1985

DIDACTICAL PHENOMENOLOGY OF MATHEMATICAL STRUCTURES

HANS FREUDENTHAL

DIDACTICAL PHENOMENOLOGY OF MATHEMATICAL STRUCTURES

D. REIDEL PUBLISHING COMPANY

A MEMBER OF THE KLUWER ACADEMIC PUBLISHERS GROUP

DORDRECHT / BOSTON / LANCASTER

Library of Congress Cataloging in Publication Data

Freudenthal, Hans 1905–
 Didactical phenomenology of mathematical structures.

(Mathematics education library)
Bibliography: p.
Includes index.
 1. Mathematics—Study and teaching. 2. Mathematics—
1961–   . I. Title. II. Series.
QA11.F769 1983   510′.7   83–3215
ISBN 90–277–1535–1

Published by D. Reidel Publishing Company,
P.O. Box 17, 3300 AA Dordrecht, Holland.

Sold and distributed in the U.S.A. and Canada
by Kluwer Boston Inc.,
190 Old Derby Street, Hingham, MA 02043, U.S.A.

In all other countries, sold and distributed
by Kluwer Academic Publishers Group,
P.O. Box 322, 3300 AH Dordrecht, Holland.

Printed in The Netherlands

TABLE OF CONTENTS

EDITORIAL PREFACE

The launch of a new book series is always a challenging event — not only for the Editorial Board and the Publisher, but also, and more particularly, for the first author. Both the Editorial Board and the Publisher are delighted that the first author in this series is well able to meet the challenge. Professor Freudenthal needs no introduction to anyone in the Mathematics Education field and it is particularly fitting that his book should be the first in this new series because it was in 1968 that he, and Reidel, produced the first issue of the journal *Educational Studies in Mathematics*. Breaking fresh ground is therefore nothing new to Professor Freudenthal and this book illustrates well his pleasure at such a task. To be strictly correct the 'ground' which he has broken here is not new, but as with *Mathematics as an Educational Task* and *Weeding and Sowing*, it is rather the novelty of the manner in which he has carried out his analysis which provides us with so many fresh perspectives. It is our intention that this new book series should provide those who work in the emerging discipline of mathematics education with an essential resource, and at a time of considerable concern about the whole mathematics curriculum this book represents just such resource.

ALAN J. BISHOP
Managing Editor

A LOOK BACKWARD AND A LOOK FORWARD

Men die, systems last. Immortality is assured to those who build their name into a system. Although even immortality is not what it used to be any more, and I did not crave for it, I once a day set my mind on writing my first systematic work, after a few that can rightly be called chaotic. The result has been the most chaotic of all — so chaotic that when the reader expects a preface he has to wait for Chapter 6. Moreover the work is incomplete. When unexpectedly Chapter 18, 'Text and Context', showed the first symptoms of elephantiasis, I cut it off, appointed Chapter 17 to be the last, and exiled the remainder to a separate book, which is very likely to become still-born. Let me add that my secretary and collaborator for almost 25 years, Mrs. Breughel read and wrote the last line of the illegible Dutch manuscript of this book the day before she retired.

But a manuscript like the present deserves a look backward by its author, which at the same time should be to its reader a look forward.

A common theme of the greater part of my publications in mathematics education has been: change of perspective; in particular what I called *inversion* and *conversion*, a mathematical virtue, practised and cherished from olden times. No mathematical idea has ever been published in the way it was discovered. Techniques have been developed and are used, if a problem has been solved, to turn the solution procedure upside down, or if it is a larger complex of statements and theories, to turn definitions into propositions, and propositions into definitions, the hot invention into icy beauty. This then if it has affected teaching matter, is the *didactical* inversion, which as it happens may be *anti-didactical*. Rather than behaving antididactically, one should recognise that the young learner is entitled to recapitulate in a fashion the learning process of mankind. Not in the trivial manner of an abridged version, but equally we cannot require the new generation to start just at the point where their predecessors left off.

Our mathematical concepts, structures, ideas have been invented as tools to organise the phenomena of the physical, social and mental world. *Phenomenology* of a mathematical concept, structure, or idea means describing it in its relation to the phenomena for which it was created, and to which it has been extended in the learning process of mankind, and, as far as this description is concerned with the learning process of the young generation, it is *didactical phenomenology*, a way to show the teacher the places where the learner might step into the learning process of mankind. Not in its history but in its learning process that still continues, which means dead ends must be cut and living roots spared and reinforced.

Though I did not use the term explicitly, didactical phenomenology already played a part in my former work. In the present book I stress one feature more explicitly: *mental objects* versus *concept attainment*. Concepts are the backbone of our cognitive structures. But in everyday matters, concepts are not considered as a teaching subject. Though children learn what is a chair, what is food, what is health, they are not taught the *concepts* of chair, food, health. Mathematics is no different. Children learn what is number, what are circles, what is adding, what is plotting a graph. They grasp them as *mental objects* and carry them out as *mental activities*. It is a fact that the concepts of number and circle, of adding and graphing are susceptible to more precision and clarity than those of chair, food, and health. Is this the reason why the protagonists of concept attainment prefer to teach the number concept rather than number, and, in general, concepts rather than mental objects and activities? Whatever the reason may be, it is an example of what I called the anti-didactical inversion.

The didactical scope of mental objects and activities and of onset of conscious conceptualisation, if didactically possible, is the main theme of this phenomenology. It was written in the stimulating working atmosphere of the IOWO.* So it is dedicated to the memory of this institution that has been assassinated, and to all its collaborators, who continue to act and work in its spirit.

* The Netherlands 'Institute for Development of Mathematics Education'.

CHAPTER 1

AS AN EXAMPLE: LENGTH

1.1–1.11. PHENOMENOLOGICAL

1.1–3. *What is Length?*

1.1. "Length" has more than one meaning. "At length", "going to the utmost length", "length and width" include in their context "length" in different meanings. The one I am concerned with becomes clear if along side the question

> what is length?

I put a few other questions:

> what is weight?
> what is duration?
> what is content?

"Length", "weight", "duration", "content" are *magnitudes*, among which length has its special status.

If I use the word length in the sense, made more precise here, I mean length of something, of a "long" object. "Length" then is synonymous with "width", "height", "thickness", "distance", "latitude", "depth", which are related to other dimensions or situations. For the sides of a "lying" rectangle one prefers "length" and "width", for a "standing" one, "width" and "height".

1.2. Without stressing it, I have turned my question "what is length?" towards an answer such as "length of . . . is . . .". This is a typically mathematical turn: transforming apparently isolated terms into symbols of functions. The question

> what is "mother"?
> what is "brother"?
> what is "neighbor"?

are more easily answered according to the pattern

> mother of . . . is . . .
> brother of . . . is . . .
> neighbor of . . . is . . . .

More precisely:

> mother of x is she who has born x,

1

brother of x is every y such that y is a male and x and y have the same
   parents,
neighbor of x is every y such that x and y live beside each other.

Afterwards "mother" can also be defined in an "isolated" way:

   x is mother if there is a y such that x is mother of y.

   Linguistically "man", "stone", "house" belong to the same category as
"mother", "brother", "neighbor" — as nouns they enjoy a substantiality, though
that of "mother", "brother", "neighbor" differs from that of "man", "stone",
"house". "Being mother", "being brother", "being neighbor" get a meaning
only by the — explicit or implicit — addition "of whom". In "they are brothers",
"they are neighbors" the additional "of . . ." seems unnecessary but is not: they
are brothers or neighbors of each other.

1.3. Back to "length", interpreted in "length of . . ." as a functional symbol:
a function that talks about "long objects" how long they are, though not necessarily numerically specified, as in

   the length of this bed is 1.90 m.

Functional value may be vague: long, very long, short, very short, and so on.
The reason why I neglect these values now is that I will start by focusing on a
phenomenology of *mathematical* structures. Are "long", "very long", "short",
"very short" not mathematical concepts? Such questions will be answered later
on; in order not to complicate things, I delay the answer.

1.4. *Magnitudes* *

Before continuing let me consider the terms mentioned earlier. All of them aim
at functions:

   weight: weight of (a heavy object),
   duration: duration of (a time interval),
   content: content of (a part of space).

Let me introduce abbreviations:

   $l(x)$: length of x,
   $p(x)$: weight of x,
   $d(x)$: duration of x,
   $v(x)$: content of x,

* *Directed* magnitudes will incidentally be considered in Sections 15.9–12. Otherwise
"magnitude" is always understood in the classical way. In this context "rational" and "real"
always means "positive rational" and "positive real".

where x is something that can properly be said to have a length, weight, duration, content.

We again pose the question of the possible values of the function l (and of p, d, v). Not "long short", "heavy light", "big small", respectively, but since we speak mathematics, more precise values. This does not oblige us to state something like 1.90 m, 75 kg, 7 sec, 3 m^3, expressed in the metric system, or in any *a priori* system of measures. This is a liberty we can profit from to get deeper insight. Indeed, it appears that we can go rather far without accepting any special system of measures.

Let us call the

values of l lengths,
values of p weights,
values of d durations,
values of v contents,

and the

system of lenths L,
system of weights W,
system of durations D,
system of contents V,

and look for their properties.

1.5. *Adding Lengths*

The first thing we notice is that we can *add* lengths even before conceiving them numerically. How is it done? Given two lengths α and β, we provide ourselves with two "long objects" x and y with lengths

$$l(x) = \alpha, \qquad l(y) = \beta,$$

respectively, and compose them (in a way that asks for detailed explanation) into a new "long object" $x \oplus y$. This object has a certain length, consequentially named $l(x \oplus y)$. It was our intention to define the sum of lengths α and β and by definition we put

$$\alpha + \beta = l(x \oplus y),$$

that is to say

1.5.1 $\quad l(x) + l(y) = l(x \oplus y).$

In other words

> the length of the composite equals the sum of the composing parts.

As regards this kind of definition one has to pay attention to one point: For the lengths x and y we have chosen representative "long objects" x and y, respectively, with lengths as prescribed. Instead we could have chosen other representatives, say x' and y', thus such that again

$$l(x') = \alpha, \qquad l(y') = \beta,$$

which would lead to a composite $x' \oplus y'$. In order for the definition 1.5.1. to be meaningful, we must be sure that

1.5.2 $l(x' \oplus y') = l(x \oplus y),$

in other words, that

> the length of the composite does not depend on the choice of the representatives.

I have to take care that my way of combining "long objects" fulfills this condition.

In a similar way this ought to be true of weights. Given two weights α and β the sum of which I propose to define, I take two "heavy objects" x and y with weights α and β, respectively, compose them into a new "heavy object" $x \oplus y$ and define

$$p(x) + p(y) = p(x \oplus y).$$

Again, replacing x and y by x' and y' with the same weights, respectively, must not change the weight of the composite. This requirement looks self-evident, and it is so for a good reason, indeed: we would never have focused on length, weight, and so on if this condition were not fulfilled.

A second remark: If composing is meant to lead to defining the sum, it must be carried out in such a way that the components do not overlap. Suppose I want to add a length α to itself in order to define the length $\alpha + \alpha$. Then for each of the summands I need *another* representative, thus

$$l(x) = \alpha, \qquad l(y) = \alpha,$$

in order to get

$$\alpha + \alpha = l(x) + l(y) = l(x \oplus y).$$

So I cannot manage with one representative for each length. Fortunately with lengths it is rather easy to provide oneself with two, three, or more representatives of the same length; instruments like a ruler can repeatedly be applied. In the case of weights and so on, the difficulty of obtaining enough representatives looks greater, but this is a point we are not concerned with here.

In carrying out the operation $\oplus$ as imagined in the various cases, order does not play a role and, as a consequence, addition of lengths, weights, etc., obeys the laws of commutativity and associativity:

$$\alpha + \beta = \beta + \alpha,$$
$$(\alpha + \beta) + \gamma = \alpha + (\beta + \gamma).$$

The first property stated in the systems L, P, D, and V of lengths, etc., therefore is:

I.   A commutative and associative operation (+) of addition in L, and so on.

1.6. *Order of Lengths*

Adding will later be joined by subtracting; that is, "the smaller from the bigger". But "smaller" and "bigger" are ideas we have not yet come across. They will now be considered.

Relations like "smaller bigger" belong to the so-called *order relations*: any pair of elements α, β of L is in exactly one of the situations

1.6.1 $\alpha < \beta,$ $\alpha = \beta,$ $\beta < \alpha$

and for three of them, $\alpha, \beta, \gamma \in L$,

1.6.2   if $\alpha < \beta$ and $\beta < \gamma$ then $\alpha < \gamma$

holds (the so-called *transitivity*).

Such an order relation can now be defined in L by means of the addition. We express the property that

by adding, something can become only larger,

in a formula

1.6.3 $\alpha < \alpha + \kappa$

for any lengths α and κ. This immediately ensures transitivity 1.6.2. Indeed, if $\alpha < \beta$ and $\beta < \gamma$, then there is a κ and a λ such that

$$\beta = \alpha + \kappa, \qquad \gamma = \beta + \lambda,$$

so

$$\gamma = \beta + \lambda = (\alpha + \kappa) + \lambda = \alpha + (\kappa + \lambda),$$

and thus

$$\alpha < \gamma.$$

The first requirement, 1.6.1, on an order relation is a bit trying. It means

1.6.4   If $\alpha \neq \beta$, then there is
          either a κ with $\beta = \alpha + \kappa$,
             or a λ with $\alpha = \beta + \lambda$,

though not both together.

For a moment I call two "long objects" x, y directly comparable if either x can be considered as a composing part of y or y as a composing part of x. Then 1.6.4 can be translated as follows into the language of "long objects":

1.6.5   Given two "long objects" x, y, then I can find directly comparable "long objects" x', y' such that $l(x) = l(x')$, $l(y) = l(y')$, and however I choose x', y' in this way, one thing is true:
either x' is a composing part of y',
or y' is a composing part of x'.

The second property we have stated for the systems L, W, etc. is:

II.   The definition $\alpha < \alpha + \kappa$ for all lengths α, κ determines a total order in L, etc.

1.7. *Multiplying Lengths*

If we repeatedly add the same length, then the resulting lengths can be denoted as

$$1\alpha = \alpha,$$
$$2\alpha = \alpha + \alpha,$$
$$3\alpha = \alpha + \alpha + \alpha,$$

and so on; in general

$$n\alpha = \alpha + \ldots + \alpha \text{ with } n \text{ summands.}$$

Laws like

$$
\begin{aligned}
&(m + n)\alpha = m\alpha + n\alpha, \\
&(mn)\alpha = m(n\alpha), \\
&n(\alpha + \beta) = n\alpha + n\beta, \\
&\text{if } \alpha < \beta \text{ then } n\alpha < n\beta,
\end{aligned}
$$

1.7.1

are obvious.

From adding we have now derived multiplying elements of L, etc., by positive integers, that is, elements of $\mathbf{N}^+$. As an inverse of this operation one has dividing, which means:

Given a length α and an $n \in \mathbf{N}^+$, then the equation

1.7.2 $n\beta = \alpha$

has a solution β. There is only one such β, since if

$$n\beta' = \alpha,$$

then

$$\beta < \beta' \quad \text{or} \quad \beta = \beta' \quad \text{or} \quad \beta' < \beta.$$

In the first and third case this would result in

$$\alpha = n\beta < n\beta' = \alpha, \qquad \alpha = n\beta' < n\beta = \alpha,$$

respectively, which is impossible, and leaves us with

$$\beta = \beta'.$$

The solution β of 1.7.2 gets the name

$$\beta = \frac{1}{n}\alpha.$$

Thus $\frac{1}{n}\alpha$ is defined by

1.7.3 $\quad n\left(\frac{1}{n}\right)\alpha = \alpha.$

This then is our third property of lengths, etc.:

III.   For every $\alpha \in L$, etc., and every $n \in \mathbf{N}^+$ there is one $\frac{1}{n}\alpha \in L$, etc., such that

$$n\left(\frac{1}{n}\alpha\right) = \alpha.$$

The following laws for dividing are easily verified:

1.7.4 $\quad \dfrac{1}{m}\left(\dfrac{1}{n}\alpha\right) = \dfrac{1}{mn}\alpha,$

$$\frac{1}{n}(\alpha + \beta) = \frac{1}{n}\alpha + \frac{1}{n}\beta,$$

$$\text{if } \alpha < \beta, \text{ then } \frac{1}{n}\alpha < \frac{1}{n}\beta.$$

1.8. *Rational Multiples of Lengths*

Multiplying and dividing elements of L, etc., by elements of $\mathbf{N}^+$ can be combined. One puts

1.8.1 $\quad m\left(\dfrac{1}{n}\alpha\right) = \dfrac{m}{n}\alpha,$

which results in multiplying of lengths by positive rational numbers. Since, however, a rational number can be denoted in various ways,

$$\frac{m}{n} = \frac{km}{kn},$$

we have to make sure that the definition 1.8.1 is valid; that is, we have to prove that

$$m\left(\frac{1}{n}\alpha\right) = km\left(\frac{1}{kn}\alpha\right).$$

This is indeed true. According to 1.7.4

$$\frac{1}{kn}\alpha = \frac{1}{k}\left(\frac{1}{n}\alpha\right),$$

thus

$$km\left(\frac{1}{kn}\alpha\right) = mk\left(\frac{1}{k}\left(\frac{1}{n}\alpha\right)\right) = m\left(\frac{1}{n}\alpha\right).$$

Thus we can multiply every length, etc., by every positive rational number $r \in Q^+$. We easily find the laws, for $r, s \in Q^+, \alpha, \beta \in L$, etc.:

1.8.2
$$\begin{aligned} &(r + s)\alpha = r\alpha + s\alpha \\ &r(s\alpha) = (rs)\alpha \\ &r(\alpha + \beta) = r\alpha + r\beta \\ &\text{if } \alpha < \beta, \text{ then } r\alpha < r\beta. \end{aligned}$$

1.9. Real Multiples of Lengths

Starting from one length, etc. say α, we can form all its rational multiples. They constitute a set $Q^+\alpha$. In $Q^+\alpha$ two arbitrary elements are rational multiples of each other. So $Q^+\alpha$ cannot possibly exhaust what we imagine the system of lengths to be. Indeed, the diagonal and side of a square are not rational multiples of each other. However, $Q^+\alpha$ does exhaust the system of lengths, etc. "aproximately". One knows about a property, the so-called Archimedean axiom:

IV. Given an $\alpha \in L$, etc., then there is no element of L, etc., bigger than all of $Q^+\alpha$, and no element of L, etc., smaller than all of $Q^+\alpha$.

I now take an arbitrary $\beta \in L$, etc. It does not necessarily belong to $Q^+\alpha$, but according to IV it must lie "in between". For each $r \in Q^+$

1.9.1 $\qquad r\alpha < \beta \quad$ or $\quad r\alpha = \beta \quad$ or $\quad \beta < r\alpha$

holds. I now want to represent β as a real multiple of α,

$$\beta = u\alpha, \qquad u \in R^+$$

in such a way that the order fits, that is

$$\text{if } u < v \quad \text{then} \quad u\alpha < v\alpha \quad \text{for} \quad u, v \in R^+,$$

In particular for $r \in Q^+$,

1.9.2
$$\begin{aligned} &\text{if } r < u \qquad \text{then } r\alpha < u\alpha, \\ &\text{if } u < r \qquad \text{then } u\alpha < r\alpha. \end{aligned}$$

How to find such a u? Well, 1.9.1 causes a partition of $r \in Q^+$ into three classes (the second can be empty, or can consist of one element if β is a rational multiple of α). Such a partition is called a Dedekind cut:

the lower class: the $r \in Q^+$ with $r\alpha < \beta$
the upper class: the $r \in Q^+$ with $\beta < r\alpha$,

where at most one $r \in Q^+$ can escape this division. Now there is a real $u \in R^+$ that "causes" the cut, that is to say

if $r\alpha < \beta$ then $r < u$.
if $\beta < r\alpha$ then $u < r$.

If now we put

$$\beta = u\alpha,$$

we fulfill the requirements of 1.9.2.

It has been shown that

of two given elements of L, etc., each is a positive real multiple of the other.

We can now conclude with the property

V.   For each $\alpha \in L$, etc., and each $u \in R^+$, there is an $u\alpha \in L$, etc. Similarly to those of Q^+ one can formulate laws for $u, v \in R^+$ and $\alpha, \beta \in L$ etc.:

$$(u + v)\alpha = u\alpha + v\alpha$$

1.9.3 $u(v\alpha) = (uv)\alpha$

$$u(\alpha + \beta) = u\alpha + u\beta$$

if $u < v$ then $u\alpha < v\alpha$.

1.10. *Length Measure*

Let us break off the exposition and not insist on a systematic approach to magnitudes.

For instance we could continue with a numerical treatment of magnitudes: A measuring unit (m, kg, sec, m³, or suchlike) is chosen in order to express each length, and so on, as a positive real multiple of the unit. Then each length, etc., is represented by a measuring number and according to its generation we find the fundamental rule

under the composition $\oplus$ the measuring numbers are added,

from which follows among others

the longer, heavier, . . . object has the bigger measuring number.

1.11. *What is Lacking Here*

The preceding was an example of phenomenology; namely, for the mathematical structure "magnitude". Or rather, it was a fragment of such a phenomenology. No attention was paid to measuring; connections between different magnitudes should have been considered; and finally, what has not been mentioned at all is that length is ascribed not only to "long objects" but also to broken and curved lines. How broken lines, say the perimeter of a triangle, should be dealt with is easy to guess. Curved lines are a different case. The classical way is approximation by broken lines but I shall skip it here in order to resume it later on.

1.12–1.29. DIDACTICALLY PHENOMENOLOGICAL

The preceding was not *didactical* phenomenology. In order to stress the difference I started with phenomenology as such. But also in the sequel didactical phenomenology will often be preceded by phenomenology as such, to create a frame of concepts and terms on which the didactical phenomenology can rest.

The difference between phenomenology and *didactical* phenomenology will soon become apparent. In the first case a mathematical structure will be dealt with as a cognitive product in the way it describes its — possibly non-mathematical — objects; in the second case, it will be dealt with as a learning and teaching matter, that is as a cognitive process. One could think about one step backwards: towards a genetic phenomenology of mathematical structures, which studies them in the cognitive process of mental growth.

One might think that a didactical phenomenology should be based on a genetic one. Indeed I would have been happy if, while developing the present didactical phenomenology, I could have leaned upon a genetic one. This, however, was not the case, and the longer I think about the question, the more I become convinced that the inverse order is more promising. In the sequence "phenomenology, didactical phenomenology, genetic phenomenology" each member serves as a basis for the next. In order to write a phenomenology of mathematical structures, a knowledge of mathematics and its applications suffices; a *didactical* phenomenology asks in addition for a knowledge of instruction; a *genetic* phenomenology is a piece of psychology.

All the psychological investigations of this kind which I know about suffer from one fundamental deficiency: investigations on mathematical acquisitions (at certain ages) have involved the related mathematical structures in a naive way — that is, they lack any preceding phenomenological analysis — and as a consequence, are full of superficial and even wrong interpretations. The lack of a preceding *didactical* phenomenology, on the other hand, is the reason why such investigations are designed in almost all cases as isolated snapshots rather than as stages in a developmental process.

1.13–1.25. COMPARISON OF LENGTHS

1.13–14. *Length Expressed by Adjectives*

1.13. Many mathematical concepts are announced by adjectives. Adjectives belonging to length are: "long, short", but also "broad, tight", "thick, thin", "high, low", "deep, shallow", "far, near", "wide, narrow", and finally also "tall, sturdy, diminutive, insignificant". Of course the ability to distinguish such properties precedes the ability to express them linguistically. For the adult it is — at least unconsciously — clear how these expressions are related to the same magnitude, length, and he often presupposes children to be well acquainted with this relation. Researchers in this field are often not aware of this difficulty. It is not farfetched to ask oneself how the child manages to develop a knowledge of these connections. A disturbing factor is the overarching of this complex of adjectives by "big and little", which can serve so many aims (up to "big boy" and "little girl").

Bastiaan (5; 3) asks how big is a mole. When I show with my hands a mole's length, he insists "no, I mean how high". He is compelled to differentiate "big". Clearly he is conscious of the fact that both cases mean a length.

The insight that both expressions mean a length is not at all trivial, for instance, that a *high* tree, if cut, is *long*. As a matter of fact, even adults may have problems with the equivalence of distances in the horizontal and the vertical dimensions, at least with regard to quantitative specification.

How does the connection within this complex of adjectives come into being? How is the common element constituted? If I may guess, I would attribute a decisive role to the hand and finger movements that accompany such statements as *that* long, *that* wide, *that* thick, and so on (likewise *that* short, and so on) — movements that can turn in different directions and possess different intensities but always show the same linear character. (Compare this with the mimic expressions of embracing, which may accompany "that much", but also "that thick", and with the mimic and acoustic expression of lifting belonging to "that heavy".)

The common element in this complex of adjectives for length is possibly not yet operational in all young school children; as a matter of consciousness it may even be absent in many older ones. Acquiring it and becoming conscious of it are an indispensable condition for mathematical activities.

1.14. Around such adjectives as "long" there is a complex of relational expressions like:

longer, longest, as long as, less long, not as long as, too long, very long.

Here again the ability of distinguishing precedes that of linguistic expression (for instance, something cannot pass through a hole because it is *too thick*; the smaller cube is placed upon the bigger). Inhibitions work against using comparatives and superlatives — "big" is used where "bigger" and "biggest" are meant.

The adjectives of the last list aim at comparing objects with respect to length. This activity develops long before what mathematicians call the order relation of lengths is constituted, not to mention becoming conscious of the order relation. The constitution of an order relation in whatever system includes at least the operational functioning of transitivity, that is, drawing factual conclusions according to patterns like

> a as long as b,
> b as long as c,
> thus a as long as c

and

> a shorter than b
> b shorter than c
> thus a shorter than c,

which of course does not mean the ability of verbalising or even formalising transitivity.

Contradictory Piaget, P. Bryant* showed that young children (from the age of four onwards) possess an operational knowledge of transitivity. On the other hand, I reported** on third graders who could apply the transitivity of weights in seesaw contexts but were not able to understand a formulation of transitivity.

Little if any information on the development of the concept of length can be drawn from traditional research. Thought on this subject is obscured by such terms as "conservation" and "reversibility", which are supposed to cover the most divergent ideas, and by a faulty phenomenology.

1.15. *Congruence Mappings*

One of the mathematical notions that have been absorbed by "conservation" in order to be mixed together with quite different ones is

> invariance under a set of transformations.

As an aside I will illustrate this notion by a number of examples:

> The *number of elements* of a set ("cardinality") is invariant under one-to-one mappings.
> *Convexity* of a plane figure is invariant under affine mappings.***
> *Parallelism* of lines is invariant under affine mappings.***

---

* P. Bryant, *Perception and Understanding in Young Children*, London, 1974, Chapter 3.
** *Weeding and Sowing*, p. 255.
*** A reader not acquainted with the concept of "affine mapping", may read instead: parallel projection.

The difference between the *surfaces* of a *sphere* and of a *ring* is invariant under arbitrary deformations (the surface of a ring cannot be deformed into that of a sphere).

Lengths of line segments and *measures of angles* of pairs of lines are invariant under congruence mappings of the plane or the space (movements and reflections, with glide reflections included).

The length ratio of line segments and *measures* of angles of pairs of lines are invariant under similarities.

The property of being a *regular pentagon* is invariant under similarities.

The property of being a *cube with side 1* is invariant under congruence mappings.

The property of a plane figure to represent the *digit 2* is invariant under movements (though not under reflections).

The *shape* of a figure is invariant under similarities.

Both the *shape and size* of a figure are invariant under congruence mappings.

The expression *congruent* is well-known; congruent figures are, as it were, the same figure laid down in various places. In mathematics this concept is made more precise by that of a congruence mapping, which extends to the whole plane or space; then figures are called congruent if they can be carried into each other by congruence mappings.

The simplest figure is the line segment. "Equal" rather than "congruent" line segments is an older terminology. The now prevailing terminology reserves "equality" to coincidence; that is, actual identity. Yet congruent line segments are equal in a sense; that is, with respect to length. And conversely: line segments with the same length can be carried into each other by congruence mappings.

1.16. *Rigid Bodies*

Line segments are mathematical abstractions. They are connected with the former "long objects" via the phenomenon of the rigid body. A rigid body can be displaced, and provided it is not badly belaboured, it remains congruent with itself under this operation. Rigidity is the physical realisation of the property we called invariance of shape and size under movements. The fact that in geometry we consider by preference properties that are invariant under movements is related to the dominance of rigid bodies in our own environment – molluscs would prefer another kind of geometry.

I am pretty sure that rigidity is experienced at an earlier stage of development than length and that length and invariance of length are constituted from rigidity rather than the other way round. Rigidity is a property that covers all dimensions while length requires objects where one dimension is privileged or stressed. However, stressing this one dimension may not lead to restricting the preserving transformations. If lengths are to be *compared*, the free mobility of rigid bodies must play its part. The mobility must be fully exploited; all movements must be allowed, not only the most conspicuous translations, but also rotations, in order to compare "long objects" in all positions. The shape of a body or the stressing of one dimension as its length may not result in restricting the mappings under which rigidity expresses itself as invariance. Adjectives like "high, low" within the complex of terms that indicate length can exert an influence to restrict the set of transformations; "high, low", stressing one direction in space, can lead to restricting the set of transformations to those that leave invariant the vertical direction — displacements along and rotations around the vertical — a restriction that would impede the overall comparison of line segments and "long objects".

1.17. *Similarities*

Side by side with the congruence mappings I repeatedly mentioned the similarities. The latter play a part in interpreting visual perception. "What is farther away, looks smaller" (at least at big distances); this is a feature unconsciously taken into account by a perceiver and sometimes made conscious to himself — a curious interplay which has been studied many times. If a rigid body moves away, its shape as understood by us remains invariant; visually conceived the rigid bodies are invariant even under similarities, while the similarity ratio depends on the distance between object and perceiver.

Nevertheless just this fact can contribute a great deal to the mental constitution of rigidity. The invariance suggested by the continuous behaviour of some striking characteristics might provoke the attribution of more invariances, in particular those of size and length.

1.18–20. *Flexions*

1.18. The rigidity of rigid bodies has to be understood with a grain of salt. Though its wheels and doors can turn independently, a car can globally and under certain circumstances be considered as a rigid body. Another extreme case is clay, which by mild force can be kneaded and deformed. In defining rigidity all depends on what you call "not badly belaboured". A liquid or a gas can be given some other shape without using any force, but according to the degree of rigidity more or less strong forces are needed to deform a rigid body. More or less rigid parts can be movable with respect to each other, such as in the case of animal bodies, while certain arrangements of the parts with respect to each other may be privileged, such as the state of rest, which can be congruently copied ad lib. It is that privileged state in which length measures of animal bodies are defined. The heights of, say, two people are compared while they are standing; we are convinced that they do not change when they sit down, and we know that they will show anew the former relation if they rise again. We also judge that if they sit down and the taller person looks smaller, the difference must be ascribed to longer legs — something we can reconsider under the viewpoint of addition of lengths.

1.19. What comes about here is another principle of invariance of length, that is to say, invariance under a kind of transformations other than planar or spatial congruence mappings. It is transforming "long objects" by plying or bending them with a negligible effort: two objects to be compared are laid side by side or one on top of the other while certain deformations are allowed. Typical examples of this are measuring instruments other than the ruler and the measuring stick — for instance, the measuring tape, the folding or coiled pocket-rule — but a more primitive device used to measure lengths, the piece of string, should not be forgotten. It shows marvellously two ways of comparing lengths: in the tight state it measures a straight length, and fitted around a curvilinear shape it measures a circumference.

As opposed to the rigid bodies considered earlier, I will call these objects *flexible* the admissible deformations of these objects being called *flexions*. Flexions are reversible — this is an important feature. Moreover, flexible objects possess one or more privileged states. Among the privileged states there might be one in which the object is straightened and used as a measuring instrument: the measuring tape, the folding pocket-rule, the coiled-rule, and again the piece of string that can be stretched with a little force and that in this state resists further stretching. One's own body is of the same kind; in order to have it measured, one jumps to one's feet (though not to one's toes). Similarly, one measures the length of a stalk or reed or a stair-carpet: by stretching. Or of a car antenna: by pulling it out. A sheet of paper is flexible, though there is a well-defined state of maximal stretching. Plastically deformable substances such as clay are again different, a "long object" made of clay, if carefully handled, can be considered as flexible, though a kneading transformation is no flexion.

1.20. Where can we put the flexions mathematically? The mathematical counterparts of the rigid bodies (which may be moved without being badly belaboured) were the geometrical figures subjected to movements in the plane or in space, transformations that map everything congruently; in particular every line segment whatever its length or direction might be. If our objective is *measuring lengths*, this requirement is exaggerated; in order to serve for measuring, the "long objects" need display this invariance in the length direction only. Only in the length direction should the object be rigid; there is no need for rigidity in the other dimensions. This kind of object is mathematically idealised by what is called curves — curves which are described by a moving point or appear as boundaries of a plane figure. Of course curves which are — entirely or partially — straight are also admitted: straight lines and broken lines. It is these *mathematical curves* that are subjected to *mathematical flexions*. What does this term mean? If it refers to curves, I am concerned with one dimension only

− no width and no thickness − and in this one dimension they shall be rigid.
The arc length, which as a measure replaces straight length, should be invariant
under flexions. Mathematically, flexions are defined as mappings of curves that
leave the arc length invariant.

But what do we mean by the arc length of a curve? The answer looks obvious:
straighten the curve while not stretching it and read the arc length on the
final straight line segment. Well, isn't it a vicious circle? What do we mean
by straighten without stretching? No stretch − this just means that the arc
length must be preserved, but arc length still has to be defined. As a matter of
fact, it is curious that I prohibited stretching only, and kept silent about shrinking, but of course the mistake you can make when straightening the object,
is pulling too hard and stretching. This shows once more that the alleged clarity
of the straighten-out definition of arc length rests not on visual but on kinesthetic
intuition.

Yet another definition of arc length deserves to be considered. In order to be
compared, curves are rolled upon each other. In particular, in order to measure
the length of a curve, it is rolled upon a straight line. Rolling yes, but of course
skidding is forbidden. But what does it mean mathematically: no skidding? That
the pieces rolling along each other have the same (arc) length. This again closes
the vicious circle.

There is no escape: In order to define flexions mathematically, we must first
know what arc length is, and arc length must be defined independently with
no appeal to mechanics.

How this is to be done, I have already mentioned. First, one defines the
length of a polygon − that is, a curve composed of straight pieces − as the sum
of the lengths of those pieces. Given a curve, it is approximated by "inscribed"
polygons, that is, with their vertices on the curve. The smaller the composing
straight pieces, the better the curve is approached. In this approximation process
one pays attention to the respective lengths: as the curve is approached by the
polygons, the lengths converge to what is considered as the length of the given
curve. Not only should the total curve get an arc length by this definition,
but also each partial curve, and it is plain (though the proof requires some attention) that these lengths behave additively: if a curve is split into two partial
curves, the length of the whole equals the sum of the lengths of the parts. It is
now clear what we have to understand by a mapping that preserves the arc
length (by a flexion): not only should the total arc length be left invariant.
but also that of each part.

It is strange that an intuitive idea like *invariance of arc length* and *straightening without stretching* requires such a cumbersome procedure in order to be
explained mathematically. The reason is now obvious: when trying attempts
at explaining arc length mathematically, we are compelled to renounce our
mechanical experiences. It is particularly intriguing that *physically* I can compare
two flexible objects by flexion or the borders of two plane shapes by rolling the
one upon the other, before I start *measuring* length, whereas our *mathematical*

definition of flexion presupposes arc length, which includes the whole measuring procedure and even the addition of lengths.

1.21. *Rigidity and Flexibility*

We have been concerned with two kinds of mappings:

> congruence mappings in plane of space, and
> flexions of curves.

Both are mathematically defined by the invariance of length, though the first requirement cuts deeper than the second if the view is fixed on curves and arc length.

The fact that congruence mappings and flexions leave length invariant is implicit in their definition. In physics the counterpart of mathematical congruence mappings and flexions is the movement of rigid bodies and the bending of flexible bodies, but whether in physical practice something is (at least approximately) a rigid body or a flexible body and which physical operations are allowed if length should (at least approximately) be preserved are physical facts, depending on experiences we have somehow acquired. This acquisition of experience starts rather early, certainly as early as in the cradle. It is empirical and experimental, and though this experimenting starts, as Bruner asserts, in an *enactive* way, in the course of development it is supported more and more by representative images of what is recollected or pursued (the *ikonic* phase), and it becomes more and more conscious in order to be verbalised (the *symbolic* phase). In the context of the phenomenon of "length" a phenomenological analysis is required to state and to distinguish invariance under congruence mappings and flexions, but anyway it is clear that the related learning process starts in the enactive phase (with no representative images and unconsciously, that is in the most effective way) and that pieces of it can be made conscious in the learning process.

Bastiaan (3; 9) finds a glass marble on the foot path: "If I push hard, it would roll into the street". It does happen. The marble rolls under the tyre of a car parked at the curb. Bastiaan cannot reach it. I show him a little stick. By sight he judges: "It is not hard enough." It is a soft stick, but nevertheless he succeeds.

This example does not concern using rigid or flexible objects to compare lengths. What matters here is experiences with mechanical properties of things. At a certain moment in his development a child judges at sight whether something is "hard" enough to be applied as a lever to exert a certain power (ikonic phase), and he even finds words — hard enough — to express this fact (symbolic phase).

I do not have the slightest idea how this complex of mechanical properties becomes mentally constituted; an able physicist, observing children, could discover a lot of things in this field. There is one conjecture which I dare pronounce: that rigidity precedes flexibility. The environment strongly suggests the rigid body as a model. Surprising experiments show that under conditions of incomplete information about kinematical phenomena there is a strong tendency to interpret them as movements of rigid bodies.

As a consequence I think that length is first constituted in the invariance context of congruence mappings – that is, connected to rigid bodies – and only at a later stage gets into that of flexions – that is, of flexible objects. This can happen if the child sees lengths compared or even measured by flexible instruments – fitting ("is the sleeve long enough?") and measuring with a tape.

In any case it is crucial to pay attention to the double invariance context of length.

1.22. *Make and Break*

I hesitated – unjustly as it will shortly appear – as to whether I should augment the two kinds of transformations that show invariance of length (that is, congruence transformations and flexions) with a third, which I would call

break–make transformations:

a "long object" is broken into pieces and remade.

The "long object" may be a stick that is factually broken, or a string that is cut, or a train of blocks that is split into two or more partial trains. In the first two examples remaking will not yield a complete restoration of length even if carried out carefully, with some loss in the second case if the partial strings are *tied* together. In the third case the restoration can be complete though it need not be: the parts can be put together in another order, and this can even be visible if the particular blocks are distinguished by length, colour, or other characteristics.

It is a meaningful and non-trivial statement that under break–make transformations length is invariant. It is meaningful if it is the original and final state that are compared, disregarding the intermediate ones. Indeed, how should we formulate the question if the intermediate states are to be admitted? "Do they remain as long together?" If "together" means adding lengths, this question is premature at the stage of simply comparing lengths, and if "together" means "taken together" the question aims at comparing the initial with the – now also mental – final state, which is no news.

Whenever the break–make transformation reproduces the initial state, the question "are they of the same length?" is trivial. Or rather, the answer reveals only whether the child that was questioned has remembered the initial state and is able to compare an actual and a mentally realised state with each other. If the final state is not wholly identical with the initial one, the answer also reveals whether the child knows which characteristic matters if length is meant. These two abilities will be reconsidered later.

Insight into length invariance under break–make transformations can be split into two components:

first, that under breaking (partitioning) and making (composing) "long
objects" are transformed into long objects, and

second, that in composing "long objects" length is not influenced by the
order of the composing parts.

Actually, these two insights form the basis for measuring lengths and will reappear in that context. If the second insight is to be placed into the context of invariance of length with respect to certain transformations on "long objects", instead of break–make transformations we could better use the term

permutation of composing parts.

I can now explain the hesitation I felt before writing this section. Break–make transformations or permutations of composing parts as a third kind of transformations look logically and phenomenologically superfluous. Within a phenomenology of magnitudes, and particularly length, as sketched in the beginning, the break–make transformations (permutations of composing parts) and the associated invariance of length can be derived from the congruence mappings, flexions and their invariance properties. But this derivability is a consequence of coupling the comparison of lengths with measuring, which is genetically and didactically premature. It is true that composing "long objects" occurs in that phenomenology as a special operation, indicated by $\oplus$, but the context in which it occurs is length rather than comparing length; namely the formula

$$l(x \oplus y) = l(x) + l(y).$$

$\oplus$ occurs there as a logical rather than geometrical and mechanical operation. $x \oplus y$ appears as something that is uniquely determined by x and y, whereas for break–make transformations it is essential that x and y can be put together in various ways and however composed, yield objects of the same length.

1.23–24. *Distance*

1.23. Up to now in our didactical phenomenological analysis we have considered length as a function of concrete objects (possibly replaced by their mental images). This, however, does not cover all cases of length. Length as distance between A and B answers the question "how far is B from A?" In a purely formal sense "how far?" is quite another interrogative than "how long?" In "how far ... ?" two points occur as variables, whereas in "how long is this object?", the object is the only variable. Length is a function of whole objects, whereas distance is a function of two points "here" and "there". We are so accustomed to the procedure which connects both of them that we can hardly imagine the early stage where we must have acquired it by a learning process and ask ourselves whether this connection is as obvious for children as it is for us.

If "how far?" is to be reduced to "how long?", a "long object" must be placed between A and B, between here and there. So, if A and B are railway stations or stopping places along a highway, the rail connection or the stretch of highway may be considered as concrete "long objects" whose distance is asked for. In general, if there exists a concrete path between A and B, their distance is the length of the path; if there are more such paths, it should be stipulated which one is meant. But how far is it from the front room of my ground floor to the rear room on the first floor? From here to across the canal, if no bridge is visible? From here to the sky? Only from the context can it be understood what is meant. In the context of geometry, mechanics, and optics the distance is measured along a straight line; in the context of spherical trigonometry and in the context of (surface of air) navigation, along arcs of great circles, "geodesics" or shortest paths as determined by straightened strings on curved surfaces. Of course, with this remark I do not mean spherical trigonometry or navigation should have been studied or exercised in order to decide that lengths should be measured along geodesics; contexts like this develop long before they are made conscious. The value of rectilinearity is suggested to the young child, enactively, if he is called to come straight in your open arms, the ikonically by all the horizontal and vertical straight lines in his environment, and symbolically by straight lines in schemas and by the word "straight line". The part played by rectilinearity in the constitution of "length" remains unconscious until it is explicitly discussed. Straightening flexible objects if lengths are to be compared may still be an automatic act — for instance, automatic imitation — and there might be children who as automatically put between two unrelated points a mental "long object", an imagined ruler, or a string in order to interpret distance as length. Well-known experiments where children get disoriented as soon as a screen is placed between the two points may prove how important this act of inserting a "long object" can be for reducing "how far?" to "how long?". But whatever these experiments mean, if some judgment about the distance of unrelated points must be motivated, one cannot but make explicit the necessity of rectilinear connections. From this moment onwards the significance of rectilinearity for the concept of length becomes more and more conscious — another connection between length and rectilinearity will be indicated later on.

1.24. How does a child learn what matters if lengths are to be compared? Sets of objects of the same kind but of different length may play an important part: big and little spoons (and equal ones), long and short trains (and equal ones), high and low trees (and trees of equal height). The objects are compared at sight if they are lying parallel and side by side; in order to be compared they are brought into such a position, physically or mentally, as rigid bodies, by congruence mappings. This requires comparing physical with mental objects, and mental ones with each other. Memory for length initially functions in a rather rough way, it seems. Remembering length during long periods remains a difficult task. As for myself, I am often surprised that relations of length differ greatly from what I remember they should be. Comparing objects side by side gains precision in the course of development: the ruler is laid close to the line to be measured, while observing the prescription to aim perpendicularly to the line. The connection between "length" and "distance" is stressed, and the weight is shifted to "distance" if one of the objects to be compared, or both of them, bear marks by which the ends of the objects to be compared can be indicated. Comparing can be done indirectly, using the transitivity of the order relation; for instance, by taking distances between fingers of one hand, between two hands, between the points of a pair of compasses, or between two extant or intentionally placed marks on a long object, and carrying them from one place to another. With all these methods length as a function of long objects is replaced by distance as a function of a pair of points. It already starts with showing "that big" or "that small" with fingers or hands, although in its exaggerated appearance this gesture is more an emotional expression of "awfully big" or "miserably small" than a true means to compare lengths. More refined methods of comparing lengths are based on geometry and will be dealt with in that context.

1.25. *Conservation and Reversibility*

Before extending the analysis of *measuring* lengths I tackle the question already touched in Sections 1.12 and 1.15: how psychologists interested in the development of mathematical concepts deal with such concepts, in particular length. The investigations, started by Piaget, show the following pattern. The general problem is to acquire knowledge about the genesis of such fundamental concepts as number, length, area, shape, mass, weight, and volume. Subjects are shown groups of objects which agree with respect to one or more of these magnitudes (the same number of chips in a row, reeds of the same length, and so on) and are asked to state that they agree with respect to the characteristic A (number, length, or suchlike). Then one of the objects of the group is subjected to a transformation that according to adult insight does not change the characteristic A while other characteristics may be changed (for instance, changing the mutual distances of the chips in the row or bending the reed). After this operation the subject is asked whether the characteristic A has remained unchanged; if this is affirmed, one speaks of *conservation*, and the subject is classified as a "conserver". Psychologists are reasonably unanimous about the average age of conservation of the various characteristics, whereas people who have some didactical experience with children usually judge these ages absurdly high. The large percentages of non-conservers in psychological experiments are achieved by a particular strategy: The transformation that should be ascertained to conserve A is intentionally chosen so that it changes another characteristic B so drastically that the attention is diverted to B (for instance, if A is cardinal number or mass, a striking change of length, or if A is length, a striking change of position or shape). What is actually being investigated is whether the subject is able to separate these characteristics sharply from each other and how strongly he can resist attempts at misleading him. Built-in misleading is in general characteristic of the psychological, as opposite to the didactical, approach.

By no means should the question be rejected as to the stage of development at which children master invariances of certain magnitudes. On the contrary, it is a merit of Piaget's to have been the first to have formulated such problems. The problem, however, is obscured by the use of such terms as "conservation"; often the researchers themselves have no clear idea of the kind of transformations with respect to which the so-called conservation should be established. For each experiment designed to take in young children, one can contrive a more sophisticated version to embarrass adults. For instance, show a person two congruent paper clips and ask him whether they are equally long; the question is of course affirmatively answered. Then unfold one of them, straighten it out, and repeat the question. Whatever he answers can be wrong. It depends on what the experimenter meant. An adult subject would react to the question by asking, "What do you mean?" (In our terminology, length invariance under congruence mappings? Or under flexions?) Young children in the laboratory are not likely to ask counter questions. The fact that they do not ask proves that they are intimidated (in the terminology of the psychologist, "put at their ease") — their critical behaviour being eliminated by situational means.

For a good experimental design it is indispensable that experimenter and subject have a clear idea of the kind of transformations with respect to which invariance is to be established. Perhaps psychologists would answer that then the fun goes out of it, as the chance of getting wrong answers would be minimised. So much the better, I would say. Such a result would better agree with the opinions of children's capacities held by didacticians.

Of course this does not mean that all problems are disposed of. I could enumerate a lot of developmental problems that from the viewpoint of a sound phenomenology are interesting enough. For instance I would like to know whether constituting rigidity mentally precedes length, whether length invariance under congruence mapping and length invariance under flexions help or impede each other, what role is played by similarities in the mental constitution of length, and how the equivalence of "long" and "far" is acquired. So there are many more questions I would like to have answered. The most urgent question, I think, is about the significance of the break—make transformations for so-called conservation (not only of length). If I may trust my own unsystematic experience, I would consider them as crucial. Yet in order to answer such questions, a quite different design of experiments is required than that of snapshots, registering which percentage of subjects at a certain age do "conserve". Also required is a more positive mentality than that of tricking children into making mistakes.

Another vague term that is often used in that kind of research is "reversibility". Originally it was related to answers given by subjects when they motivate pronouncements on conservation. For instance, one of two strings of equal length is made crooked while the other remains straight; the subject is asked whether they are still equally long. If it is affirmed, the subject is asked to give reasons. If he answers "If straightened out, it is again the same", he shows "reversibility"; that is, the capacity to mentally reverse the transformation, which is considered a good argument for equality of length. Of course, it is no argument at all, and though it is interpreted by the experimenter as such, it was probably not meant that way by the subject. From the equality of initial and final states nothing can be derived about intermediate ones. If the subject had said, "They are equal because I got the one from the other by mere crooking", the answer would have been as good as, or even more to the point than, the argument of reversal. The subject, however, would not have been counted among the true conservers, because he lacked reversibility.

This "reversibility" as a proof for "conservation" is the original meaning, but subsequently it has been used in many other and mutually unrelated senses. There are, however, also researchers who reject the reversibility argument. They postulate standard answers that have to be given in order to establish conservation. Then the question "why is this as long as that?" must not be answered by a material argument but by a formal one, if the subject is to be classified as a conserver; he should answer something like "because they have the same length". To the question "why do they have the same content?", it must be "because they include equal parts of space". It goes without saying that such investigators are even farther away from meaningful mathematics.

The lack of insight into the difficulties with the equivalence between "long" and "far" has already been mentioned. Often they are increased by a stress on intentionally misleading connecting paths — a pattern in the plane that suggests a system of paths or two points on the rim of a round table that invite marching along the edge — where the experimenter had, of course, meant straight paths.

These details may suffice. I would certainly not judge that all the investigations I have in view are worthless, but many of them suffer from wrongly placed stresses. The method of snapshots need not be rejected but in order to be applied it requires a background theory — or at least ideas — about the intermediate development. Such theories do exist, but they are so vague and general that anything can be fitted to them and they do not provide criteria for attributing relevance to certain questions or complexes of questions.

What is lacking here can be made clear parabolically. Let us assume somebody is investigating the development of flora during the year by snapshots. On trees and brushes he notices various kinds of buds at various stages. In the next snapshots he identifies leaves and petals at the same places. Later on the former have remained whereas the latter have been replaced by fruit. Then the fruit, and finally the leaves too, have disappeared. He has not paid attention to stamens, pistils and insects and does not know where the fruit and leaves went. Perhaps he does not even know that the leaves and flowers were locked up in the buds. His phenomenology was utterly fragmentary, he did not know what

he had to look for, and there is a good chance that he will wrongly interpret
what he has seen. Perhaps terms like *growing, blooming, bearing fruit* are lacking
in his vocabulary, or they mean states rather than processes. Ideas about development would have given him a greater chance of noticing essentials.

1.26–29. MEASURING LENGTHS

1.26. *Yardsticks*

Measuring length requires instruments — measuring sticks or rules. At first the
measuring instrument will be smaller than the thing to be measured. Remarks
to the contrary in the psychological literature rest on misapprehensions about
measuring, or on artificial experiments.

The first yardstick I see used by children is the step. For a long time they
do not care whether all steps are equally long. Almost always they count one
step too many (the zero step as one). From the beginning it is clear that fewer
steps mean a shorter interval, though it is not as clear that composition of
intervals goes along with addition of numbers of steps. At about the same time
as measuring distances by steps, or somewhat earlier, one notices the activity of
jumping over a certain number of pieces in patterns of tiles in order to see how
far one can jump. I do not claim that this is really a measuring activity, though
this kind of jumping may influence measuring by steps.

Bastiaan (4; 10) spontaneously measured the width of a path by steps. "This is six further".
I show him I can do it in one step. He does the same with two steps. He continues measuring
by pacing.
    Bastiaan (6; 5) has made a large construction of roads, bridges, walls and tunnels in a
sandpit. In order to make a drawing of the construction he measures distances with his
two forefingers parallel at a fixed distance (about a decimeter), proceeding with the left
forefinger in the hole made by the right one.
    Bastiaan (almost 7; 6) measures distances with a span between thumb and little finger
which he knows is one decimeter.

Measuring with a measuring instrument means laying down the instrument
congruently a number of times until the length to be measured is exhausted.
If the object to be measured is a distance between two points, rectilinearity
of continuation must be practised as the measuring instrument is repeatedly laid
down. It is surprising that even 12 years olds may neglect this. If the straight
line between the two points is blocked, the path is partially replaced by a
parallel one. It is a remarkable fact that usually parallelism is better observed
than is the rectilinearity of the continuation in the non-blocked case. Indeed,
the latter is more difficult. To do this reasonably, one has to develop a certain
technique, which requires more geometrical insight than — unfortunately — is
being taught in the primary school.

There is a rich variety of yardsticks. Most of the traditional length units are
taken from the human body: inch (which means thumb), finger, palm, foot,

short and long ell, yard, step, double step, fathom; for larger distances the stadium (= 100 fathoms = 600 feet), the Roman mile (= 1000 double steps), an hour's walk. The so-called metric measures are related by powers of 10: metre, kilometre, centimetre, millimetre, micrometre, picometre. At variance with them: light year, parsec.

1.27. *Change of Yardstick*

If the object to be measured is not exhausted by applying the yardstick congruently a number of — say n — times, the problem arises of what to do with the remainder. In many cases one will resign oneself to the fact that a bit is left or is lacking, which means that the object is a bit longer or a bit shorter than n times the unit. Likewise the case where the remainder looks to be about half, one-third, or two-thirds the unit is not problematic. For greater precision a more systematic procedure is required. Two systems are familiar: common and decimal fractions. A less usual variation is binary fractions (or fractions with another base). A most natural system, now obsolete because of its complexity, is continued fractions, as I have explained elsewhere*. If a_1 is the measuring unit and a_0 the object to be measured,

$$a_0 = p_1 a_1 + a_2,$$

then the remainder a_2 $(< a_1)$ is used as a new unit,

$$a_1 = p_2 a_2 + a_3,$$

and so one goes on, expecting that eventually the division will terminate, that is

$$a_{n-1} = p_n a_n.$$

Then a_n is a common measure of a_1 and a_0, and by reckoning backwards, one will find, say

$$a_0 = r a_n,$$
$$a_1 = s a_n,$$

which implies

$$a_0 = \frac{r}{s} a_1.$$

It is an advantage of this procedure that it involves a systematic search for a a denominator, provided a_0 is truly a rational multiple of a_1; that is, if the procedure indeed stops. But this need not happen. Then the procedure has to be stopped at a certain stage, the remainder is neglected, and the length of a_0 is expressed approximately in terms of a_1.

With the methods of decimal fractions one is saved the trouble of finding

---

* *Mathematics as an Educational Task*, p. 203.

a suitable denominator. The measuring unit is again and again divided into ten equal parts (even if such a partitioning is not yet marked on the measuring instrument), and one has only to see how often the subdivided unit goes into the remainder. It is a disadvantage of the decimal method that even simple fractional lengths such as $\frac{1}{3}$ of the unit can only be indicated approximately.

Length is one of the concepts by which common and decimal fractions can operationally be introduced. This subject will be resumed in the chapter on fractions.

1.28–29. *Measuring Length at an Early Stage*

1.28. Terms that should occur early in measuring length are "double", "three times", "half", and "a third". It struck me that 5–6 year olds who reasonably understood length did not know these terms, or at least, not as related to length; the dominance of the adjective "big" seems to block applying "double" and "half" to the linear dimension.

Bastiaan (5; 3), at a certain moment during a straight walk at the other side of our canal between two bridges at a large distance from each other, does not understand the question "Are we half-way?", but later spontaneously indicates the point where the "middle starts" (that is, the second half).

Terms like "half full" and so on (of a glass) function earlier and better.

Additivity of length is still a problem at this age. A long object is paced off anew after it has been lengthened by a second object. It is not noticed that the second pacing gives another length for the first piece.

One should realise that these are not trivial things – knowing

how lengths are composed,
that the results are again lengths,
that pieces of lengths are again lengths,
that the length measure of a part is smaller than that of the whole, and
that length measures behave additively under composing.

1.29. The length of flexible objects is measured after straightening. The circumference of curved figures is measured by means of a flexible object – a string – laid along side. It can also be done by rolling the curve upon a straight line. It is not at all trivial that this yields the same result. The length arising from rolling a circle is grossly underestimated by children, and even by adults.

Conversely, rolling a wheel can be used to measure linear distances (expressed by the number of revolutions of a bicycle wheel or a measuring wheel).

Geometrical knowledge can lead to more sophisticated methods of measuring distances. Some of them are possible at an early age. We will reconsider this question later.

Reading and designing maps with distance data does not necessarily presuppose acquaintance with ratio.

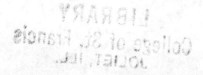

The relation between distances and the times needed to cover them does not necessarily presuppose an acquaintance with velocity.

Climbing stairs can be put into relation with distance.

The distances in a network of streets are accessible early.

CHAPTER 2

THE METHOD

2.1. *Aspects of Phenomenology*

I started with an example to be used as a subject matter which I can appeal to when I explain my method. I chose "length" because it is both a rich and relatively easy subject.

First of all, what of the terms "phenomenology" and "didactical phenomenology"? Of course I do not mean "phenomenology" in the sense that might be extracted from the works of Hegel, Husserl, and Heidegger*. Though the clearest interpretation I can imagine is that by means of the example of chapter I, which is to be continued in the following chapters, nevertheless it is worthwhile trying something like a definition.

I start with the antithesis — if it really is an antithesis — between *nooumenon* (thought object) and *phainomenon*. The mathematical *objects* are *nooumena*, but a piece of mathematics can be experienced as a *phainomenon*; numbers are *nooumena*, but working with numbers can be a *phainomenon*.

Mathematical concepts, structures, and ideas serve to organise phenomena — phenomena from the concrete world as well as from mathematics — and in the past I have illustrated this by many examples**. By means of geometrical figures like triangle, parallelogram, rhombus, or square, one succeeds in organising the world of contour phenomena; numbers organise the phenomenon of quantity. On a higher level the phenomenon of geometrical figure is organised by means of geometrical constructions and proofs, the phenomenon "number" is organised by means of the decimal system. So it goes in mathematics up to the highest levels: continuing abstraction brings similar looking mathematical phenomena under one concept — group, field, topological space, deduction, induction, and so on.

Phenomenology of a mathematical concept, a mathematical structure, or a mathematical idea means, in my terminology, describing this *nooumenon* in its relation to the *phainomena* of which it is the means of organising, indicating which phenomena it is created to organise, and to which it can be extended, how it acts upon these phenomena as a means of organising, and with what power over these phenomena it endows us. If in this relation of *nooumenon* and *phainomenon* I stress the didactical element, that is, if I pay attention to how the relation is acquired in a learning–teaching process, I speak of *didactical*

---

* Is it by accident that — with Habermas included — the names of the most pretentious producers of unintelligible talk in the German philosophy start with an H?
** *Mathematics as an Educational Task*, in particular, Chapters II and XVII.

28

phenomenology of this *nooumenon*. If I would replace "learning–teaching process" by "cognitive growth", it would be *genetic* phenomenology and if "is ... in a learning–teaching process" is replaced by "was ... in history", it is *historical* phenomenology. I am always concerned with phenomenology of mathematical nooumena, although the terminology could be extended to other kinds of nooumena.

2.2. *The Part Played by Examples*

The piece of phenomenology with which Chapter I began was clearly an *a posteriori* constructed relation between the mathematical concept of length and the world of long objects structured by an operation of composing, ⊕. Length was interpreted as a function on this world. I did not analyse how I arrived at this function. Although this was indispensable, I omitted it because I had to tackle this question in the didactical phenomenological section and I wanted to avoid duplication. But as a consequence the didactical phenomenological section contains pieces of pure phenomenology, such as Section 1.15 about the congruence mappings and Sections 1.18–19 about the flexions. Likewise in the sequel I will not clearly separate phenomenology and didactical phenomenology from each other. As promised in the preface I would not sacrifice readability to systematics.

Where did I look for the material required for my didactical phenomenology of mathematical structures? I could hardly lean on the work of others. I have profited from my knowledge of mathematics, its applications, and its history. I know how mathematical ideas have come or could have come into being. From an analysis of textbooks I know how didacticians judge that they can support the development of such ideas in the minds of learners. Finally, by observing learning processes I have succeeded in understanding a bit about the actual processes of the constitution of mathematical structures and the attainment of mathematical concepts. A bit — this does not promise much, and with regard to quantity it is not much, indeed, that I can offer. I have already reported a few examples of such observations, and I will continue in the same way. I do not pretend that at this or that age this or that idea is acquired in this or that way. The examples are rather to show that learning processes are required for things which we would not expect would need such processes. In the first chapter I showed a child suddenly confronted with the necessity to differentiate "big" according to various dimensions, a child placing "far" into the context of "long" and learning about the connection between "half" and "middle". I am going to add another story, which happened a few hours after the event where "half" and "middle" were tied to each other:

Bastiaan's (5; 3) sister (3; 3) breaks foam plastic plates into little pieces, which she calls food. He joins her, takes a rectangular piece, breaks it in about two halves, lays the two halves on each other, breaks them together and repeats the same with a three-layered combination – the fourth piece was already small enough.

I do not know where I should place this observation, whether I should classify it as mathematics, say geometry, or whether it belongs to general cognitive behaviour. I report this observation because I think it is one of the most important I ever made because it taught me a lesson on observing. I do not know whether the age of 5; 3 is an early or a late date for this kind of economic breaking; I do not know either whether Bastiaan imitated or adapted something he had observed before. I know only one thing for sure: that what he did is important and worth being learned. For myself it is fresh material to witness that in no way do we realise all the things children must learn. If I look at what people contrive to teach children, I feel inclined to call out to them: do not exert yourself, simply look, it is at your hand.

Why do people not look for such simple things, which are so worth being learned? Because one half of them do not bother about what they think are silly things, whereas those who do bother are afraid to look silly themselves if they show it. *Weeding and Sowing* is full of such simple stories. I told them in lectures. I do not care whether a large part of the audience interprets my reporting as senility, provided that by my example a small part of the audience is encouraged to follow suit – this, indeed, requires courage.

2.3. *Enactive, Ikonic, Symbolic*

Above I used Bruner's triad "enactive, ikonic, symbolic". Bruner* suggested three ways of transforming experiences into a model of the world: the enactive, the ikonic, and the symbolic representation. Corresponding to the dominance of one of these, he distinguishes phases of cognitive growth.

Bruner's schema can be useful. It has been taken over by others, and its domain of application has been extended, in particular towards the attainment of concepts in learning processes, where similar phases are distinguished. Later I will explain my objections to the idea of concept attainment as such, although I would not oppose the extension of Bruner's triad to concept attainment. As a matter of fact, in Bruner's work there is an example that shows how the three ways of representation can be extended to concept attainment: *enactively* the clover leaf knot is a thing that is knotted, *ikonically* it is a picture to be looked at, and *symbolically* it is something represented by the word "knot", whether or not it is accompanied by a more or less stringent definition.

There is a well-known pleasantry: ask people what "spiral" stairs are. All react the same way: they make their forefinger mount imaginary spiral stairs. Of course, if need be, they would be able to draw them. Does this mean that they are in the enactive or in the ikonic phase? Of course not. For the concept in question they possess a symbol, the words "spiral stairs", though if a *definition* is to be produced, one would have more or less difficulty in passing from the enactive or ikonic to the symbolic representation.

* *Studies in Cognitive Growth* (Edited by J. S. Bruner), Toward a Theory of Instruction, 1966, pp. 10–11.

Consider the number concept "three" and the geometrical concept "straight". Before the child masters these words, he can be familiar with what they mean: clapping his hands thrice and running straight to a goal if it is suggested to him (the enactive phase); sorting out cards with three objects or straight lines pictured on them (the ikonic phase). Mastering the word *three* (or *straight*) means he is in the symbolic phase, since "three" as a word is a symbol for the concept three (or "straight" is for *straight*). But likewise the three dots on dice can be a symbol; for instance, in playing the game of goose. A child that counts intelligently is in the symbolic phase even if this counting is accompanied by moving counters on the abacus. Adding on the abacus is enactive only for a moment. After the first experience it has become symbolic, though the symbolism differs from that of the written digits. The Roman numerals are as symbolic as the Arabic ones. Notches and tallies to indicate numbers belonged to the symbolic phase, even before people invented numerals — they are as symbolic as Roman and Arabic numerals. The cashier in the supermarket who prints amounts of money is neither enactively nor ikonically busy. A little child who claps his hands in joyfulness expresses his feelings symbolically even if he cannot yet pronounce the word joy. As early as kindergarten, children accept a drawing of a dance position where dancers are represented by strokes rather than manikins. If the doors of the men's and ladies' rooms are distinguished by plates of figures in trousers and skirts it does not mean that the decorator imagined the users to be in the ikonic phase; he did so because this difference is differently symbolised in the hundreds of languages that mankind speaks and writes — moreover the plates themselves are already symbols.

With these examples I intend to say that in learning–teaching situations, which are our main interest, Bruner's triad does not yield much. Bruner's domain of application is the psychology of the very young child, and in this period the phases can meaningfully be filled out.

2.4–5. *Concept Attainment and the Constitution of Mental Objects*

2.4. I would like to stress another idea, already stressed in my earlier publications. Let me start with a semantic analysis of the term "concept". If I discuss, say, the number concept of Euclid, Frege, or Bourbaki, I set out to understand what these authors had in mind when they used the word "number". If I investigate the number concept of a tribe of Papuans, I try to find out what the members of this tribe know about and do with numbers; for instance, how far they can count.

It seems to me that this double meaning of "concept" is of German origin. The German word for concept is *Begriff*, which etymologically is a translation of Latin "conceptus" as well as "comprehensio" and which for this reason can mean both "concept" and "(sympathetic) understanding". "Zahlbegriff" can thus mean two things, number concept and understanding of number; "Raumbegriff," concept of space and geometrical insight; "Kunstbegriff," concept of art and artistic competence.

Actually, in other languages too "concept" is derived from a word that means understanding (English, *to conceive*; French, *concevoir*) which, however, does not have the misleading force that the German *begreifen* has. I cannot say whether it has been the influence of German philosophy – in particular, philosophy of mathematics – that created the double meaning of number concept, of space concept, and in their train as it were, of group concept, field concept, set concept, and so on. At any rate the confusion has been operational for a long time and has been greatly reinforced by the New Math and by a rationalistic* philosophy of teaching mathematics (and other subjects) which in no way is justified by any phenomenology. It is the philosophy and didactics of concept attainment, which, of old standing and renown, has gained new weight and authority in our century thanks to new formulations. In the socratic method as exercised by Socrates himself, the sharp edges of concept attainment had been polished, because in his view attainment was a re-attainment, recalling lost concepts. But in general practice the double meaning of concept has been operational for a long time. Various systems of structural learning have only added a theoretical basis and sharp formulations. In order to have some X conceived, one teaches, or tries to teach, the concept of X. In order to have numbers, groups, linear spaces, relations conceived, one instills the concepts of number, group, linear space, relation, or rather one tries to. It is quite obvious, indeed, that at the target ages where this is tried, it is not feasible. For this reason, then, one tries to materialise the bare concepts (in an "embodiment"). These concretisations, however, are usually false; they are much too rough to reflect the essentials of the concepts that are to be embodied, even if by a variety of embodiments one wishes to account for more than one facet. Their level is too low, far below that of the target concept. Didactically, it means the cart before the horse: teaching abstractions by concretising them.

What a didactical phenomenology can do is to prepare the converse approach: starting from those phenomena that beg to be organised and from that starting point teaching the learner to manipulate these means of organising. Didactical phenomenology is to be called in to develop plans to realise such an approach. In the didactical phenomenology of length, number, and so on, the phenomena organised by length, number, and so on, are displayed as broadly as possible. In order to teach groups, rather than starting from the group concept and looking around for material that concretises this concept, one shall look first for phenomena that might compel the learner to constitute the mental object that is being mathematised *by* the group concept. If at a given age such phenomena are not available, one gives up the – useless – attempts to instill the group concept.

For this converse approach I have avoided the term *concept attainment*

---

* In the 18th century sense of *a priori* concepts epistemology.

intentionally. Instead I speak of the constitution of mental objects,* which in my view precedes concept attainment and which can be highly effective even if it is not followed by concept attainment. With respect to geometrically realisable mental objects (square, sphere, parallels) it is obvious that the constitution of the mental object does not depend at all on that of the corresponding concept, but this is equally true for those that are not (or less easily) geometrically realisable (number, induction, deduction). The reader of this didactical phenomenology should keep in mind that we view the nooumena primarily as mental objects and only secondarily as concepts, and that it is the material for the constitution of mental objects that will be displayed. The fact that manipulating mental objects precedes making concepts explicit seems to me more important than the division of representations into enactive, ikonic, and symbolic. In each particular case one should try to establish criteria that ought to be fulfilled if an object is to be considered as mentally constituted. As to "length" such conditions might be

> integrating and mutually differentiating the adjectives that indicate length, with "long, short",
> comparing lengths by congruence mappings and flexions,
> measuring lengths by multiples and simple fractions of a measuring unit,
> applying order and additivity of measuring results, and
> applying the transitivity of comparing lengths.

2.5. In opposition to concept attainment by concrete embodiments I have placed the constitution of mental objects based on phenomenology. In the first approach the concretisations have a transitory significance. Cake dividing may be forgotten as soon as the learner masters the fractions algorithmically. In contradistinction to this approach, the material that serves to mentally constitute fractions has a lasting and definitive value. "First concepts and applications afterwards" as it happens in the approach of concept attainment is a strategy that is virtually inverted in the approach by constitution of mental objects.

---

* Fischbein calls them *intuitions*, a word I try to avoid because it can mean inner vision as well as illuminations.

CHAPTER 3

SETS

3.1–9. *Sets in Advanced Mathematics*

3.1. *Sets as an Aim in Itself*

"Sets as an aim in itself" is a restricted, almost sterile, and hardly popular domain of mathematical research. Terms characterising it are *cardinality, continuum hypothesis, well-ordering, transfinite ordinals, alephs*. Up to the middle of the present century there was a certain need in algebra and analysis for transfinite ordinals, which subsequently were eliminated by means of the so-called Zorn's lemma.

For a while, general set theory was a subject of profound axiomatic research; in particular, the continuum hypothesis was in the focus of attention.

3.2. *Sets as Substrata of Structures*

Wherever sets in advanced mathematics are not an aim in themselves, they fulfill various tasks. For instance, they serve as substrata for structures — a metric space, a group, a category is a set with a number of properties. (Sometimes the term "set" is verbally replaced by "class" in order to avoid certain paradoxes of set theory.) Structuring the substratum set can happen in various ways, for instance:

> a set becomes a metric space by putting a distance function upon its pairs of elements,
>
> a set becomes a group by prescribing a certain operation between its elements, and
>
> a set becomes a category by imposing certain mappings between the member sets.

If sets serve as substrata for certain structures, they are in general not subjected to drastic set theory operations. *Sub*sets are formed, mainly to introduce substructures; substrata of structures are *mapped* on each other to define mappings of the imposed structures; set theory *products* are formed to get structure products of the imposed structures; and the substratum set is partitioned, for instance, as a set of *equivalence classes*, in order to derive new structures.

3.3. *Sets as a Linguistic Tool*

In an even weaker way, sets occur virtually as only a linguistic tool where some predicates are replaced by their extension. For instance, to avoid repeating the clumsy predicate "n times differentiable", one introduces the set C^n of n times differentiable functions and expresses "f is n times differentiable" by $f \in C^n$. A more recent acquisition is the conventional set symbols $\mathbf{N}, \mathbf{Z}, \mathbf{Q}, \mathbf{R}, \mathbf{C}$, which replace the predicates "... is a natural, integral, rational, real, complex number", respectively.

A certain operation on sets (though important as such — namely, forming the power set $\mathscr{P}(A)$ of A (that is, the set of subsets of A) — functions purely as a linguistic tool if the fact $C \subset A$ (C is a subset of A) is expressed by $C \in \mathscr{P}(A)$, or the fact that Ω is a set of subsets of A is expressed by $\Omega \subset \mathscr{P}(A)$ or even by $\Omega \in \mathscr{P}(\mathscr{P}(A))$. Sometimes the empty set occurs as a purely linguistic tool; for instance, if the non-existence of solutions of a certain equation is expressed by the emptiness of the set of solutions. Interpreting a mapping from A to B or a relation from A to B as a subset of the product $A \times B$ is also a purely linguistic use of a set theory operation (the set theory product). A more essential use of set theory language will be dealt with later.

3.4. *Sets in Topology*

Operating genuinely and explicitly with sets happened first in topological contexts. Not with the substrata of limits and convergence that are sequences and series rather than sets; nor even in the case of the maximum of a function was the set of its values originally made explicit. This changed when the need was felt to use and to define the upper bound of a function (or functional) that has no maximum — a phenomenon important in history because of Dirichlet's principle. In this case the formulation is indeed easier if the set of values is made explicit. A similar case: in order to prove that a continuous function vanishes somewhere in the interval between a negative and a positive value, one conceives the set of points where the function is not positive, and then takes the upper bound of this set. Upper bounds, lower bounds, upper limits, lower limits, and accumulation points are indeed an opportunity to explicitly introduce sets and to operate with them. In real and complex function theory the need is felt to consider open and closed sets, interior points and boundaries of sets — the drawings illustrating such concepts reinforce the set theory context. In "set theory topology" the adjective expressively indicates this context, although in "algebraic topology" it is no less influential.

3.5. *Measures*

Area, volume, and measure are functions on sets which, as it happens, can call for *explicit* sets on which they are defined. As long as it is simple "figures" to which an area or volume is to be ascribed, there is no urgent need for making the underlying set explicit; even the part of the plane delimited by the "horizontal" axis, a function graph, and two ordinates, whose area is expressed by an integral, need not be explicitly described as a set. The need really arises as soon as areas and volumes are to be attached to more or less arbitrary sets, that is in measure theory. Here *algebra of sets* becomes operational: unions and intersections (not only finite ones) and complements. Measures, that is, functions on sets with certain additivity properties, are an important organising tool for many mathematical phenomena. Probability is one of them.

3.6–7. *Solution Sets for Conditions*

3.6. Traditional geometry knew an explicit procedure for introducing sets as solution sets for conditions. Instead of *sets* one called them *loci* — the circle with centre M and radius r is the locus of the points at distance r from M. Among the set theory operations the intersection played an important part, arising from the combination of two or more conditions.

3.7. Considering within some structure the sets of elements maintaining a certain relation with a given element is an important principle in algebra, too. The multiples of 4 in Z form a subset: the set of numbers divisible by 4. The intersection of multiples of 4 and multiples of 6 is the set of multiples of 12 — a connection between intersection and least common multiple.

In algebraic structures one often focuses on subsets closed with respect to some operations. A set theory pattern, which is typical, may be illustrated by an example taken from group theory: Let G be a group and A a subset of G; one asks for the subgroup of G generated by A, that is, the group obtained constructively by starting from A and applying over and over the operations of product and inverse. However, the subgroup of G generated by A can also be obtained in one blow: defining it as the smallest subgroup of G containing A, or in still another way as the intersection of all subgroups of G containing A. The same pattern works similarly in rings, fields and similar algebraic structures.

We meet with sets when considering divisibility properties. The multiples of a given element a in a commutative ring R form the (principal) *ideal* generated by a. Starting with a set A in R and forming all linear combinations of elements of A with coefficients from R, one gets the *ideal* generated by A. Ideals I in R are characterised as non-empty sets of R with the property: $a, b \in I \wedge c \in R \to a - b \in I \wedge ca \in I$. The ideal generated by A in R is the smallest ideal of R containing A, or otherwise, the intersection of all ideals of R containing A. In Z every ideal is a principal ideal; this is equivalent to the existence of a greatest common divisor of a subset A of Z. For instance, the ideal of Z generated by by 4 and 6 is also generated by their greatest common divisor 2. Ideals have been invented and the theory of ideals has been developed to master divisibility in such rings where the existence of the greatest common divisor is not assured, that is, where not every ideal is a principal ideal. Ideals were in history the first explicit occurrence of sets and set theory methods in algebra.

3.8. *Zorn's Lemma*

Applying set theory in advanced mathematics, as mustered up to this point, is light or even very light guns. The heaviest — and then downright heavy — is Zorn's Lemma. Its use is in eliminating the transfinite ordinals of olden times, and the way to prove it is similar to that of formerly providing a set with a well-ordering. Zorn's Lemma runs as follows:

Let Ω be a non-empty set of sets (partially) ordered in the natural way, that is, by means of inclusion. A chain Θ is a subset of Ω such that for each pair A, $B \in \Theta$ we have either $A \subset B$ or $B \subset A$ (thus a chain is *totally* ordered). Suppose that the union of each of its chains also belongs to Ω. (Ω is closed with respect to forming unions of chains.) Then — Zorn's Lemma asserts — Ω possesses maximal elements, that is elements $X \in \Omega$ such that if $X \subset Y$ then $X = Y$, for every $Y \in \Omega$.

By means of Zorn's Lemma one can prove, for instance, that each group G possesses maximal proper subgroups: Let $a \in G$ not be the unit element of G; take for Ω the set of all subgroups of G that exclude a. Let Θ be a chain in Ω; then the union of Θ is obviously a subgroup* that excludes a, and thus belongs to Ω, which shows that the condition of Zorn's Lemma is fulfilled. According to Zorn's Lemma Ω possesses a maximal element, which is a maximal proper subgroup of G.

This proof is characteristic of applications of Zorn's Lemma. A concept from this complex of ideas that is crucial in modern algebra and analysis is that of *filter*:

Let R be a set. A filter F of R is a set of subsets of R with the properties:

> the empty set is not a member of F,
> $A \in F \land B \in F \rightarrow A \cap B \in F$, and
> $A \in F \land A \subset C \rightarrow C \in F$.

By means of Zorn's Lemma one shows that each filter of R is contained in a maximal filter, different from R (also named an ultrafilter).

If $a \in R$ then there is one maximal filter of R containing the one-element set $\{a\}$, namely the set of all sets of which a is an element. But otherwise as soon as R is infinite, maximal filters are almost pathological objects, beyond intuition and construction: each maximal filter F of R possesses the perplexing property:

> for each $A \subset R$ either $A \in F$ or $R \backslash A \in F$.

---

* Indeed: If $x, y \in \bigcup_{H \in \Theta} H$ then $x \in H_1$, $y \in H_2$ for certain $H_1, H_2 \in \Theta$; now $H_1 \subset H_2$ or $H_2 \subset H_1$, and in the first case $x, y \in H_2$, thus $x^{-1}y \in H_2$, thus $x^{-1}y \in \bigcup_{H \in \Theta} H$, and similarly in the second case. This line of reasoning is characteristic of applications of Zorn's Lemma.

3.9. *Cardinality*

All sets occurring in the applications shown hitherto had some structure, which was at stake in the set theory operations performed on them. This is also true of Zorn's Lemma. Indeed, Ω was a set of sets and by this very fact bears the structure of pairs of member sets being part of each other.

The absolutely unstructured set still possesses one characteristic: cardinality. Sets that can be mapped one-to-one upon each other have by definition the same cardinality, and if A can be mapped one-to-one into but not onto B, A is by definition of smaller cardinality than B. This then operationally defines cardinality and order in cardinalities. After this definition, it does not matter much whether cardinalities themselves are introduced as mathematical objects. In fact, it can be done by putting sets "having the same cardinality" into one "class", which is named their cardinality.

First of all, cardinality should be appreciated as a historical-philosophical phenomenon: the courage to extend an elementary concept like number, which apparently needs no analysis, to infinite sets and to defy the seeming paradox that an infinite set can have the same cardinality as some true subset. Technically viewed, cardinality appears to be important for five reasons:

Firstly, equal cardinality of sets and subsets can be exploited in a positive sense.

Secondly, the countability of the sets of integer, rational, and even algebraic numbers — at first sight unexpected — allows unexpected constructions in this field.

Thirdly, the uncountability of **R** guarantees in a simple way the existence of non-algebraic numbers and in a more general way that by the difference of cardinality one set can be distinguished as a *true* extension of another.

Fourthly, the unexpected phenomenon that forming the product of an infinite set by itself does not increase its cardinality and, as a consequence, that line segment, square, cube, and so on have the same cardinality, is the source of the problem of how to distinguish dimensions, which has been solved by paying attention to more structure, namely topological structure.

Fifthly, the well-known drawer principle: if a set A is being mapped in a set of lower cardinality, at least two elements of A have the same image.

3.10. *Sets as a Purely Linguistic Phenomenon*

The reader may ask why I have elaborated so much on the phenomenon of sets in advanced mathematics, at the risk of supplying no more than verbal information. In fact what I did was not much more than to lift a tip of the magic veil spread over set theory by innovators in the past, who claimed to have shown that set theory, hitherto a privilege of advanced mathematics, could be successfully taught in primary school and even kindergarten. After the "back to basics" reaction it is still or even more necessary to analyse those pretentions because they are characteristic of an approach to designing mathematics education that lacks any phenomenological background.

Set theory as meant by those innovators was quite another subject than is meant in advanced mathematics. In today's school mathematics sets are not an organising device for mathematical (or non-mathematical) phenomena, but an aim in themselves. Intersection, union, complement, and power set are not introduced when and where the subject matter asks for these organising devices. Instead a subject matter has been created to exercise and train these operations. It should be noticed that this is not an unusual way of creating school teaching matter. Rather than developing set theory schemas as organising tools from subject matter that asked for such organising and schematising, empty boxes are taught and, in order to appease one's didactical remorses, filled with (false) concretisations. Sets originating this way always remain within the palpably concrete sphere or are purely linguistic phenomena. Collecting a finite number of objects in a set, which as a mental object nobody has asked for, only in order to apply set operations on it, is one aspect of this false concretisation. Another is the so-called Venn diagram; a third, logical blocks. Later we will stress that the proper problem with sets is to grasp and recognise them *abstractly*.

Set as a purely linguistic phenomenon expresses itself in this kind of school mathematics in particular by the cult of creating sets in extension. Every predicate can be transformed into a set — indeed, the set of things sharing the predicate. From ". . . is red" one can form the set of red things; from ". . . has long hair and wears spectacles" the set of long-haired spectacle-wearers, which in turn is the intersection of the long-haired people and the spectacle-wearers. Or closer to mathematics: from ". . . is more than 7 or less than 3" one forms the set of the numbers more than 7 and less than 3, to be represented as the union of the set of numbers > 7 and the set of numbers < 3. Or "divisor" can be transformed into the set of pairs $\ulcorner x, y \urcorner$ such that x is a divisor of y. In advanced mathematics sets are created in extension where they are needed. In the kind of school mathematics we have in view, sets are training matter, and what is trained are things to which only side attention is paid in advanced mathematics. Consider the parallelism between logical and set theory operations:

$$\text{to} \wedge \text{(and)} \quad \text{corresponds} \quad \cap \text{(intersection)},$$
$$\text{to} \vee \text{(or)} \quad \text{corresponds} \quad \cup \text{(union)},$$
$$\text{to} \neg \text{(not)} \quad \text{corresponds} \quad R \backslash \ldots \text{(complement)}.$$

The last line exhibits what is wrong with this parallelism: in order to have "not" correspond to complementation, one must know with respect to which domain the complement has to be taken. Here I indicated it by R. In general it is known under various names: basic set, universe, choice set, reference set. It is a conception occurring in school mathematics only; in normal mathematics it is entirely unknown and this for more profound reasons than the difficulty with the negation.

True mathematics is a *meaningful activity* in an *open domain*, rather than a haphazard one in an *a priori* fixed reference set. I admit that in set theory axiomatics, it is postulated that from every object one can make a one-member set, that for every set of sets the union exists, that every set has its power set, and that every predicate has its extension. In the school mathematics I sketched, one feels obliged to exert oneself to realise all these possibilities. But in true mathematics there is not the slightest need for, say, the union of the set of natural numbers that are divisible by three and not the power of a prime number, with the set of finite subsets of a three-dimensional vector space, with the set of all sunsets. Sets are formed and used where they are needed.

With regard to the parallelism of logical and set theory operations, I would not exclude the need — at a certain moment — to illustrate an equivalence like

$$\neg(p \wedge q) \leftrightarrow \neg p \vee \neg q,$$

or perhaps even

$$(p \vee q) \wedge r \leftrightarrow (p \wedge r) \vee (q \wedge r),$$

by its set theory counterpart. As soon as these logical equivalences provide for real wants, nobody will raise objections. Even then, exaggerations and wrong concretisations should be avoided. By wrong concretisations I mean concretising the corresponding set theory equalities

$$R\backslash(A \cap B) = (R\backslash A) \cup (R\backslash B)$$
$$(A \cup B) \cap C = (A \cap C) \cup (B \cap C)$$

dogmatically by means of Venn diagrams rather than by rectangular partitions

	B	R\B
A	$A \cap B$	$A \cap (R\backslash B)$
R\A	$(R\backslash A) \cap B$	$(R\backslash A) \cap (R\backslash B)$

Fig. 1.

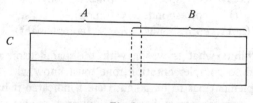

Fig. 2.

(Figures 1 and 2). It is pedantry to go further if there is no reason for it, and in particular trying to extend this parallelism to "→" and "⊂" — such is the whole ritual of sets as a purely linguistic phenomenon.

3.11. *The Good Reference Set*

A few lines above I stated: "True mathematics is a *meaningful activity* in an *open domain*, rather than a haphazard one in an *a priori* fixed reference set." Curriculum developers and textbook authors apparently believe they offer teachers and pupils firm ground under their feet as soon as they restrict mathematical activities to a fixed reference set. This, they think, prevents possible surprises that might be disappointing. Fixed reference sets have been a source of confusion − in particular in probability (which we are obliged to anticipate here) − and have provided rather than firm ground under the feet, a swamp of contradictions. In traditional probability, one can also learn what should really be done if reference sets are tried, and this I will explain here.

Take a die. The six possible results of a throw form such a reference set R. Related propositions are of this kind: the result x of a throw is even, or is < 3, or is > 6. Such propositions determine subsets of R; these subsets form a "Borel system", where one can play set theory algebra and form unions, intersections, complements.

Yet this reference set is not of any real importance. In fact, one promptly passes to considering throws with $2, 3, \ldots, n$ dice, or rolls of one die rolled $2, 3, \ldots, n$ times, or together with the die a coin showing head or tails, and so on. This requires forming ever and ever new reference sets, arising from the old ones by forming products; that is, products of different sets, powers of the same, and products with a finite (or even an infinite) number of factors. One lives, as it were, with an open reference set which at any moment, if need be, can be enriched by adding new factors or impoverished by neglecting factors. A colloquium lecture on a probability theme may start with the announcement that the lecturer supposes the reference set (more precisely the probability field) so large that every stochastic variable he will introduce is meaningful if conceived as a function on this set. It is just by means of stochastic variables that the probabilist can bring any reference set he means to the fore. For instance, the result of the nth throw of a certain die is considered a stochastic variable x_n defined on a probability field, which has among its factors at least the nth throw of that die though it can have many more factors.

It is a point of view that works very well wherever reference sets are at stake. Let us illustrate this. Consider the following two reference sets:

R_1, the set of flowers with predicates like
... is red, ... is multicolored, ... is tall,

R_2, the set of animals with predicates like
... is a mammal, ... lives in Africa, ... is extinct.

As soon as both are to be placed within the same context, the correct reference is not obtained by throwing R_1 and R_2 on one heap but by forming their product

$$R = \ulcorner R_1, R_2 \urcorner.$$

In this way any proposition p on R_1 can be considered as one on R, which is not influenced by the second component; similarly any proposition q on R_2 can be considered as one on R, which is not influenced by the first component. By this interpretation any proposition

$$x \in A$$

about R_1 ($\supset A$) is jacked up to a proposition

$$\ulcorner x, y \urcorner \in \ulcorner A, R_2 \urcorner$$

and any proposition

$$y \in B$$

about R_2 ($\supset B$) to a proposition

$$\ulcorner x, y \urcorner \in \ulcorner R_1, B \urcorner.$$

The proposition

the flower is red

on flowers is jacked up to a proposition on

$\ulcorner$flowers, animals$\urcorner$

by stating that

in the pair $\ulcorner$flower, animal$\urcorner$ the flower is red,

which of course is equivalent to

the flower is red,

and this means that we can dispense with the more complicated form and even need not introduce it.

In order to make the connection between propositional logic and set algebra, indicated in Section 3.10, we have only to ensure that all propositions under future consideration (or all predicates) deal with the same set of objects — that is, for the predicates, have the same domain. This is achieved by means of the product of the domains of all propositions occurring in the context, jacking up — at least in theory — each proposition to one about the new domain, the reference set of the given context. Extending the context involves enriching the reference set by new factors. Restricting the context means omitting factors of the reference set; that is, projecting it on a poorer one, with a restricted system of factors.

Rather than an organisation of mathematics, the present explanation organises a system of propositions on reality by mathematical means, in particular, by set theory. It is less relevant for mathematics than it is for relating mathematics to its applications. The most embracing reference set which, starting from reality, can come under consideration might be dubbed: world. In this sense "world" is neither a set of space and time points, nor a set of objects in space-time, nor any physical datum whatever, but the reference set of the context dealing with reality. "World" is the product of an awful number of factors such as "flowers", "animals", "colours", "throws of coins", "throws of dice", "moments", "points", "persons", "feelings", "thoughts", and whatever one may think of.

If I pronounce some proposition such as

>the flower is red,
>the animal lives in Africa,
>the die shows a six,

I occupy myself with one factor, one facet of the "world", while disregarding the others. If I pronounce the three of them together, I am grasping more of the "world". Yet this is still a small reference set. Realistic contexts require a much broader reference set — actually but also potentially by their openness, by the potentiality of extending.

The reason why this phenomenological analysis is less relevant for mathematics than for applications (for instance, probability, where it originated) is to be found in the peculiarities that distinguish mathematical language from the vernacular, in which applications are usually formulated. I shall return to these peculiarities, but I will anticipate the most essential one: The variables of mathematical language are omnivalent in principle; letters can indicate anything, whereas any restriction of domain must be made explicit. Such variables are rare in the vernacular — the variable "something" is akin to mathematical variables. Domains of variables in the vernacular are most often extremely restricted; "flower" can be applied only to flowers, "now" only to moments. Yet this also offers us an advantage: the linguistic symbol promptly displays the reference set, the facet of the "world", to which it is restricted, and by this fact doubts about the reference set of a context are virtually excluded. It is the same situation which we meet with stochastic variables, which are not variables in the usual mathematical sense. If I denote the result of the next throw with a die by $\underline{x}$, then the proposition

$$\underline{x} = 3$$

has a different meaning from

$$x = 3$$

in a mathematical context. In a mathematical context the variable x must be bound in some way in order to occur in a proposition, but regarding the proposition $\underline{x} = 3$ I can directly pronounce judgments such as "... is true" or "... has the probability $\frac{1}{6}$".

3.12. *Elementary Structures*

From a didactical phenomenological point of view implicit and explicit uses of sets are to be distinguished. As a matter of fact for mental objects to function well it is not required and often not even beneficial that they are made explicit. This is particularly important for the didactical phenomenology of sets as mental objects because, unlike numbers and geometrical objects, the vernacular knows no terms to indicate sets in general. Nevertheless sets as substrata of structure must play a role early in cognitive development even though they are neither recognised nor isolated.

Sets, as they emerge in spontaneous development, are heavily structured. This is not surprising. Attention is more strongly attached to structures than to what lacks structure, and the inclination towards structuring that which lacks structure dominates that of destructuring structure, at least as far as building rather than destroying activities are concerned.

As examples I shall indicate such structures. First, the *structure of succession* or *file*, placing objects besides or behind each other, for instance to make trains for riding or putting blocks upon each other, perhaps according to size. Or the same in a rhythmic-acoustic or rhythmic-gymnastic way, combined or not with files of solid objects. Counting in the early meaningless manner — or in a wrong sequence — shows the same pattern. Creating from a given file a new one by regular skipping (for instance when jumping in a tile pattern) is a more sophisticated way of structuring, as is the regular inserting of new elements into a file.

By being provided with periodicities (have beads of different colour regularly alternating in a string), the file is transformed into a *repetition structure*. With repetition structures one can didactically distinguish: passive recognising, imitating, and describing. Repetition structures are developed not only linearly but also in the plane and the space — it would be of interest to know in which stage the potential infinity of the repetition structure becomes conscious. Akin to the repetition structure is the *cyclic structure*, arranging objects cyclically — persons around a table or a centre or the mental structures of the day, week and year cycle.

More topological structures are: the *detour structure* (the child makes a detour after which it meets the adult guide again), the *border structure* (walking on borders), the *enclosure structure* (a true or symbolic enclosure of oneself or somebody else by means of drawn or otherwise marked enclosures), the *barrier structure* (blocking the path with the arms stretched), the *hide-and-seek* structure, the *look-around-the-corner structure*, the *labyrinth structure*, the *island structure* (objects are declared to be islands within a surrounding sea).

This list of general structures can be complemented by more or less specialised ones: the box with blocks that cannot be packed arbitrarily, the puzzle that must be composed in a definite manner (that is, where the parts are mutually closely related); the family in general, or in varying realisations, structured according to generations, sexes, kinship, the content of the toy-cupboard, structured with respect to a lot of relations, the seating accommodation in a car or bus, a pack of cards.

In appropriate places I will reconsider these structures, which can be processed into a wealth or didactical material. The enumeration was meant only to evoke an idea of how strongly the early cognitive possession is dominated by structures. Sets as substrata of these structures are in a few cases easily recognised, in other cases it is harder or even difficult. I mean: for the author and his readers. Does the child notice the substratum set? One should ask rather whether it would be relevant to him.

3.12a. *Order Structures*

As far as elementary structures are concerned, those that lend themselves at the earliest stage to mathematising seem to be the order structures — linear and cyclical order (with partial orders as a rather deranging phenomenon). They, too, are obtained from richer structures by impoverishing — in a particularly effective way. At the start this richer material is of a quite concrete character, that is, felt by hands, feet, senses. In a developmental line the first order relations might be

> spatial inclusion,
> spatial succession.

Where

> temporal succession

is to be placed is not easy to say. Everybody knows the difficulties that many — though not all — little children have with "yesterday" and "tomorrow": they use the same word for both of them, and as by preference, it seems, "tomorrow". What is wrong here? The mental grasp of the time direction, the idea of something like a future that in fact is (still) non-existent? Is the preference for "tomorrow" to be explained by the greater frequency of the word?

Daphne (3; 8) gets a number of sweets. Mother says: "This we will keep for tomorrow." Daphne: "And this for yesterday, and that for Thursday". Mother: "Thursday is today, isn't it?" Daphne: "Then for Friday".

Order relations are expressed in

> comparing spatial and temporal dimensions and quantities,

characterised by comparatives such as

> longer, shorter, thicker, higher, older, more, fuller, and so on,

by prepositions such as

> in front of, behind, after, between, and so on,

by adverbs such as

> first, then, more, less, and so on.

— I would not be able to say in which succession. Farther away, though perhaps only partially, is

> comparing intensities

by means of

> heavier*, hotter, sweeter, dearer, nicer, and so on.

Order relations are

> perceived,
> stated,
> actively imitated,
> created,

for instance, if the child follows a prescribed path, from one stone or step to the next, or if, more or less consciously he chooses his path, jumping from one stone to another in the neighborhood, climbing from one branch to the next.

Much concrete material — paths, stairs, strings, straight and crooked linear objects, nests of objects — suggest, a

> global order — that is, comparing totally,

though under less concrete circumstances — with temporal or intensive criteria — this global suggestion can be lacking. Whereas in the first case the global order is

> accepted and — at most — analysed afterwards,

it happens in the other cases that the global order is

> synthesised from the local one by pairwise comparison.

The link between local and global order then is

> transitivity

— two local mutual situations when combined yield a third, which is mathematised by

$$a < b \land b < c \rightarrow a < c.$$

Transitivity, if not imposed by definition, is an empirical feature. Cognitively viewed, the complex of order relations as it develops itself is a complex of justified — and sometimes unjustified — analogies, linguistically supported by comparatives and superlatives such as

---

* Developmentally "weight" starts as an intension.

      big, bigger, biggest,
      sweet, sweeter, sweetest,
      much, more, most,

and so on. Putting moments, weights, intensities, and so on, on a number line may arouse the suggestion of global linear order.

    Necessary — and perhaps even sufficient — for the constitution of

      one mental object "ordered set"

seems to me the

      operational mastery of transitivity*

and the

      operational inverting of order.

    Necessary for the constitution of

      the mental object "linear order"

seems to me the

      operational mastery of endo- and isomorphisms of orders

and the

      rejection of only partial, as it were, total order.

    Our exposition has increasingly focused on linear order. *A priori*

      cyclic orders

are

      globally

given;

      spatially

— around the block, around the tree, the closed string of beads —

      temporally

— the day cycle, the week cycle, the merry-go-round, continued repetitions. A restricted grip on the global character is within the scope of the mental object possible by means of

      local analysis

* Possibly as early as the age of four, as shown by P. Bryant, contradicting Piaget. See Chapter 1.14.

– an ordered triple determines the difference between a cyclic order and its converse
– neighborhood relations determine the cyclic order,
but formalising

>what corresponds to transitivity

is far away in the conceptual domain of cyclic order.* However, for constituting cyclic order as a mental object, the

>operational mastery of endo- and isomorphisms, and
>operationally inverting

look even more important than in the case of the linear order.

In the planar geometrical context the cyclic order is enriched by

>right and left turn,

which will be considered at appropriate places.**

3.13. *The Elimination of Structure*

Recognising the substratum set requires thinking away the structure based on it. This can be quite difficult if the structure has arisen from a natural order or topology as is the case with most of the examples of general structures. It may be easier, though still difficult, with structures generated by classification. The colours in a supply of beads are more easily thought away than the structure of the human family. It is easiest when the elements are not, or hardly, distinguished from each other, and unrelated (chips of the same kind, a flock of sheep, the reverse side of a pack of cards) or – another extreme – if the whole looks like a hotchpotch, which does not invite structuring.

Didacticians who undertook to make children familiar with explicit sets have, as it appears, struggled with the problem of how to eliminate structure in order to evoke the suggestion of an unstructured set. The most natural method – indistinguishable objects – has been the least common, probably because of a wrong concretisation of the requirement "that the elements of a set should be different". The most usual paradigm of the bare unstructured set is the hotchpotch – the elements are being put into a Venn diagram or between braces in such a way as to kill any suggestion of order, higgledy-piggledy, and if they are numbers or letters, criss-crossed in the diagram, and in a chaotic order between the braces. All that could be experienced by the child as order, is set at defiance – the order in the family where every member knows his place at table, has his own bed to sleep in, the order in the tidy toy-cupboard, in the flower garden – so as to evoke the set deprived of any structure. It is suggested

---

* Cp. *Mathematics as an Educational Task*, p. 472 sq.
** Sections 10.3 and 14.12–16.

that as soon as objects figure as elements of a set, they are freely movable in space — an ontology based on a wrong phenomenology. Indeed, in the great majority of cases the elements of sets are *mental* objects, and as soon as point sets are discussed, the pupil is asked to relearn what he has been taught to unlearn: that for the individual elements of point sets the only thing that matters is the place.

In spite of the chaotic order in the Venn diagram or between the brackets, the suggestion that the place does not matter is still not strong enough. On the contrary, teacher and pupil must explicitly be informed about it, as they must about the possibility that — contrary to the verbal suggestion — a set is allowed to consist of one or even of no element. Why is it allowed? By prescription, since this is the only way to introduce such objects as one-element sets and the empty set as long as there is no need for them.

The didactical means by which children are forced to constitute structureless sets are kill or cure medicine, like the horse-drench. It is impossible to check whether and how they function, because in fact the constitution of the mental object is skipped and replaced by a verbalism that does not cover any mental object. These verbal products are afterwards subjected to purely linguistic operations, as demonstrated earlier.

3.14–16. *Equality*

3.14. The approach through blinkers is the most disturbing feature of these horse cures. If sets are to be constituted close to reality a broader context is required. "Close to reality" does not mean Venn diagrams of pictures, but a living context.

School texts often explain: A set is determined if for each object it is known whether it is element of the set or not. This is misleading. If it were literally true, I would not be able to speak of sets of natural numbers without precisely enumerating all their elements. It is as misleading as the suggestion that in order to become elements of a set, objects must be stripped of their position.

The decisive feature, however, is what equality means in the case of sets. In order to speak about sets, one must know criteria of equality. Do two pictures represent the same family? Does this picture represent the family across the street? Are these the same marbles that were here last week and that are now in the bag? Do A and B have the same sorts of trees in their gardens? (Notice that rather than — concrete — trees, I said sorts.) Do A and B have the same school marks?

Of course, in order to answer such questions a broader context is required. We should first know something about the pairwise equality of the elements belonging to such sets, that is, know what equality means with simple objects. Afterwards we can look for the composition of the sets in question.

By "broader context" I mean formally the ontological question of what is equality, that is, criteria in general, not necessarily restricted to sets. Not absolute criteria, but criteria determined by the context itself. As a context I now choose, as regards subject matter, closeness to reality, and as regards didactics, young children. I am going to sketch a broad theme.

3.15. Imagine a number of pictures of the same church building, taken from different directions, under different angles, perhaps one before the restoration, one with scaffolding, one after the restoration. Is it the same church? – a question which can be discussed, affirmed, and denied. The result does not matter. What really matters is the insight that criteria are required to answer such questions and that there is nobody to offer you the criteria.

In such a discussion analogies will be cited as arguments. A man in different clothing, with one tooth more or less, a woman who got a different name by marriage, a mother-to-be with a child in her belly – are they different persons? "I feel a new man", "Saul became Paul", "the saltpillar called Lot's wife" – themes of endless discussions.

The most likely final result is: The church building is continually the same, but "the church building previous to restoration", "the church building under restoration", and "the church building after restoration" are different things. Mathematically formalised: From C (the church building) and t (time) we get the pair $\ulcorner C, t \urcorner$, thus besides the object C we have new objects $\ulcorner C, t_1 \urcorner$, $\ulcorner C, t_2 \urcorner$, $\ulcorner C, t_3 \urcorner$ – all different. In a similar way the other examples can be dealt with: the church building according to different aspects, Saul–Paul before and after Damascus, and so on.

If the church building, or Paul, is considered to be a simple (not composite) object, then the church building previous to, during, and after restoration, Paul before and after Damascus, are composite (mental) objects. The way of composing, however, by mental pair formation, of course differs from that of composing a family, a construction box, a jigsaw puzzle, the seating accommodation in a bus or car, a pack of cards. Composing in the latter sense is how sets are composed of elements. Of course, as was stressed earlier, in all such examples, in order to get to the substratum sets, the structure has to be eliminated, which is not always easy. Sometimes this can be done by means of criteria of equality. The family remains unchanged when its members move, the construction box and the jigsaw puzzle are the same however their parts are arranged (though the orderly packed state can be considered as a special object), the pack of cards is not changed by shuffling; the seating accommodation in the bus is the same whether occupied or not.

Yet all these examples show much more structure, which cannot be done away with by these operations. Family, construction box, puzzle, quartet game, seating accommodation keep their individuality not only as sets but also as structures (unless parts get lost or are changed).

Let me illustrate this equality of structure by another example. Take a ludo board* such as you can buy in a shop – a quite simple structure with a

* A cyclic arrangement of squares with four entrances on which four gamblers move their men according to certain rules.

set of squares as set substratum. Take a second specimen, exactly congruent. Is it the same object? No, not as a concrete ludo board. But you can consider both of them as pictures of the same — abstract — ludo board. As a matter of fact, the boards may even be of different manufacture, of different size, in other colours and nevertheless serve as images of the same object. (If by chance they have different numbers of squares, one should consequently distinguish ludo games with different numbers of squares, as with the game of draughts, where there is the international and the Polish—German variant.) The concrete boards appear as pictures of the mental board in the same way as you can make various snapshots of a family or consider a construction box, a puzzle, a pack of cards, the seating accommodation in a bus as an abstract pattern, which can be realised in various — congruent or non-congruent — ways. The term "picture" which I used, evokes the idea of a two-dimensional image, while other examples suggest visual or instrumental realisations, but reciting the same poem in various ways or singing a song can also be interpreted as a picture of a more abstract object. A more comprehensive term than picture would be model* — the concretisations are models of the same abstract structure.

The structure of the ludo board is what is called in mathematics a directed graph. The square fields can be thought of as being replaced by their centres (the nodes of the graph) and neighboring ones connected by arrows in the direction they are passed through in the course of the game. This is a simplified topological image of the usual ludo board. In the background there is an even more abstract structure, lacking any geometric appeal: the set P of nodes and on it the special relation $R(x, y)$ (there is an arrow from x to y). However, the set substratum can also be chosen differently: along with the set P one considers the set Q of arrows, while adding the relations $R_1(x, y)$ (node x is the tail of arrow y) and $R_2(x, y)$ (node x is the point of arrow y). Choosing the set substratum of a structure is a matter of taste and practice.

I should add that I have analysed the structure of the ludo *board* only, not that of the *game*, which is much more involved. Then a set of pawns should be added, a concept of position — which is a relation between nodes and pawns —, a rule as to how a position can change into another, data on the starting position and a rule on the consequences of a final position. Here I have restricted myself to the *ludo board*.

I will now reconsider the other examples of structures. First of all, the family, or rather a particular family, say, of two adults, father and mother, three children, boy, girl, boy. This structure looks more variegated than that of the ludo board. There is a linear order on it according to age, there are classifications according to generations and sex, and there is a genealogy. Another structure is *the* family, of which the particular families are models. It must allow for more

---

* As explained in *Weeding and Sowing* (p. 130 sq.) this is one of the diametrically opposed meanings of "model", that is, as *after-image*. Model as *pre-image* would just be the abstract ludo board, the abstract construction box, after which the concrete ones have been manufactured, or the poem or song on paper as conceived in the author's mind.

than two generations, for more complex and disconnected genealogies. The structure of the construction box is determined by classes of geometric shapes, perhaps also by functional classes. The jigsaw puzzle reminds one of a graph — representing the particular pieces by means of nodes and neighborhood by means of joining line segments — although such a structure would be too rough to allow one to reconstruct the puzzle from the graph, which does not determine the jigsaw puzzle structure sharply enough. The true structure is the geometric division of the rectangular point set according to the jigsaw pieces. If I focus on this structure, I only neglect what is stuck or painted on the pieces.

3.16. When the set theory frenzy reached the Netherlands, it was the stamp collection that was raised to the rank of a paradigm for sets — indeed, the plain meaning of the Dutch word for set is collection. So one could even argue why all elements of a set should be different: a serious philatelist does not insert more than one specimen of a postage stamp into his collection. Following this argument the 50-cent stamps in stock at midnight on 1 August, 1975 at the Utrecht Central Post Office would not constitute a set.

One can distinguish — at least — two different concepts of postage stamp: the concrete printed and gummed piece of paper, and the sort of which this piece of paper is a specimen. (Different philatelists may adhere to different notions of sort; some of them know more sophisticated sorts than do others, but this does not matter here.)

What then is a stamp collection? The set of printed pieces of paper? Yes and no. The collection of philatelist X grows, is transformed by exchange, but does not change its identity. Well, this is not a new point of view; we can account for it by speaking of X's collection at time t. Is this really the set I mean? If in this set I replace the specimen of a certain sort actually or mentally by another of the same sort, it remains the same stamp collection although the set has changed.

The philatelist does not collect pieces of paper but sorts. He does not collect so as to own a sort. The stamp collection is not a set of sorts. In a stamp collection sorts are represented by specimens — not all sorts are represented and perhaps some sorts are represented by more than one specimen.

In the gardens of A and B are the same sorts of trees — we met this example earlier. In all $Y-Z$ bookshops you can buy the same books, at all $M-N$ ice cream stands you can buy the same ice creams, the countries P and Q have the same fauna. Of course, a book as a concrete object can sit in one bookshop only, two ice creams are never the same, and one lion is not another.

The concept that fits these situations is *assortment* rather than set, the same assortment of stamps, trees, books, ice creams, animals, respectively. One considers the set Ω of all stamps, trees, books, ice creams, animals, respectively. This set is divided into sorts, which form a set Σ. An assortment is a set of sorts, a subset of Σ.

It seems a rather weak concept: such an assortment could be empty, indeed. In itself, such an assortment is poor: a set with a bit of structure — its elements being sorts. In the stamp collection case it is no more than a list of the respective sorts. However, this set bears a strong external structure by its relatedness to the set Ω of all stamps and the various stamp collections of the other philatelists. The assortment gets its significance from the fact that it is an assortment of the set Ω, partitioned into sorts. In this way it becomes meaningful to ask whether two philatelists share the same assortment of stamps (although their collections are separate), or whether the one is larger than the other, or whether they overlap, or have nothing in common (different countries or ages).

Consequently, a stamp collection is a set (of pieces of paper) plus an assortment; the set is, as it were, externally structured. A group or a linear space can be studied in itself. On the other hand, one can stress the relations of the group, or the linear space with others. Then one acts in what is called the category of groups or linear spaces; there the external structure of the particular group or space expresses itself by the so-called morphisms of the category.

3.17. *Structures as Mental Objects*

Except in artificial examples and exercises, sets are usually endowed with, and are dependent on, structures and can be grasped through these structures only. As a substratum a set becomes explicit if the structure is recognised and consciously eliminated. This has been the meaning of the preceding discussion. The examples of structures we chose were taken from common experience. Artificial ones need not be rejected provided one realises their deficiencies. Logic blocks, for instance, suffer from four ills: All is prestructured, structure is restricted to classification, a closed (even finite) mathematical universe is mirrored in order to exercise set algebra, and finally, all relevant predicates always determine one unique element together. These four ills are the expression of a systematism that, like unfortunately all systematisms, attracts didacticians of mathematics. (Of course, reasonably applied, logic blocks may be a useful tool.)

Recognition of structures and – through the structures – of substratum sets is a schematising abstraction, which in the didactical process deserves concrete support, though not so strong a support as is supplied by logic blocks. One is better advised to draw pictures that reflect structures and substrata schematically, for instance, along with the ludo board the schematising graph. For classifying according to one characteristic, the most suggestive picture is the set of bags (Figure 3) and in the case of double classifications two sets of crossing bags (Figure 4). A pack of cards is schematised as the product of

$$\{2, 3, 4, 5, 6, 7, 8, 9, 10, J, O, K, A\}$$

and

$$\{C, H, S, D\}.$$

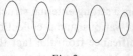

Fig. 3.

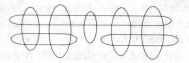

Fig. 4.

There are many ways to represent assortments schematically: by means of a check list, by a schematic page of a stamp album where some places are filled in and others are not, by a page of a stamp catalogue with some items checked. Yet with all these representations the set Ω remains in limbo. In order to remedy this defect one can combine these pictures with that of the classification.

We will encounter more of this kind of schema. Again we stress that they *do not concretise concepts* – as do Venn diagrams – but are *derived from real structures by schematising abstractions* and have to serve for *constituting mental objects*. Whether and how fast and by which stages one passes to *constituting concepts* is another question. Stages would be characterised by their verbal machinery. One can restrict oneself to simply introducing words like "set", "group", "ordered set", "graph", "directed graph" for this or that, in the way one speaks about numbers and addition, one can denote particular structures by letters, and in formalising the notation of a structure one can even go so far as to distinguish formally the substratum and the structuring relation (such as with an ordered set $\ulcorner S, < \urcorner$). It depends on the level of formalising whether these are pure names used to call up objects or whether one intends to describe mental operations with the objects by formal operations with their names. From traditional arithmetic instruction one knows that formalising is possible at an early age – the first formalised activity a child is taught is column arithmetic. It is also known that premature formalisms can be pernicious. One should profit from these experiences in the didactical subject matter that is dealt with here; introducing formal machinery only where it covers mental objects and where it is required to describe and facilitate operations with these objects. With the view on this issue fixed, I will review a few set theory concepts.

3.18–3.20. NUMEROSITY AS A SET THEORY OBJECT

3.18. *Constitution of Cardinal Numbers*

At present I will not deal with the constitution of the cardinal numbers; this will be done in Chapter 4. Yet as far as constitution of sets and operations on sets are concerned, something must be anticipated. In the constitution of numerosity (or cardinal number) I distinguish the following activities:

Eliminating from structures with the same substratum the structuring component in order to arrive at substratum sets,

and in this connection transforming the inclusion relation into an order relation ("less" instead of "contained in").

Using isomorphisms of structures with different substratum to compare different sets.

Using transitivity of equality and order (of numerosity number).

Intentionally I did not add:

structure-free comparison of sets,

which by set theory is, as it were, ordained for comparing cardinalities, but as important as it might be at the level of concept formation it is irrelevant and ineffective for the constitution of mental objects — which will be shown immediately. Meanwhile I shall soften this strong assertion. There is one exception: whether one set has *many* more elements than the other can be decided with the naked eye (or ear), without structuring the sets. There are early developmental stages in which a short sequence of constituted small numbers leads to an indefinite "many" (which can be expressed verbally by a *definite* numeral). Then the development progresses by differentiating "many". The mental constitution of "many" could, at least partially, be performed by a procedure of structure-free comparison, and in this way structure-free comparing would influence, via the motor of "many", the development of numerosity as a mental object.

I review the three activities which at the start of the present subsection I pointed out as being essential for the constitution of numerosity. In order to clarify what I said, I review some well-known experiments: Two rows of objects are laid down in such a structure (Figure 5) that by means of the greater

· · · ·
· · · · · ·

Fig. 5.

length of one row the larger number of objects in that row is made visible: the isomorphism of one of the structures with a substructure of the other and transitivity are ostensively used. Then the longer one is compressed to make both of them equally long. Thus in order to maintain the order of numerosity, one has to eliminate the change of structure and again apply transitivity. Such experiments are considered by psychologists as tests of "conservation", they do not, however, unveil the rock on which the so-called non-conservers foundered.

What these three activities mean for the constitution of numerosity number will be discussed later. At present we are concerned with the part played by sets – implicitly or explicitly.

Of course, one may first ask: "do they really play a part?" As I have stressed repeatedly,* numerals as cardinal numbers – indefinite ones included – are a different linguistic category than adjectives: if I want to say about a football team not that each of its members but the team itself is good, it is not enough to say "eleven good football players"; I shall rather attach the adjective "good" to the object "eleven": a good eleven, or a good team. Numerals, however, are by their very essence related to collectives of things, the particular football player cannot be "many" or "eleven", but it should be: eleven football players. This collective, however, is not explicitly indicated as it is in the case of the football team: the numerals stamp by their mere use the object they are attached to as a collective.

Well, it is possible that in the first developmental stages the numerals – as well as the non-verbalised mental numerosities – are concerned with structures rather than with sets. At least with regards to the number 2, this is a quite natural assumption. At an early stage the child's attention is drawn to natural pairs: two eyes, two ears, two hands, the pair of parents, sun and moon, twins. Is the "two" at this moment already mentally constituted or is it a new step that the "two" is recognised in unstructured sets of two? Indeed, the pairs we just cited are more or less strongly structured, the first three, in particular, by "right and left"; but also externally structured because a union of pairs of eyes, and so on, displays a natural classification structure which extends to the pair of parents; "sun and moon" show a clear internal structure; in the case of the ("identical") twins it is weak or absent.

Natural triples are less conspicuous, natural quadruples, however, are, for example the legs of many animals, and tables, the wheels of cars – strong internal structures in the first and third cases, and external ones in all of them. Natural quintuples, such as the fingers of the hand, are imposed explicitly on the child as means of constitution of a number. This imposition may not be too successful, as a consequence of the strong structure of the system of fingers, stressed by nursery rhymes, and impeding rather than favouring the constitution of number: it is always the same triple or quadruple of fingers that is lifted if children accompany the answer to the question of their age by this obligatory gesture.

Structures may impede the constitution of number but they are also, as seems to me, indispensable in instigating this process. I repeat the red currants story:

At the rectangular table Bastiaan (4;3) is seated opposite his younger sister, father opposite mother, grandpa opposite grandma. At the dessert of red currants he suddenly lifts his little spoon in the greatest agitation and ejaculates "So many are we!" Indeed they were six.

* For instance, *Weeding and Sowing*, p. 215 sq.

I asked him: "Why?" and he answered "I see it so", and then "two children, two adults, two grandpa and grandma". Possibly the six currants lay on the spoon in the same configuration of six as we occupied around the table, but this I could not see. At that time Bastiaan was still quite unsure with numbers and obstinately refused to count. This event distinctly marks in his development the constitution of numerosity numbers.

This is one of the observations that taught me how important for the development of mental numerosities is the structured comparison of sets. Adults impose a special kind of structured comparison: by means of the ordered set of natural numbers, N, used to count out a given set S. This asks for two remarks: On the one hand, for *comparing* sets, N is not indispensable. Often, yet not always, N is useful as a mediator between sets that are to be compared. There are cases, examples of which we shall deal with later, where thanks to mutually related *structures* on the sets that are to be compared, the intervention of N is inappropriate. On the other hand, the intervention of N is in general insufficient to settle the cardinal number of a set. Unless this number is very small, settling the cardinal number of a set requires the use of an extant structure or the creation of a new one, in order to be sure that by the counting process S is one-to-one exhausted. The elements of V might be given in a spatial or temporal linear order, as a string of beads, as a passing train, as strokes of a clock. If this is lacking, an order must be imposed on V. How it should be imposed is not trivial and is a matter of learning. It can be done by a kind of *coordinate system*: a line from the upper left to the upper right is mentally constructed, a second line below it, and so one continues to the lower right corner. A variation is the so-called ploughing scheme; first line from left to right, second from right to left, and so on, alternating. In a horizontal and in a three-dimensional field, one may start in the rear and progress to the front, or follow the inverse way. The use of polar coordinates also occurs — spiralling from the centre to the exterior, or the other way round.

Structuring by analysing according to coordinates can be promoted in the learning process by means of examples in which such structures are conspicuous. The learner is seduced into strengthening them where they are weaker, and to imposing them when they are absent. Another way of structuring is by means of *classification*: if the elements of V are different in shape, colour, or size, these characteristics can be used for classification. Neighborhood as a classifier can lead to local groupings in S, mentally or physically marked by enclosures.

It is my intention to stress the significance of structures in the mental development of numerosities and numerosity and for the factual determination of cardinal numbers. Though the cardinal number is a characteristic of the structureless set, its development and application can be effective only through structures (at least with regard to definite numbers, rather than indefinite ones, which are indicated by words like "many" and "few"). The didactical consequences are clear. One should not try to have the constitution of numerosities depend on structureless sets, as system fanatics prefer nor on the counting-out structure of N adhered to by parents, when teaching their children the number sequence.

I restrict myself to this remark. I did not aim at a phenomenology of number. This will be developed later. Then methods of determining numerosities in the reality will be systematically reviewed.

3.19. *Cardinals of Sets Connected by a Structure*

At this point I will discuss structures that may serve to determine cardinals and relations between cardinals.

First of all, comparing cardinals of sets that are mutually connected by a structure, which may be realised by a more or less explicit mapping between the sets S_1 and S_2. Examples:

> the nose of x; V_1 a set of people, V_2 their noses — by counting noses the number of people in V_1 is determined;
>
> ballot paper assigned to x; instead of the votes cast for a person, one counts the ballots;
>
> hooks and eyes;
>
> ticket for seat x in the theatre; the unoccupied seats are counted by means of the unsold tickets;
>
> counting out a set by numbering its elements by means of numerals or written numbers.

A more sophisticated relation between two or more sets that can lead to a comparison according to cardinality is alternating arrangement. The cards of a pack are equally distributed among four persons A, B, C, D by dealing one card (or any number of cards) at a time while alternating cyclically, starting with A. In a directed string of beads where colours A, B, C, D follow each other cyclically, there are equally many of each colour provided the string starts with A and finishes with D. Such structures determine mappings where each element (except the last) is mapped upon the next, and these mappings settle the equality of numbers.

With planar structures similar phenomena can be perceived: A chess board has as many black as white squares. The same holds for any like structure of m by n squares if m or n is even. For $m = 2$ (Figure 6), the squares in each row

Fig. 6.

are mapped upon those in the other to shows that there are as many white squares as black. From this the general assertion follows for all even m by splitting the structure into strips of two lines. The same method shows that for odd m and n the numbers of black and white squares differ by one.

A similar phenomenon is the planar tesselation into equilateral triangles of alternating colour or other regular and regularly coloured tesselations (Figure 7 and 8). At a glance one states that globally each colour is equally frequent,

Fig. 7.

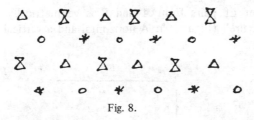

Fig. 8.

though it requires caution to cut out of the infinite pattern a finite piece where indeed all colours occur equally often.

There are cases where the cardinal equivalence of sets is shown by a one-to-one mapping rather than by counting out. The mappings are natural ones inspired by geometry rather than arbitrary ones like those used with the structureless sets pictured in Venn diagrams. According to my experience pre-school children are able to recognise the equality globally albeit without motivating it; they are also able to construct equipartitions of stocks of similar objects by alternately dealing them out to persons or laying them on separate piles. Young school children are even able to reason about it.

3.20. *Cardinalities of Sets in a Union Structure*

The formulas

$$\#(A \cup B) = \#A + \#B \quad \text{for } A \cap B \text{ empty}$$

and

$$\# \lceil A, B \rceil = \#A \cdot \#B$$

and similar ones with more terms describe in a formalised way facts that are basic to certain methods of counting and comparing sets. These methods are not at all self-evident or trivial.

When second graders in groups of two were asked to count large quantities of objects (sticks, chips, thumb-tacks, paper clips, and so on), those who had not followed the experiment program in the first grade did not hit spontaneously on the idea of sharing the work, whereas those who had followed it, immediately shared the work.

It is one thing, if a set is divided into subsets, to act according to the addition formula; it is another thing, even if the formula is mastered, to divide a set intentionally in order to use the formula. For all the structures we are going to deal with, it is the same experience: using given structures or introducing new ones can require abilities of different levels. Everything depends on how distinct or how vague the structure is.

3.21. *Cardinalities of Sets in a Rectangle Structure*

The abstract set of pairs from A and B is geometrically visualised by the "rectangle structure" (Figure 9): A horizontal and a vertical order structure is

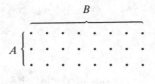

Fig. 9.

imposed on A and B, respectively. Almost as effective with small sets is the rhythmic model of pairs, such as those known, for instance, from reciting the number sequence (twenty-*one*, twenty-*two*, twenty-*three*, ... , thirty-*one*, thirty-*two*, ... , forty-*one*, with strong stresses at the end) or from reading a matrix (*a-one*-one, *a-one*-two, *a-one*-three, ... , *a-two*-one, *a-two*-two, ... , with stresses on the first number).

But even if the product structure is visualised or rhythmically scanned, its recognisability can be vague. In a meeting room with m rows of n chairs each, arranged in straight columns, the product structure is clearly visible. If the chairs of subsequent rows shift the width of half a chair (as do the stars in the American flag), the two intertwining product structures must be unravelled before applying the product formula. Or the product structure can display holes or other irregularities, which are first filled up or mended to make the product structure clear. The number of type-writer touches on a sheet of paper is structured according to lines, but the column order structure of touches, though indicated on a rail, is rather weak.

In the case of equicardinality of the two factors the product structure is reinforced; in its geometric visualisation as a square the equality of cardinals is easily recognised. The square structure invites one to conceive other structures as parts of it. For instance the "border structure" (Figure 10), which can be understood as the difference of two square structures, and the triangle structure (Figure 11), completed to a square structure, which in turn can be interpreted as the union of two congruent triangle structures intersecting in the diagonal, leading to the numerical evaluation

$$n^2 = \Delta + \Delta - n,$$

thus

$$\Delta = \frac{1}{2}(n^2 + n).$$

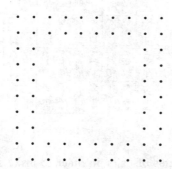

Fig. 10.

Fig. 11.

All these geometrical visualisations — rectangle, square, triangle — need not be realised in the euclidean geometrical way; the pattern may be deformed affinely or even topologically, provided the structure according to rows and columns remains more or less clear. Instead of a linear order the columns can be endowed with a cyclic one; then the product is borne, as it were, by a cylinder; or if the rows are also cyclically ordered, by a torus. The column elements can be the faces of a die; the product set then consists of sequences of faces of dice, placed in a row.

On the other hand the euclidean structure can play an essential part if sets are to be compared. With a row or a tower of congruent objects the length or the height, respectively, can be a criterion of the cardinality. A large number of such objects can be counted by arranging them in rows or towers of 10, where the number 10 is verified only once by counting, and assured subsequently by congruence.

If the rectangular structure of a set of pairs is not given *a priori*, its mental image can be supported by visual images. If from *m* boys and *n* girls all pairs boy-girl are to be formed, the actual pairs cannot be present at the same time.

However, with *n pictures of each boy* and *m pictures of each girl* all *pictures of pairs* can be formed (Figure 12); the rectangular pattern then serves to

Fig. 12.

arrange the $m \cdot n$ pictures of pairs, in order to make sure that none is missing and none is counted twice. Once this has been understood, the pictorial component can be dropped, and a diagrammatic rectangular schema, stripped of pictorial features can take over, which in turn is replaced by a mental schema.

3.22. *Cardinalities of Sets in an Equipartition Structure – m Baskets with n Eggs Each*

I stressed elsewhere* that in many cases the arithmetical product operation cannot – or not without reinterpretation – be justified by the set theory product. In *m baskets with n eggs each* the resulting $m \cdot n$ eggs suggest a set product, which in fact is absent. The structure involved is a mapping structure, which I will name *equipartition*. The set C of eggs is mapped into the set A of baskets – egg x lying in basket y – and in this mapping f each element of A has the same number of originals ($\#f^{-1}a = \#f^{-1}b$). The corresponding formula is

$$C = \#A \cdot \#f^{-1}a \qquad (a \in A).$$

In the present case the equipartition can somewhat artificially be restructured as a product: In the original set of each element of A (that is, in each basket) the elements (that is, the eggs) are arbitrarily numbered from 1 to $\#f^{-1}a$; by this procedure the particular sets $f^{-1}a$ $(a \in A)$ are mapped one-to-one upon each other, such that C is represented as the product of

$$A \text{ by } \{1, 2, \ldots, \#f^{-1}a\}.$$

In other cases the particular $f^{-1}a$ $(a \in A)$ might be naturally related. The set of legs of n cats is structured on one hand by the cats and on the other by the set {left foreleg, right foreleg, left hindleg, right hindleg}.

* *Mathematics as an Educational Task*, pp. 189, 250.

If on each of its n pages a book contains two pictures, one above and one below, the number of pictures can be obtained as well by n times two pictures, as by two times n pictures, namely, n upper and n lower pictures. Many examples in the early phase of learning multiplication are of the kind n times m where n is large and m is small. As long as the tables of multiplication are not yet sufficiently memorised and multiplications are performed as (n times) repeated additions, such multiplications can be arduous. Teachers are inclined to impose the commutativity of multiplication and its application. Instead, one can proceed conceptually by strengthening the structure to such a degree that it is seen as m times n — in the present case, n times two pictures as n upper and n lower pictures, thus twice n pictures.

In other cases it can be utterly useless to strengthen the equipartition as to form a product structure. If in a club A of n persons a president and a secretary are to be chosen and the number of possibilities is asked for, the set of "boards" is structured according to president and to secretary. Let f be the mapping that maps each board on its president; then the original sets $f^{-1}a$ (for various a) cannot meaningfully be mapped upon each other with the aim that the equipartition is strengthened to become a product structure; attempts at doing this can only cause confusion. Didactically the equipartion is needed as a structure of its own.

The example of the last paragraph still admits of another structuring: the ordered pairs formed by *different* elements of A (nobody can be both president and secretary at one time), which leads to the number

$$\#A \cdot \#A - \#A = n^2 - n = n(n-1).$$

However, if more factors are at stake (president, secretary, treasurer) this filling up to full products becomes laborious.

3.23. *Roads Model of the Product*

There is no use in imposing the rectangle image of the set theory product by every means, and certainly not if in the long run this model would push aside or even suppress more effective models. In the example I have often cited of "three roads from A to B, two roads from B to C, how many ways from A through B to C?" an image is suggested that fundamentally differs from the rectangular one. Of course, one can help the pupil recast it into the rectangular pattern, but the rhythmical structure of the description of ways is certainly more effective. Anyway, it is desirable that the pupil encounters the road structure — once disentangled — as a structure of its own that embodies the set theory product. Whereas visualising set products by rectangles can be continued to at most three factors (visualised by boxes instead of rectangles), visualising by means of road structures knows of no restrictions on the number of factors. The road structure, initially a problem situation that is elucidated by the set product, is raised to the rank of a model of its own, the *road model* of the set product. (Figure 16) It is an extremely flexible model; it admits of an arbitrary number of factors. It invites the embodying of products with "equal" factors by running to and fro between *A* and *B* (*n* equal factors by running *n* times) (Figure 13). It also invites problems like that of president and secretary

Fig. 13.

(from Section 3.22) to be embodied by running to and fro while not repeating roads already used. The set of sequences of length *k* out of a set with *n* elements is embodied by running *k* times while roads used earlier are prohibited.

3.24. *Tree Model of the Product*

Consider the task of colouring flags, the upper bar by three colours, the middle by two others, the lowest again by two others, all prescribed. After a number of failures, the task is structured by the pupils according to "upper bar, middle bar, lower bar". This can be an action structure ("first colouring all flags with an upper bar black", and so on) if the flag patterns to be coloured are given. In the process of describing, a verbal structure, possibly rhythmically accentuated, can develop. Schematising produces the *tree model* (Figure 14).

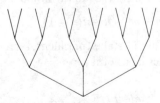

Fig. 14.

A variant of the tree model is the *substitution model*, which can produce beautiful patterns (Figure 15).

The tree model, which is useful in many applications, does not account adequately for the set product, since the equality of branches of higher order is not visualised. By identifying the nodes on the same level one can pass from the tree model to the road model (Figure 16).

3.25. *Equipartitioning Relation*

A relation between two sets *A* and *B* can be visualised as a subset of the rectangular image of $\ulcorner A, B \urcorner$, although this is hardly illuminating. A more effective illustration is by means of the set of connections between the related elements of *A* and *B* (Figure 17), and in the case of *A = B* by directed connections (arrows).

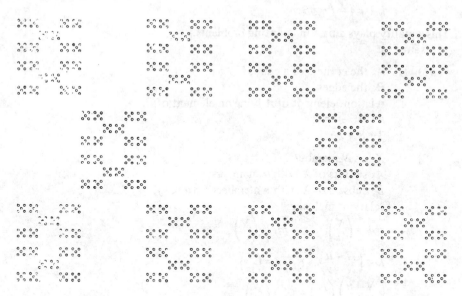

Fig. 15.

Fig. 16.

Fig. 17.

In the case of the full relation — each element of A related to each of B — this gives an image of the set product of A and B. Although this picture is less appropriate as a model of the set product, it is desirable that pupils recognise the set product therein.

In combinatorics an important part is played by what I would call an *equipartitioning relation*: A relation between A and B such that each element of A is related to k elements of B, and each of B to l elements of A. Then in the model $k \cdot \#A$ connections leave from A, and $l \cdot \#B$ from B, which leads to the equality

$$k \cdot \#A = l \cdot \#B.$$

This equality plays a part in counting problems.
  Examples:

(1) A: the corners of a cube,
          B: the edges,
          relation: element of A lying on element of B,
          $k = 3, \quad l = 2,$
          $3 \cdot 8 = 2 \cdot 12.$

(2) X has N members,
          A: subsets of X with u members,
          B: subsets of X with v members $(u < v),$
          relation: inclusion.

$$\#A = \binom{N}{u}, \qquad \#B = \binom{N}{v},$$
$$k = \binom{N-u}{N-v}, \qquad l = \binom{v}{u},$$
$$\binom{N-u}{N-v}\binom{N}{u} = \binom{v}{u}\binom{N}{v}.$$

It is worthwhile signaling the different character of these examples. The sets
A and B of the first example (and similar examples) as well as the relation
between both of them can be given visually. In the second example – even if
u and v are numerically specified – the sets A and B are given purely verbally
(though the description can be simplified by speaking of u-tuples and v-tuples
from X rather than of subsets with u and v elements, respectively). Notwithstanding this verbal definition, the sets A and B might be mentally constituted
in this context, thanks to the visual power of the relation of inclusion between
elements of A and B.

3.26. *Structure of the Power Set*

A similar phenomenon is shown by $\mathscr{P}(X)$, the set of subsets of X. Again it is a
verbal definition, which can become operational only in rather strongly formalised
mathematics. If on a lower level $\mathscr{P}(X)$ is to be constituted as a mental object,
the rather strong and visually active structure of $\mathscr{P}(X)$ must not be neglected.
This structure arises from the inclusion between the elements. For small X the
power set $\mathscr{P}(X)$ can be considered as a graph, directed by inclusion (Figure 18

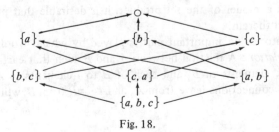

Fig. 18.

with $X = \{a, b, c\}$), which visualises the structure of $\mathscr{P}(X)$ schematically. By the transition from $\mathscr{P}(\{a, b, c\})$ to $\mathscr{P}(\{a, b, c, d\})$ an induction can be prepared, leading to the formula

$$\#\mathscr{P}(X) = 2^n \quad \text{if } \#X = n.$$

3.27. *Model of the Set of Subsets with u Elements*

Let us reconsider the set A of u-tuples from a set X with N elements.* To constitute A mentally, in particular with a view to determining $\#A$, there are other methods than that indicated at the end of Section 3.25. I will enumerate some of them.

(1) One introduces the set C of $(u - 1)$-tuples from X and the relation of inclusion between A and C, which is equipartitioning.

(2) One introduces the set D of sequences consisting of u different elements of X (earlier embodied, for instance, by means of the tree or the road model of Figures 14 and 16), and the "forget" mapping of D on A (forgetting the order) that maps a sequence into its substratum set.

(3) One introduces an order in X, interprets an element of A as a choice sequence of length N consisting of u choices of "yes" (belongs to A) and $N - u$ choices of "no" (does not belong to A), and visualises this by means of a Galton board (Figure 19), with $N + 1$ stories, where "no" means a fall to the left, and

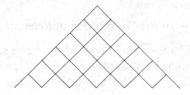

Fig. 19.

"yes" a fall to the right. All choice sequences with u "yes" choices finish at the uth place of the lowest story (counted from the left, starting with 0). So the set A is embodied by the set of all paths on the Galton board finishing at the uth place of the lowest story.

3.28. *Drawer Principle*

The last examples especially show the importance of an extant or imposed structure for the constitution of sets as mental objects. Elsewhere** I have investigated the difficulties pupils can have with the problem we generally describe as the drawer principle. It appears that pupils who have no difficulties with problems of the type

---

* N and u should be taken as definitely fixed numbers.
** *Weeding and Sowing*, p. 210 sq.

> Are there in any school class children that have their birthdays in the same month?

can be greatly embarrassed by the type

> Are there two people in the world with the same numbers of hairs on their heads?

or

> Are there two match boxes with the same number of matches?

The analysis showed as the source of the difficulty not the cardinal aspect of number but the constitution of the decisive sets. In the first example the relevant sets are the children in a class and the months of the year, which are immediately recognised. This immediateness is lacking in the other examples, although they look isomorphic to the first one. Of course, the set of people and the set of matchboxes are not problematic, whereas the other — superficially viewed, the set N — has at first sight not enough relevant structure to be helpful. But in order to be helpful an external structure must be imposed on N — as the set of possible numbers of hairs on heads or of possible numbers of matches in match-boxes — if the drawer principle is expected to work. In problems where the cardinal aspect of numbers plays a part, it can easily happen that by the data of the problem N is externally structured and that recognising this structure is the very thing that matters.

3.29. *Formal Machinery for the Constitution of Relevant Structure*

With the last example I have ventured beyond the frame set out in the beginning of 3.19 — to discuss structures that play a part in determining cardinalities and relations between cardinalities. The drawer principle is instead a method for gaining information about certain mappings from the inequality of cardinal numbers. I now take up the thread from the end of Section 3.17, where I promised to define the minimal formal machinery needed to constitute as mental objects such sets as appeared in the paradigmatic examples I displayed. Obviously this machinery will comprise much less than the means by which I presented my examples, although even this representation could have been formalised much more heavily. I could also have opted for less formalising. Instead of N and u I could have said 20 and 4. It would have been extremely difficult to avoid the letters X, A, B for sets; the presentation would probably have become incomprehensible or cumbersome if I had tried it. I could have avoided the symbol # though not the word "cardinal number" or something like it. I could no more easily have renounced words like "set", "union", "product", and so on, because it was just their didactical place that was being discussed. In teaching, however, and particularly in teaching young children, one can at most aim at doing something with sets, unions, products, and so on, rather than describing their status.

In Section 3.18 I gave and illustrated criteria that aimed at the cardinal comparison of more or less clearly structured sets. The terminology used when exercising this ability at pre-school age, is characterised by such words as "as many as", "more", "less". As many black as white, as many stars as rounds, or more, or less. If not in kindergarten, then in the first grades, one can afford to follow such a statement by the question "Why?". After each black a white, after each star a round — this could be the answer, and the pupil could make such statements more precise by the language of arrows. This, indeed, would be a new symbolic means, enriching the pupil's language paradigmatically. It should be used to describe or delineate a visible mapping rather than — as happens in modern methods — to construct an arbitrary, artificial, unnecessary mapping. Is there any reason or need to introduce verbal constructions like sets of whites, blacks, stars, rounds and to speak of mappings of such sets? I do not think so. "The set of all whites has the same number of elements (or the same cardinal, or cardinality) as that of the blacks" is a weighty expression by which mathematical language is intentionally dissociated from the vernacular — a dissociation of language that at least suggests, if not instigates, or even puts the seal on, a dissociation of realities.

The additivity of cardinal number in the union of disjoint sets is mathematically the principle that defines addition of cardinals, and is *a posteriori* a fact to be stated and an expedient in counting. "x chips and y chips together is $x + y$ chips" (with of course, definite numbers instead of x and y) is the minimal linguistic means to verbalise this fact. In set language it is done by the formula

$$\#(A \cup B) = \#A + \#B.$$

At the level of, say, primary school it is the only opportunity to use the symbol $\cup$ (there are somewhat more opportunities for $\cap$). Between the formulation in the vernacular and its formalisation by means of the above formula many intermediate stages can be distinguished, like "if I form the union of disjoint sets, their numbers are added". What good can be gained for understanding addition by teaching a formalisation that transgresses the vernacular? Nothing, as far as I can see. The didactical problem of the arithmetic operations is — besides teaching algorithms — a problem of application, which I shall reconsider. The pupil must learn, for instance, that an addition is required to know how much money he had yesterday if he has this much now and spent that much in the meantime, and that he should add 7 to his own age to know the age of somebody 7 years his senior. Sets do not help much in this case. Even though I can make a set of seven marbles, with seven florins or seven years this is an artificial if not meaningless concern.

Compared with the addition formula there are better reasons for explicitation of the multiplication formula

$$\#\ulcorner A, \bar{B} \urcorner = \#A \cdot \#B.$$

This formula does not spring from a mere definition – at least if multiplying is generated as repeated adding. There is no doubt that sets of pairs can and should be constituted as mental objects at an early age and that the product formula can and must function. Should the formula be made explicit and, if so, how? I say "yes", and add "in a paradigmatic way", though one paradigm is not very likely to suffice.

From 4 boys and 3 girls I can form 4 · 3 boy–girl pairs.
4 roads from A to B and 3 from B to C is 4 · 3 from A via B to C.

Should one go further? Replace 4 and 3 by letters? Replace "this much . . ." by "the number of elements of the set of . . ."? Anyway, it would be less artificial than with the addition formula since the members of the pair in the multiplication formula can indeed naturally be interpreted to form a set, at least, up to a certain limit. The product formula leads via the rectangle model of the set of pairs to the area formula for rectangles. As far as I know, even the most fanatical advocates of sets did not dare to interpret length and width as *sets* of the number of times a measuring unit could be laid down along the one side and the other, which indeed is didactically absurd. Formalising the area expression stays, as far as I can ascertain, always in the sphere

4 horizontal by 3 vertical pieces is 4 · 3 squares,

which is the same vernacular expression as in the earlier examples, or more tightly formalised,

area = length times width,

which is a simplifying rather than complicating way and clearly lacks any association with sets.

In the case of equipartition mappings is there any need to go beyond vernacular formulations like

5 boxes with each 12 eggs is 5 · 12 eggs

or in the case of equipartition relations beyond

3 edges from each of 8 corners and 2 corners on each of 12 edges is 24 pairs ⌜edge, corner⌝,

(of course with schematic illustrations)? Yes. In the case of equipartition relations it can pay one not to be satisfied with the sloppy motivation of "thrice counted" but to understand and to have understood that behind this "multiple counting" is the counting of related pairs.

3.30. *Need for More Formalising*

The sets most often discussed in the present chapter have the peculiarity that they can be visualised (or otherwise embodied) in a way that is both more honest and more subtle than Venn diagrams. It is exactly for this reason that premature formalising can be dispensed with, at least if the examples where the sets occurred had sufficient paradigmatic force.

When and in what measure can a need to formalise arise? I must anticipate this question, which will be asked and answered later in a broader context. If I ask it, I mean not only formalising but also transgressing the limits of the vernacular by adaptations of it that, as seen from the vernacular, look like jargon. The sets considered in the present chapter bear, or can be indicated by, meaningful names, such as "ludo board", "paths finishing at the uth node". Sometimes they were denoted by letters, but this happened because I had to deal with a variety of such examples side by side and to distinguish them efficiently and economically.

Formalising is a linguistic activity, and the need for formalising is in the first instance a need for a better means of communication (communicating with both others and oneself). In particular, far-reaching requirements are to be fulfilled by the formalism as soon as oral, or often even ostensive, communication should give way to written and other kinds of communication that lack ostensive means. However, not only do languages have a communicative function, but they also serve as closed, more or less automatically functioning, systems. In a more advanced stage of formalising, the need is felt to guarantee this automatism: formalising is made subservient to developing algorithms.

As far as the communicative aspect of formalising is concerned, one can distinguish: formalisations of vocabulary, some of a simplifying, some of an elucidating character, and formalisations of syntax, mostly of a systematising character. An example of the first sort: the reinterpretation of the vernacular "quadrilateral" and introduction of terms like *square, rectangle, rhombus, parallelogram, trapezoid*. An example of the second sort: the convention that squares are to be counted both among the rectangles and among the rhombuses. Another example of the first sort: indicating *definite* objects by letters, and of the second, using letters to indicate *variables*.

Historically letters for variables were first used in euclidean geometry, where they were obviously needed for communication: as soon as geometrical constructions and reasonings are to be laid down in writing, ostensive means of expression like "this point" and "that point" do not work any more; they are replaced by conventional variables, indicated by letters. This situation becomes more involved if the exposition is illustrated by a drawing: if the same letters are placed at certain points on the paper they also have the appearance of constant proper names of these points. On the other hand, this formalisation in geometry has an algorithmic offshoot: a straight line through A and B does not get a brand-new name but an algorithmic one, AB. But this does not lead to developing any even weakly automatised algorithm.

A formalisation of great value and momentous consequences, which we are so well acquainted with that we do not experience it as such any more, is the creation of our familiar system of number notation. In contradistinction to the ancient Egyptian system and the so-called Roman numerals, which, as it were, literally translate the numerals of the vernacular, the Indian—Arabic system is a translation of the abacus language and was distinctly motivated historically by the need to transfer the advantages of the abacus expression onto the powder table or writing paper. The Indian—Arabian figures have become an inseparable piece of our written language (beyond the frontiers of any particular written language); nevertheless they have remained a foreign element in it, as appears from the fact that they are taught in the arithmetic lesson rather than in the language lesson. This is so for a good reason, because their flawless syntax is very unlike that taught in the language lesson. Formalising the number language by means of the Indian—Arabic notation is highly motivated by its algorithmic sequel: Indian—Arabic arithmetic. Even the notation of operations has then been formalised, both within the vernacular and by special symbols beyond the vernacular.

Fixation of a spoken language in writing is quite another sort of formalising — it is a more primitive, less specialised activity and does not invite algorithmisation. It is certainly not by accident that teaching both formalisms starts at the same age and that they are very much alike regarding their fertility in education. This is a question which will be reconsidered later.

The function of formalising the number system is primarily communicative, and only by the algorithms of column arithmetic is it made subservient to guaranteeing the reliability of algorithms. Too often, this component is identified with mathematics. Modern mathematics in the primary school is readily interpreted as formalising for the sake of algorithms. It is the traditional view that at the age of obligatory instruction the child becomes susceptible to learning formalisms (though not to learning formalising), but it is a mistake to believe that this is also the moment of formalising for the benefit of automatising. Innovators were not satisfied with the formalisms taught according to tradition at the lower levels of the primary school. A large number of new formalisms were invented for this stage of development, but their communicative function is often doubtful, and the algorithmic one remains within the narrow limits of little games. Up to now, only formalisms with a strongly communicative character have proved successful at low levels. An example: the arrow language for arithmetic operations and their inversions and for symbolising operations on the number line. No one ever tried to go beyond this limit to teach the denoting of mappings by letters at an early age. In general, to decide at which developmental stage this would be feasible would depend on the total state of formalising. The most natural, and experimentally tested, formalism are letters for variables representing numbers or magnitudes, which possibly now starts too late. With respect to sets, which were the subject of the present chapter, I can discover a need for formalising and formalisms only at a quite advanced stage.

CHAPTER 4

NATURAL NUMBERS

4.1–4.9. PHENOMENOLOGICAL

4.1. *According to Peano*

A great deal of what would be expected here, has been presented elsewhere earlier.* In order to avoid useless repetition, I am going to summarise it within the context of what has to be added.

The most usual introduction of natural numbers at an advanced level is Peano's axiomatics:

A set **N** ("natural numbers") with

(1)   an element $0 \in \mathbf{N}$ and

(2)   a one-to-one mapping f ("successor of . . .") of **N** into **N**,

such that

(3) $f\mathbf{N} = \mathbf{N}\setminus\{0\}$

(thus: each element has precisely one successor and — up to 0 — precisely one predecessor by means of f) and

(4) $(0 \in M \subset \mathbf{N}) \wedge (fM \subset M) \rightarrow M = \mathbf{N}$

(thus: with the same f the set **N** is minimal with respect to (1) and (3)).

Instead of using subsets, (4) may also be formulated with predicates as the "principle of complete induction":

(4′)   Let E be a property of natural numbers such that
$E(0)$ and
$E(n) \rightarrow E(f(n))$ for all $n \in \mathbf{N}$,

then

$E(n)$ for all $n \in \mathbf{N}$.

If we let

$M = \{n \in \mathbf{N} \mid E(n)\}$

then under the suppositions of (4′)

$0 \in M$

---

* *Mathematics as an Educational Task*, Chapter XI.

and because of

$$n \in M \to E(n) \to E(f(n)) \to f(n) \in M,$$

it follows that

$$fM \subset M,$$

and because of (4)

$$M = N,$$

so

$$E(n) \quad \text{for all } n \in N.$$

Thus $(4) \to (4')$, and conversely it can be argued that $(4') \to (4)$.

In **N**, operations can be introduced inductively, the addition by means of

$$a + 0 = a$$
$$a + f(n) = f(a + n) \quad \text{for all } n,$$

which defines generally the sum

$$a + b$$

(though the proof is not as obvious as it looks). Commutativity and associativity of the addition are inductively proved. Similarly, the multiplication:

$$a \cdot 0 = 0$$
$$a \cdot f(n) = a \cdot n + a.$$

The order relation in **N** is no problem at all.

From here a bridge can be built to the finite cardinals: putting

$$\#\{x \in N \mid 1 \leq x \leq n\} = n,$$

one has to prove that indeed all these cardinals are differnt. These then are the finite cardinals, and any set with such a cardinal is called finite. Moreover, one can prove that for finite sets M_1, M_2,

$$\#(M_1 \cup M_2) = \#M_1 + \#M_2 \quad \text{if } M_1 \cap M_2 = O,$$
$$\#\lceil M_1, M_2 \rceil = \#M_1 \cdot \#M_2,$$

which means that the inductively defined and the set-theoretically defined operations coincide.

In Peano's axiomatics **N** accounts for the counting sequence, f for the counting act, and (4) for the idea, not easily verbalised, that the counting act successively exhausts the counting sequence. By the axiomatisation the counting sequence has been frozen as it were; **N** is static; time and action look as though they are eliminated.

On the algorithmic plane the counting sequence is, as it were, concretised by means of the decimal notation: in

$$n = a_k \ldots a_1 a_0$$

each a_i may be one of the symbols $0, 1, 2, 3, 4, 5, 6, 7, 8, 9$. The counting act changes these symbols in a well-defined way. Infinity seems to be encapsulated, but it again appears in the subscript k (the number of digits minus 1 if $a_k \neq 0$), which fortunately is of a lower order than n itself.

4.2. *According to Euclid*

The phenomenology at the background of Peano's axiomatics is the counting process in the time flow; the numerals one, two, three, . . . should properly be read as first, second, third, ... Another phenomenology, which is more original, is accounted for in Euclid VII. Before Peano, and even years later, it has influenced theoretical expositions on the number concept.

Euclid says

Unit is that after which a thing is called one.
Number is a set composed of units,

and still in the *Encyklopädie der mathematischen Wissenschaften*, founded and edited by Felix Klein, we can read*

Counting things means considering them as of the same kind, taking them together and assigning to them individually other things, which are also considered as of the same kind. Each of these things to which by counting other things are assigned is called a *unit*, and each of the things which by counting are assigned to the others, is called *one*. The result of counting is called *number*. Because of their being of the same kind, as regards the units and the ones, respectively, the number does not depend on the order according to which the ones are assigned to the units. (Translation by the author.)

Hilbert in 1904** still tried to base number theory on a phenomenology of this sort, and his later formalist approaches are to be understood from the same phenomenological viewpoint.

From Euclid to Schubert a number consists of units ("ones" according to Schubert) — a mental object composed of simple objects of the same kind. Number as a cardinal is explicitly mentioned by Euclid only in the case of the "one" (though in fact the "one" is no "number"): by means of the unit one can say of a thing that it is one thing. Euclid could have added that by means of the two-ness, three-ness, . . . two, three, . . . things can be named. Euclid

---

* I A 1 (1898), H. Schubert, *Grundlagen der Arithmetik*. In his criticism of H. Schubert's contribution ('Über die Zahlen des Herrn H. Schubert', Jena, 1899), G. Frege stands on somewhat firmer ground than in that of Hilbert's *Grundlagen der Geometrie* (Jahresbericht DMV 12 (1903), pp. 319–324, 368–375), but as to insipidity both criticisms are well matched. Frege's lack of understanding both of phenomenology and of axiomatics has proven symptomatic of the logistic attitude in general.
** Verhandlungen des III. Internationalen Mathematikerkongresses in Heidelberg, 1904. From 3rd to 7th ed., Appendix VI of *Grundlagen der Geometrie*.

kept silent about the way this designation is to be established. Various authors have filled in this gap with various formulations.* In order to count a heap of things, one has to see them as a set of things of the same kind — cows and horses as a herd of animals, because after the numeral a noun is expected — and to these objects other objects, likewise all of the same kind, are assigned — counters on the beadframe, or chips, or tallies, or mental equivalents of them, which together form numbers.

At the time that Schubert wrote his article, such a consideration was still the normal one, though it should have been obsolete since in the course of the nineteenth century complete induction had become the most striking property of **N** and had been formulated as such by Peano. This then is the ordinal approach to natural numbers, preferred today, which leads in the straightest way and with a stress on what is essential for mathematics to the mathematical use of numbers as arithmetical objects. It is a disadvantage of this approach that it does not match a phenomenology where a number *consists* of units, and an even greater disadvantage that numbers serving to *count something* and being (cardinal) *numbers of something* come at the end of a long and arduous course of reasoning.

4.3–8. *According to Cantor, Frege, Russell & Whitehead*

4.3. Number as (cardinal) number of something is Cantor's approach. Cantor says:**

"Potency" or "cardinal number" of M we call the general concept which by means of our active thinking power is created from a set M in the way that we abstract from the quality of its different elements m and the order of their presentation.

Since each single element m, if its quality is disregarded, becomes a "one", the cardinal number $\bar{\bar{M}}$ (of M) is itself a definite set composed of nothing but ones, which exists in our mind as the mental image or projection of the given set M. (Translation by the author.)

This is followed by the definition of equivalence:

Two sets M and N are called equivalent, denoted by

$$M \sim N \quad \text{or} \quad N \sim M,$$

if it is possible by a law to put them in a relation with each other, where to each element of one of them corresponds one and only one of the other.

And later on:

that two sets M and N have the same cardinal number if and only if they are equivalent.

The "if" is motivated by the fact that whenever the cardinal number of M is at stake, the position and the character of the elements of M does not matter

* For instance E. Schröder, *Lehrbuch der Arithmetik und Algebra*, Leipzig, 1873.
** *Math. Annalen* **46** (1897), pp. 481–483 = *Gesammelte Abhandlungen*, pp. 282–284.

and consequently the elements of M can be replaced with the corresponding "ones" of N according to the correspondence. The "only if" is motivated in an even more curious way,

that between the elements of M and the different ones of their cardinal number $\bar{\bar{M}}$ there exists a mutually one-valued correspondence ... Thus we can say that $M \sim \bar{\bar{M}}$. Likewise $N \sim \bar{\bar{N}}$. So if $\bar{\bar{M}} = \bar{\bar{N}}$, then $M \sim N$.

We see the mathematical construction entwined with a phenomenology that continues Euclid's: the cardinal number of M consists of units, albeit of more than a finite set.

This cannot do any harm, one should say, because the only thing that matters is the definition of equivalence. It suffices to know when sets are equipotent; it does not matter what the cardinal number of a set is. This is *intentional* concept building specialised on cardinal number. I assign the same cardinal number to two sets as soon as they are equivalent.

4.4. Cardinals are also numbers in the sense that you can perform operations on them defined as follows: the addition by means of the union, the multiplication by means of the set product:

$$\#M_1 + \#M_2 = \#(M_1 \cup M_2) \quad \text{if } M_1 \cap M_2 = O,$$
$$\#M_1 \cdot \#M_2 = \#\lceil M_1, M_2 \rceil.$$

This is — for finite sets — simpler than in Peano's approach, and the various laws for these operations are much more easily proved.

Powers is another feature obtained by set operations

$$(\#M_1)^{\#M_2} = \#(M_1{}^{M_2}),$$

where $M_1{}^{M_2}$ is defined as the set of mappings of M_2 into M_1.

4.5. Notwithstanding the concession made in the last paragraph of Section 4.3, the question of what the cardinal number of a set is has been answered up to a point. The answer, which stems from G. Frege and Russell & Whitehead, is a paradigm for what is called *extensional* concept formation. All sets equivalent with M must possess the same cardinal, thus cardinality of M is defined as the thing common to them. But what is the thing common to them? Set theoretically viewed it is the class of all sets equivalent with M. (A remark: the class of all sets equivalent with M is an object that must be handled carefully, as appears from various paradoxes; the term "class" instead of "set" is meant as a warning sign.) At Frege—Russell's standpoint one is in a world far away from Euclid—Cantor's phenomenology. For non-mathematicians this definition is so extraordinary that, for instance, Piaget read the Frege—Russell definition as though a cardinal number were a class of equivalent units.

4.6. Here we are not interested in cardinals in general, but especially in number, that is, finite cardinals.

Cantor* starts with a set E_0 with one element and assigns to it the cardinal 1. He adds another element to get a set E_1, to which he assigns the cardinal 2:

By adding fresh elements we get the series of sets

$$E_2 = (E_1, e_1), E_3 = (E_2, e_2), \ldots,$$

which in an unlimited sequence deliver successively the other, so-called finite, cardinals, denoted by 3, 4, 5, ... The auxiliary use we made of the same numbers as subscripts is justified by the fact that a number is not used in this sense until it has been defined as a cardinal.

The last sentence shows that the finite numbers are not presupposed, but truly defined.

All cardinals obtained by this process are called finite. The set of the finite cardinals ($N\backslash\{0\}$) has the (countably) infinite cardinal $\aleph_0$. Each infinite set M (that is, whose cardinal is not finite) contains a countably infinite subset, which arises as follows: Take an element a_1 of M; since M is infinite, $M \neq \{a_1\}$. Hence there is an $a_2 \in M$ with $a_2 \neq a_1$, since M is infinite, $M \neq \{a_1, a_2\}$. So it goes on, to produce an infinite sequence $a_1, a_2, \ldots$ of different elements of M.

4.7. Cantor's presentation of natural numbers is as naive as Euclid's. The only difference is that the numbers are not statically present as in Euclid. They are produced in the course of time, and counting out a countably infinite subset from an infinite set is also done in time. Cantor's new idea is not his presentation of natural number but the first clear explanation of what it means for a number to be the number of a set of things.

On the first point, Frege has explored a less naive approach: He took all sets Φ of cardinals such that

> Φ contains the 1
> with each cardinal Φ contains the one that is 1 larger,

and of all these sets he took the intersection — thus the smallest — and this was the set of finite cardinals. This method to describe in one blow infinite constructions previously projected in time has since become paradigmatic. At the end of his work Frege confesses that his plan became unsettled by paradoxes that had been discovered in the meantime.

Russell & Whitehead proceeded more carefully: At this point they do not consider sets of cardinals. They define sets Φ according to Frege's prescription, but their elements are finite sets rather than cardinals:

> Φ contains the empty set,
> Φ contains with any set X the set $X \cup \{X\}$,

---

* *Math. Annalen* **46** (1897), pp. 214–215 = *Gesammelte Abhandlungen*, pp. 289–290.

and among all these sets they determine the smallest. The example sets for the finite cardinals then are

$$O, \{O, \{O\}\}, \{O, \{O\}, \{\{O, \{O\}\}\}, \ldots.$$

It looks, as a satirist observed, like a library with

in the first volume, nothing
in the second, the table of contents of the first,
in the third, the table of contents of the second,
and so on – .

These are the units from which the Russell & Whitehead numbers are composed.

The existence of this – or some – infinite set must explicitly be postulated according to Russell & Whitehead.

4.8. Another point where Cantor proceeded naively was in counting out a countably infinite subset from an infinite set. In order to describe this process in one blow, one imagines that assigned to *any* non-empty subset X of M is one of its element $\varphi(X) \in X$. The desired countably infinite subset A of M consists of consecutively the elements

$$a_0 = \varphi(M), \quad a_1 = \varphi(M\backslash\{a_0\}), \quad a_2 = \varphi(M\backslash\{a_0, a_1\}), \ldots.$$

More precisely, by induction

$$A_0 = \{\varphi(M)\}$$
$$A_{i+1} = A_i \cup \{\varphi(M\backslash A_i)\} \quad \text{for all } i \in \mathbf{N},$$

and finally

$$A = \bigcup_{i \in \mathbf{N}} A_i.$$

Or, without explicit induction,

$$\{\varphi(M)\} \in \Phi$$

and

$$\text{if } X \in \Phi, \quad \text{then } X \cup \{\varphi(M\backslash X)\} \in \Phi,$$

and among these sets Φ, take the smallest (the intersection of all of them).

The indispensable instrument for this definition is the function φ, which to every non-empty part X of M assigns an element of X – a function whose existence is obtained by the so-called axiom of choice.

4.9. *According to Dedekind*

A countably infinite set can be mapped one-to-one on a proper subset; for instance, the A is mapped on $A\backslash\{a_1\}$ just obtained by an f such that $fa_i = a_{i+1}$. Thus every infinite set can be mapped one-to-one upon a proper subset –

the M which we have just considered, by mapping A according to f and the remainder identically.

This characteristic property of infinite sets, that of being equivalent with certain proper subsets, was R. Dedekind's* starting point for a quite original theory of finite sets.

In the theories we have summarised up to now the finite cardinalities were inductively generated, though finally the inductivity became somehow encapsulated. In Dedekind's method the induction is self-generated. Dedekind defines

> a set is infinite if it is equivalent with a proper subset,

and of course

> a set is finite if it is not infinite.

He starts with some infinite set M — his proof that such a set exists is not convicing.

Since M is infinite, there is a one-to-one mapping f of M on a proper subset. Let $a \in M \backslash fM$. A subset N of M is defined by

(1) $a \in N$,
(2) $fN \subset N$,
(3) N minimal with respect to (1) and (2).

Substituting N for $\mathbf{N}$, a for 0, and f for the successor mapping, one gets exactly the system of properties that defines $\mathbf{N}$. Moreoever — without the axiom of choice — one has succeeded in finding in any infinite M a subset equivalent with $\mathbf{N}$, thus a countably infinite subset.

4.10–11. *Tested Didactically–Phenomenologically*

4.10. If all or part of the preceding analysis is considered too high level and consequently out of place, I would object that it is subject matter found in some modern textbooks, for instance, for the 6th grade. There it looks like a closed system, though in fact it is, as it were, tiptoeing in wooden shoes, displaying subtleties in a rudely defective context. I have discussed this elsewhere.**

It will have become clear from my exposition that with each profundity we get further away from the phenomenology of number as it is naively experienced. The definitions quoted from Euclid interpret a directly-from-the-abacus abstracted consciousness about numbers. They are, however, not operational in the number theory to which they are meant to be the introduction. Complete induction was repeatedly applied from antiquity onwards, sometimes even in a profound way, but not until Pascal was it formulated. Complete induction turned up in

---

* *Was sind und was sollen die Zahlen?* 1887, 1893.
** *Mathematics as an Educational Task*, Chapter XI.

ever more examples, and was ever more consciously used, but not until two and a half centuries after Pascal did the key position of complete induction for the concept of natural number become clear. Likewise it became clear that complete induction could serve to define the arithmetical operations and prove their properties. The most modest interpretation of Peano's axiomatics is to consider it as descriptive. Anyway, it is operational, wherever induction plays a part in mathematics. A step further is to ascribe a definitory character to it, that is, defining what natural numbers are; then Peano's axiomatics is only operational as long as one moves within the foundations of mathematics.

As far as Cantor's definition of cardinal numbers describes number as numerosity and draws attention to its invariance under one-to-one mappings, it is a phenomenology of the naively experienced number; if it is meant as defining number, it belongs to the foundations of mathematics; if, moreover, Frege's and Russell & Whitehead's formulations are used, it is research in foundations by methods that have arisen from a profound criticism in the foundations of mathematics. Dedekind's definition of infinite set, produced by a clever change of perspective, presupposes a well-stuffed arsenal of mathematical strategies.

The definitions of addition and multiplication by complete induction can be seen as an adequate phenomenological description of the naive process of adding by "counting forth" and of the naive process of multiplication by repeated adding (producing multiplication tables); interpreted as definitions, they belong to the foundations of mathematics, and the same holds for the laws of addition and multiplication, whereas, when pronounced for numerosity rather than for counting numbers, they are obvious — the need for proofs arises only in systematising the counting number approach by Peano's axiomatics.

In the numerosity approach the definition of addition describes adequately what happens if quantities are taken together, though it is expressed at a linguistically high level; in order to be practically operational, it must be completed with the counting-forth definition. The definition of multiplication in the numerosity approach adequately accounts for the rectangle model of multiplying, which is an indispensable complement of multiplication defined by repeated addition. For the power definition in the numerosity approach (cf. Section 4.4) traditional school mathematics does not possess an analogue, though creating one would be feasible and worth paying attention to.

4.11. Numbers, counting, and arithmetical operations are first of all a means to organise phenomena where quantities play a part. All theories of natural numbers are rooted in these means of organisation. But all theories go beyond that. Mathematics is characterised by a tendency which I have called *anontologisation*: cutting the bonds with reality. This tendency is entirely justified. It is, however, the result of historical and individual development and cannot be supposed to be innate to the learner's mind. Even less can the learner be presumed to be susceptible to anontologised mathematics. Attempts at instilling it lead to false concretisations.

Auxiliary tools in the process of anontologising are: change of perspective from description to definition, where observed properties of mathematical objects are used to define the objects in order to detach them from their origin; and the replacement of intentional by extensional concept formation. Neither tool is simply present in the mind. They require development and exercise.

If a pupil who can identify rhombuses at glance isolates from the multitude of properties of the rhombus a few characteristic ones, which bring about a formal definition of rhombus, he performs a local change of perspective and a first step on the way to anontologising mathematics. With number, the same step is much longer because number is more familiar to him and the necessary change of perspective would be more global in character. Even more difficult is extensional concept formation by equivalence, which I paid attention to elsewhere*. This restricts to narrow domains teaching theories of natural number.

4.12. *Phenomenologically Too Low a Level*

One current in the nineteenth century, which I have neglected up to now, was characterised by a superficial kind of formalism: the natural numbers considered as symbols, sequences of digits, with operations performed according to merely conventional rules. In a sense it is not as mad a phenomenology as one would think. Rather it is to the point as far as it describes a kind of instruction in arithmetic and the attitude fostered by this instruction: an attitude of viewing numbers as symbols and doing arithmetic according to conventional rules.

Nevertheless, this phenomenology is unsatisfactory. It does not account for the relatedness of numbers and arithmetic with reality, where numbers and numerical operations have a meaningful counterpart. Moreover, this phenomenology does not carry farther than understanding the most primitive arithmetic. The activity described by this phenomenology is that of a mechanical (or electronic) calculator rather than that of a human calculator who can handle the instrument and is obliged to understand at a higher level the aim of the numbers and operations. It cannot be denied that in the nineteenth, and even for many years into the twentieth century there was a need for human calculators who experienced the numbers and the operations at this primitive level, and this need justified socially a corresponding kind of instruction. With the rise of better perfected calculating machinery, and its mass production, the need vanished for people educated according to those principles. If numbers and numerical operations are taught today, the instruction can no longer amount to programming a computer, to which numbers and operations are indeed meaningless and conventional.

The primitive formalism of the nineteenth century has given away in the twentieth century to a more profoundly understood formalism, by which not only elementary arithmetic but the whole of mathematics is interpreted as a

---

* *Mathematics as an Educational Task*, pp. 31–32, 213–214.

language with a conventional syntax. However, the mathematician does not then move within this language but reflects on it — reflections that can be syntactical or semantical. The same holds for all kinds of formal languages which are extensively studied today; they are "spoken" only by and with computers.

4.13. *Developmental Phenomena*

From the concept of natural number, which has occupied us up to the previous section I now turn to the mental object or objects that can play a part in acquiring the concept. I use the plural "mental objects" because I must distinguish at least between particular natural numbers and the natural number as mental objects, but certainly also between numerosity numbers and counting numbers as well as between numerosity number and counting number. For acquisition of the concept "natural number" even more is required: constituting certain relational patterns between natural numbers.

Developmental facts about the constitution of numbers and number are, as far as I know, scanty. Small cardinals seem to be constituted early — as people assert, 2, 3, 4 at the ages 2, 3, 4, respectively. In Section 3.18, I asked what part structures play in this. One could add many more questions: to what degree the numbers aimed at are bound to certain objects or kinds of objects or to certain representations, whether and how the constituted numbers are expressed verbally or enactively, what in a certain stage of restricted constitution impedes or stimulates the constitution. According to my experience the continuation, at least as far as verbalising is concerned, has the character of differentiation: a numeral used in the sense of "many" starts playing a more specific part, in order perhaps to be replaced by another that takes over the meaning of "many".

The majority of children learn counting before constituting cardinal number; counting is reciting numerals, initially in an arbitrary order, later in the right order though with gaps, finally without gaps. In this period they also learn counting *something*; it is arbitrarily pointing to the objects, before it becomes a systematic procedure; first graders can still have difficulties with one-to-one exhausting even if they know that this is the way to do it. The connection between counting and cardinal number can be lacking in spite of verbalising. I know of only one case of a child that started counting only after the complete constitution of the numerosity number (cf. 3.18).

4.14. *Conservation*

With a view to this lack of factual data on the development there is no way out except the phenomenological method. In fact, Piaget took it and, influenced by mathematical ideas, chose conservation as a criterion of what I have called constitution of number. For the cardinal number as defined by Cantor, conservation or — in mathematical terms — invariance with respect to certain transformations is indeed an important characteristic.

I have already discussed "conservation" critically (Section 1.25). The term itself is confusing because it does not indicate the transformation with respect to which invariance is postulated. In the case of length it could be, according to my enumeration, congruence transformations, flexions, and break-and-make transformations or permutations of parts. Numerosity looks simpler: everything that has to do with transformations can be brought under the concept of one-to-one mapping. It can, indeed, but this does not mean that it must. Many ideas reduced to one when logically possible can refer to ideas that belong to different abilities and to abilities that are being learned under different circumstances. Moreover, we noticed already in Sections 3.18—27 that the constitution of cardinals may depend on structures and isomorphisms, rather than on sets and one-to-one mappings.

Up to the present day nobody can say whether the stress Piaget put on conservation was justified, whether indeed certain invariances characterise the constitution of certain mathematical objects, whether this might be true of all mathematical objects for which Piaget developed conservation criteria, for some of them, or for none. I guess that in principle Piaget chose the right way, but I believe he deserved to be followed more critically on this way in every detail than he has been in fact.

Another doubtful element in Piaget's method, and a matter of principle, is the use of temporal cross-sections instead of longitudinal observation (which he rarely performed) The unavoidable drawback of that method is that the observed conservation phenomena are possibly only criteria of the constitution of mathematical objects rather than developmental phenomena on the way to it.

4.15. *Invariances With a Single Set*

With all these provisos I am going to follow Piaget in the phenomenological search for invariances that might matter in the process of constituting the numerosity number. I first mention four of them, all related to one single set rather than to comparing two sets:

> Invariance
> > over time,
> > under change of standpoint,
> > under shake transformations,
> > under break-and-make transformations.

Some comments on these terms, in particular the first:

A child (4; 8) counts her fingers. "Five". And how many did you have yesterday? "I have forgotten". Another child (4; 6) laughs at this question.

This example shows that the invariance over time is not at all self-evident. This insight is acquired earlier or later by different children. It is no logical fact.

Indeed, the invariance depends on conditions. The child knows circumstances under which something becomes more or, less — plants that germinate, flowers that drop, families that increase or decrease, drops that split and unite.

> If something is added, it becomes more,
> if something is taken away, it becomes less,
> if nothing happens, it remains equal,

these are principles for all kinds of magnitudes, discrete and continuous. The only problem is to know whether nothing happened. This knowledge depends on a vast body of factual experience. Adults are hardly able to imagine that and how they acquired this knowledge in their own development.

Invariance under change of standpoint is a similar experience. A set moved to another place and observed there, or observed in the same place by another observer, must possess the same number. This too is not self-evident. It can mean a discovery to look for a cause for why under certain circumstances a set observed from another standpoint can be more or less.

The shake transformations again are concerned with one set: cookies on a plate can be moved, flowers in a vase rearranged, sheep in a flock run — and at least mental continuity ties the initial to the final stage and guarantees the invariance. Like the invariance over time and under change of standpoint, invariance under shake transformations has never properly been tested with children in Piaget-like experiments as far as I know. First of all, in the classical test for conservation of number the relevant shake transformation is obscured by a built-in misleading cue: the discrete set of objects is presented in a such way that by the very fact of presentation not only the number is defined, but also the length, which is intentionally not kept invariant. Moreover, the shaken set is compared with an unshaken one, which produces a new difficulty.

In the cases of duration, change of standpoint, and shaking, one could maintain that they concern conservation of set rather than of number: the set of fingers, cookies, animals remains the same over time, under change of standpoint, after shaking, and so does of course its number as a property of the set.* I have no objection against seeing it this way. I do not believe it matters much. The case of the break-make transformation, however, is different. It is breaking and rebuilding a set. The reason why I consider these transformations separately is similar to the reason I gave in the case of length. I should add that it was in considering "length" that I first became aware of this kind of transformations and their importance.

Breaking divides a set into two or more whether the pieces are taken apart or merely separated by true or symbolic walls. Remaking brings them together or removes the separation. Meanwhile the objects do not cease to belong together,

---

* I could have considered these invariances also with respect to lengths; then the invariance of set would have had as its analogue the invariance of long object.

which can be expressed in the intermediate question, "Are there still as many *together*?"

Break—make transformations seem to me to be the most effective means of developing invariance as a general feature, for number as well as for length. The learning process could take the following course: A small number of objects is removed from the set in order to be added back again — possibly at another place. This procedure is continually repeated, while the question is asked, "Are there now more, or less, or the same?"

If information on the development of conservation of number is wanted, these invariance phenomena should be tested separately: "How many were there yesterday?" "How many are they if you look that way?" "How many are they now?" (after shaking) "How many are they now together?" (after breaking). A child that is not familiar with these invariances will not be able to answer immediately; he will rather try counting anew.

A child that cannot yet count sets could be asked the same questions in the version "Are there as many?" instead of "How many are there?" Or one should ask him to estimate the number before and after.

4.16. *The Counting Number*

It has often been noticed that many children count before having constituted number as a mental object. Let us distinguish

> counting, that is, reciting the number sequence;
>
> counting something, that is, in the counting process connecting the numerals with the set that is counted out or produced;
>
> interpreting after counting the counting result as number of the counted or produced set.

Counting may or may not be accompanied by insight into the structure of the counting system.

Various aspects of this insight are tested with questions like

> what follows 3?
> what follows 23?
> what follows 9?
> what follows 29?
> what follows 99?
> what precedes 4?
> what precedes 24?
> what precedes 10?
> what precedes 30?
> what precedes 100?
> which is earlier, 6 or 9?
> which is earlier, 26 or 29?
> which is earlier, 29 or 36?

Children who lack this insight count — silently — anew from 1 up to the desired place, as do many adults, if the order in the — less structured — alphabet is at stake.

Insight into the structure of the system of numerals is made easier or acquired by

> reading and writing numbers,
> producing the number sequence by writing.

This ability does not depend on "counting something" and on the relation to numerosity.

> Counting by tens, by hundreds

allows one to get on quicker — it plays a part in some games. House numbers suggest

> counting by two.

A difficult job is

> counting back

if it is not supported by acquaintance with the written image. Likewise

> counting back or forth a given number of steps from a given number

does not in principle depend on "counting something" and on the relation with numerosity, though this counting activity will usually be motivated, or be accompanied by "counting something".

The objects to be counted can be

> visible, palpable, audible, kinesthetic, rhythmic,
> all of these combined,
> movable, fixed,
> demonstrable, observable, mental.

The counting activity can be accompanied by movements of

> fingers, eyes, hands, feet, or other limbs.

The counted objects can be

> indicated, marked, separated.

The numbers can be firmly attached to the objects to be counted by

> mapping the number sequence on the number line

and

> operations on the number line.

Then it means

counting steps on the number line,

and

combining sets of steps into jumps on the number line.

Illustrative are the

simultaneous* counting of steps taken
simultaneous* counting of steps on stairs,

which can be

real or symbolised,

in general,

simultaneous* counting of rhythmic events.

A usual "mistake" in counting steps (taken, or of stairs) is taking the zeroeth for one. Possibly

in counting rhythmic events, counting something

arises or gets experienced: rhythmic movements of the forefinger. Whereas

correct simultaneous* counting of a rhythmic event (for instance the rhythmic counting process) is a matter of coordination,
correct counting out is a matter of organisation,

that is,

organisation by setting apart or marking the counted objects,

or if the objects to be counted cannot be manipulated,

organisation by means of present structures,
or by structuring.

By this organisation

skipping and multiple counting

must be prevented.

It is a well-known experience with children at a certain stage that, when asked "how many?" they count, say, 1, 2, 3, 4, 5, without answering the question properly, and count anew if the question is repeated. So they do not consider the counting result as that characteristic of the set which we call its number. The step to this can perhaps be provoked by replacing the question

how many are they?

---

* That is, simultaneously with the acts performed.

with the task

> give me . . . beads

from a set of beads. If now the child reacts by counting, he is compelled to interpret the counting result as the number of beads. Continuing this line, one can charge the child with

> picking m beads out of a set of n beads, with $m < n, m = n, m > n$,

and in this way

> according to the feasibility try to have the cardinal order relation constituted.

The strongest stimulus to constituting numerosity number in this context is structural isomorphisms: if I have successfully counted the eyes of six persons, I do not need to count anew their ears. The insight is created that

> isomorphically structured sets possess isomorphic counting structures,

hence

> isomorphically structured sets lead to the same counting result,
> counting a part leads to a smaller counting result.

This insight can be generalised while abstracting from counting: without counting, the child draws the conclusion that a group of people have as many ears as eyes. This leads to the insight that

> isomorphic structures have already *potentially* isomorphic counting structures

and to the possibility of asserting that

> on account of structural isomorphism of sets, they have as many elements,
> on account of inclusion one set has fewer elements.

If three sets are considered where the first is isomorphic with the second by another kind of structure than with the third, it can still be concluded that the second and the third have equally many elements. It can by replacing structures of a *different* kind by counting structures, thus of the *same* kind. From here the

> use of transitivity of "equally many" and "less than"

can develop.

By this course one can imagine the numerosity number constituted as a mental object: starting from the counting number and eliminating it while preserving its possibility.

4.17. *The Structure of the Counting System*

The *mathematical* structure of the counting system is the successor relation with well-known properties that allow induction and, consequently, the defintion of N. But in genetic and didactic phenomenology the counting system is endowed with a richer structure, from which N is derived by a theoretic impoverishment.

First of all, genetically: Acquiring numbers is, in general, preceded by acquiring numerals, which as a system is — with a few exceptions at the start — well-structured by tens, hundreds, thousands, and so on, and most often in its acquisition is integrated with the total vocabulary. But however this process of acquisition may run, there can be no doubt that acquiring the mathematical structure of N is possible only by way of acquiring the stronger structure with which N is endowed, the decimal system — from counting forwards and backwards to the numerical operations. In true mathematics the decimal structure of N does not count and, as is easily understood, for good reason. There are no good reasons, however, why no attention is being paid in developmental research, as far as I know, to the decimal structure. There is a strong tendency to read the developmental — and sometimes even the didactical — phases from the structure of mathematical science, and by preference from one particular structure of science, Bourbaki's system. Some developmental psychologists, such as Piaget, have even raised this parallelism to the rank of a principle. With regard to the organisation of mathematical subject matter, this view has been expressed, if not earlier, then anyway with new aplomb, at Royaumont (1959) and Dubrovnik (1960). For very good reasons the decimal structure of N does not appear in any scientific structure of mathematics except as a curiosity that is just mentioned. In such genetic or didactic research as is steered or even dominated by the idea of structure of science, there is no inducement to pay attention to the stronger structure of N. This explains the lack of interest I alluded to earlier. Yet whoever is influenced by genetical or didactical phenomenology feels the stronger structure of N as a *conditio sine qua non* and as a new argument against the idea that spontaneous and guided development are determined by the structure of science.

One additional remark: Even if the decimal structure is rejected as irrelevant within the scientific structure of mathematics, one cannot bypass the fact that the principle of building fixed numbers of units to introduce higher units should certainly fit somewhere into the scientific structure (cf. equipartition by mappings and relations in 3.22 and 3.25). It is rather the restriction to one system of bundling that looks unmathematical. Yet even structuring by bundles is in general neglected in developmental research.

In the decimal structure of N two components can be distinguished,

> the (decimal) bundling,
> the positional arrangement of the bundles.

Both have been known from olden times to all civilisations, the oldest included

— the Sumerian—Akkadian and the Egyptian. The sexagesimality of cuneiform texts was a sophistication of professional arithmetic, artificially imposed upon an originally decimal system, but anyway it included the positional idea. In Egyptian arithmetic we find the decimal bundling, but the positional idea seems lacking. There can be no doubt, however, that operations on numbers, in particular on big ones, were performed from olden times on some kind of abacus; and however an abacus is constructed, it certainly embodies the positional idea. On cuneiform tablets the numbers were also positionally *written*. It is the decimal *digit* system that is of more recent date, but it was preceded by a decimal abacus, which was gradually superseded by the decimal digit system when writing material became widely available. Unfortunately — one would add from a didactical point of view.

For the mathematical layman the decimal counting structure is an indispensable element of the counting system. I doubt whether the man in the street ever disregards the decimal structure, being an inessential element of the counting system — not until true mathematics is at issue is there any need for such an elimination. Quite a few know that other counting systems will also do and that, for instance, computers work in the binary system. This knowledge, however, is most often not deeply rooted or it is directed in the wrong way: the impression can arise that some number system — in whatever base — is required; that such a supplementary structure as decimality is indispensable. The large part decimality plays in arithmetic puzzles indicates another mentality than is appreciated in mathematics.

From a practical rather than theoretic perspective the extra structure of N, decimality, is valuable and even indispensable. This holds for the didactical phenomenological perspective too. The extra structure is a powerful means to actually master the bleak successor structure of N. One comes to grips with the counting system by decimalising it, or at least one imagines that one does; and with regard to mere routine, it is certainly true. Performing the arithmetical operations and estimating "orders of magnitudes" is organised by means of the decimality. Not until mathematics is carried on to a more advanced level, do the idea and the need arise to eliminate the particular number system from the phenomenal structure of the counting system in order to get the truly mathematical N.

It is well known how in counting large quantities of objects of the same kind, decimality is concretised: objects are laid down in rows of ten, which are combined in squares of ten by ten, to systems of tens of such squares, and so on. Or towers of ten or one hundred such objects are built, for instance, when counting money. This principle is also systematised by such material as blocks, rods of ten blocks, plates of ten rods, blocks of ten plates.

Such material concretises the bundling only — it neglects the positional order. This latter component is fully embodied by the abacus, an old instrument rediscovered in Western Europe. Whereas this kind of instrument maintained itself in Russia and the Far East, it disappeared in the Western world after the rise of Indian—Arabian arithmetic — strangely enough also in instruction, which relied on the counting frame with 100 beads, each of the same value. As a didactic tool the abacus enjoys an increasing popularity. If I am not mistaken, its reintroduction was due to Maria Montessori.

Unlike the Eastern abacus with its intermediate units of 5, 50, and so on, our instructional abacus is purely decimal, as is the Russian one. In the Russian abacus the units, tens, hundreds, ... follow up each other in the vertical direction. Our instructional abacus, however, follows the order of the written positional system. It is fabricated with 10 or 20 beads on a line or a bow; if it is 20, then 10 are of one colour and 10 are of another. Another variant is the pictorial abacus which allows for an arbitrary number of chips between vertical lines. On such abaci numbers admit of a multiplicity of representations; in the standard representation, corresponding to the digital numeral, the number of beads or chips used is minimal.

A variant of the decimal abacus is the so-called minicomputer, where in each position the numbers from 0 to 10 are represented according to the binary system.

After these technical details on the abacus and before dealing with its use in arithmetical practice, I will briefly tackle a question of principle. When I mentioned how calculating on the abacus was superseded by written column arithmetic, I heaved the sigh "unfortunately". Why "unfortunately"? Is it not a blessing we would be silly to waive writing numbers as one likes it, beside and below each other when compared with working in the shackles of the abacus, however flexibly it might be constructed and used?

Yes, it is a blessing for those who can afford it. Writing the digits neatly below each other is the precondition for the functioning of the positional system, and by "neatly" I mean not only a calligraphic fact but mentally observing the positional idea. And this holds for decimal *fractions*, too. A pupil who calculates excellently in N and gets into difficulties with decimal fractions proves that he has not yet grasped the essentials of the positional system.

The abacus, in whatever form, compels the learner to reflect again and again not only on what units, tens, hundreds, ... are, but also on what tenths, hundredths, ... are, and how the ones arise from the others. It looks like a shackle — calculating on the abacus. Indeed it is for those who have outgrown it. But the chance to outgrow it should be granted to anyone who has no other chance, who cannot do without it, who is otherwise left with the choice: number chaos.

4.18. *Invariances With Two or More Sets*

In Section 4.15 aspects of conservation were discussed with regard to a single set. Numerosity, however, does not become an essential characteristic and instrument until more sets are compared.

Recall the enumeration in 3.18 of activities which together might represent the constitution of numerosity number:

Eliminating the structuring component from structures with the same substratum in order to arrive at substratum sets,

and in this connection transforming the inclusion relation into an order relation ("less" instead of "contained in").

Using isomorphisms of structures with different substratum to compare different sets.

Using transitivity of equality and order (of numerosity number).

The structures meant here can be present in the material in order to be discovered and used afterwards, or they can also be imposed intentionally by the user. The structure most usually imposed in this context on unstructured or insufficiently structured sets is the counting structure, though as the examples of Sections 3.18–30 show it is certainly not the only one, and often it is operational only through a one-, two-, or three-dimensional order structure or some other structure which is needed to make systematic counting feasible.

In Section 4.16 I sketched how the elimination of the counting structure might develop; the elimination could be triggered by other, more striking structures that supersede the counting structure — as an example I took the relation between numbers of eyes and ears when determining them for a group of people.

My phenomenological sketch of the constitution of numerosity number via the counting number allows for variants and short cuts. In principle it is possible, and though it seems exceptional, in fact it happens, that numerosity number is not constituted via counting number but precedes it.

In both cases the invariance of the numerosity number under one-to-one mappings, though *mathematically* the most prominent and even constituting property of the concept of cardinality, is an *a posteriori* phenomenon if compared with the invariance under isomorphism of structures, and the same holds for invariance of the order as established by the cardinals. Constitution of those invariances requires that structures be impoverished to sets and isomorphisms to one-to-one mappings. The invariance principles to be acquired are: if the mapping f of A in B is

one-to-one, onto	then $\#A = \#B$
one-to-one, not onto,	then $\#A < \#B$,
onto, not one-to-one,	then $\#A > \#B$,

and conversely,

if $\#A = \#B$ and f one-to-one,　then f upon,
if $\#A = \#B$ and f upon,　then f one-to-one,
if $\#A < \#B$,　then f not onto
if $\#A > \#B$,　then f not one-to-one.

Modern textbooks often teach and test such properties by means of pairs of Venn diagrams. Such exercises are false concretisations which are likely to block the mental constitution of these properties, or at least to thwart their application. Recognising relevant sets and mappings is the very thing that matters in applications. I have concluded this from experiments with the drawer (or Dirichlet) principle, which, indeed, is an efficient test of the last property I mentioned above.

4.19. *Comparing Cardinals Numbers by Estimate*

Earlier I indicated the part played in the constitution of cardinals by indefinite numerals such as "many" and "few". As far as they express rough estimates, "more" and "less" are of the same character, if they mean "many more" and "much less".

The comparison by estimate is based on a — most often unconsciously handled — pattern, which shall be made explicit with a view to didactics: Two sets of objects, *A* and *B*, are compared

> while *A, at least as densely* distributed as *B*,
> covers a *larger space* than *B*,

or

> while *A, more densely* distributed than *B*,
> covers the *same space* as *B*.

A conflict arises if

> *A*, more densely distributed than *B*,
> covers less space than *B*.

When applying this pattern,

> the spaces covered by the sets are globally, at sight, compared

whereas

> the densities are locally compared

and

> the homogeneity of the distribution is globally judged.

A special case is the following:

> *A* and *B* are densely packed within equal spaces
> whereas the particular objects of *A* cover less space than those of *B*.

A variant:

> The spaces covered by *A* and *B* are replaced sets *A'* and *B'*,
> respectively, which are known to have the same number of elements
> or the one is known to have more than the other.

Another variant:

The spaces are replaced with time intervals
and the elements of A and B are events.

In general, the principle explained here can be formulated as follows:
Consider a finite set X with all its subsets. X is internally or externally structured in such a way that subsets apart from their *number* # possess a certain *character* k which can be described extensively; that is, by means of a magnitude that behaves additively under composition:

$$k(Y_1 \oplus Y_2) = k(Y_1) + k(Y_2).$$

This is called a k-homogeneous set.
Indeed if k is thought to be a

volume (area, length),
weight,
duration,

thus $k(Y)$ the

volume (area, length) covered by Y,
weight of the combined members of Y,
duration of the combined events of Y,

then

$$\delta(Y) = \#Y / k(Y)$$

is the number of members of Y per unit of

volume (area, length),
weight,
time,

independently of Y ($\subset X$), which justifies the expression "k-homogeneous set".

$$\delta(X) = \#X / k(X)$$

is what was called

density

in the introductory examples. Here, too, we can speak of a *distribution density* with respect to k:

$$\delta(X) = \#X / k(X).$$

Let A and B now be two k-homogeneous sets. The principle we wish to formulate, is

if $k(A) > k(B)$ and $\delta(A) \geq \delta(B)$, then $\#A > \#B$,
if $k(A) \geq k(B)$ and $\delta(A) > \delta(B)$, then $\#A > \#B$,

and the conflict is expressed by

$$k(A) > k(B) \quad \text{and} \quad \delta(A) < \delta(B).$$

The k-inequalities in the premises can globally be tested, the δ-inequalities locally and perhaps by samples.

The assertions based on applying this principle can be strengthened, such as

more than the double,
more than half,

and so on, if the estimations of k and δ allow it — an extension we shall consider when dealing with ratio and proportion (Chapter VI).

4.20–28. *The Constitution of Addition as a Mental Act*

4.20. For many years it was a habit to indicate the operation of set union by a plus-sign — thus $A + B$ instead of $A \cup B$ — and even now there are authors, in particular in measure theory, who stick to that notation. It happened first in topological algebra that one abandoned the old notation: in additive structures one needed $A + B$ to indicate the set of $a + b$ with $a \in A$ and $b \in B$.

The plus sign for the union formation was of course inspired by the close connection between union and sum, witnessed by

$$\#(A \cup B) = \#A + \#B \quad (\text{if } A \cap B = O)$$

and in fact used to define addition. If m and n are to be added, one provides oneself with two disjunct sets A and B of which m and n are the cardinal numbers

$$m = \#A, \quad n = \#B, \quad A \cap B = O,$$

forms their union and puts

$$m + n = \#(A \cup B).$$

It is easily shown that the result does not depend on the choice of A and B, that is, if

$$\#A = \#A', \quad \#B = \#B', \quad A' \cap B' = O,$$

then

$$\#(A \cup B) = \#(A' \cup B').$$

At the lowest level this then is the way to perform addition; the learner who has to add m and n creates the required sets of fingers, beads, strokes or whatever.

But it can happen as well that the sets are given in advance: he is asked to add not m and n, but the numbers of objects presented in some way, say

5 cars and 3 cars together,

where the cars can be real cars, five on this side of the road and three on the other side, or drawn, or suggested by a story, or some of the cars may be realised in one way and the others in another. Well, if their reality is not palpable enough, the calculator can provide himself with substitutes A' and B', equivalent to the sets A and B, which he is able to take together; A' and B' can be sets of fingers, or beads, or strokes. In all these cases the numbers to be added are clearly recognisable as cardinals of sets, and their addition as reflecting the union operation, even though the sets themselves may be inaccessible, and uniting them cannot actually be performed.

4.21. Problems do not arise until

the terms to be added are not plainly recognisable as cardinal numbers of sets

or

the addition is not plainly recognisable as reflecting the union operation of sets.

In

5 marbles and 3 marbles

the related sets and the operation are clearly recognisable. They are less so in the case

John has 5 marbles, while Pete has 3 more, how many does he have?

Pete's set is not obtained by taking two *given* sets together. One should rather consider the imaginary set of Pete's marbles split into two sets, one set of marbles equivalent with John's and another set of the 3 marbles he has more than John, which can be done in many ways. The formula

$$\#(A \cup B) = \#A + \#B$$

is again being applied, though with $A \cup B$ prescribed and to be split into A and B.

Another type of less clear recognisability of the related sets is

John has 5 marbles, yesterday he had 3 more, how many did he have?

Here a "lost" set must be mentally reconstructed in order to be united with a present one.

In

John won three marbles today, how many more does he have than he had yesterday?

nothing is really to be added, though an addition is suggested by "won" and "more", and the situation is even more troubled by the foggy set of John's marbles yesterday. Though looking more complicated,

Yesterday John won 5 marbles and today 3 marbles, how many more does he have than he had yesterday?

is probably an easier case, since one has really things to add. But what about replacing "won" with "lost" and "more" with "fewer"?

This kind of example could be extended with many more where the sets are somehow recognisable although the operation is harder to identify — in particular, as a consequence of misleading linguistic cues, such as in

John has lost three marbles, he still has 5, how many did he have originally?

4.22. This stock of examples can be multiplied ad lib., but we restrict ourself to this choice in order to signal another type of addition problems in which it can be properly said that no sets are distinguishable, and constructing sets from the objects occurring in the problem is perhaps less desirable. In

5 steps (of stairs) and 3 steps,
5 days and 3 days,
5 km and 3 km,
5 florins and 3 florins,
5 times and 3 times,

one can hardly speak of sets consisting of 5 and 3 elements, respectively. Whereas in the former group of examples the sets are still constructible in the domain of the objects, in the latter it is hard, if not impossible. Images that are to represent 5 or 3 marbles can be quite realistic, for instance, in a Venn diagram, but an explanation that 5 or 3 chips should be understood as being 5 or 3 days is nothing more than a verbalism.

In such examples new problems of recognition can be created by the way in which the addition is suggested:

John is 5 years, how old will he be 3 years from now?
Today is 5 January, what date will it be 3 days from now?
It is 5 o'clock, what will the time be 3 hours from now?

with all kind of variations which can arise, for instance, from a change of perspective.

4.23. How is the arithmetical knowledge about addition that was acquired by uniting sets transferred didactically to this kind of problem? Possibly the numerical performance of the act of addition is simply suggested on the basis of the genuine set operation via the abstracted arithmetical problem 5 + 3. This is certainly the case with such patterns as 5 times and 3 times, where every indication is missing concerning what should be present, happening, done, taken three times. Though originally the numbers were cardinals of sets, and operations with numbers were operations with cardinals based on operations with sets, in these problems they are stripped of all substance in order to become numerals and algorithmic operations on numerals ("computing number").

It is not astonishing that such didactical procedures, if applied again and again, create hosts of underachievers: children that fail on any kind of word problem. According to this didactics the transition from

     5 marbles and 3 marbles

to

     5 days and 3 days

is performed via

     $5 + 3$,

thus the transition occurs algorithmically and without any experience as to what adding days can really mean. How can and should this be improved?

4.24. Earlier I drew attention to how closely numerosity and counting number are interwoven. $5 + 3$ is defined cardinally, but from olden times it has been calculated ordinarily. Set theory fanatics tried to fight this didactically with Venn diagrams, but fortunately with little success thanks to the natural inclination to operate by counting. The result of $5 + 3$ is obtained by counting 3 steps forth from 5 onwards: to the 5 beads on the abacus one adds 3 — one, two, three — both terms as well as the sum being defined by concrete sets. Or: starting with the mental 5, one counts on — 6, 7, 8 — while thumb, forefinger, middlefinger, one after the other, are raised in order to steer the deployment of the second summand. Or: the same, without the fingers, while the activity of adding is steered by rhythmic or visual images.

One could teach children the elementary additions in the same way as the tables of multiplication are often taught, by intentional memorising. Perhaps here and there it really happens this way, but certainly not as a rule. Memorising addition is unintentional; the addition tables are learned by performing additions again and again. Yet memorising is not the sole aim of performing the operations. By the process of adding the addition is experienced as meaningful and understood in such a way that, if needed, it can be made explicit. By this meaningful activity on summands more complex additions are prepared where the summands and the sum are not represented by sets. In brief, by the activity of adding the addition is constituted as a mental activity.

Let us review our examples. 5 steps and 3 steps — with feet on the fifth step, it goes on — one, two, three mentally — to the sixth, seventh, eighth — counted aloud. 5 days and 3 days — turn three days (pages) further in the calendar. 5 km and 3 km — the signpost shows 5 km to the point you came from and 3 km to the point you are going to, and tells you to count from point 5 further to point 8. To the 5 florins, represented by a banknote or by 2 coins of 2½ florins, three florins are added — 6, 7, 8 — which are mentally counted as 1, 2, 3.

4.25. These examples are distinguished from the former ones not only with regard to the materialisation but also structurally. One is no longer taking the union of two unstructured sets. We are not faced, as in the case of the marbles, with five and three disconnected steps, days, km, florins, times, which are raked together. They are objects that follow each other in space or time, not as elements of a set but as paces on a road, in a process that in turn suggests and perhaps elicits a counting process.

I already stated that counting processes are embedded in time and possibly also in space. By pronouncing the number, by indicating objects, points in time and space are isolated in the continuous medium, and this can be done more or less sharply — a string of beads the size of a dove egg or a mustard seed, densely threaded or loosely spread. Counting queuing cars — they follow each other at a distance and are counted one by one. Counting the wagons of a freight train, a continuous stream articulated by the buffers — the rear buffers, of course, because with the rear buffer, the ordinal number of the wagon is "accomplished", as the odometer of a car accomplishes the kilometers. Counting when playing at hide-and-seek — numerals are stretched over an interval, and after 100 it is: I am coming.

One counts marbles, but likewise intervals: pronouncing 21 such that it takes a second. It depends on the intention: counting spatial or temporal phenomena or articulating the spatial or temporal stream. Of course, one can maintain that mathematically it is all the same: the disconnected marbles, the beads loosely or tightly on the string, the sequences of spatial or temporal intervals — all of them are sets, all have cardinal numbers, and adding them is based on taking the union. This is true and it proves how universal mathematical concepts are. If you assign to each kilometer of a trip its end, to each year the stroke of the clock that completes it, to each m^3 of water passing the bridge of the Meuse its last drop, to each ton of flour streaming out of the mill its last particle, you obtain a mapping where the successive intervals are mapped on a sequence of points — a one-to-one mapping of something that looks continuous but by this mapping is cut into a discrete looking sequence of slices. This then is the mathematical justification of a procedure by which natural numbers are used to count in a continuous medium; but though it is mathematics it is nothing but the unconscious background made explicit.

How old are you? "Four" the child says, raising four fingers and knowing that at some precise moment a fifth will be added. Before the natural number is constituted as a characteristic of sets in the discrete realm, it is already applied in the *continuous* one — that is, to magnitudes. The examples I mentioned — length, time, monetary value — are paradigmatic. Continuous climbing is articulated by steps, the stream of time by the torn-off pages of the calendar, the road by km-posts, money by coins and notes, "so many times" means so many times the same — it is these articulations that are counted. In traditional teaching, magnitudes are delayed until the children are ready to learn common and decimal fractions. This reservation is justified by nothing but a pseudo-didactical systematism. The first step in analysing a magnitude, where measuring the magnitude is articulated by the natural multiples of a unit, is possible and desirable at an early age; counting can and must immediately be transferred from discrete quantities, represented by sets, to magnitudes. Modern textbooks start measuring much earlier than tradition allows, but unfortunately this kind of measuring is not yet sufficiently integrated with the operations on natural numbers.

4.26. The device beyond praise that visualises magnitudes and at the same time the natural numbers articulating them is the number line, where initially only the natural numbers are individualised and named. In the didactics of secondary instruction the number line has been accepted, though it is often still imperfectly and inexpertly exploited; in primary education it makes progress little by little. The progress is slowed down on the one hand by Venn diagrams, on the other hand by rudimentary material like Cuisinaire rods. It seems to me a disadvantage of the number line that it is so easily drawn and that it cannot be sold together with the textbook as teaching material.

The Cuisinaire rods — which in fact have come down from Fröbel — were, once introduced, a large step forward: a translation of natural numbers into lengths, and of operations on natural numbers into operations on lengths. The articulations in the continuous stream are being concretised, the intervals coagulated and embodied in coloured rods. The lengths are torn from their context "length"; numbers is the peak of what the rods can represent — no other magnitudes.

The number line eclipses the Cuisinaire rods in many respects: The virtual infinity is better expressed by the number line. The number line knows no compulsory scale; number lines on different scales — on the blackboard and on paper — are immediately identified, notwithstanding their incongruency. And what is most important: in manipulating Cuisinaire rods the route from the visual to the mental realm is diverted by an irrelevant motorism. Later we shall consider what relevant motorism is and how it applies to working on the number line.

Does it look strange that I deal with these matters under the title "The constitution of addition as a mental act"? Counting and adding are closely knit in the constitution of number. Counting is again and again adding one, and additions are performed by counting. This then characterises the arithmetic of the number line and its didactics which we are engaged in.

The number x stands on the number line where x is accomplished. What x?

x cm of the ruler,
x km of the road,
x cars in a row,
x books on a stack,
x pages of a book (tens only)
x ticks of a clock,

x hours, days, years (on the time axis),
x grams or kilograms,
x cc of the measuring glass.

Yes, accomplished. The ruler or road is accomplished by scanning or pacing it with real or imaginary fingers or feet, the queue runs out, the books are fetched from and put into the stacks, the book is leafed through to page x, the time is accomplished and the hand of the clock or the spring balance accomplishes the interval x. The number x stands where it is "full". I can move or jump to the point. And if it is a jump it is again pictured in a continuous manner by an arrow, perhaps with a "$+ x$" on it. The jump can be composed or thought to be composed from little jumps. The jump can be done from every starting point, pictured by the same arrow "$+ x$", and this then pictures the addition, in one blow, or dissolved in little jumps. The intermediate jumps have been made explicit or just blurred. The number line can bear numbers at the tens only – like a ruler – with intermediate strokes for the units, or at the fifties – like the measuring glass – with intermediate strokes for the tens, or only at the hundreds and thousands, like the kilometers along the road. In between, interpolation takes over: the 175 should be at a certain spot between 100 and 200 – second graders are able to localise it. And finally there is the line (or path) with no number or marks at all, except an origin.

How far is it from here to there on the number line? The little steps are counted. But you can also take the "from here to there" between your thumb and forefinger, carry it back to 0, and read it off. Adding n to m can be performed by counting but it can also be done in one blow: the piece that is "accomplished" at n is taken between the fingertips and carried over to m – between the fingertips or on a ruler or a strip. Addition is being performed geometrically, rather than by counting. But that means that you can dispense with numbers on the number line. The number line is mirroring a magnitude, and the geometrical shift on the number line mirrors addition for this magnitude, as it does subtraction.

This then is another way to have addition begin, with magnitudes rather than with marbles, but with magnitudes that are visualised and bound to the number line – the length on a ruler, the volume on the scale of the measuring glass, the weight on the scale of the spring balance, and (somewhat harder) the time on the time axis. Davydov* has shown how adding and subtracting with magnitudes – length, volume – can didactically precede the numerical operations in order to develop and support them. I tried it with a boy (5; 6) whose arithmetic abilities were negligible; he was able to perform meaningful operations on length, volume, and, with the spring balance, on weight, up to fair understanding of Archimedes' principle.**

* Cp. Hans Freudenthal, 'Soviet Research on Teaching Algebra at the Lower Grades of Elementary School', *Educ. Stud. in Math.* 5 (1974), pp. 391–412.
** 'Bastiaan's Experiments on Archimedes' Principle', *Educ. Stud. in Math.* 8 (1977), pp. 3–16.

Well, let it be true that the title of the present *chapter* does not seem to justify the present subject, which is magnitudes; then, however, the title of the present *section*, which is about addition, does justify it. From the beginning addition should be of a broader range than operating on natural numbers. The natural numbers are characteristics of discrete sets, adding them is rooted in throwing together such sets. The set theory medium of magnitudes is continuity, and adding there means composing. The universal model of magnitudes is the ray, primarily structured not by numbers but by congruent displacement, corresponding to addition of magnitudes. Not until a unit is chosen for the magnitude and correspondingly a "1" is placed at the ray, are addition of magnitudes and addition of numbers related to each other, then the ray is articulated by points corresponding to the natural numbers to become the number ray, and cardinal addition is translated into set theory addition.

4.27. With the idea of natural numbers on the number line we are as far from their cardinal origin as geometrical addition is from uniting sets. In this analysis the connection between numerosity and measuring number has been made didactically via the counting number and magnitudes. Earlier I explained a short cut: dividing the ray into succeeding congruent intervals and interpreting number as the cardinal of a set, to wit, of subsequent intervals. Anyway from the start onwards the natural number obtrudes itself on the learner in all its aspects. Only a system fanatic could be offended by this challenge. It is a fact that natural number has many aspects, one of which is its use and its indispensability as a measuring tool, not only in applications of arithmetic but also on behalf of its didactics.

Problems arise around addition (I have already elaborated on this theme) as soon as the summands to be added and the operation of addition are not plainly recognisable as cardinals of sets and their union, and attempts at a set interpretation are artificial or obnoxious. For this reason, as I said, the constitution of addition and learning to add should be more broadly oriented towards adding magnitudes represented on a number line, bearing initially the natural numbers only. Neither a recourse to sets nor algorithms that exploit verbal cues can be of any help if such problems are to be solved. The Venn diagram visualisation of adding is too narrow, and algorithmising prematurely involves the risk of a wrong perspective that may influence the mind in a way obnoxious to mathematics. It might be wholesome to first understand addition broadly and to visualise it in a way that can be considered as definitive from a mathematical viewpoint.

This suggests that not only adding but even counting should be accompanied by activities on the number line. One more marble from the bag to be counted is accompanied by one block joining the train of blocks and one segment on the number line. Counting by estimate is accompanied by vague indications on the number line. "More" and "less" mean directions on the number line; "this more" and "this less" make more precise how much to proceed in one direction or the other. Two parallel number lines, used for John and Pete: the marbles

John wins from Pete are transferred as a line segment from the Pete line to the John line. Transferring contents and weights from one receptacle to another is similarly represented. Concretely or symbolically adding numbers given by objects or at the fingers or the abacus beads, or by tallies is one of the accesses, which certainly should not be neglected, but it is not the only one. The other, broader one is via magnitudes and the number line.

4.28. A few lines above I stamped the visualisation of the addition on the number line as definitive. This is true as long as I restrict myself to **N**. The extension to **Z** seems artificial at an elementary level. A more natural extension is towards two dimensions, vector addition in $\mathbf{R}^2$ or in the lattice $\mathbf{Z}^2$, which then includes the extension to **Z**. This addition would be introduced geometrically, by parallel shifts of vectors, in order to be expressed arithmetically afterwards. This subject can lead into a field of free activities which as motivations may influence the more regimented, properly arithmetical, activities. Later (in Chapter 11) we will reconsider these questions.

4.29. *The Additive Structure of N*

The additive structure of **N** includes more than the act of adding. It is, as it were, the whole complex of relations

$$a + b = c,$$

possibly also expressed as

$$c - b = a \quad \text{(for } b \leqq c\text{)}$$

and supplemented by

$$a + b + c = d,$$
$$a + b = d - c,$$

and all other relations one would like to consider in this context.

On a higher level it includes experience, and on a still higher one formulated knowledge, of such properties as are

commutativity,
associativity,
equivalence of $a + b = c$ and $c - b = a$,

and many more properties of this kind.

This structure grows as the learner explores **N**, but fundamental properties such as the ones just mentioned can be experienced fully and clearly and even formulated within a quite restricted part of **N**.

The structure of **N** is partly accessible to memorising; beyond this, it is obtained and analysed by means of the algorithms of the decimal system, which will be dealt with later in the present chapter.

However, the relations considered so far yield an insufficient grasp of the structure of N. For instance, the relation

$$a + b = c$$

can be structured by prescribing c and asking for the totality of solutions $\ulcorner a, b \urcorner$, the list of splittings

$$8 = + \frac{8 \quad 7 \quad 6 \quad 5 \quad 4 \quad 3 \quad 2 \quad 1 \quad 0}{0 \quad 1 \quad 2 \quad 3 \quad 4 \quad 5 \quad 6 \quad 7 \quad 8}$$

which exhibits a striking structure of increasing and descreasing sequences and a central symmetry. Of course, splittings are also useful for the algorithm of passing over the tens when adding, but there is more to it. A structure like the one exhibited by this list invites questions of "why?", which with their answers help one to understand the additive structure of N more profoundly. One list like this is not enough; they are available for each c. Interpreted in two dimensions, in the lattice, these lists become point sets with a remarkable interior and exterior structure.

Other lists are created if

$$a + b = c$$

is viewed with b fixed as a condition for a, c, such as in

$$3 = - \frac{3 \quad 4 \quad 5 \quad \cdots}{0 \quad 1 \quad 2 \quad \cdots}$$

again with a striking structure, which asks for explanation. What characterises these pairs of points, when viewed on the number line?

The order relation, viewed in the context of addition, also belongs to the structure of N. It is obvious that adding more yields more, and subtracting more yields less, but it is not so obvious what this means for solving inequalities and for other applications; for instance, that to solve

$$x + 4 < 10$$

one is advised start with the largest solution. Much insight into the structure of N is required to solve

$$x + y < 10.$$

Other additive structures in N are arithmetical sequences

$$1, 4, 7, 10, \ldots$$

corresponding to jump sequences on the number line. Where do two sequences of this kind, the preceding and

$$2, 6, 10, 14, \ldots,$$

meet each other? The question is easily answered but to understand it requires the multiplicative structure of **N**.

The field of hundred is a structure in **N**, and in this field the arithmetic sequences show special structures. The table of addition is also a structure in **N**, symmetric with respect to the diagonal, and a chessboard distribution of even and odd numbers — why?

Properties of such structures can be traced, understood, explained. Properties of **N** can be known and applied; for instance commutativity to add the smaller to the larger number conveniently, and associativity to simplify additions algorithmically — for instance, by completing tens.

4.30. *The Constitution of Subtraction as a Mental Act*

Discussing subtraction after addition does not aim at a didactical separation, and certainly not at a succession in the genetic and didactic process. In all contexts where addition is didactically offered, subtraction is implicitly present in order to be made equally explicit.

Formally,

> subtraction results as the converse of addition,

and in fact this aspect of subtraction should not be neglected.

$$8 - 3 = 5 \quad \text{because of} \quad 5 + 3 = 8$$

is reasoning by which each subtraction can be reduced to an addition, which might be known in advance by memorising. In the case of division as the converse of multiplication, this is indeed the way to solve elementary problems after the tables of multiplication have been memorised. Subtraction is dealt with differently, and there are reasons why it is so. The inference pattern of inverting an operation is probably less familiar to a six-year old than to an eight-year old; but the main reason seems to be that subtraction is as concrete as addition, whereas division is much less so than multiplication.

In the domain of objects, subtraction means taking away, as addition means annexing. The older arithmetic books had great difficulty when picturing subtraction. Venn diagrammatici, at a loss what to do with subtraction, invented the strangest aberrations. Meanwhile new inventions bear witness to the fact that a fresh spirit has befallen developers: rigidity and unifying dogmatics have given away to a creative imagination of situations that suggest mental activities. With regard to subtraction: withering flowers, birds flying up, dwarfs running away, walls and towers breaking, and many more of this kind. The problem suggested by the picture is often not uniquely determined. If a story is to be told about the picture, it may be the style of

there were c,
b went away, $c - b = a$
thus a were left;

or

there were c,
a were left, $\qquad c - a = b$
thus b had gone;

but also

a are left
b runs away $\qquad c = a + b$
that is c together.

Yet this lack of uniqueness is not at all a didactic shortcoming.

The objects can be pictured in disorder or in a row; the objects to be taken away can be on one side of the row, on the other, or criss-crossed in between. But explicit taking away suffices as little for the mental constitution of subtraction as uniting explicitly given sets suffices for addition. "Which ones are more, and how many more?" In order to answer this question about a picture with two kinds of objects, indirect taking away is required: one set is diminished by a subset equivalent to the other.

The numerosity aspect suffices as little for subtraction as for addition to mentally constitute the operation. I desist from repeating for subtraction the arguments I displayed in the case of addition. Subtractions too should be constituted more broadly, with magnitudes geometrically interpreted on the number line.

In numerical subtractions — on the abacus, with tallies, on the number line, or mentally — one observes two methods,

taking away at the start,
taking away at the end,

and counting the remainder. "Among 8 children, 3 girls, how many boys?" can be answered by "from 4 up to 8" and counting these numbers "one, two, three, four, five" on the fingers. Or the three are taken away at the end: one away is 7, another away is 6, another one away is 5, and meanwhile thumb, forefinger, and middle finger have been raised to control the process. Children learn quickly which method is more useful in each particular case: if the subtrahend is smaller than half the minuend it is taken away at the end, otherwise at the start.

With geometrical subtractions on the number line both methods can apply: the subtrahend can be congruently cut away at the start or at the end of the minuend, and the remainder is measured again.

The geometrical concreteness of the number line is particularly useful in understanding problems like

$a - \ldots = c$
$\ldots - b = c.$

4.31. *The Constitution of Multiplication as a Mental Act*

Multiplicative terms like "double" and "times" precede multiplication as an arithmetical operation.

> I told you three times,
> walk three times up and down,
> the clock struck three times,
> when the (minute-) hand has gone around three times,
> you must sleep three more times, then it is your birthday.

It is

> three times doing, undergoing, experiencing, awaiting

something. And then in simple additions, before multiplication is on the program, it can be

> seizing three times four marbles

or simply

> three times four marbles,

and this means

> four marbles and four marbles and four marbles.

No operation — not even addition and subtraction — offers itself as naturally and is understood as spontaneously as multiplication.

Daphne (5; 1) is asked: "How many prongs does this fork have?" Two forks, three forks, four forks? Almost imperceptibly her fingers (the thumb excluded) tap on the table while she is counting on. When, at "five forks" she hesitates a bit, her elder sister whispers "20" to her grandmother, which makes Daphne angry.

Though this is not an explicit multiplication problem — the term "times" does not occur — it shows the inductive origin of multiplication as repeated addition. This then is the way products are calculated and how tables are built. Remembering our exposition on addition one should apply "so many times" in *magnitudes* early. Indeed, this is one of the functions of multiplication

> 3 kg, 3 m, 3 km (travelling)

are 3 times as much and cost 3 times as much as

> 1 kg, 1 m, 1 km;
> 3 km cycling

is 3 times as much and takes 3 times as long as

> 1 km cycling;

three of the same

weigh three times as much as

one;

and so on. Especially on the number line multiplication as repeated addition can be effective:

3 logs reach 3 times as far as 1 log,
with 3 steps (jumps) you go 3 times as far,
with 3 turns a wheel covers 3 times the path.

This implies that

3 times 4

can be realised by

3 rods, steps, jumps of 4,

subsequently performed on the number line.

So far multiplication was present as nothing but repeated addition. 3 rows of 4 marbles each may be placed

after each other;

with a change of perspective, they can be placed

in order below each other,

a two-dimensional arrangement, the

rectangle model

of multiplying. From two sets A and B the

set of pairs or product set $\ulcorner A, B \urcorner$

is formed. This restructuring reveals new aspects of the product. To calculate

$m \cdot n$

one provides oneself with two sets A, B such that

$m = \#A$, $n = \#B$,

forms the set of pairs $\ulcorner A, B \urcorner$, and puts

$m \cdot n = \#\ulcorner A, B \urcorner$,

where the result does not depend on the special choice of A, B.

Only by means of the rectangle model of the product do properties of multiplication become visible: *commutativity*, the rectangle is rotated a quarter of a turns and *distributivity*, two rectangles of equal height (or width) moved side by side. (Figures 20 and 21.)

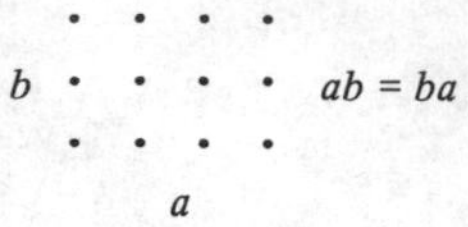

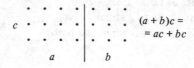

Fig. 20.

$$(a + b)c =$$
$$= ac + bc$$

Fig. 21.

In the preceding

> set of pairs and the rectangle model were used to
> restructure the set that is basic to the arithmetical product,

among other reasons in order to create insight into properties of multiplication. The converse function is accomplished by the

> set of pairs and the rectangle model as structures discovered or introduced in sets in order to calculate their numbers as products.

This function has been discussed extensively in Chapter 3.

> The rectangle model leads in a natural way to

> the area of the rectangle:

When one passes from the discrete number sequence to the number line, the discrete point-like beads are, as it were, condensed into unit rods; in two dimensions they form unit square tablets, which together cover the rectangle. Its area is expressed by unit squares after multiplying length by width as the − implicit or explicit − arithmetical operation. The part played by the rectangle model in multiplying magnitudes in general shall be discussed at another place.

All that has been said about the set of pairs and the rectangle model can be extended to the

> set of triples and plank model.

The latter shows

> associativity

and leads to the

> volume of the plank

expressed by unit cubes. Sets are

structured as triples

on behalf of calculation of their cardinals.

In Chapter 3 we displayed other product sets with two or more factors, like the

tree model, the roads model,

as showing product structures.

4.13a. *The Role of Equipartition Structures in Learning Processes on Behalf of Multiplication*

Among the models just mentioned, I forgot the one illustrated in Section 3.22 by

m baskets with n eggs each.

Not intentionally but as an oversight, which in Section 3.22 I imputed to others as a mistake. I myself had overlooked what I had earlier signaled as a serious gap in the whole didactical literature. Was I so overcome by set theory dogmatism as to forget about my own ideas? Anyway, by observing learning processes I recalled them.

Multiplication is, first of all, repeated addition and this operation can efficiently be structured by the set of pairs in the rectangle model — the set theory product — in order, among other things, to calculate cardinalities as products. This model, however, is insufficient. Not mathematically, since the eggs in each basket can be numbered in order to impose on the set of eggs the stronger structure of set of pairs with the elements

⌐basket, number of egg within its basket⌐.

It is didactically insufficient since the sophisticated mathematical restructuring can hardly be expected to arise — spontaneously or perforce — in learning processes or, if it arises at all, be made conscious in order to be made explicit and, if need be, available.

Let us consider the following sequence of problems:

(a)   A picture representing seven baskets, each with six eggs. Question: how many eggs?

(b)   A picture representing seven baskets, with the legend: in each basket there are six eggs. Question: how many eggs?

(c)   The text: I have seven baskets; in each of them there are six eggs. Question: how many eggs?

(c′)   Or more streamlined: 7 baskets, 6 eggs each; how many eggs?

It is a sequence leading from "material" and "pictorial" to "mental" and "verbal", which, for that matter, can be lengthened and refined. The factual situation which I observed in a traditional third grade, treating more modern material, was that of problem (b).*

Pupil A counted – very fast – 1, 2, 3, 4, 5, 6, with a finger at the first basket, 7, 8, 9, 10, 11, 12, with the finger at the second basket, and so on, up to 37, 38, 39, 40, 41, 42 with the finger at the seventh basket.

Pupil B counted, while his finger glided from one basket to the next: 6, 12, 18, 24, 30, 36, 42.

Pupil C did the same as B, without using his finger.

Pupil D said: 7 times 6 = 42.

(Between pupil A and B one could interpolate a variant A' who does the same as A but without using his finger as a marker.)

How would these pupils have reacted in the situations (a) and (c)? In particular would pupil D in situation (a) have used multiplication, and would pupil A have interpreted situation (c) as a multiplication or would he have failed completely? What background part did multiplication play with pupils B and C in situation (b)? Rather than using $1 \times 6 = 6$, $2 \times 6 = 12$, $3 \times 6 = 18$, ... it is easier to recite the sixes table in the singsong 6, 12, 18, ... , and meanwhile, with the finger or an eye on the basket or while counting on one's fingers, control the number of steps to be taken in the multiplier. A bunch of questions. A long term observer could have said more about them.

I can say just as little about whether some pupils saw the structure of set of the pairs in the data – one needs to know much more about their former learning processes. The structuring articulation after each six may be presumed to be operational in all pupils, A marked it with the displacement of his finger, and B and C with counting by six. The solely mental presence of the eggs was certainly a factor in favouring this structuring. It may be guessed that in situation (a) it would have been less favoured.

How would these pupils have reacted to a counting problem in a rectangle pattern (seven rows of six eggs)? Probably A would also have counted; B and C would perhaps learn by such examples to interpret a rectangle model directly as a multiplication.

More questions. I have never felt so frustrated by the lack of continuity in the learning processes I have had the chance to observe.

4.32. The Multiplicative Structure of N

The multiplicative structure of **N** is the whole of the relations

$$a \cdot b = c,$$

possibly also expressed as

* I have adapted the problem to my model terminology. Originally it was about house trailers and people.

$$c : b = a,$$

complemented by

$$a \cdot b \cdot c = d$$
$$a \cdot b = d : c$$

and all one can think about in this context.

At a higher level it includes experience, and at a still higher level formulated knowledge, of such properties as

> commutativity,
> associativity,
> distributivity,
> equivalence of $a \cdot b = c$ and $c : a = b$,

and many more properties of this kind.

The structure grows as the learner explores **N**, but fundamental properties such as the ones just mentioned can be experienced fully and clearly and even formulated within a quite restricted part of **N**.

The structure of **N** is partly accessible by memorising tables; beyond this, it is obtained and analysed by means of the algorithms of the decimal system, which will be dealt with later in this chapter.

However, the relations considered so far, yield an insufficient grasp of the structure of **N**. The relation

$$a \cdot b = c,$$

for instance, can be structured by fixing a and having b run through the number sequence; this yields for c an arithmetical sequence visualised by jumps on the number line.

Or I can fix c and find systematically its splittings into two factors, which are illustrated by rectangles with a given area.

I can also split into more factors, factorising into prime factors. Divisors, multiples, remainder classes are other means of structuring. Divisibility properties can serve to simplify multiplication.

As in the case of the additive structure of **N**, the order structure deserves attention if tied to the multiplicative structure. It is obvious that larger factors yield larger products, but this does not necessarily include the change of perspective:

> given the product, increasing one factor and decreasing the other go together

is an order reversing behaviour, which plays a part in division:

> smaller portions — larger numbers,
> bigger portion — smaller numbers.

In another way the order structure of **N** is operational in division: the number
sequence is divided by an arithmetical sequence into intervals;

> which numbers are in such an interval?
> in which interval does a given number lie?

are relevant questions.

4.33. *The Constitution of Division as a Mental Act*

The relation of dividing to multiplying is much more involved than that of subtracting to adding. It is much less symmetric in the case of multiplication and
division than it is for addition and subtraction. This discrepancy extends from
the most simple activity to the fully developed concepts. At a glance one can
tell whether a number can be subtracted from another: the smaller from the
bigger. Whether a division terminates is a surprise. Nothing like the remainder
in division exists in subtraction, or should it be $3 - 8 = 0$ remainder 5 (that is,
if the problem is about money, the amount by which one is another's debtor)?
On the other hand, at an early stage children know, besides the indivisible
object, other — continuous — objects that suggest unrestricted divisibility and
invite the extending of **N** to **R**⁺, whereas suggestions of restricted subtraction
and extension of **N** to **Z** are scarcer and weaker.

Division does not occur as universally in the function of inverting multiplication as does subtraction with respect to addition. There is no counterpart in
division to performing an act three times as shown by the former examples.
Even "half" is no good as a counterpart of "double", as witnessed by "bigger
half" in the vernacular.

Phenomenologically viewed, dividing arises in three ways: as

> continually taking away,
> distributing in equal parts,
> inverting a multiplication.

Dividing by repeated subtractions is the counterpart of multiplying by
repeated additions:

> how often can you take away a set of three from this pile;
> how many jumps of three do you need to go from here to there;
> how many times does the three-metre rod fit along the corridor;
> how many times can you scoop 3 litres from this vessel?

Being able to subtract q times d from a number or magnitude a means that the
remainder must be smaller than d:

$$a = qd + r \quad \text{with} \quad r < d.$$

The ancient mechanical calculator, which did not know multiplication tables,
performed divisions as repeated subtractions; that is to say, rather than starting

with d, starting with a $10^m d$ in order to apply the same procedure with $10^{m-1} d$ to the remainder, and so on. Likewise, in the usual long division algorithm the subtractive structure of the process is recognisable, though perhaps shortened by the knowledge of multiplication tables.

This kind of division should be contrasted with that of dispensing equal shares of equivalent objects to, say, q persons. Each gets one of the equal shares, *a qth part* — a strange terminology, if you stop to think about it, but so familiar that it looks natural.* In every language I know, the ordinal number is used to indicate how many shares it is of one share — leaving aside division by two, for which there are special expressions like one half. This terminology looks even stranger if instead of a quantity of objects divided among q persons, a number, say 12, is divided in, say, three parts. Then 4 is *the*, rather than *a*, third part of 12. It is so easy to pass from concrete sharing to abstract dividing and at the same time from *a* qth part to *the* qth part, but whether it is as easy for the learner, we simply do not know.

Distributing a small quantity in a small number of equal parts is most often an intuitive procedure, in particular if done with magnitudes, which in principle can be divided with no remainder left. It is exercised and understood early; in particular, as meting out fair shares to a number of persons. It can be done by giving them cyclically the same share until nothing is left or something is left that does not admit of dividing.

These two kinds of division were formerly distinguished as

ratio division

and

distributive division,

and separately learned and trained as such.** The question of the ratio division is

how many times does d go into a?,

that of the distributive division is

what is the qth part of a?

The difference is particularly striking if a and d are concrete numbers of the same kind:

how many persons are dealt with if each gets d out of a florins?
how many florins does each of q persons get if a florins are distributed?

Under both aspects the remainder has the same function: a remainder too small to fit what is taken away, or too small to be fairly shared.

---

* This question will be tackled once more in Chapter 5, on fractions.
** Cp. *Mathematics as an Educational Task*, pp. 252–254.

With division as the *converse of multiplication* it is different. The arithmetical sequence of multiples of d determines intervals in $\mathbf{N}$; the number a is placed in one of them. If it is between qd and $(q + 1)d$, then q is the quotient of "a divided by d". After q steps of d on the number line, there are still r simple steps left to reach a, and that is the remainder.

This is how elementary divisions are *learned*: one takes one's bearings in the frame of the table of d and situates a with respect to it. To achieve this, one has to know the multiplication tables well.

From the viewpoint of inverting multiplication the two kinds of division can be distinguished as follows: If I read

$$a = q \cdot d$$

as a equals q times d, the factors q and d are not exactly the same thing: d is the thing that is taken q times. This is strongly felt if d is not simply a number but a concrete number or a magnitude:

> 3 weeks = 3 · 7 days
> 3 dozen = 3 · 12 pieces
> 3 notes of 25 florins equal 3 · 25 florins.

The set whose product is the cardinal bears a certain structure, which means that the factors are not *a priori* commutants.

In the equation

$$a = qd$$

I can ask for q if a and d are given — ratio division —, or for d if a and q are given — distributive division. Thus, also as a converse of multiplication, division shows two facets. Well, one could do the same with the addition, distinguishing the summands and asking two questions:

> how much added to a to get c?
> to what is b added to get c?

Yet the asymmetry is not so striking in addition as it is in multiplication, where one of the factors can be concrete and the other abstract.

More profoundly viewed, *two* aspects of division are not yet enough. Concrete numbers are multiplied by concrete ones too, magnitudes by magnitudes, such as

> length times width = area,
> quantity times unit price = total price,
> time by speed = distance,
> time by rate of interest by capital = total interest,

and though such equations can also be solved with respect to different factors, it is not usual to distinguish here various kinds of division. For good reason indeed: Mathematics is powerful thanks to its universality. One can count all sets by the same sort of numbers, as one can measure all magnitudes by the same sort of numbers. Numbers do have a numerosity and a counting aspect; addition has a cardinal, an ordinal, and a measure aspect; multiplication has the aspect of repeated addition and of pair formation, and likewise division has its own variety of aspects. But in spite of this wealth of aspects, it is always the same operation – a fact that expresses itself by algorithmics. As a calculator one may forget about the origin of one's numbers and the origin of one's *arithmetical* problem in some *word* problem. But at the same time one must be able to return from the algorithmic simplicity to the phenomenal variety in order to discover the simplicity in the variety. It is the secret of aspects that they are discovered at one time and neglected at another time, and the knowledge of it is part of the constitution of the mental object or act in question.

4.34–43. *Algorithmics in N*

4.34. Algorithmics was touched on in the last section and earlier in Section 3.30. In the case of division I alluded to a very special algorithm, whose definitive version is long division. Algorithmics can, however, be understood more broadly as an organisation by which one is advised to follow prescribed rules, where each particular step requires a decision bound by certain criteria, under which the step is easily performed. For the computer such rules are more narrowly formulated than for man; the human calculator should have the liberty of replacing a multiplication by 98 with one by 100–2 in order to apply distributivity, but in a computer program the advantage of less computer time does not outweigh the disadvantage of a more involved program. It is as though a bank whose customers may have several accounts would think about saving postage by sending messages regarding the various accounts in one envelope. The savings might not outweigh the costs of a foolproof program for such a complicated system. For crossing a road pedestrians follow rules such as "first look left, then look right" (not useful on one-way streets), but this can hardly be called an algorithm. A crossroads with traffic lights is different – "cross when the light is green" looks more like an algorithmic rule.

Though algorithms may sometimes look like an aim in itself, they are not. They serve to simplify complex activities – complex up to impossibility. Insight is superseded by automatisms; that is, automatisms that are dependable even though they are controlled by little or no insight. This is what you teach computers by means of good computer programs. Man is less easily programmed. Programs are inculcated by numerous repetitions; whether people are slow or fast learners, flawless functioning is quite exceptional.

This seriously restricts the usefulness of learning algorithms, and in particular if it happens with no appeal to insight. How much time and trouble should be spent on teaching some algorithm which might be expected to be applied only rarely by the learner? In a period of non-activity the algorithmic ability might fade away or be lost. If some opportunity would then occur to apply it, it

would be forgotten or shrouded in fog. But if one has learned the algorithm by
insight, there is some chance that it can be reconstructed. "I can remember there
was a trick", but whether the reconstruction succeeds or not may depend on
how that trick has been learned.

Algorithmics, as considered here, springs from radical formalising. Mental
objects are fixed, or even replaced by, linguistic symbols — in the language of
the abacus or the digital system —, operations on mental objects are supported,
or even replaced, by strictly regimented syntactic operations on linguistic
symbols — the column arithmetic — relations between mental objects are translated into, or replaced by, regimented syntactic propositions, properties of the
operations and relations are expressed, or replaced, by formulas — but this is a
new stage of algorithmising: algebraisation.

4.35. The usual algorithmics of the operations in N rests on the decimal structuring of N, followed by a formalisation by means of the positional system. A
number $n \in N$ is dissected into a sum

$$n = a_i 10^i + a_{i-1} 10^{i-1} + \ldots + a_0 ,$$

where $i \in N$ and for the a_j the values $0, 1, \ldots, 9$ are allowed. If $a_i \neq 0$ is prescribed, this presentation of n ($\neq 0$) is unique. Positionally n is written as

$$a_i a_{i-1} \ldots a_0 ,$$

that is with a_j beads, chips, strokes in the jth column (starting at the right with
the 0th column) of the abacus, or with the digital symbol for a_j in the jth
position (starting at the right with the 0th position) of the writing material.
(I neglect abaci with intermediate units of 5, as used in the Far East).

The school abaci now available realise the columns by means of metal rods
with — mostly — 20 beads each. Representing numbers requires no more than
9 beads to a rod; the surplus is meant as a reserve for adding numbers or dissolving a higher unit into 10 lower ones. Consequently, number representation
is not unique on the school abacus; there is a most economical, the reduced
one, which reflects the digital representation with 9 beads at most on each rod.

For initial learning of the algorithms of adding and subtracting, this non-
uniqueness, compared to the uniqueness of the reduced presentation, is an
advantage. This is one of the reasons why in initial learning the abacus deserves
to be preferred above the digital representation. This does not mean neglecting
the digital presentation; on the contrary, the two tasks

to represent a written number — in various ways — on the abacus,
to note down in digits a number — arbitrarily given on the abacus,

program the learner in a natural way to

transfer 10 units as one higher unit to the left,
transfer one unit as 10 lower units to the right

in the operations of

adding

and

subtracting.

For the addition

$$8 + 5$$

first 8 then 5 beads are made visible from the hidden supply on the units rod, from which 10 are taken away and replaced with a bead in the tens rod, after which the result is read as 13.

For the subtraction

$$13 - 8$$

one bead is placed on the tens rod and three beads on the units rod, then the one on the tens is replaced by 10 on the units, after which 8 beads are pushed to the rear, leaving 5 visible.

This procedure is usually maintained for a short time only. Soon, in performing the addition, one observes pupils structuring the 5 as 2 + 3 when they push the 5 to the fore; then comes a stage where they more or less mark pushing the string of 5 beads, and finally one where they immediately unite 5 from the 8 beads with the 5 to be added, pushing them behind and replace them with one bead in the tens. Then the addition 8 + 5 is functioning mentally, and the abacus is used as an aid to memory.

With subtraction one can likewise observe the abacus structuring the mental act, though the learning process takes more time. Two ways of structuring can be observed:

$$13 - 8 = (10 + 3) - 8 = (10 - 8) + 3$$

and

$$13 - 8 = (10 + 3) - 8 = (10 + 3) - (5 + 3) = 10 - 5;$$

that is, from the 10 arisen by dissolving, the pupil immediately takes away 8, or he splits the taking away process into two steps; first 3, then 5. On the number line the first means taking away at the start, the second, at the end.

The didactical use of the abacus, as here demonstrated, is to suggest mental procedures of this kind. When observing the children's activities, the teacher can reinforce the inclination to mentalise by useful interaction and can clear away possible blocks.

There are a number of transitional stages between the abacus with the beads and the column arithmetic on paper. A pictorial abacus — columns with strokes instead of rods with beads — is one of these stages. In such a situation, where no beads are being pushed and erasing is annoying, mental acts are stimulated.

The pictorial abacus can be even more dissimilar to the genuine one: though there are still drawn columns, digits rather than strokes are put in the columns, such as

$$\begin{array}{l}18 \\ 5+\end{array} \rightarrow \begin{array}{c|c}1 & 8 \\ & 2+ \\ & 3+\end{array} \rightarrow \begin{array}{c|c}1 & 10 \\ & 3+\end{array} \rightarrow \begin{array}{c|c}1 & 0 \\ 1 & 3\end{array}$$

$$2 \quad 3$$

and similarly with subtraction.

The abacus is not the only tool used to suggest completing and reducing to tens in addition and subtraction. This can be suggested in another way on a number line where the tens are marked in a special way, or only the tens are numbered and the numbers in between are indicated by strokes or dots.

$$18 + 5$$

is structured on the abacus as

$$(10 + 8) + 5 = 10 + (8 + 5) = 10 + 13 = 23,$$

but on the number line the closeness to 20 prevails,

$$18 + 5 = 18 + (2 + 3) = (18 + 2) + 3 = 23,$$

a method also applied in mental arithmetic.

As long as algorithmics is dominated by the abacus or some intermediate form, it does not matter in which order the various columns are processed

$$\begin{array}{l}365 \\ \\ 287+\end{array} \rightarrow \begin{array}{c|c|c}3 & 6 & 5 \\ & & \\ 2 & 8 & 7+\end{array} \quad 5 \mid 14 \mid 12 \rightarrow 6 \mid 4 \mid 12 \rightarrow 652$$

is one way, but there are many more.

$$\begin{array}{l}365 \\ \\ 287-\end{array} \rightarrow \begin{array}{c|c|c}2 & 16 & 5 \\ & & \\ 2 & 8 & 7-\end{array} \rightarrow \begin{array}{c|c}& 5 \\ 0 & 8 \\ & 7\end{array} \rightarrow \begin{array}{c|c}& 5 \\ 0 & 7 \\ & 7\end{array} \rightarrow \begin{array}{c|c}& 15 \\ 0 & 78 \\ & 7\end{array}$$

can also be performed in another order. Proceeding from right to left recommends itself in the final algorithm as the method that admits of the shortest notation. The definitive algorithm is automatised to such a degree that while working on a column, one does not pay attention to the next one at the left.

In mental arithmetic it is a habit to proceed from the left to the right. Sometimes instruction builds such a strong system separation between mental and column arithmetic, determined by the horizontal or vertical position of the summands, that pupils do not grasp that both operations mean the same. Of course, numbers to be added or subtracted should not always be given in the vertical position; pupils should have the opportunity to rearrange the given numbers vertically and in order.

In the definitive algorithm haptic and visual auxiliary means are replaced by "borrowing" and "keeping in mind". The mental activity can be burdened with multiple "keeping in minds", the influence of more summands and more columns in additions, repeated borrowing in subtraction. It is even more burdened in mental arithmetic, when visual support is lacking and pupils may have forgotten the result of the first partial operation as soon as they accomplish the next. Poor achievement in column arithmetic may result from a lack not of understanding but of attention and concentration. Even though the pupils may be fascinated in general, this attention can slacken, certainly at the age when children are taught column arithmetic. Even good pupils can make a lot of errors; less able pupils can be discouraged by their failures and finally stop learning altogether.*

Learning column addition and subtraction has been viewed here as a − I do not say gradual but − step-by-step development. By preference the steps should be taken spontaneously by the learner − for the course that was sketched here, it has been shown to be feasible. As much as possible, the particular steps should be observed by the teacher and made conscious to the learner, as a means of reinforcement. Pupils who are not strictly led may develop methods of their own. Once I observed a pupil who in subtractions almost systematically rounded the subtrahend upwards:

$$
\begin{array}{c}
365 \\
\underline{287 -} \\
\end{array}
\rightarrow
\begin{array}{c}
365 \\
300 - \\
\underline{13 +} \\
\end{array}
\rightarrow
\begin{array}{c}
65 \\
\underline{13 +} \\
\end{array}
\rightarrow
78
$$

Perhaps this method is even better than the usual one. Anyway a pupil who contrives such things proves to act with so much insight that he can learn the traditional way too.

On the other hand, pupils should not be pushed to take a step on the road to algorithmisation unless they have really got there. These are general didactical principles, though particularly relevant for learning column arithmetic. Replacing insight with algorithms is a meaningful activity provided the algorithm has arisen from insight rather than having been imposed and blindly accepted. "If it does not do any good, it does not do any harm", is not a convincing argument. Algorithms should be learned by algorithmising, and this means most often by *progressive algorithmising*, which is a special case of *progressive schematising*.

Even this is not sufficient. Once a performance has been learned, the way in which it has been learned is readily forgotten. For algorithms this may mean that their sources of insight are clogged. In the aftermath of algorithms, teaching should aim at *retention of insight*.

---

* Another cause of failure, possibly even more important than lack of attention and concentration, which seems not to have received sufficient attention, is failure of short term memory. This is not the place to advise remedial teachers. My experience has shown that systematic training of short memory can be helpful.

4.36. Spontaneous algorithmisations are less easily prepared for multiplication. The operation is more difficult, and the natural motivation is not strong enough to conquer the difficulty. Multiplying is defined as repeatedly adding, and if 4 times 8 or 4 times 8 florins can be obtained by simple adding, if the price of 20 Season's Wishes at 0.50 florin a piece can be obtained by pushing a key twenty times, extra stimuli are required to store the results of multiplication in the memory. Once and for all — I came near to saying — but that is just not what happens. We know that this is not impossible, imprinting something in one's memory in one stroke — a name, a phone number, a birthday — if only we concentrate on it. But it seems that learning the tables looks more like being introduced to twenty people in a row — there are not many who can tie their attention to such a procedure. Learning the tables is a process of slow inculcation, of transfer from short term to long term memory. In column multiplication the knowledge of tables is rewarded; so column multiplication might be a good motivation to learn tables. It sounds paradoxical: The exercises by which memorising tables is motivated do not start until the learner is supposed to be familiar with them — a gap that is not easily bridged.

I do not know serious research about how to guide the memorisation of tables — I mean research on actual learning processes. Pupils develop a great many strategies in order to facilitate the ever fresh computations of the table products. One of them starts with familiar products, such as n^2 ($n = 1, 2, \ldots, 9$), which are somehow attractive, in order to go up and down. But it can happen as well that if, say, 7×8 is asked, the pupil leafs back to 6×8, which he recalls was asked earlier. Some more knowledge about pupils' strategies might lead to more effective techniques in teaching tables.

4.36a. In the original version, I turned at this point immediately to column multiplication in general (Section 4.37). Remedial work with 5th and 6th graders revealed to me a source of failure. It concerns certain apparently simple automatisms of the positional system of N (and as important for decimal fractions) which are fundamental and indispensable if the algorithms are to function at all.

I means the automatised rules

        multiplying by   10:   attaching a zero,
        multiplying by 100:   attaching two zeroes,

and so on, in general

        multiplying by 10 . . . 0:   attaching as many zeroes as there are after
                           the one.

The converses, related to division:

        division by   10:   dropping a possible zero at the end,
        division by 100:   dropping two possible zeroes at the end,

and so on, in general

> division by 10 . . . 0:   dropping as many possible zeroes at the end as
> there are after the one.

Adults I observed — from teachers in training to professors of sciences — were
as familiar with these rules as they were unable to argue them, and even to feel
any need for arguing them — a shortcoming which is perhaps more serious for
teachers in training, with a view to their future job, than it is with professors.
Some of them, perhaps even many, will have learned such rules by insight. It
looks like a typical case of clogging the sources of insight by rote exercise. I
should add that in the textbooks I consulted I met with the same lack of insight
— that is to say, didactical insight in how to stimulate useful learning processes
towards these rules by didactic sequences.

Let us start with the phenomenon that the tables for 2, 3, . . . terminate with
the round numbers 20, 30, ... It is a phenomenon experienced by learners
with feelings varying from satisfaction to astonishment. Anyway, they are
keen to reach at the terminus the safe harbour of 10 times. Would it not be
wise to make good use of this emotional concern to have them to explain this
phenomenon?

If we do nothing, the road to the above rules is paved by the empirical
induction

$$10 \times 1 = 10, 10 \times 2 = 20, 10 \times 3 = 30, \ldots, 10 \times 10 = 100, 10 \times 11 = 110, \ldots$$

thus ten times means a zero at the end. From here an easy — too easy —
generalisation leads to the multiples of 100, 1000, and so on. It is obvious,
however, that this learning process provides less insight than is possible and
desirable. What to do about it?

First of all, the phenomenon of the handsome terminus of the tables should
be explained, which may happen in two steps: first, commutativity — known
from the rectangle model — which transforms

> 10 twos into 2 tens,
> 10 threes into 3 tens,

and so on; second, coupling this to the promotion of

> 2 ones into 2 tens,
> 3 ones into 3 tens,

and so on by attaching a zero, which can be supported by the abacus or other
material. This, then, is the indispensable link in the learning process towards
the rule of multiplying by 10 while attaching a zero,

> 10 times 463 are 463 tens,

which requires the three units promoted to as many tens, the six tens promoted to as many hundreds, the four hundreds to as many thousands, which results in attaching a zero. The promotion of

> units to tens,
> tens to hundreds,

and so on, by attaching a zero thus reflects multiplication by 10.

The next step would be understanding multiplications by 100, 1000, ... as repeated multiplication by 10, in order to get the promotion of

> tens to thousands,

and so on, by attaching two zeroes, reflecting multiplication by 100, and so on.

An (intelligent) 12-year old says that multiplying a number by 10 and once more by 10 boils down to multiplying by 20. She masters the rule of attaching one zero, but without understanding it. She does not master those of multiplying by 100, 1000, and so on. She does not know what is 1000×1000. She does not know how to divide by 10. She knows almost nothing about decimal fractions.

Division by 10, 100, ... , if possible, should be made explicit in this context. This didactical sequence has its sequel in decimal fractions (cf. Section 5.24a).

A question, which will then be repeated, is whether in such a sequence the power notations

$$10^2 = 10 \times 10,$$
$$10^3 = 10 \times 10 \times 10,$$

and so on, should be used. The law of exponents is a powerful means, which, however, can didactically lead the wrong way ($10^2 \times 10^3 = 10^6$). The reader will have understood that the power notations in Section 4.37 are his business rather than the pupils', but this does not exclude a final transfer to the pupil.

4.37. Column multiplication in the decimal system is based on the knowledge of the tables, that is, the products

$$i \cdot j \quad \text{with} \quad i, j = 0, 1, \ldots, 9,$$

on the rule

$$10^p \cdot 10^q = 10^{p+q},$$

and on applying distributivity. If two numbers m, n are given decimally,

$$m = a_k 10^k + \ldots + a_0,$$
$$n = b_l 10^l + \ldots + b_0$$

the product

$$m \cdot n = (a_k 10^k + \ldots + a_0) \cdot (b_l 10^l + \ldots + b_0)$$

is built from the partial products

$$a_i 10^i \cdot b_j 10^j = a_i b_j \cdot 10^{i+j}.$$

These partial products are computed and pulled together according to some principle which assures that each gets its turn once and only once. The question of how this can be done most efficiently should be answered by the algorithm of multiplication. Let us forget the usual solutions and ask ourselves how we would tackle such problem if they were new for us.

Well, the most natural image is a two-dimensional pattern, a table with two entries: one factor is written in the ordinary way, horizontally, the units at the right and progressing to the left according to powers of 10; the other vertically, from below to above. At the crossing of the 10^i row and the 10^j column the product $a_i b_j$ is placed, omitting the factor 10^{i+j}.

	b_l	b_{l-1}		b_1	b_0
a_k	$a_k b_l$	$a_k b_{l-1}$	$\ldots$	$a_k b_1$	$a_k b_0$
a_{k-1}	$a_{k-1} b_l$	$a_{k-1} b_{l-1}$	$\ldots$	$a_{k-1} b_1$	$a_{k-1} b_0$
.					
.					
.					
a_1	$a_1 b_l$	$a_1 b_{l-1}$	$\ldots$	$a_1 b_1$	$a_1 b_0$
a_0	$a_0 b_l$	$a_0 b_{l-1}$	$\ldots$	$a_0 b_1$	$a_0 b_0$

It is these partial products $a_i b_j$ provided with the factors 10^{i+j} that should be pulled together.

Which ones among them bear the factor 10^p? One gets

$$a_0 b_0 10^0$$
$$+ (a_0 b_1 + a_1 b_0) 10^1$$
$$+ (a_0 b_2 + a_1 b_1 + a_2 b_0) 10^2$$
$$+ (a_0 b_3 + a_1 b_2 + a_2 b_1 + a_3 b_0) 10^3$$
$$+ \ldots$$

which, indeed, suggests a way of conbining: according to oblique lines under an angle of 45° from below left to above right. I think everyone would do it this way if he were allowed from this point of view to reinvent multiplication. In fact, it is the way to multiply polynomials in x,

$$a_k x^k + a_{k-1} x^{k-1} + \ldots + a_1 x + a_0,$$
$$b_l x^l + b_{l-1} x^{l-1} + \ldots + b_1 x + b_0.$$

The boxes of this pattern contain two-digit numbers yielded by the tables. Let us do it numerically

	7	6	5
7	49	42	35
8	56	48	40
9	63	54	45

and assemble

10^4	10^3	10^2	10^1	10^0
49	98	146	94	45

or vertically

$$14945$$
$$9864$$
$$49$$
$$\overline{603585}$$

which could also be done mentally, with a lot of "keeping in mind".

It is a matter of technique how this assembly along the oblique lines is performed. For instance, one could do it on the abacus. First the 45, then one column to the right 40 and 54, one more column to the right 35, 48, 63, and so on. Written down

$$45$$
$$500$$
$$540$$
$$3500$$
$$4800$$
$$6300$$
$$42000$$
$$56000$$
$$490000$$
$$\overline{603585.}$$

That is to say, one attaches as many zeroes as correspond to the oblique lines, but the zeroes can also readily be dropped. For safety one can place the corresponding power of 10 at the row, columns, and oblique lines (Figure 22), thus

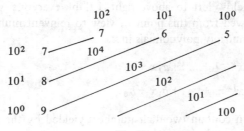

Fig. 22.

Adding according to the oblique lines can also be done mentally, which would mean writing down the result in one line. That, then, is the so-called abridged multiplication — not new — which I have presented here.

The method exposed here is mathematically the most natural, but it is not the usual one. There are reasons why it is not. First of all are historical reasons. Abacus traditions may have played as big a part as accident when the system was chosen, but once the choice had been made, it became a tradition. It is not easy to change such things. It is safer to have a teacher teach a method he has mastered than one he must learn himself and that is not convincingly better than the old one.

Didactical arguments can also be adduced in favour of the usual method, and this is very likely to tip the scale. After learning the tables and products like

$$3 \cdot 40 = \qquad 30 \cdot 40 =$$

and so on, the first true column multiplications will be

$$24 \cdot 2 = \qquad 24 \cdot 8 = .$$

The rectangular pattern shows no more than one column

$$
\begin{array}{r|l}
 & 2 \\
2 & 4 \\
4 & 8
\end{array}
\qquad
\begin{array}{r|l}
 & 8 \\
2 & 16 \\
4 & 32
\end{array}
$$

while the usual method

$$
\begin{array}{r}
24 \\
\times \quad 2 \\
\hline
8 \\
40 \\
\hline
48
\end{array}
\qquad
\begin{array}{r}
24 \\
\times \quad 8 \\
\hline
32 \\
160 \\
\hline
192
\end{array}
$$

is more perspicuous.

The next step in the learning process would be dropping the zeroes and shortening the procedure: Less on the paper and more mentally: 2 times 4 is 8, 2 times 2 is 4, 8 times 4 is 32; write down 2, keep in mind 3, 8 times 2 is 16, plus 3 kept in mind is 19. The result is written in one line, without intermediate steps.

This then determines the sequel. The next step is

$$
\begin{array}{r}
24 \\
\times \quad 80
\end{array}
$$

reduced to

$$
\begin{array}{r}
24 \\
\times \quad 8
\end{array}
$$

which had been mastered earlier and is augmented by a zero. Similarly

$$\begin{array}{r} 24 \\ \times\ \underline{800} \end{array}$$

is dealt with.
Then

$$\begin{array}{r} 24 \\ \times\ \underline{82} \end{array}$$

invites splitting into

$$\begin{array}{r} 24 \\ \times\ \underline{2} \end{array} \qquad \begin{array}{r} 24 \\ \times\ \underline{80} \end{array}$$

brought into one pattern

$$\begin{array}{r} 24 \\ \times\ \underline{82} \\ 48 \\ \underline{1920} \\ 1968 \end{array}$$

where finally a zero may be dropped.

This is in principle the general method, but it does not mean hurrying to apply it directly to

$$\begin{array}{r} 789 \\ \times\ \underline{765} \end{array}$$

which initially is better dissolved into

$$\begin{array}{r} 789 \\ \times\ \underline{5} \end{array} \qquad \begin{array}{r} 789 \\ \times\ \underline{60} \end{array} \qquad \begin{array}{r} 789 \\ \times\ \underline{700} \end{array}\ .$$

Perhaps this even helps better one's understanding of

$$\begin{array}{r} 789 \\ \times\ \underline{705} \end{array}$$

It is a quite natural course, this way. Viewed through the rectangular pattern it is assembling the partial products according to columns, 0th column, 1st column, and so on, and within the columns upwards. But this table of partial products is not made explicit. I confront the two methods with each other without drawing any conclusion. I myself do it according to the method I learned at school; in spite of many efforts I never succeeded in chasing it away in favour of the more efficient abridged method.

4.38. In a learning process directed towards reading the ordinary algorithm, the rectangle model of multiplication need not be excluded. It is entitled to be taken seriously because of its convincing geometric power as well as its significance for the product as a means to calculate areas. Its use has many times been proposed, and it has recently been elaborated in the Wiskobas curriculum.*

The product $m \cdot n$ is visualised by a fabric with n warp and m woof threads (Figure 23). The number of crossings can be counted and calculated more or less adroitly. The threads are taken together in bundles of ten. Thus, for instance $24 \cdot 8$ becomes the Pattern of Figure 24, and by a similar, though more complicated, pattern $24 \cdot 82$ is illustrated. Counting the crossings adroitly is a way to structure the activity.

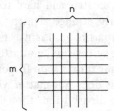

Fig. 23.

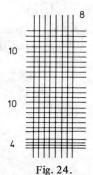

Fig. 24.

From the abacus the suggestion comes to combine ten thin threads into the thick thread, which for $24 \cdot 82$ means (Figure 25) a way to structure the counting activity even more sharply.

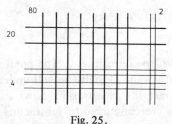

Fig. 25.

* Leerplanpublikatie 10, Wiskobas Bulletin 8, nr. 5–6, Nov. 1979.

The method can be extended to factors with more than two digits, although this can hardly be recommended. Instead, genuine algorithmisation should have started in the meantime and even made fair progress. Finally the geometric picture will fade away. In the long run, visualising may not interfere with algorithmising. But if need be, the picture can be called in, in particular when the area of the rectangle is dealt with. Then the fabric becomes, as it were, a braided mat, the threads become strips, horizontal and vertical; say, one mm for the thin ones, and one cm for the thick ones. This is not absurd — fabric threads too have a thickness and can be close enough to fill an area.

4.39. Among the four arithmetical operations, division is of course the most difficult. Its algorithm is complicated and hard to memorise. To make it even harder, it contains a particularly strange element that has no analogue in the other algorithms: estimating partial quotients. Few pupils attain reasonable proficiency and accuracy in long division, and after a lapse of time, with little opportunity to practice, the algorithm is soon forgotten.

For learning the algorithm it does not matter whether dividing is understood as repeated taking away or as distribution, though the first approach is more appropriate to exploring self-reliantly and describing the applied procedure.

Dividing 56789 by 3

then means subtracting 3 again and again, though these subtractions can be performed at a larger scale, say, 10000 threes at the first step in the present example

$$
\begin{array}{rl}
56789 & \\
30000 & 10000 \ \text{times} \\
\hline
26789 & \\
24000 & 8000 \ \text{times} \\
\hline
2789 & \\
2700 & 900 \ \text{times} \\
\hline
89 & \\
60 & 20 \ \text{times} \\
\hline
29 & \\
27 & 9 \ \text{times} \\
\hline
2 &
\end{array}
$$

together 18929 times, with the remainder 2. The pupil will initially fail to take away the largest possible amount. Finally he will understand that the decision how many times to subtract the divisor, boils down to an elementary division problem. Large dividends are conducive to subtracting as much as possible in one stroke.

The above pattern does not differ too much from the definitive one. One writes the partial quotients above rather than at the side, while dropping the zeroes to get the familiar pattern.

In this example the divisor was a one-digit number. The real difficulties arise with longer divisors. This extension should not be attacked unless long division with a one-digit divisor has been reasonably mastered algorithmically. Technical difficulties can persist even though the principle has been understood. In view of the mass of calculations required in long division, it may be taken as a wonder if no mistakes slip in. One out of three long divisions correct may be a discouraging experience, yet in fact this is the normal achievement. If it is even worse, there is a big chance that the pupil stops altogether. Such long divisions are of no practical use any more; if it is serious, one uses the calculator.

Apart from the mass of calculations there is another factor that makes long division in general difficult. In the case of one-digit divisors the multiplication tables suffice to find the partial quotients: when dividing by 7, all one has to do is locate a number < 70 between the rungs of the ladder for 7. This convenience is lacking in general, say, with the divisor 47. The table of 47 is not something to be memorised. If 331 is to be divided by 47 one has to scan the multiple of 47 that is just below 331. Rounding the divisor is one way to try it; in the present case: dividing 331 by 50, or rather 33 by 5. It goes six times. 50 goes 6 times in 330 with a large remainder. $47 < 50$, so the correct partial quotient must be at least as large as 6. But 6 times $47 = 282$, which is 49 less than 331. Thus 6 was not enough. It should be 7, thus

$$331 : 47 = 7 \text{ rem. } 2.$$

A long reasoning with a lot of computations in order to get this result, which with a longer dividend is only the first step. Trial and error, and finally the experience of failing — a new experience of uncertainty that is not to the benefit of the work.

4.42. Divisibility too is a structure of N which is also algorithmically approached.

There are algorithmic rules for divisibility by 2, 3, 4, 5, 6, 8, 9, 10; that for 11 is relatively simple, but that for 7 is not worth memorising. The three-folds of 37 are funny. The multiples of 142857 show nice regularities which have profound roots.

The rules, particularly, for divisibility are well-known, but very few people feel the need for asking "why?". The rules are considered to be empirical facts. Divisibility by 9 can be elucidated by the abacus: Transferring a bead from one rod to another makes a difference that is a multiple of 9; for instance

$$10000 - 1000 = 9000.$$

Transferring all beads to the rod of the units yields the total of the digits, which differs from the original number by a multiple of 9. Division by 9 leads to the same remainder with both of them.

In order to see whether a number is prime or to factor it, one tries the prime divisors 2, 3, ... systematically; at $\sqrt{N}$ one is allowed to stop.

The greatest common divisor of two numbers is delivered by the algorithm, named after Euclid.

4.43. Without much ado we have based the algorithmics of N on the decimal structure of N. I think it should be done this way. Innovators like to do a lot with structures on other bases. They make one believe that it is mathematics if one moves into another positional system. It is, however, only a slightly different algorithm rather than an expression of mathematics. Some of them assert that the principle of the positional system is better understood if it is embodied more than once — by a variety of systems and not only in base 10. There is, however, not the slightest indication that they are right, though I do not exclude other bases for remedial use. Unorthodox positional systems are rather a symptom of innovation by new subject matter. If compared with mathematics resulting from pondering more profoundly the subject matter and its relations to reality, unorthodox positional systems are a mere joke. Jokes are a good thing in instruction. It is good didactics to motivate pupils by jokes, and an unorthodox positional system may even be a *good* joke. Of course, other bases can have their own significance, in particular base two, that is, for computers. A context that justifies other bases mathematically will be touched on in Section 5.27.

CHAPTER 5

FRACTIONS

5.1–2. *The Title*

5.1. It is not a slip of the pen — "fractions" rather than "positive rational numbers" in the title of the chapter. It looks old-fashioned, this terminology. To the present view rational numbers are the proper mathematical objects that are meant here. This view is correct, as a consequence of how the mathematician interprets his formulae. If a and b are numbers,

$a + b$ is *not* the assignment "add b to a",

rather it is again a number, to wit the sum of a and b. If this is understood,

3 + 2 is again a number,

which more briefly can be written "5", though if you like it you may write "$\sqrt{25}$" as well, or $\log_{10} 10^5$.

3 + 2 = 5

then should not be read

if I add 2 to 3 I get 5,

but

3 + 2 and 5 are the same thing —

sometimes also formulated as

"3 + 2" and "5" are different names of the same thing,

such as, for instance,

"Amsterdam" and "capital of the Netherlands"

are names of the same thing.

On the left and right of the equality sign, the *same* object occurs. Likewise in

$$\frac{2}{3} = \frac{4}{6} = \frac{6}{9} = \dots$$

there is talk of again and again the same thing, only represented in various ways; and this thing is a *rational number*. Well, one can agree to prefer the way $\frac{2}{3}$, and in general, for every rational number, the expression by means of a fraction where numerator and denominator have the common divisor 1, the simplified

133

fraction; as one prefers for the number 5 the expression 5 rather than $3 + 2$, $10 - 5$ and so on, though the others are equally well admissible. There is, however, a difference: "5" is not only the *preferred* name of the number 5, it is its *first* name, the name by which it has been introduced to me, and under which I first made acquaintance with it, whereas "$3 + 2$" and "$10 - 5$" are aliases by which I can also call it up. "$\frac{2}{3}$", however, is only the simplest name of a certain rational number, and I would not even be able to say about many rational numbers under which name I first met them. This then is the reason why the various fractional expressions of the same rational live so much more their own lives, and why they are known under a special name: *fraction*.

But whatever one may feel about it, the mathematical object that matters is the rational number rather than the fraction. Nevertheless, I put the word "fractions" into the title, and I did it intentionally. Fractions are the phenomenological source of the rational number − a source that never dries up. "Fraction" − or what corresponds to it in other languages − is the word by which the rational number enters, and in all languages I know it is related to breaking: fracture. "Rational number" evokes much less violent associations; "rational" is related to "ratio", not in the sense of reason but of proportion, of measure − a learned context, and much more so than "fraction".

5.2. In fact, fractions have much to do with ratio, and I hesitated about whether I should not place the word "Ratio" under "Chapter 5". Not as a substitute for "Fractions" but as the subject that deserved priority − priority for didactic reasons but also on behalf of the exposition. I delayed "Ratio" to Chapter 6, though repeatedly in the present chapter I shall anticipate it. From the start I have struggled with problems of priority while I wrote this book, and I can only hope that the damage caused by that struggle looks bearable. As a matter of fact, I have turned the present chapter inside out several times. It is the wealth of phenomena mastered by fractions and ratio that caused the trouble. In order to write a phenomenology I have to pay attention to all these phenomena; organising them too systematically may mean simplifying so much that it infringes on the phenomenological task.

It cannot be denied that the didactics of fractions is characterised by unifying trends. As a rule, natural numbers are approached on a variety of tracks. If it is the turn of fractions, pupils are supposed to be so advanced as to be satisfied with one approach from reality. To my view, this wrong assumption is the reason why fractions function much worse than natural numbers and why many people never learn fractions.

It is my intention to present fractions in their full phenomenological wealth − I only hope that I do not drown myself in this ocean.

5.3. *Fractions in Everyday Language*

5.3.1   *half as* (by analogy with equally as, twice as, . . .)

followed by

> ..., much, many, long, heavy, old, ...

compares quantities and values of magnitudes.

Less usual

> *a third as, two thirds as*
> ..., much, many, long, heavy, old, ...

5.3.2 *two and a third times as*

> ..., much, many, long, heavy, old, ...

is as it were an extension of

> *twice as*
> ..., much, many, long, heavy, old, ...*

Yet

> *one third times as*
> ..., much, many, long, heavy, old, ...

can hardly be considered as belonging to everyday language.

5.3.3 *half of, third of, fourth of, ...*

describes a quantity or a value of a magnitude by means of another. The indefinite or definite article adds nuances

> ..., a (one), the
> *half of, third of, fourth (quarter) of, ...*
> a   the
> cake, way, travel, hour, pound, money, million, ...

So does

> ..., a (one), the
> *half of, third of, fourth (quarter) of, ...*
>   seven
> cakes, hours, pounds, millions, ...

Multiples can be formed

> *two thirds of, three fourths (quarters) of, ...*
>   a (one), the
> cake, way, travel, hour, money, million, ...

---

* In many other languages I would be able to add "little", "short" and "few" to these lists.

5.3.4   half a . . . , half the . . .

is used in the same sense.

5.3.5   From the noun or numeral to the measuring number expressed by
figures

$$\frac{1}{3}, \ \frac{2}{3}, \ 5\frac{1}{4}$$

m, kg, l, sec, bottle, million.

5.3.6   A strange phenomenon — I think in all languages — is

time or *times*

after fractions. In Section 5.3.2 we already met with

. . . times as . . . .

"Times" belongs to multiplication (cf. Section 5.31). With a natural number m
it occurs in

m times doing, undergoing, experiencing, waiting for,

something, for instance

m times seizing n marbles ($m \cdot n$ marbles),
m times laying down a measuring stick (m times as long),
m times turning the key in the hole,
m times around the clock, the race track, the Earth,
m times rolling of a wheel,
m times swinging back and forth.

At a certain moment fractions are allowed for m. This linguistic extension is
more easily understood if m is a mixed number:

2   times as long

brings

$2\frac{1}{2}$ times as long

in its wake.

$\frac{1}{3}$ time as long

looks unnatural, but the whole number in

$2\frac{1}{3}$ times as long

suggest that this means

$$2 \quad \text{times as long and } \frac{1}{3}.$$

In cyclic or otherwise periodic processes

$$\frac{1}{2} \text{ time}, \quad \frac{1}{3} \text{ time}, \quad \frac{2}{3} \text{ times}$$

might be self-explaining without the whole numbers. In

$$\frac{1}{2} \text{ time}, \quad 2\frac{1}{2} \text{ times turning the key in the hole},$$

$$\frac{1}{3} \text{ time}, \quad 2\frac{1}{3} \text{ times around the clock, race track, Earth}$$

$$\frac{1}{2} \text{ time}, \quad 2\frac{1}{2} \text{ times swung back and forth}$$

$$\frac{1}{3} \text{ time}, \quad 2\frac{1}{3} \text{ times rolling of a wheel}$$

the fraction suggests an action whose last phase has only partly been performed. If this is applied to the movement of measuring — for instance the use of a measuring tape —

$$\frac{1}{3} \text{ time as long}, \quad 2\frac{1}{3} \text{ times as long}$$

becomes clearer as a process of fitting a measuring tool periodically, where the last phase is only partially performed.

5.3.7   The more natural terminology is

3 *times* . . .

$$\frac{1}{3} \text{ of} \dots$$

and also

$$\frac{2}{3} \text{ of} \dots$$

applied to a

number, quantity of objects, divisible object, value of a magnitude,

such as

7,   30 marbles,   a cake,   5 kg.

but arithmetic and mathematics are better served with one term only. In exceptional cases the *times* is replaced with *of*, as in

3 packages *of* 5 kg each,

but this is another *of* than after a fraction. For the sake of uniformity and following an old tradition

> *times* takes over from *of*.

In textbooks this is most often simply prescribed:

$$\frac{1}{3} \text{ } time \quad means \quad \frac{1}{3} of.$$

In the preceding we sketched natural ways from

> *of*   to   *times*:

One way is from

> 2 times

via

> $2\frac{1}{3}$ times

to

> $\frac{1}{3}$ time;

the other is the cyclic or periodic process:

> 2 times around the clock,
> $\frac{1}{2}$ time around the clock.

Later on I will deal with this question again when multiplication of fractions is discussed.

5.3.8   In another way

> *of*   or   *out of*,   or   *in*,   or   *to*

suggests a fraction in

> 3 (out) of (in, to) every 5   (people living in cities),
> 5 (out) of 100   (5%)
> 35 miles to the gallon,
> a scale of 1 to 1000.
> one chance in a hundred,
> 3 out of 5 parts.

In a mixture:

>   3 parts salt and 2 parts pepper.

5.3.9   A stronger looking terminology

>   *every third* lot wins,
>   *every fifth* man is Chinese.

It seems that this is the origin of the ordinal numbers as a means to indicate denominators of fractions: Counting 1, 2, 3, . . . , 10 to count out *the tenth*; all these "tenth" people or objects together form *a* (*one, the*) *tenth* of the whole. Thus the *tenth* part is in fact the last of all of them. In an obsolete terminology nine parts means $\frac{9}{10}$, the remainder that is left if the tenth is counted out. "Decimate" originally meant counting out the tenth (to be shot).

5.4. THE FRACTION AS FRACTURER

5.4.1. *Causing Fractions*

We have already explained how magnitudes are divided, with or without a remainder. In order to divide substance, measured by magnitudes, many methods are available: fracturing can be

>   irreversible, or reversible, or merely symbolic.

The equality of parts is judged

>   at sight or by feel,

or by more sophisticated methods. One of them is

>   folding in two

in order to halve,

>   folding in three

in order to divide in three equal parts;

>   repeatedly folding in two and three

leads to more fractions.
    Heavy objects are halved by

>   weighing the parts in one's hands or on a balance,

while repeatedly correcting the lack of equilibrium. Similarly

>   comparing and correcting

play a part if in general a substance measured by magnitudes is to be distributed;
for instance, a liquid over a number of congruent glasses, where the heights of
the liquid are then compared.

Planar and spatial figures or objects as well as large amounts are sometimes
distributed with regard to area or volume while using

> congruences and symmetries;

for instance, the round cake into congruent sectors, which can be done

> at sight or by feel.

In all these examples I disregarded proper measuring. I aimed at drawing
attention to more primitive methods. In the mental constitution of all kinds of
magnitudes, meting out fair shares seems to me an important link — more important than what is investigated under the title of conservation by psychologists.
As far as I know, developmental psychologists have paid hardly any attention
to this aspect. I have observed many times that 7 to 8-year olds are able to
estimate one half or a third of an irregular area to be coloured; by this ability
they are mastering an important component of the mental object "area", whereas
knowledge of the formula for the area of a rectangle as shown by 10 to 12-year
olds need not mean progress — on the contrary it can equally well mean retro-
gression. Earlier I stressed the importance of break-make transformations for the
development of magnitudes as mental objects.

5.4.2. *Whole and Part*

In the most concrete way fractions present themselves if a whole has been or is
being

> split, cut, sliced, broken, coloured

in equal parts, or if it is

> experienced, imagined, thought

as such. In this complex of phenomena we will try a classification, illustrated
by examples.

The whole can be

> discrete   or continuous,
> definite   or indefinite,
> structured or lacking structure,

which means extremes with a variety of transitions in between.

The attention can be directed to

> one part, a number of parts, all parts.

The parts themselves can be

connected or disconnected.

The way of dividing can be

structured or unstructured.

5.4.3. *Examples – Definite Whole (Figures 26–32)*

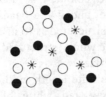

Fig. 26.

Fig. 27.

Fig. 28.

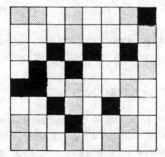

Fig. 29.

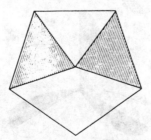

Fig. 30.

Out of a bag of marbles – discrete definite whole – I have taken a tenth; my attention is fixed on this tenth, and perhaps also on the remaining nine tenths.

From the same marbles in front of me, lying or rolling, structured as a sequence, I took arbitrarily a tenth – an unstructured choice – or the first tenth part or every tenth of them – a structured choice.

Out of 60 beads in a bowl $\frac{1}{6}$ are red, $\frac{1}{3}$ white, $\frac{1}{2}$ blue – a discrete definite whole, structured according to colour, unstructured in space. The same as a string – structured as a whole, and if the beads follow each other regularly, say 1 red, 2 white, 3 blue, also structured as to distribution.

A lottery – discrete definite whole, structured by numbering – the attention is fixed on the parts that gain prizes.

A strip – continuous, of definite length, with a linear structure – with one or more segments coloured, say, red–white–blue – connected parts, or disconnected – here and there some red, white, blue spots – divided with or without structure.

The same with the circular disc – continuous, definite, cyclically structured, divided into sectors, which separately or taken together represent parts (roulette, spinner, sector diagram).

The same with more or less structured geometric figures:

A square with or without an underlying squared paper structure, or a regular or irregular polygon with or without an underlying lattice structure, regularly or irregularly divided.

The edges or faces of a cube, combined into parallel quadruples and pairs, respectively.

Curvilinearly bounded planar domains, or spatial domains bounded by curved surfaces, regular or irregular, divided in various ways.

5.4.4. *Examples – Indefinite Whole (Figures 33–34)*

Mankind – discrete indefinite whole – divided according to blood groups, where attention can be paid to one or more of them; the whole can further be taken as unstructured, or as structured according to sex, race, geographical distribution, and so on.

Fig. 31.

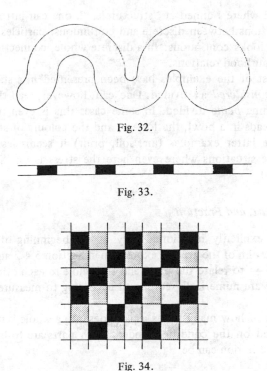

Fig. 32.

Fig. 33.

Fig. 34.

A string of beads of indefinite (possibly infinite) length – discrete, indefinite, linearly structured – and as a string of finite length, divided according to rank numbers or colours, which can be placed in a structured or unstructured way.

A strip of indefinite (possibly infinite) length – continuous, indefinite, linearly structured – and as a strip of finite length with coloured segments or spots, structuredly or unstructuredly distributed.

A wall or a tiled floor – continuous, of indefinite extension, structured in a pattern of bricks or tiles – divided according to colour, gloss, pictures, material – disconnected pieces, structuredly or unstructuredly distributed with or without structure.

The air – continuous, indefinite, a structureless whole – divided into gases, oxygen, nitrogen, and so on, connected parts, structureless distribution.

The soil – continuous, indefinite, a structureless whole – divided according to categories of use – disconnected parts, structureless distribution.

Print in a certain language – discrete, indefinite, a structureless whole – divided according to letter symbols, structureless distribution.

Time – continuous, indefinite whole – structured according to various criteria.

Remark. All these examples are to be taken with a grain of salt. Where I called something "structured", it is possible to neglect the structure if it does not matter, and where I aimed at "structureless", one can introduce structure. There are transitions between discrete and continuous: particles can be so small that the whole looks continuous. In a discrete whole, connection can be built up from neighbourhood relations.

   Remark. Most of the examples have been presented in a static way: something *is*, or *is considered* as divided. One can, however, read them also in the sense of something *being* divided. In some cases this is even more natural, as it is with the beads in a bowl, the lottery, and the colours of strips and planar domains; in the latter examples (air, soil, print) it seems less plausible, but one can imagine situations where even here the stress is on "is being" rather than "is" divided.

5.4.5. *Whole, Part, and Fraction*

Fractions were explicitly mentioned only at the beginning of Section 5.4.2, though it was the aim of the general exposition in Section 5.4.2 and the examples in Section 5.4.3–4 to relate the parts and the whole to each other by fractions. Parts and whole are numerically compared according to measures that can vary greatly.

   The question of how many times a part goes into a whole is meaningful only if one has agreed on the condition under which parts are to be considered as equivalent. The criterion can be

   number

or

   value of a certain magnitude.

This will be elaborated later.

   In spite of the many sided classification and the wealth of possible examples, the approach to fractions from the point of view of "part–whole" is much too restricted not only phenomenologically but also mathematically – this approach yields proper fractions only. The traditional didactics of arithmetic restricts itself to this approach, mostly even in the narrow sense of dividing a cake. After these concrete cake divisions – with proper fractions only – the learner is immediately introduced to dividing abstractly presented quantities and values of magnitudes; with arbitrary decrees like "$\frac{1}{2}$ times means the same as $\frac{1}{2}$ of"; and with arithmetical rules a straight way is leveled to the rational number. Some innovators inserted a stage of fraction operators as inverses of multiplication operators. This could have been a progress were it not that even they are satisfied with too small a basis of orientation.

   Pupils with a knack for digesting algorithms learn to operate on fractions anyhow, pupils who are less or not at all gifted in this specific way learn it by trial and error or not at all. After one or two years of fractions, some pupils master the algorithms though they have no idea what fractions mean and what you can

do with them; others do not even know the names of the particular fractions.
The phenomenological poverty of the approach seems to me largely responsible
for this didactic failure.

5.5. FRACTIONS AS COMPARERS

5.5.1. *Comparing Concrete Objects*

The traditional didactics overlooks the fact that the concreteness of fractions
does not stop with breaking a whole into parts. As the linguistic analysis of
Section 5.3 showed, fractions also serve in comparing objects which are

>   separated from each other

or are

>   experienced, imagined, thought as such:

>>   in this room there are half as many women as there are men,
>>   the bench is half the height of the table,
>>   the street is $2\frac{1}{2}$ times as wide as the footpath,
>>   John earns half as much as Pete,
>>   copper is half as heavy as gold.

Comparing is performed according to certain criteria,

>   directly and indirectly.

   Directly: the objects which are to be compared are brought close together, or
are in some other way considered, as though the smaller were part of the bigger,
by which strategy the fraction as comparer is reduced to the fraction as fracturer
of one concrete object.
   Indirectly: a third object, say a measuring stick, mediates between the two
objects to be compared by being, or regarded as being, transferred from the one
to the other.
   The above examples admit of another formulation:

>   the number of women in this room is half the number of men,
>   the height of the bench is half the height of the table,
>   the width of the street is $2\frac{1}{2}$ times that of the footpath,
>   John's income is half of Pete's,
>   the (specific) weight of copper is half that of gold.

Rather than

>   objects with respect to number or magnitude value

we now compare

>   numbers or magnitude values themsleves.

It looks like too much sophistication to make this distinction, and in the unsatisfactory phenomenology of psychological research as well as in traditional didactics it is disregarded. In our phenomenological analysis it is not superfluous. One should fully realise that comparing with respect to number or magnitude value precedes comparing numbers or magnitude values themselves, and that the former remains, or should remain, imminently present in the latter as long as fractions are to be more than a formalism.

5.5.2. *Fraction and Magnitude*

Earlier we explained how distributing into three equal parts can take place: with small quantities at sight, with larger ones by alternately taking away equal parts, or algorithmically by division, as the inverse of multiplication. If the division terminates, no new problem turns up. If not, then in realistic problems the question arises of what to do with the remainder. If its division is feasible, then the *mathematical* distribution problem and its relation to the *real* one have changed. It is no longer a finite set that is distributed; the finite set model does not fit the real distribution problem any more. For instance rather than six — discrete — loaves of bread that are distributed, it is bread, in a quantity that is thought to be arbitrarily divisible and according to a rule that states when quantities can replace each other in order to be considered equal.

The mathematical model that fits this task of distribution, is *magnitude*. It was already discussed in Chapter 1, and it is a subject to be dealt with once more. Meanwhile we repeat the essentials:

> To constitute a magnitude in a system of quantities requires:
>
> an equivalence relation, which describes the conditions for replacing objects (for instance quantities of a certain substance) with each other and which leads to *equality* within the magnitude,
>
> a way of taking together objects (quantities), which leads to an *addition* in the magnitude,
>
> the unrestricted availability of objects with the same magnitude value (that is, in the same equivalence class), which makes *addition unrestrictedly possible*,
>
> the possibility of dividing an object into an arbitrary number of partial objects that replace each other, which leads to *division by natural numbers*.

Multiplication by natural numbers is a derived operation defined by repeated addition; "*n*th part of" becomes the inverse of "*n* times". By composing multiplications and divisions with each other one gets *multiplications by rational numbers*.

If we restrict ourselves to *mathematics* only, then in order to define what magnitude is we could be satisfied with postulates on addition and division. In a phenomenological approach we must start with objects which by an equivalence relation are required to form classes representing magnitude values. The un-restricted availability of such objects in each class is indeed indispensable. I stress this point which via a defective phenomenology has produced a defective didactics of fractions.

Our exposition shows an asymmetry between multiplication and division. The operator "nth part of" can be applied to the object before it can to the magnitude value. An nth part can be a concrete part of a given something. On the other hand, "n times" cannot be realised by means of the given object; one has to call in others, perhaps arbitrary ones, whereas "nth part" can be realised within the object and only the *choice* of the part is arbitrary.

This asymmetry is so striking that no phenomenology can be allowed to dis-regard it and no didactics of fractions may pass over the results of this analysis. It is, however, just the point where the traditional didactics of fractions shows its defects, for which a defective phenomenology is likely to be made responsible. The fraction as part of something is of such a convincing and fascinating con-creteness that one is easily satisfied with this one phenomenological approach and forgets about all others. In all examples, whether visualised or not, one restricts oneself to fracturing. The nth part is exclusively seen or imagined *within* the whole — something that would not be feasible with "n times". Phenomenologically this approach leads to proper fractions (< 1) only. The insufficiency would appear as soon as mixed fractions (> 1) are taught, but when this point is reached, mathematising fractions and the operations on fractions are already in full swing if not completed; the required extension to mixed fractions is simply dragged along in the stream of mathematising, or accomplished purely formally without any phenomenological bonds. Expressions such as $1\frac{1}{2}$ are paper work, unrelated to reality, which is still visible in the proper fraction.

The "fraction as fracturer" is not only too narrow a start, it is also one-sided. It is strange that all attempts at innovation have disregarded this point. Modern phenomenological analysis has carefully approached the concept of magnitude; the part played by equivalence and fractions has been recognised, but this phenomenological analysis has never taken a didactical turn. In particular, it has not been realised that the didactics of magnitudes cannot be built on that of fractions, which in turn require magnitudes to be approached didactically-phenomenologically. The "fraction as fracturer" can be described by a quite restricted equivalence concept; it does not require any more than dividing something into n equal parts. But in the didactic reality an equivalence of broader scope is needed, as well as the unrestricted availability of objects in every equivalence class. So far this need has not been recognised in the didactics of fractions and in the choice of didactical models.

5.6. *Aspects of the Fraction*

Let us summarise the contents of Sections 5.4—5 formally and replenish it.

As the mental stress is on

acting or stating

the fraction appears

in an *operator* or in a *relation*

– halving versus "half as big".
    Both the fraction operator and relation can work on and relate to each other, respectively,

objects

with respect to certain characteristics (number, length, salary, weight ...) – "half of the stick, the bench is half the height of the table, and so on – or

quantities and magnitude values

– this length is half that, this weight is $2\frac{1}{2}$ times that.
    If the objects to be compared are

part and whole

or are considered as such, the fraction appears in the

fracturing operator or relation.

    If they are

separated,

it is better to speak of the

ratio relation.

    If it is about quantities and magnitudes, the fraction occurs in the

ratio operator,

which transforms a number, length, weight into another one.
    From the ratio relation as stated between objects one can pass to the ratio operator, which acts on quantities and magnitudes, by an intermediate stage, the fraction in the

transformer,

such as "mapping a half scale", "stretching $2\frac{1}{2}$ times". This operation is performed

on the object itself,

though not by breaking, but by

mapping and deforming.

If we leave the concrete sphere around fractions step by step we arrive at the

fraction as measurer

preceding a unit — $2\frac{1}{2}$ in $2\frac{1}{2}$ kg, $2\frac{1}{2}$ m, $2\frac{1}{2}$ cc, $2\frac{1}{2}$ bottle — or without a unit, as is the case with $\frac{1}{2}$, $2\frac{1}{2}$, . . . that measure segments

on the number line;

the fraction operator as

inverse of the multiplication operator;

and the

fraction as rational number.

As we explained earlier, the traditional didactics knows the fraction only in the fracturing operator, from which it passes straightforwardly to the end of the sequence: the fraction as rational number.

5.7. *The Fraction in an Operator*

Among the aspects in Section 5.6 we meet an operator aspect on three occasions, that is,

the fracturer

which claims to act on *concrete objects* by breaking them into equivalent parts,

the ratio operator,

which puts *magnitudes into a ratio* with each other,

the formally defined fraction operator

in the *number field*.

The differences look sophisticated, but didactically they are not so — the medium in which the fraction operator acts is being stripped of its concreteness in a stepwise manner. Initially it acts on the objects cited concretely, while their magnitude aspects are the factors that check the fairness of the distributive procedure. Next, the magnitudes themselves are objects, while the concrete objects measured by them are disregarded or passed over in silence. There is a remarkable intermediate stage, the transformer, which, as it were, preserves the substance while changing the magnitude values proportionally. Finally, the fraction operator acts in the pure number domain, where it satisfies the need for inverses of multipliers.

In the ratio *relation* the ratio *operator* is, as it were, coagulated, from an operation to a relation between the object operated on and the result. The fraction as measuring number, as spot on the number line, and finally as rational number is the result of applying the fraction operator to a unit. In all aspects of the fraction, the operator aspect is felt. In a didactics of fractions it should be appreciated accordingly, and in modern approaches it is in fact done. Unfortunately, this is allied with misconceptions, expressed in such formulations as the "fraction as operator". Logically such interpretation is of course, feasible – number and vectors, too, can be interpreted as operators. Elsewhere* I have shown the didactical rocks on which this logic must founder. The interpretation of the fraction as an operator is untenable, as is the involved terminology. One badly needs the fraction as a *number*, which for that matter may have arisen by applying a fraction *operator* to a unit. This means that in the *fraction operator* one must distinguish the *operator* from the *fraction*. The operator with a fraction *in it* cannot afford a second self in the form of the fraction *as* an operator.

It is a fact that the operator aspect is more important for fractions than it is for natural numbers. In the constitution of the mental object "natural number" the growing together of the cardinal and ordinal root is decisive, and only after natural numbers have been constituted are they used in operators such as "three more than", "three less than", "three times (as much as, as many as)".

Fractions, however, show the operator aspect from the start, which justifies a didactics which calls itself – by exaggeration – the operator interpretation of fractions.

An operation known as early as natural numbers is distributing. If a finite set of equivalent objects is distributed into three equal parts, say, among three persons, each part is a third, that is, a third of the whole – a strange terminology whose troubled source I have uncovered in Section 5.3.9 – yet we are so accustomed to this strange use of the ordinal numeral that we are not aware any more of its curiosity, let alone inclined to protest it or to ask ourselves why year after year hosts of pupils do not grasp it.

5.8. *Models of the Ratio Relation (Figures 35–38)*

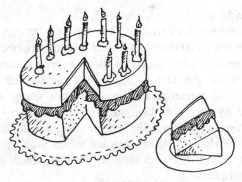

Fig. 35.

* *Mathematics as an Educational Task*, pp. 260–262.

Fig. 36.

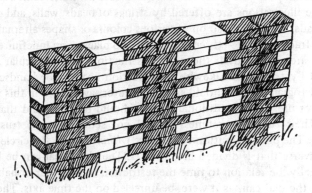

Fig. 37.

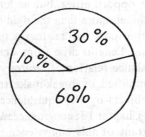

Fig. 38.

The universal model of magnitude is "positive number" visualised on the number line as length, although other models may be equally useful didactically, in particular if fractions are concerned: area, volume, weight, time, to mention a few. Lengths and areas have their own visualisations; with some precaution volumes, too, can be graphically visualised, though proper spatial visualisations, which at the same time can be rather palpabilisations, are highly recommended.

Weights can be visualised linearly, on the scale of a spring balance or on the beam of an old fashioned steelyard where a running weight is displaced; time is visualised on the time axis, unreeling the clock dial, as it were. Each of these models deserves our attention since it can be useful in the ratio relation.

I did not mention here the classical model for fractions, the pie distribution, not to mention more recent ones, such the fractions boxes. In the didactical practice they may certainly not be skipped. The pie distribution is the predecessor of general sector divisions of the circle that are applied as statistical sector diagrams and in roulettes and spinners. As didactical models for fractions they are especially effective if various sectors are to be taken together in order to make assertions about "*m* out of *n* parts" or "*p* parts of this against *q* parts of that". On a spinning top, mixing *p* parts of one colour with *q* parts of another colour to get a certain colour shade is an effective illustration of mix ratios. Likewise, one can mix liquids in a given ratio and illustrate the mixture in a sector diagram. Handsome illustrations are offered by strings of beads, walls, and other patterns where beads, stones, and so on of various colours or shapes alternate regularly in a certain fraction ratio — three white and two black — an indefinite whole where no limits are suggested. If the subject is fractions, the particular shares will be expressed by fractions. Likewise, the fraction box can be handsomely used to display histograms, but I should say that I never saw it used in this way.

Whoever uses these traditional models should bear in mind that they do not suffice. Their rude concreteness should not seduce him to trust this narrow approach. The pie distribution takes place *within* the pie; the circle to be divided is the universe that is divided into sectors. The clock dial can be handled more smoothly: by the relation to time the restriction to one hour or half a day can be removed; the dial can, as it were, be unreeled on the time axis. The fraction box is the most restricted tool; it resists not only extending but also refinement. The drawn rectangle has more opportunities, but as long as the rectangle is only *sub*divided, it is not worth much more than the rigid fraction box.

Lengths and areas are the most natural means to visualise magnitudes with respect to teaching fractions. Lengths arise from straight long objects by means of congruence as an equivalence relation; if *arbitrary* long objects are admitted, congruences have to be amplified by break-make transformations or flexions. Areas arise from planar objects by the equivalence relation of area equality, which will be dealt with in Chapter 13; congruences and break-make transformations contribute to the extent of this equivalence class. In the process of pie division the circle sectors are compared by congruence, which should guarantee the equality of area or volume.

Line segments are the most simple visual representatives of magnitude values. Two magnitude values in a fraction relation are easily visualised by two line segments in the same ratio (Figure 39); in order to make the ratio comparison easier, relevant parts can be marked; the representing line segments are by preference taken parallel. This, however, is not the only way. Two trees beside each other (Figure 40) that are in fraction relation, which can be stated by

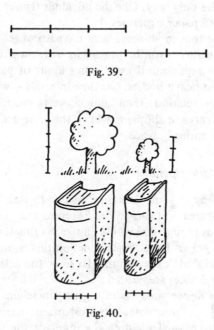

Fig. 39.

Fig. 40.

measuring or using intermediate scales, two books with thicknesses in fraction ratio, ages on the time axis; weights on the scale of the spring balance are other examples. Most of these representations show more than one linear extension, which means that the other extensions can also be discussed. They are not as thin as pure lengths, indeed.

Thinner lengths can be stylised by low rectangles, strips which are the same in one extension and variable in the other (Figure 41). In order to systematise this and to facilitate comparison, one may draw the rectangle on a squared paper background (Figure 42), where comparing is reduced to counting. But again,

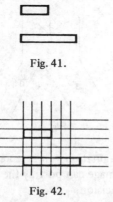

Fig. 41.

Fig. 42.

this should not be the only way. One should admit figures which overlap or run counter to the squared paper structure.

I stress once more that in all these cases the pairs of geometric objects — line segments, planar domains — can be present in their own right to embody fractions or they can be representatives of other kinds of pairs of objects — two trees, two books, two heavy bodies, two time intervals — which are to be understood in their fraction relation. Then quite concrete couplings can arise: weight and price on the balances in shops, weight *in* the scale and length on the beam or *on* the scale of the spring balance.

5.9. *Models of the Ratio Operator*

In the most natural way, "$\frac{3}{5}$ of" is realised by two figures, one of which is $\frac{3}{5}$ of the other in length or area. Yet this procedure represents "$\frac{3}{5}$ of" unsatisfactorily as an operator. It is as though one would illustrate a function not by a *graph* but by *one point of the graph*. For a linear function this is, indeed, enough, but in no way does it satisfy our expectations. To show the action of "$\frac{3}{5}$ of" in its whole domain, other devices are needed.

The most popular device today is to suggest a machine — in the present case it would be the "$\frac{3}{5}$ of" machine. It is most often merely a verbal suggestion illustrated by a conventional picture. The input of the machine is numerical data, which however can also be represented geometrically. The machine itself does not show any structure, geometrical or otherwise. It is a "black box". As far as my experience extends, textbook authors, teachers, and pupils use these machines merely verbally, with no relation to any concretised fraction operation. It is my impression that the machines owe their origin to attempts at introducing the *concept* of function rather than functions as mental objects; the false concretisations which are then unavoidable have adopted here the form of a pseudo concretisation: a verbal suggestion.

More concreteness is provided by the picture of flow distribution in order to embody fractions (Figure 43). As a matter of fact, the magnitude flowing

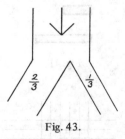

Fig. 43.

in and out, exists only in imagination — it is replaced, as it were, by an indefinite time —, but the branching image can portray the fractional part (and its complement) with a geometrical precision.

Whichever model is chosen, one is free to interpret it arbitrarily, for instance, the flow image as length, weight, money, and so on.

5.10. *Mapping Models of the Ratio Operator*

A complete geometrical as well as global picture of the fraction operations is obtained as soon as they are genuinely interpreted by geometrical operations. If in order to do so, lines are mapped, there are a few possibilities, all of them affine mappings (Figure 44).

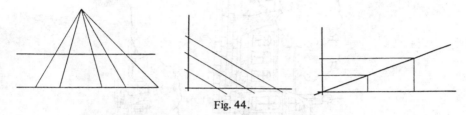

Fig. 44.

central projection of parallel lines (lamp shadow),

parallel projection of, say, orthogonal lines (sun shadow),

composition of two parallel projections (such as used in the graphic representation of a linear function).

Performing the geometrical constructions in detail can be both advantageous and disadvantageous: all the details become consciously clear, but the procedures are protracted.

A more attractive way is to use

planes, rather than lines, that is, *projection planes*.

The detailed constructions are even more difficult to perform, but they can readily be dispensed with if the pictures are differentiated to show clearly which points correspond to each other in the original and the image (Figure 45). What I mean is two pictures beside each other, one an enlargement or reduction of the other, where the same ratio relation can be stated for each particular detail. The same can be done in three dimensions by building models in different scales.

A danger one should anticipate if one uses such two- and three-dimensional representations is the possible confusion of length, area, and volume ratios. Nonetheless even if it is ratio of lengths that matters, planar figures are to be preferred as means of representation because of their more global expressiveness; in order then to stress length, one can rely on two artifices:

as planar figures one chooses narrow strips, which are transformed according to length only, while places are distinguished by means of ornaments,

or one takes plain two-dimensional parts, which are transformed according to both extensions, and to which one attaches drawings that suggest one extension, such as worms, snakes, whips, spectacle frames.

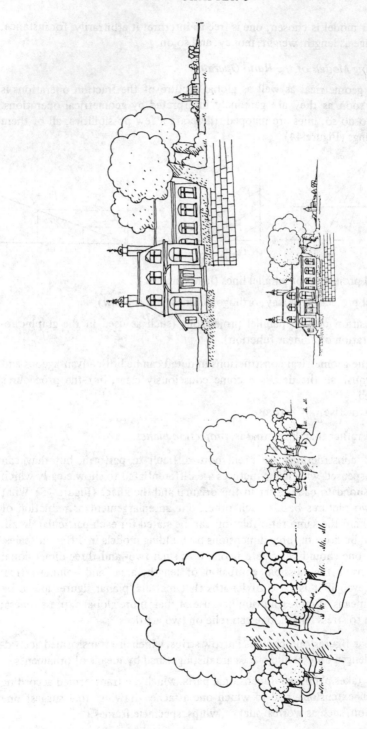

Fig. 45.

5.11. *Mathematical Theory of Rational Number from the Point of View of the Ratio Operator*

It is well-known how rational numbers are introduced, starting with natural numbers (or integers): one considers pairs ("fractions") of integers with a non-vanishing second member and prescribes an equivalence relation

$$\ulcorner m, n \urcorner \sim \ulcorner m_1, n_1 \urcorner \leftrightarrow mn_1 = m_1 n;$$

rational number are then the equivalence classes of these pairs. The arithmetical operations are defined appropriately for the pairs, and accordingly for the equivalence classes.

I now sketch how it is done if the multiplication operator is chosen to start with, and an *a priori* genetic rather than *a posteriori* axiomatic way is followed. Fractions then are not the result of a definition; instead they are discovered and described.

We consider a magnitude S and within S multiplications by natural numbers ($\neq 0$), which form a set M, with composition as an operation in M. M then is a

commutative semigroup
with identity and a
cancellation rule: $a \circ x = a \circ y \rightarrow x = y$.

Such semigroups can in general be extended to groups, which is easily proved.

In the present case it is even easier because the elements of the semigroup are given as multiplications within a magnitude S. I display the sequence of steps (l.c. italics stand for natural numbers $\neq 0$):

(1)   "k times" is a one-to-one mapping of S into itself.

(2)   The inverse of "k times" is called "kth part of".

(3)   All the "k times" form a set M; the "kth part of" a set M^{-1}.

(4)   (k times) $\circ$ (m times) = km times.

(5) M is closed and commutative under composition.

(6)   Given a set T and one-to-one mappings φ, ψ of T on itself, then from

$$\varphi \circ \psi \circ \psi^{-1} \circ \varphi^{-1} = \text{identity}$$

one concludes:

If φ and ψ commute, then φ and ψ^{-1} do so also, as well as φ^{-1} and ψ^{-1}, moreover $(\varphi \circ \psi)^{-1} = \psi^{-1} \circ \varphi^{-1}$.

(7)   Applying (6) on S instead of T and two elements of M instead of φ, ψ, one gets

$M \cup M^{-1}$ with composition as its operation is commutative.

(8)   From the last part of (6) it follows that

(nth part of) ∘ (mth part of) = (nmth part of).

(9)   One defines

$$\left(\frac{m}{n} \text{ of}\right) = (m \text{ times}) \circ (n\text{th part of}),$$

which according to (7) can also be written

$$= (n\text{th part}) \circ (m \text{ times}).$$

Here $\frac{m}{n}$ is not yet meant as a symbol for a rational number. It is rather an arbitrary symbol, expressed by means of m and n.

(10)   The multiplication rule

$$\left(\frac{m}{n} \text{ of}\right) \circ \left(\frac{k}{l} \text{ of}\right) = \left(\frac{mk}{nl} \text{ of}\right)$$

is derived from (9), (8), (7).

(11) $$\left(\frac{k}{k} \text{ of}\right) = (1 \text{ time})$$

follows from (8) and (2).

(12)   The cancellation rule

$$\frac{mk}{nk} \text{ of} = \frac{m}{n} \text{ of}$$

follows from (10) and (11). This allows one to introduce rational numbers as classes of fractions.

(13)   The $\left(\frac{m}{n} \text{ of}\right)$ form a set N, which according to (11) is closed and commutative.

(14) $\left(\frac{m}{n} \text{ of}\right)$ is a one-to-one mapping of S onto itself with $\left(\frac{n}{m} \text{ of}\right)$ as its inverse.

(15) N is a commutative group of one-to-one mappings of S onto itself.

It looks awfully complicated, though it reflects nothing more than the occurrence of rational numbers in multiplication operators; addition is lacking, and the rational numbers are not yet freed from their operator formulation. However, the preceding sequence should not be understood in the way that any of its steps would be made explicit, except if it is done paradigmatically. If we take a closer look at what is didactically required in this line of thought, then we get the following sequence:

the mental object object "one-to-one mapping", albeit specialised to stretchings and shrinkings of the number ray,

the mental activity of composing and inverting mappings,

the recognition of "k times" (for paradigmatic k) as a one-to-one mapping, and the identification of certain mappings as "k times".

the view of, and identification of, the inverse of "k times" as "kth part of" or "$\frac{1}{k}$ of" (known as such from the division task),

the mental composition of "k times" and "m times" (for paradigmatic k and m) and the recognition of the result as "km times",

the mental composition "nth part of" and "mth part of" (for paradigmatic n and m) and the recognition of the result as "mnth part of",

the definition and recognition as a mapping of "$\frac{m}{n}$ of" as composed of "m times" and "nth part of", in arbitrary order,

the mental composition of "$\frac{m}{n}$ of" and "$\frac{k}{l}$ of" and the grasp of the multiplication rule,

the grasp of the cancellation rules,

inverting "$\frac{m}{n}$ of" into "$\frac{n}{m}$ of".

The only steps of the mathematical analysis that do not figure in this imaginary didactical sequence are those where commutativity is ascribed to certain pairs of mappings. In most cases this property is so obvious that to make it explicit would cause confusion. The only case where it is required to do this is the commutativity of "m times" and "nth part of".

It is perhaps surprising that in the mathematical analysis the inverse of "m times" is not immediately called "$\frac{1}{m}$ times" but "$\frac{1}{m}$ of" – mathematically viewed, nomenclature is not bound by any rules. We did so, because – as has been added in parentheses – the inverse of "m times" must be first identified with the familiar and visually rooted "mth part of", and labelling this inverse by "$\frac{1}{m}$ of" requires a motivation which must be prepared carefully.

The preceding approach can hardly be compared with that of introducing rational numbers as equivalence classes of number pairs; the approach by operators follows a didactical sequence, whereas the one using equivalence classes accounts formally for an already acquired arithmetic ability.

To what degree can the sequence described above and justified mathematically be realised? Well, this is a badly formulated question. As a matter of fact, this sequence must be implicit in any didactics of fractions – it is rather a check list. The proper problem is that the sequence is fleshless. Restricting oneself to this list would be a hobby, inspired by a mistaken hunt for purity of method. The sequence is fleshless, its basis is too narrow. It is walking with blinkers which, for that matter, do not sufficiently protect one against disturbances.

Both the mathematical and the didactical sequence start with mappings (multiplications) in a magnitude. This magnitude must be specified somehow, and the most obvious specification is length, visualised as a number ray (or number line), which we may suppose to be familiar to the pupils. There multiplications by natural numbers are readily recognisable mappings, as are their products and inverses. This intuitive recognisability, however, is insufficient; communication requires verbalisation, which initially might be ostensive, but gradually should be refined by means of relative and functional linguistic devices.* A system of arrows from (variable) x to mx to show "m times" remains locked in the ostensive sphere. More sophisticated linguistic devices are required; for instance,

indicating a point by A,

its "m times" image by mA,

its "$\frac{1}{n}$ of" image by $\frac{1}{n}A$,

its "$\frac{m}{n}$ of" image by $\frac{m}{n}A$,

which boils down to plotting positive rational scales on the number ray and putting them into a mutual relation.

This could be a quite useful exercise in detail if it were not for the fact that the number ray is already familiar to the pupils as the infinite ruler, where the natural numbers are lodged, perhaps even intercalated by some fractions. This presence cannot be obscured. In fact, it even gets systematised: by applying the operation "$\frac{1}{m}$ of" on the natural numbers, which may be assumed to be prefigured on the number ray, all the rational numbers on the number ray come into being. This would not be inconvenient were it to occur rather later. As it is, the rational numbers now fulfill a double task: numbers lodged on the number line as well as linguistic parts of ratio operators. Of course, in the long run this is unavoidable, and at a certain moment this consequence *must* be accepted and made conscious; but then one *must* be able to choose this moment such that the consequence *can* be made conscious in order that the rational number can play its double role well and in an undisturbed fashion.

This, however, is not the major objection against the fleshless didactical sequence. Intentionally I had the fractions marching on a broad phenomenological front. The phenomenological wealth should be put to good use. The steps isolated in the mathematical sequence should be taken not *in abstracto* but in a variegated context. Even if each of the steps could be taken paradigmatically, one should not expect that the *didactical* sequence in which I try to realise the *mathematical* one didactically automatically contains the wanted paradigms.

* *Weeding and Sowing*, Chapter IV, Section 15.

I wish to add that besides the didactic realisation of the mathematical sequence room must be created for

adding and subtracting fractions,
isolating the fraction as part of the ratio operator,

replacing "$\frac{m}{n}$ of" by "$\frac{m}{n}$ times".

Even then I would not yet have accounted for the algorithmisation or formalisation of fraction arithmetic.

I am now going to sketch a rich didactical sequence for the arithmetic of fractions.

5.12–13. *A Rich Didactic Sequence for the Arithmetic of Fractions*

5.12. Eight bottles of beer, three persons and each of them gets his fair share — a ten-year old girl reacted to this problem by setting up a long division and then reproaching me that it did not terminate. To my answer "yet they did share it", she reacted as though she had awakened from a dream — suddenly she noticed more things between heaven and earth than are dreamt of in the arithmetic lessons she had had so far.

She drew sketches of eight bottles beside each other, divided each of them in three parts, gave each person eight thirds — in fact she did not know this word, but said "little bottles" — and because it was suggested by the total problem she gave the leftmost to A, the rightmost to C, and the part in between to B.

Possibly there would be pupils who assign all the lowest thirds to A, the middle ones to B, and the highest to C. "Can it be done otherwise?", may be asked. The children track down a rich variety of solutions. (Permutations disregarded there are 280, but it is not the aim of the question to find all of them.)

The same problem can be posed with other numbers. It is practically instructive to deal with the following ones next to each other:

24 bottles and 5 persons,
26 bottles and 5 persons.

In a visual context the children learn

with respect to an m-partition changing wholes into $m$ths (for small m), using additive splittings k into $k_1 + \ldots + k_i$ with a view to getting additive splittings $\frac{k}{m}$ into $\frac{k_1}{m} + \ldots + \frac{k_i}{m}$, exercising, in particular, the splitting off of wholes.

The initial notation is k $m$ths; the notation $\frac{k}{m}$ is of a later date. If wholes are split off, the initial notations would be $1 + 1 + \frac{2}{3}$, $2 + \frac{2}{3}$, to finish with $2\frac{2}{3}$.
  The aim is to

transfer addition, subtraction, order relation from N isomorphically to

$$\frac{1}{m} N \left(x \to \frac{x}{m} \right).$$

while slackening the visual bond, this can be supported by tables like

0	1	2	3
$\frac{0}{7}$	$\frac{1}{7}$	$\frac{2}{7}$	$\frac{3}{7}$

0	1	2	3	4
$\frac{0}{12}$	$\frac{1}{12}$	$\frac{2}{12}$	$\frac{3}{12}$	$\frac{4}{12}$

$$= 0 \qquad = \frac{1}{6} \qquad = \frac{1}{4} \qquad = \frac{1}{3}$$

At this point simplifying fractions with denominators like 12, 24, 60 can be practiced.

Then tables are again visualised on the number line where corresponding points are joined (Figures 46 and 47).

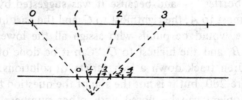

Fig. 46.

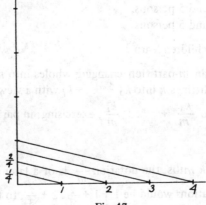

Fig. 47.

Multiplications can be prepared as repeated additions; after posting

$$\frac{2}{3} + \frac{2}{3}$$

$$\frac{2}{3} + \frac{2}{3} + \frac{2}{3}$$

$$\frac{2}{3} + \frac{2}{3} + \frac{2}{3} + \frac{2}{3}$$

ask the question: "How can you say this in other ways?"
Cautious examples of division:

$$\text{half of } \frac{2}{3}, \quad \text{of } \frac{4}{3}, \quad \text{of } \frac{6}{3},$$

and even more cautiously

$$\text{a third of } \frac{2}{3}, \ldots$$

5.13. The beer is distributed among couples and after the distribution among both members of each couple — a somewhat more compact sequence than the former, which aims to

transfer addition, subtraction, order from $\frac{1}{n}\mathbf{N}$ to $\frac{1}{pm}\mathbf{N}\left(x \to \frac{1}{p}x\right)$ and to understand the isomorphism $x \to \frac{1}{pm}x$ as the product of the isomorphisms $x \to \frac{1}{m}x$ and $x \to \frac{1}{p}x$.

The same situation visualised by tree or flow models yields Figure 48.

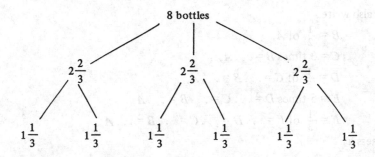

Fig. 48.

5.14 A picture of a flock of sheep; the farmer sells one out of three (that is, $\frac{1}{3}$). Strike them out. What is left? If it had been 120, how many were sold, how many left?

$$\frac{1}{3} \text{ of } 120 = \qquad \frac{2}{3} \text{ of } 120 =$$

The field of hundred: Colour $\frac{1}{5}$ of the squares red. Can it be done differently? Colour $\frac{2}{5}$ red. Can it be done differently? Find beautiful patterns!

The same with a wall of bricks — indefinite whole.

A lottery with 1000 lots. One out of five wins. How do you fix which ones? One of the five gets at least its stake back, one third of these gets double its stake. How many?

There are 10 first prizes — that is one out of ...?

Take a strip; fold it in two, three. What part is the folded strip of the original one? Fold it such that it is one sixth.

Strips below each other in a visual ratio $m : n$. If one is worth A, how much is the other?

The aim of these problems is

> Recognising and evaluating cases of the function $x \rightarrow \dfrac{1}{m} x$ in numerical wordings and visualisations.

5.15. So far $\dfrac{m}{n}$ has occurred systematically as a number only in measures like $\frac{2}{3}$ bottle, $\frac{2}{3}$ strip. The following aims at

constituting, constructing, recognising the function $x \rightarrow \dfrac{m}{n} x$.

> Here is the tree A.
> Draw tree B half as tall as A.
> Draw tree C three times as tall as B.
> Draw tree D one third of tree C.
> Draw tree E five times as tall as D.
> Draw tree F one third of E.

I can also write

$$B = \tfrac{1}{2} \text{ of } A,$$
$$C = 3 \text{ times } B = \ldots A,$$
$$D = \tfrac{1}{3} \text{ of } C = \ldots B = \ldots A,$$
$$E = 5 \text{ times } D = \ldots C = \ldots B = \ldots A$$
$$F = \tfrac{1}{3} \text{ of } E = \ldots D = \ldots C = \ldots B = \ldots A$$

I possess a

lens $\bigcirc$ through which I see everything 3 times as big

and a

lens $)($ through which I see everything $\frac{1}{5}$ as big.

I look at the flower through both of them in a row (Figure 49).

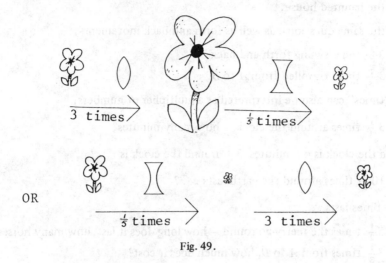

OR

Fig. 49.

What is the result in both cases?
A variety of examples should serve to

exercise the composition of the function $x \rightarrow mx, x \rightarrow \frac{1}{n}x$ in an arbitrary
number, visually supported as well as numerically isolated, for instance,
evaluating $\left(\frac{1}{3} \text{ of}\right)$ (5 times) $\left(\frac{1}{6} \text{ of}\right)$ (2 times) applied to lengths and
numbers, and recognising it as $\frac{5}{9}$.

5.16. The following serves to

replace "$\frac{m}{n}$ of" by "$\frac{m}{n}$ times".

As has been mentioned earlier, most of the textbooks do not care to motivate
this equivalence. It is annoying that "three times" is a natural operation, as is
"$\frac{1}{3}$ of", whereas the vernacular does not account for their similar character.
There is, however, one opportunity, as noticed earlier, where everyday language
– as far as I know, every language – admits of the passage from "$\frac{m}{n}$ of" to
"$\frac{m}{n}$ times", that is, in cyclic processes:

turn the key $2\frac{1}{2}$ times in the keyhole,
the big hand has gone $3\frac{1}{3}$ times around the clock,
the satellite has turned $10\frac{2}{5}$ times around the earth – where is it now?
the merry-go-round has turned $5\frac{1}{2}$ times,
so has the big wheel – where are you then?

Irregular circuits at the fair,

the roller coaster,
the haunted house,

allow the same questions, as well as forth and back movements,

$3\frac{1}{4}$ times swung forth and back,

$3\frac{1}{3}$ times travelled from A to B.

This "times" can also be interpreted as a multiplier of numbers,

$3\frac{1}{3}$ times around the clock — how many minutes?

around the clock is 60 minutes, $3\frac{1}{3}$ around the clock is . . . ?

$10\frac{2}{5}$ times around the earth lasts . . . ?

— one times lasts . . . .

$5\frac{1}{2}$ times the merry-go-round — how long does it last, how many horses?

$3\frac{1}{3}$ times from A to B, how much does it cost?

And the winding stairs

$5\frac{2}{3}$ times around, how many steps?

From the cyclic to the periodic processes,

$3\frac{1}{2}$ hundred times ticking (typing, turning of the odometer, jumping)

and to rolling a wheel

how far after 1, 2, 3 times around, $1\frac{1}{2}$, $2\frac{1}{3}$ times?

This leads in a natural way to

$1\frac{1}{2}$ times, $2\frac{2}{3}$ times a given length,

which can also be given numerically.

So a natural language of "$1\frac{1}{2}$ times", "$2\frac{2}{3}$ times" is established. As examples so far I have taken mixed fractions, which I think is didactically important. In $4\frac{2}{5}$ times the whole part 4 suggests what operation should be performed, and $\frac{2}{5}$ is dragged along. In the progression of the didactical sequence, however, proper fractions should become more frequent.

5.17. Now

"$\frac{m}{n}$ of" and "$\frac{m}{n}$ times"

are side by side; finally they shall

be identified with each other.

This can be done by applying them where both are meaningful (Figure 50):

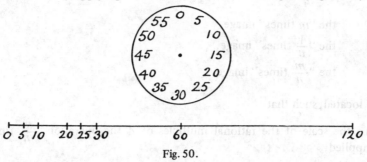

Fig. 50.

$\frac{1}{2}$ times 60 (round the clock) $= \frac{1}{2}$ of 60 (the line segment)

$2\frac{2}{3}$ times 60 (round the clock) $= 2\frac{2}{3}$ of 60 (the line segment).

It is critical that

the identification is made conscious

in order to recall it if mistakes occur. Likewise

consciously: $\frac{2}{3}$ of 1 $= \frac{2}{3}$ times 1 $= \frac{2}{3}$.

Identifying the fraction in the fraction operator and the fraction as a rational number is, however, delayed. $\frac{5}{3}$ was introduced as

"5 times $\frac{1}{3}$ of" or "$\frac{1}{3}$ of 5 times".

It shall now become

consciously: "5 times $\frac{1}{3}$ times" or "$\frac{1}{3}$ times 5 times".

In general, other examples shall be repeated in order to

consciously replace "$\frac{m}{n}$ of" with "$\frac{m}{n}$ times".

5.18. Given a point A on the ray, the rational scale of the $\frac{m}{n} A$ is systematically constructed – the expression is read as $\frac{m}{n}$ times A.

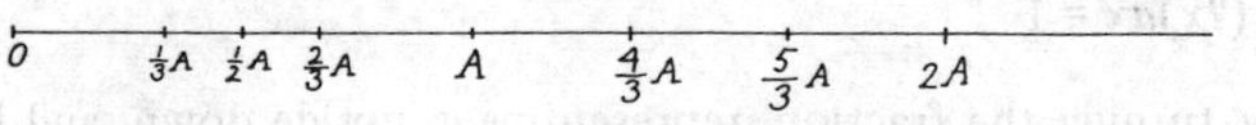

Fig. 51.

In this picture (Figure 51)

to a point $\frac{p}{q} A$

the "m times" image,

the "$\frac{1}{n}$ times" image,

the "$\frac{m}{n}$ times" image

can be located, such that

to the scale of the rational multiples of A the operation "$\frac{m}{n}$ times" is applied,

in order to validate the formula paradigmatically

$$\frac{m}{n} \cdot \frac{p}{q} A = \frac{m \cdot p}{n \cdot q} A.$$

From this, by chopping A the

multiplication formula

$$\frac{m}{n} \cdot \frac{p}{q} = \frac{mp}{nq}$$

is

paradigmatically made conscious.

Exercises in special cases such as

$$\frac{m}{n} \cdot \frac{n}{m} = 1,$$

$$\frac{m}{n} \cdot \frac{n}{l} = \frac{m}{l}$$

are included.

5.19. It seems natural to have the division of fractions join this sequence, namely

as the inverse of the multiplication

$$(?x)ax = b.$$

In the case $b = 1$ this problem was solved at the end of Section 5.18:

$$(?x)ax = 1$$

is solved by turning the fraction, representing a, upside down, and in order to solve

$$(?x)ax = b,$$

this result has to be multiplied by b. This, however, does not answer the problem of division of fractions didactically. Any hint that dividing is somehow related to reality is lacking in this approach.

Interpreting $b : a$ as a

distributive division,

that is, as a partition of b into a parts is equally meaningless unless a is an integer. It is more to the point to understand $b : a$ as a

ratio division,

answering the question

how many times does a fit into b?

for instance, if both of them are visualised as lengths. But then it is even more to the point to ask this question honestly in the context of ratios, which we will enter in the next chapter. Let us presuppose this context for a moment as a didactical precondition, with the operational conclusion:

the divisions $b : a$ and $bc : ac$ ($c \neq 0$) are equivalent,

that is, have the same solution. This is indeed an important principle, which does not become meaningful until fractions are at stake — it would not hold for divisions with a remainder.

Of course this principle can also be motivated if division is understood as the inverse of multiplication:

$$ax = b \quad \text{and} \quad acx = bc \qquad (c \neq 0)$$

do have the same solution x. As well as in the context of "ratio" the principle can also be motivated with simple approaches, such as

$\frac{2}{3}$ fits into $\frac{4}{3}$ as often as 2 into 4

$\frac{2}{3}$ fits into 6 as often as 2 into 18,

$\frac{2}{3}$ fits into $\frac{7}{6}$ as often as 4 into 7,

$\frac{2}{3}$ fits into $\frac{8}{5}$ as often as 5 into 12.

The gist of this principle is that

reducing division of fractions to that of *integers*
via fractions with *equal denominators*

— a procedure that is formally equivalent to

multiplying with the divisor turned upside down,

though it is didactically better motivated.

5.20. Adding, subtracting, and comparing of fractions are supported by the image of the number line, such as that prepared in Section 5.12:

understanding visually the mutual situation of

$$\frac{1}{4} N \quad \text{and} \quad \frac{1}{6} N;$$

finding for paradigmatic p and q one r such that

$$\frac{1}{p} N \quad \text{and} \quad \frac{1}{q} N$$

are comprised in $\frac{1}{r} N$.

Adding, subtracting, and comparing are, according to Section 5.12, performed within one $\frac{1}{n} N$, which is produced in each particular case.

5.21. Combinations of additions and multiplications are exercised in flow models (Figure 52):

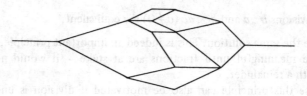

Fig. 52.

5.22. Finally the attainments are exercised in enlargements and reductions.

Pictures I, II, III of the same object,
transition from I to II by multiplication with factor a,
transition from II to III by multiplication with factor b,
from I to III?

(a, b paradigmatic fractions).

5.23. In the traditional didactics of fractions the multiplication is tied to the rectangle pattern rather than to the fraction operator. In our particular didactic sequence we chose the fraction operator; in this *structure* the rectangle model is not easily accommodated. This does not mean that it should be neglected. It can be linked to 5.20 if it does not come earlier, though certainly not in the restricted traditional form of mere *sub*dividing one rectangle.

The following didactic sequence is based on a previous treatment of areas of rectangles and similar figures (Figure 53). It is embroidered onto the pattern of

the cartesian product of $\frac{1}{n} N$ and $\frac{1}{q} N$.

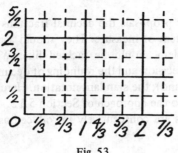

Fig. 53.

The standard problem is:

calculate the area of a rectangle with sides $\frac{m}{n}$ and $\frac{p}{q}$.

This can be joined by a sequence:

given a rectangle, find others with the same area.

These rectangles are constructed with a common SW corner and assembled in tables (Figure 54)

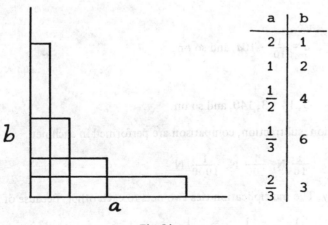

Fig. 54.

Another version,

given a rectangle, find rectangles with a times its area.

This sequence will be reconsidered in the section on area of Chapter 13.

5.24. *Decimal Fractions*

As early as in Stevin's* proposals the decimal fractions have been closely connected to a decimal system of measure. They should again be dealt with in that
context and then problem should be tackled like: Why both common and
decimal fractions; the precision and the rounding of decimal numbers; percentage
and permille (per thousand); the standard notation. Here we restrict ourselves to
subjects closely related to the approach of Section 5.12.

The decimal fractions are consecutively introduced as elements of ever finer
nets

$$\frac{1}{10} \, N \; (\frac{1}{10} \cdot 31 \text{ written as } 3, 1),$$

$$\frac{1}{100} \, N \; (\frac{1}{100} \cdot 314 \text{ written as } 3, 14),$$

$$\frac{1}{1000} \, N \; (\frac{1}{1000} \cdot 3141 \text{ written as } 3, 141),$$

and so on. These transitions are parallelled by those from mm to cm, dm, m,
..., and from g to dg, hg, kg, and so on.

The connections are again made between the layers

$$N, \frac{1}{10} \, N, \frac{1}{100} \, N, \frac{1}{1000} \, N, \ldots$$

that is

$$a = \frac{1}{10} \cdot 10a, \text{ and so on,}$$

thus

$$3, 14 = 3, 140, \text{ and so on.}$$

Addition, subtraction, comparison are performed in each net

$$\frac{1}{10} \, N, \frac{1}{100} \, N, \frac{1}{1000} \, N$$

separately. The multiplication ties two nets to each other. Because of

$$\frac{1}{10^p} \, a \cdot \frac{1}{10^q} \, b = \frac{1}{10^{p+q}} \, ab \qquad (a \in N, b \in N)$$

one gets

$$\frac{1}{10^p} \, N \cdot \frac{1}{10^q} \, N \subset \frac{1}{10^{p+q}} \, N$$

* Simon Stevin of Bruges in his booklet *La Disme* (1585).

The algorithm of multiplication is augmented with a

rule about placing the decimal point.

In divisions one takes care, according to 5.19, that dividend and divisor belong to the same net; that is, transforming the problem into the form

$$\frac{1}{10^m}\, a : \frac{1}{10^m}\, b,$$

which is equivalent to

$$a : b.$$

5.24a. *A Didactic Remark*

In remedial teaching and observations at teacher training institutions (cf. Section 4.36a), it occurred to me that the usual didactics, which aims at teaching rules for the place of the decimal point, can lead to a block of insight and of the need for insight. Once these rules have been formulated and learned, it is almost impossible to correct wrong applications by an appeal to insight. If they are needed, such rules should be the terminus of a development, which cannot be accelerated artificially. The rules should be understood on several levels.

The lowest is to start with the explanation that to the left of the decimal point are the wholes and that the decimal point is followed on the right by the tenths, hundredths, and so on, and preceded at the left by the units, tens, hundreds, and so on. Multiplying by 10 and dividing by 10 change units into tens and tens into units, respectively. This can be illustrated by an abacus with a decimal point. Equally useful is a ladder of refinement

.

.

.

1000
100
10
1
0.1
0.01
0.001

.

.

.

which can be related to the measures in the metric system. Multiplying and dividing by 10, 100, 1000, . . . are experienced as an action on this ladder. This prepares mutual multiplying (positive and negative) powers of 10. It may be asked when the usual notation for powers of 10 should be introduced (cf. Section 4.36a). However written, multiplying and dividing (positive and negative) powers of 10 should precede the formal introduction and the training of multiplying and dividing decimal fractions in general. The reduction of multiplications and divisions in this domain to those in N by means of extracting powers of 10 deserves to be preferred above memorising rules about placing the decimal point.

5.25–26. *Decimal Development*

5.25. The division of decimal fraction is by this means reduced to that of integers, that is, to what is called the

$$\text{development of } b : a, \text{ or the fraction } \frac{b}{a} \text{ in a decimal fraction,}$$

which can be of infinite length.

So far decimal fractions have been dealt with as fractions with powers of 10 as denominators, which means that a division or a fraction is to be transformed from

$$\frac{b}{a} \text{ to } \frac{e}{10^n}.$$

In order for this to be possible, the fraction in its simplified form must possess a denominator that is a divisor of a power of 10; that is

the denominator may possess no prime factors other than 2 and 5.

Other fractions do not admit of such a – finite – development.
    Transforming

$$\frac{b}{a} \text{ into } \frac{e}{10^n}$$

is performed by means of a division

$$10^n b : a = e,$$

based on

$$b : a = \frac{1}{10^n} (10^n b : a),$$

that is to say, at the dividend and finally at the quotient one passes to $\frac{1}{10^n}$ as new units before performing the division.

In fact this happens successively:

after the first division of b by a the units of the remainder are changed into tenths, with which the division is continued; the new quotient, being a number of tenths, is put in the first position to the right of the decimal point;

the remainder is changed into hundredths, with which the division is continued; the new quotient, being a number of hundredths, is put in the second position to the right of the decimal point; and so on.

If the denominator has no prime factors other than 2 and 5, the procedure terminates with the result wanted. In other cases, an infinite decimal fraction comes into being,

$$\frac{1}{3} = 0,33 \ldots$$

$$\frac{1}{7} = 0,142857142857 \ldots$$

What is mathematically relevant here differs much from what has been dealt with so far in this didactical phenomenology. It belongs to number theory and infinite series, which, however, does not exclude a phenomenological approach that fits into the present frame.

5.26. There is at this moment no need to place the infinite development of fractions into the frame of infinite series or, for that matter, into that of infinite decimal fractions in general. This can be resumed later. There is equally little need to appeal to number theory in order to explain the periodicity of the development. It is done in a more elementary way.

A division by n produces at every particular step a remainder which, consider as an integer, is a number $<n$. So among the first n partial remainders there are at least two equal ones. There is a first time in the sequence of remainders that a remainder equals a previous one. Let us assume it is the jth that equals the ith. But then the whole procedure runs from the jth onwards as it did from the ith, that is, the piece from

the ith to the $(j-1)$th quotient

repeats itself periodically. The decimal development of a rational number

eventually becomes periodic.

It can be

purely periodic

or the period is preceded by an initial segment. Examples of the first kind;

$$\frac{1}{3} = 0,3 \ldots \qquad \frac{1}{7} = 0,142857 \ldots ;$$

of the second kind;

$$\frac{1}{6} = 0,166 \ldots \qquad \frac{1}{35} = 0,285714285714 \ldots$$

How can we predict what happens?

The examples suggest: The decimal development of the — simplified — fraction is

purely periodic or not

according whether the denominator n

does not   or   does

have a prime factor 2 or 5.

This appears to be correct: The period of the development of $\frac{m}{n}$ comes into being when for certain i and j the remainders after the ith and jth division are equal, which means that

$10^i m$ and $10^j m$ leave the same remainder,

when divided by n. In other words,

$(10^j - 10^i)m$ is divisible by n.

If n does not have prime factors 2 and 5, this implies that

$(10^{j-i} - 1)m$ is divisible by n,

thus

$10^{j-i} m$ and m leave the same remainder,

when divided by n. Thus the period starts immediately after the decimal point.

Conversely: Take a purely periodic development, with a period of, say, length l. Let the period itself, considered as a natural number, be c. Thus

$$\frac{b}{a} = c\left(\frac{1}{10^l} + \frac{1}{10^{2l}} + \ldots\right).$$

The expression e between the parentheses can be calculated as follows

$$10^l e = 1 + e,$$

thus

$$e = \frac{1}{10^l - 1},$$

thus

$$\frac{m}{n} = \frac{c}{10^l - 1}.$$

Now $10^l - 1$ certainly does not have prime factors 2 or 5, nor has n. Thus

if the development of — simplified — $\frac{m}{n}$ is purely periodic, the denominator n does not have prime factors 2 or 5.

5.27. *Other Bases*

With respect to becoming acquainted with and working in positional systems other than the decimal one, the arguments of Section 4.43 can be repeated, though a certain difference is worth mentioning. In general, one may not expect that a change of basis creates more insight, even with regard to terminating and not terminating developments. If it has been understood which denominators in the decimal system lead to infinite developments, why they are finally periodic, and which cases are purely periodic, the transition to a new base g can open new perspectives. The divisors of 10 are replaced with those of g, and this has different consequences according to whether g is a prime number, the power of a prime number, or otherwise composite. It depends on the total instructional situation and in particular the special group of pupils concerned whether the insights acquired in such a course are worth the trouble of introducing other positional systems.

CHAPTER 6

RATIO AND PROPORTIONALITY

6.1. *A Preface in Between*

A first version of the present chapter has been the first specimen of didactical phenomenology that I produced – in 1973 in German. The immediate cause was a theoretical exposition by an educationalist on instructional objectives, where as a paradigmatical example the author dealt with ratio. He chose this example because in the larger work from which it was borrowed, it was a subject that could be covered by just one objective. I have repeatedly argued that formulating instructional objectives should be preceded by observing such learning processes as could reveal what is being, and thus what should be, learned; and that for observing learning processes as well as for educational development an indispensable precondition is a didactical phenomenology. At that time instructional objectives, however, were distilled from prevailing textbooks and test collections. In order to show how much is lost by this approach, I seized upon "ratio" as an example to explain what didactical phenomenology is or should be. Whoever reads what I wrote at that time – and it does not sound much different in the present version – will be struck by a tightness of style that was not my habit. With hindsight I should say that this style was conditioned at least as much by the special subject as by my intention to write a specimen of didactical phenomenology – later on I will give reasons for it.

In May 1975 I lectured in Berlin. It was the first time that I met Christine Keitel, with whom I had already corresponded. I told her about a manuscript, which was later published under the title *Weeding and Sowing*, and about a didactical phenomenology of fundamental mathematical concepts, which would be my next undertaking. I promised her a work written in a rigorous scientific style, with no regard to legibility. Christine implored me "don't" with an inflection, as though she meant "you are not obliged to". For a long time these words preyed on my mind.

In the summer of 1975 the chapter on ratio existed, as well as a provisional sketch of "Fractions", but the first chapter had still to be started. "Sets", of course. I struggled with it but I did not succeed. The subject was refractory and the tight style, which I had mastered successfully in "Ratio" deserted me. I did not write a single line.

I made up my mind. No sets. Numbers – no. Geometry – no. Finally I chose "Length", and after a short while the chapter was conceived in detail. (It shows gaps, it should be rethought.) But again I could not write it – that is, not in the tight style of Ratio – an ideal that fettered my mind. Should I straightjacket a subject that was not created for it, only to have it look in a way that was not its nature?

Moreover, I was not obliged to. I need not assume scholarly sounding language to raise expectations of depth. I do not start a career where such a language would mean a recommendation, and I do not feel happy providing work for generations to come who would fathom depths in unreadable work. My spiritual portrait is established, and my ideas on what is scientific need no correction. *Simplex veri sigillum* — I translate it as: what is true may be said in plain language.

I knew what to do, but I still did not know how to do it. I read and reread "Ratio". It looked good and well-written. It was clear and the style was honest. Why could I not write the same way on "Length"? I tried formulations, to no avail. Why, I must free myself of this model. I knew "Ratio", as it were, by heart. I must close this drawer of my mind and open another.

So I decided to move from German to Dutch. I would write the phenomenology in Dutch, in order to translate it afterwards. Language is an infection. Dutch is the only prose in which I never tried profundity.

After I had taken the decision on the language, fresh arguments emerged. The phenomenology had been started and was intended to help the developers at my side in their everyday work and in discussions about it; nobody else would profit from it in the short run — translated or not. It was meant for our colloquial talk and would be written in our colloquial language.

This was a preface in the wrong place. Another will be written when I look back on this work.

6.2. *The Logical Status of Ratio*

Belatedly I understood that the logical status of ratio is far above those concepts discussed so far. I also understood why I should separate ratio from fractions.

Ratio is a function of an ordered pair of numbers or magnitude values. So are sum, difference, product, and quotient, but they are so in an algorithmic sense: there is a recipe to figure out the function value assigned to a particular pair, or at least to act as if you had — indeed, what do you figure out if you answer 3 : 4 by $\frac{3}{4}$?

Ratio can also be figured out: transformed into a quotient, that is reading

as 3 is to 4

as

3 divided by 4,

but this is the rape of ratio. Then ratio is deprived of what it makes valuable as ratio.

Ratio is a function of an ordered pair of numbers or magnitude values. But what about the values of this function? Again numbers, values of magnitudes? One can interpret it this way, though it is the wrong way. Indeed, this would identify ratio with quotient. It is the meaning of ratio to speak about equality (and inequality) of ratios without knowing how large the ratio is, to be able to meaningfully say

$$a \text{ is to } b \text{ as } c \text{ is to } d$$

without anticipating that

$$a \text{ is to } b$$

can be reduced to a number or magnitude value

$$\frac{a}{b},$$

which then for

$$c \text{ is to } d$$

is the same:

$$\frac{a}{b} = \frac{c}{d}.$$

With the opportunity offered by numerosity and length I stressed that the recognition of equality and inequality, of bigger and smaller, phenomenologically precedes the operation of adding and measuring — it is a pity that this simple fact is impaired in its credibility by wrongly interpreted conservation principles. If ratio should be taken as seriously as numerosity and length, then equality and inequality, bigger and smaller, should play a similar role. Anyway the phenomenological exploration should uncover the same roots.

If these suppositions are confirmed — they will — then the logical status of ratio in its phenomenological context would be paraphrased as follows:

> ratio is an equivalence relation in the set of ordered pairs of numbers (or magnitude values), formally indicated by
>
> $$a : b = c : d$$
>
> if the pair $\ulcorner a, b \urcorner$ is equivalent to the pair $\ulcorner c, d \urcorner$.

We do not formulate the (axiomatic) postulates to be fulfilled by this particular equivalence relation.

It is a fact that after choosing a unit e the equivalence class of the pair $\ulcorner a, b \urcorner$ can be expressed by one number (one magnitude value), namely *that u* for which

$$a : b = u : e,$$

but this approach is an *a posteriori* insight, which in fact matters only if e does not depend on any arbitrariness (for instance, if it is the numerical unit). *A priori*, ratio depends on two data, and consequently each proposition on ratios — proportionality — depends on four of them.

This complex behaviour was what I meant when in the first paragraph of the present section I placed ratio, as regards its logical status, high above other concepts dealt with before them. Quotients and fractions are a means to reduce this complication, to lower the logical status, at the expense — as it happens — of insight. One may doubt whether fractions can be insightfully taught if insight into ratio is lacking — it is this doubt that influenced the composition of the chapter on fractions. The influence could have been stronger, but I did not dare integrate the phenomenological analysis of fractions and ratio. Should I not follow the chapters "Fractions" and "Ratio and Proportion" by a chapter "Fractions, Ratio and Proportion"?

The logical status of ratio which I explained here implies that "ratio and proportion" is more intensive mathematics, mathematics on a higher level than what has been discussed so far. This fact, I think, influenced the tight style of my first example of didactical phenomenology. By the choice of subject, the most mathematical at an elementary level, I found my mathematical bread buttered on both sides. Rather than by my desire to write a didactical phenomenology, the tight style was suggested by the choice of the subject. The attempt to imitate it with other subjects was badly motivated and doomed to failure.

The reader will have to content himself with this chaotic alternation of styles. It is rooted in subjects and views on subjects, rather than in states of mind.

6.3. *Ratio as a Relation In and Between Magnitudes*

In order not to overburden the exposition of the most relevant ideas, I start with a few concepts, terms, and notations. I will use a rather loose language, with a minimum of formalisation. For instance, I will speak of equal sizes if *objects* of equal size are intended, of equal distances, weights, times where I should properly say: paths of the same length, bodies of the same weight, intervals of the same duration. It can even happen that I speak of the ratio of two objects, where it should be the ratio of size, volume, or weight of the objects, or of the ratio of two metals in an alloy rather than of that of their masses.

I start with a heavily mathematised example,

uniform motion:

(1)   in equal times equal distances are covered,

which is equivalent to

(2a)   distances are in proportion to times,

as soon as motion is assumed to be continuous, as it should be;

(2b)   distance is proportional to time

is only another wording of (2a), and

(3)   the distance is a linear function of time

is again another formulation, as is

(4)   speed is constant,

though it looks different.

A brief comment:

From (1) it follows:

> in twice the time twice the distance is covered,
> in thrice the time thrice the distance is covered,

and more generally

> in n times the time, n times the distance is covered.

Let

$$s = f(t)$$

be the distance as a function f of the time t. We just noticed:

$$f(nt) = nf(t) \quad \text{for } n \in \mathbb{N}.$$

Replace t with $\frac{1}{n}t$. Then

$$f(t) = nf\left(\frac{1}{n}t\right),$$

which read

$$f\left(\frac{1}{n}t\right) = \frac{1}{n}f(t),$$

yields

> in $\frac{1}{n}$ of the time $\frac{1}{n}$ of the distance is covered.

If in the last formula t is replaced with mt ($m \in \mathbb{N}$) one gets

$$f\left(\frac{m}{n}t\right) = \frac{1}{n}f(mt) = \frac{m}{n}f(t).$$

Thus for each (positive) rational α:

> in α times the time α times the distance is covered.

Continuity guarantees the same for real instead of rational α.

Take two times t_0, t. Put

$$\alpha = t/t_0.$$

Then

$$f(t) : f(t_0) = f(\alpha t_0) : f(t_0) = \alpha f(t_0) : f(t_0) = t : t_0,$$

which is the formulation (2a) or (2b). This can also be written

$$f(t) = (f(t_0)/t_0)t,$$

which is formulation 3, or

$$f(t)/t = f(t_0)/t_0.$$

which is formulation 4.

There are two magnitudes concerned here: time and length; and a function f that assigns a length to a time, namely the *length* of the path covered in the *time* interval. The ratios considered here are those of pairs *in one and the same system* (time *or* length); the ratios in one system are required to equal the corresponding ones in the other — this is the postulate of the uniformity of motion.

We designate

ratios formed within a system as *internal*

to distinguish them from the external ones that are discussed below.

If t_1, t_2 are times and s_1, s_2 corresponding paths, the postulate of uniformity says

$$s_1 : s_2 = t_1 : t_2.$$

If we are tempted to interchange the middle terms, we get

$$s_1 : t_1 = s_2 : t_2,$$

again the equality of two ratios, albeit ratios of path to time.

We designate

ratios between two systems as *external*.

The uniformity of the motion is now expressed by the postulate

the ratio "path to time" is constant.

Ratios can also be interpreted as

quotients.

In this interpretation

the internal ratio is a number,
the external ratio is a magnitude,

that is, in the present case of uniform motion,

the quotient of path and time: speed.

The whole reasoning, in particular interchanging the middle terms in a proportion, is quite familiar to us. I ask myself whether we sufficiently realise that it need not be as obvious to the learner. Former arithmetic instruction was quite conscious of this jump. Rather than bridging the gulf, one invented two kinds of division, ratio division and distributive division. Together with this twin monster the former awareness of this problem seems to have vanished, and since no-one today is conscious of the mental jump from internal to external ratios, nobody raises the question as to whether it could not be too big for the learner.

The geometrical tradition of Greek antiquity allowed formulations only with *internal* ratios; algebraic operations or magnitudes were allowed only in a complicated geometrical setting. It is a drawback of Greek geometry that, because of the lack of external ratio, interchanging the middle terms in proportions in general was not allowed and had to be circumvented by means of complicated procedures. The ancient tradition has maintained itself in theoretical sciences for long times. Outstanding examples of this habit are Kepler's second and third laws:

> in equal times the radius vector from the Sun to a planet sweeps equal areas;

> the squares of the times of revolution are in the same ratio as the cubes of the long axes of the orbits.

This tradition pervaded the theoretical sciences longer than it did commercial and technical mathematics, where direct, non-geometrised algebraic operations and, in particular, external ratios were admitted earlier; even today pure mathematicians often show little understanding for calculations with magnitudes.

I used uniform motion as a paradigm. The generalisation may be left to the reader. It will be clear enough what I meant by

> internal ratio (within a magnitude)

and

> external ratio (between two magnitudes).

It is equally obvious that in mapping magnitudes,

> invariance of internal ratios,

and the equivalent

> constancy of external ratios

means

> linearity of the mapping;

in our example,

> uniform motion is a linear mapping of time on path.

The linearity of a mapping *f* is correspondingly defined in two ways:

> implicitly (postulatory),

> > to the sum corresponds the sum, $f(x + y) = f(x) + f(y),$

explicitly (algorithmic),

$$f(x) = \alpha x \quad \text{for all } x \text{ and a certain } \alpha.$$

Once more: all this is so obvious that as mathematicians we do not worry about it any more, but let us not expect that it passes by mere diffusion from our unconsciousness to that of our pupils.

Things are even more involved: A uniform motion has all time intervals of equal length, wherever they might be, mapped upon equal path intervals. It is not explicitly mentioned that the composite of two connected time intervals is mapped onto the composite of the corresponding path intervals because it is implicit in the idea of motion. The same holds for other pairs of magnitudes, such as volume and weight of some substance. If, however, f is a function that maps magnitudes onto each other, I have already abstracted from the particular time and path intervals (and similar ones); they have been superseded by lengths and durations (and suchlike). So I am obliged to require explicitly that, just as in the sphere of objects where composites correspond to composites, so in the sphere of magnitudes sums correspond to sums. In a more formalised setting I could have formulated this more sharply, but I would avoid too much formalism.

6.4. *Expositions and Compositions*

Ratio must be viewed in a broader context than that of relations within and between magnitudes. I want to sketch it by such disparate examples as:

6.4.1   a set of animal species with their average weights (or other quantitative characteristics),

6.4.2   a set of flight connections with their prices (or distances),

6.4.3   a set of countries with their numbers of population (or their areas),

6.4.4   a set of articles with prices (or weights),

6.4.5   the set of components of an alloy with their masses,

6.4.6   the set of age classes of a population with their numbers,

6.4.7   the set of categories of soil use of a nation, with the corresponding areas,

6.4.8   the set of diseases with the number of cases of each one,

6.4.9   the set of pairs of points of a plane with their mutual distances.

The *common* feature in these examples is

a set, in general indicated by $\Omega, \Omega', \ldots$ in the sequel,

and

a function, in general denoted by $w, w', \ldots$ in the sequel, which accepts values of a certain magnitude.

Between the first four (6.4.1–4) and the following four (6.4.5–8) there is a profound *difference*:

In the first group the elements of Ω are

> objects in a primitive sense, while Ω is defined by common traits of its elements (species of animals, flight connections, countries, articles),

in the second group the elements of Ω are

> classes of a universe, formed according to certain criteria that are important for that universe (ages in a population, and so on).

In the first group

> the function w describes internal properties of the elements of Ω,

in the second group

> the function w describes the size of the class (not necessarily a whole number, cf. 6.4.5).

I will call, quite arbitrarily, the first and second kind, respectively,

> expositions,
> compositions,

The ninth example, a not unimportant one, is wholly different from the preceding ones; in Section 6.5 we will return to it.

Expositions and compositions differ in how they are used. Usually they occur in couples. Anticipating a general formulation, I will explain this by examples.

> *Couples of expositions*:
>
> Ω a set of countries,
> w the function that assigns to each country the number of its inhabitants,
> w' the function that assigns to each country its area;

the ratio w to w' (population density) is variable: a country has "in proportion" the same (a larger, a smaller) number of inhabitants.

> Ω a set of filled plastic bags in a supermarket, on which are indicated:
> the price w,
> the weight w';

the ratio w to w' (unit price) is variable, for bags containing the "same" article, it will be the same; on these w and w' are linearly dependent.

> *Couples of compositions*:

We consider two alloys with the "same" components. The components of the alloys form two sets

> Ω and Ω'

with the corresponding mass functions

w and w',

for instance,

30 kg bronze consisting of 20 kg copper and 10 kg tin,
65 kg bronze consisting of 40 kg copper and 25 kg tin.

Both Ω and Ω' are the set

{copper, tin}.

In the two alloys we have, correspondingly

$$w(\text{copper}) = 20 \text{ kg}, \qquad w(\text{tin}) = 10 \text{ kg},$$
$$w'(\text{copper}) = 40 \text{ kg}, \qquad w'(\text{tin}) = 25 \text{ kg}.$$

In general the ratio w to w' can vary; if w and w' are linearly dependent, it is the "same" alloy.

Two populations — The Netherlands and The Philippines — are partitioned into age classes.

Ω and Ω'

are formed by the age classes

$\{[0, 1), [1, 10), [10, 20), \dots \}$.

w and w'

are the number of people in the respective classes.

In one population there are "in proportion" fewer babies, more aged people, and so on, than in the other.

The case of a couple of *expositions* consists of

one set Ω;

with two functions w, w' on it;

whose — mostly external — ratio is considered.

The case of a couple of *compositions* consists of

class partitionings Ω and Ω' of two universes, attained according to the same principle and identified in a natural way with each other;

with two functions w, w' on it,

whose — mostly internal — ratios are considered and perhaps compared.

6.5. *Constructs*

We pass to the example 6.4.9. It shows

a set Ω based upon a strong — preferably geometrical — structure Σ with a measure function.

In our particular case Σ was a planar figure, for instance, the whole plane, Ω the set of pairs of points, w the distance.

Other possibilities would be:

Ω the set of plane curves with w as the arc length;
Ω the set of rectangles with w as the area.

I will designate such a system Ω, w as

a construct,

or more precisely

a Σ-construct

(if Σ is the structure on which it is based).

Constructs, too, are used in couples, Ω, w and Ω', w', where it can happen that $\Omega = \Omega'$ and $w = w'$.

A couple of constructs:

Ω is the set of pairs of points of a planar figure Σ,
Ω' is the set of pairs of points of a planar figure Σ',
w and w' are the corresponding distance functions.

Moreover, there is a mapping

f of Σ in Σ'

which extends itself in a natural way as a mapping

f of Ω in Ω'.

A property of f that may be relevant, is

similarity.

As with uniformity of motion in Section 6.3, similarity can first be characterised by the condition

f maps pairs, with the same mutual distance, on pairs with the same mutual distance,

or

f conserves equality of distance,

or — a richer, but equivalent, formulation —

f maps pairwise congruent on pairwise congruent figures,

or

f preserves congruence.

This formulation does not yet involve ratio, but under the — natural — condition of continuity this characterisation is equivalent to

f preserves ratios, that is,

$$w'(f\alpha) : w'(f\beta) = w(\alpha) : w(\beta) \quad (\alpha, \beta \in \Omega).$$

This is

preservation of *internal* ratios,

taken in Ω and in Ω' respectively. As in Section 6.3, in the case of magnitudes, f being a

similarity

can be expressed by the

constancy of *external* ratios:

$$w'(f\alpha) : w(\alpha) = w'(f\beta) : w(\beta) \quad (\alpha, \beta \in \Omega).$$

Another example:

Ω a set of line segments, w the length function,
Ω' the set of squares on these line segments, w' the area function.

The use of this couple is obvious.

6.6. *The Occurrence of Ratios in Sections 6.3–5 Compared*

In Section 6.3 ratios occur in and between magnitudes, in Section 6.4 in and between expositions and compositions, in Section 6.5 in and between constructs.

The cases 6.3 and 6.5 resemble each other because of the underlying strong mathematical structure, whereas in Section 6.4 these structures are weak. Sections 6.3 and 6.5 also have in common that

proportionality and similarity

can be defined

without involving ratio

purely by

preservation of equality or congruence.

In the case of Section 6.4 this is not possible or it would require complicated reasoning.

Compared with Section 6.3 the case 6.5 has the advantage that

congruence of figures can be visualised more strongly than equality of magnitude values.

These distinctions will be seen to have important didactical consequences.

6.7. *Ratio in Similarities*

In our phenomenology the stress is now shifted to didactics.

Ratio as a concept and even as a mental object requires a considerably high developmental level. For all that, the feel and the look of ratios occur remarkably early in development. According to Piaget, topological concepts should precede euclidean ones. We anticipate that this holds at most for such spatial relations as inclusion, exclusion, and overlapping, but these are relations which no mathematician would consider topological as psychologists do. The acceptance of truly topological properties — that is, stating equivalence by means of one-to-one continuous mappings — is certainly not an attitude that can be placed in early childhood; it is much too sophisticated to be expected of little children. Piaget and researchers who repeated his assertions or experiments were seriously confused. From the inability of little children to draw circles and squares so neatly that they reasonably differed from each other, they drew the conclusion of topological predominance. Yet at an early age children are able to distinguish clearly circles and squares, which is the only thing that matters. It is true that children judge drawings in books or made by adults by other criteria than their own production — a kind of system separation which is worth studying closely.

There can be no doubt, however, that children recognise early the different sizes of objects, and their being larger or smaller. It is equally certain that they can handle similarity as an operational equivalence. I would even go so far as to assert that congruences and similarities are built-in features of that part of the central nervous system that processes our optical perceptions. The immediate reidentification of objects after a rotation (of the object or the perceiver) and after a change of distance presupposes something in the brain like a computer program for the elimination of this kind of mapping — it is riddle to me what such a program looks like; its existence, which I do not doubt, is like a miracle to me.

At a young age a child recognises drawings and models of animals, furniture, cars, bicycles, ships as images of these objects — it does not matter on which scale, and whether they are pictured side by side on different scales. "How big is a whale really?", a child can ask, convinced that the picture, except for the scale, is faithful. Well, sometimes whales are sketched by drawings in one line, but even the difference between a photograph and a characteristic sketch is grasped early.

Weighings with a spring balance, performed by Bastiaan (5; 6), were indicated by him on a horizontally typewritten "spring balance" with a different scale. He noticed inessential deviations in the figures (1 instead of 1) but neither the difference in orientation nor scale. The typewritten image was structurally faithful.

After a sequence of sunny days Bastiaan (6; 1) sees clouds again, and says: "It will rain". I tell him: "No, these are very high clouds, where no rain falls out; rain clouds are low and dark." He: "What height are these clouds? I (exaggerating): "10 thousand metres." He: "And rain clouds?" I: "Thousand metres." He (showing to the ground): "So if we are here and this (showing a height of about 30 cm) is rain clouds, then this (shows about 1 metre) is no rain clouds."

Without any hesitation children accept that objects at the blackboard are drawn ten times as large as on the work sheet, that the number line at the blackboard has a unit of 1 dm compared with that of 1 cm on the work sheet. They accept number lines where the same interval means a unit, or ten, or hundred, side by side. Children would, however, immediately protest structural modifications that violate the similarity of the image:

> what is mutually equal in the original,
> should be mutually equal in the image,

which as we know, implies

> the invariance of internal ratios,

characterising mappings as

> similarities.

Children become familiar at a young age with these

> ratio preserving mappings

as we shall call them, if they see planar or spatial figures pictured — paintings, copies of paintings, models of buildings. Systematic deviations from this mapping principle are noticed; for instance,

> the use of different scales in different directions,
> the use of different scales for different figures,
> the use of different scales for parts of the same figure.

This, however, is not done by making the scales explicit, but with formulations like

> the head is much too large — that is, if compared with the trunk,
> this is much too long — that is, if compared to the width —

objections regarding the lack of similarity though with no explicitation of ratios.

It requires more insight into geometrical relations to adduce other criteria, such as:

> what is a right angle in the original,
> should be a right angle in the image.

With this feeling or eye for similarity, as I have termed it, the child is of course still far away from similarity as a mental object, let alone as a concept. I indicate a number of intermediate stages:

> Recognising ratio preservation or non-preservation of mappings,
> Constructing ratio preserving mappings,
> Resolving conflicts in the construction of ratio preserving mappings,
> Operationally handling,

formulating,
relating to each other:
    criteria for ratio preservation, such as
        preservation of equality of lengths,
        preservation of congruence,
        preservation of internal ratios,
        constancy of external ratio,
        preservation of angles,
and deciding about the necessity and sufficiency of such criteria.

In the first steps of this sequence

ratios do not occur explicitly,

later on

equality of ratios (internal and external) becomes explicit,

and finally

ratios themselves are made explicit.

It is a sequence similar to those observed with length and other magnitudes. The strong visualisation is an advantage of the geometrical context of ratio compared with other contexts. What matters didactically is the

gradual verbalisation of visual reasoning.

Most often the contexts of ratio are not visual but are accessible to visualisation. Early familiarity with ratio-preserving mappings is a support to visualising such contexts of ratio as are not *a priori* visual. This, however, requires that the visualised ratio is somewhat loosened from the context of global similarities. In order to build a bridge from non-visual to visual ratios, the strict visualisation by similarity must be weakened.

Similarity, as mathematically understood, is a mapping that extends over whole planes. In each visualisation one is satisfied with linear or plane figures that by their size and structure suggest the whole plane: pieces of the real world and their pictures, wallpaper patterns, and other structures that can be continued inside and outside. For all activities and their various levels which are enumerated,

a rich structure of original and image

and the fact, or at least the suggestion that the one is the image of the other,

are if not required, then at least advantageous.

Too little structure may be an obstacle to the visual recognition of similarity of figures — this holds even for adults. In a wealthier structured material, the more or less algorithmic criteria for the recognition of similarity can be isolated and exercised following the above course of activities, in order to become operational in a less structured material.

It is a small step from the fact to the suggestion that something is somehow the image of something else. The lack of any suggestion may be an impediment to even think of similarities unless the dispensability and restoration of such a suggestion has been prepared in a learning process.

If pictured rectangles are to be compared with regard to similarity, the structuring addition of the diagonals alone can transform failure into success. Path lengths on ground-plans and in reality are more easily compared than circumferences of bare rectangles. The insight that all circles are similar can more easily be acquired with structured circles, such as our various coins, which on the obverse side are even similar in surface details. The bare circle is not a good medium to reveal the internal ratio of circumference (or distance covered by rolling it) to diameter. The approach via the external ratios and circumferences using different (and similarly structured) circles is more useful.

I resume this exposition with a list of activities:

> Transferring what has been
>   exercised, recognised, made explicit
> with respect to ratios and ratio preservation
>   in a richly structured context
>   into a less or poorly structured context.
> On behalf of ratios and ratio preservation
>   enriching a poorly structured context,
>   introducing a geometrical structure in a non-geometrical context,
>   translating a non-geometrical context into a geometrical one,
> understanding and using contexts that might be geometrical or not as geometrical images of each other.

6.8. *Relatively*

Whereas one can go a long way with ratio-preserving mappings without verbalising all that can be seen, experienced, constructed as ratio, other contexts require an early verbalisation (albeit not of ratio) of such ideas as

> relatively (or comparatively).

As a mental object this may be supposed at the end of the kindergarten age.

> This chocolate is sweeter because it contains — relatively — more sugar.
>
> A flea can jump — relatively — higher than a man.
>
> An air travel to South America is — relatively — more expensive than one in Europe.
>
> In the Netherlands there are — relatively — more bikes than in Germany.

In the explicit formulation the word "relatively" can be lacking, since it is clear what is meant. The terms

> relatively more, as much, less

can be given various shades of meaning

> from roughly qualitative to precisely quantitative.

In particular, to establish "more" or "less", estimations may suffice, though they can be refined by additives like

> much, very much, a bit.

"Relatively" lacks a relation term, which can be

> obvious or explicitly added.

For instance:

> If compared with the number of inhabitants there are more bikes in the Netherlands than there are in Germany.

A possible sequence of levels:

> understanding that orders (larger and smaller, more and less) can be relativised (relatively larger, smaller, more, less)

> understanding "relatively" in the sense of "in relation to . . .", with the criterion of comparison filled in at the dots,

> using meaningfully "relatively" and "in relation to",

> completing "relatively" to "in relation to . . . " in a context,

> knowing operationally what "relatively" and "in relation to" mean in general,

> explaining what "relatively" and "in relation to" mean in general.

There is a number of stages

> from roughly qualitative to precisely quantitative,

where finally according to the subject the criterion is

> internal or external ratio.

Tasks where such activities can take place can have a visual character:

> houses, people, trees on different scales
> — which ones belong together, and why?

> a group of persons, and on another scale, clothes
> — what belongs together?

walls on different scales and of different thickness
— which ones are thicker?

meadows with flowers, ponds with frogs, skies with clouds
— where are there relatively more?

Other senses can play a part:

a large orchestra that produces relatively soft sounds.

6.9–11. *Norming*

6.9. A few examples will be given to introduce the complex of techniques, wrong use of techniques, attitudes fostered (or rather, not fostered) by these techniques — a complex I designate by *norming*:

If we imagine the earth as a pin's head (1 mm diameter), the sun appears as a sphere with the diameter 10 cm at a distance of 10 m.

The scale reduction is meant to visualise drastic ratios; one chooses a familiar unit to start with; it does not matter what the scale is.

If the development of life on earth is thought to have happened in a day, man appeared one minute ago and human culture started a second ago.

It looks much like the first example: a time reduction which can be illustrated by a linear drawing. For the largest component "day" has been chosen as a unit, whereas in the first example it started with the pin's head.
    The examples

one out of five children born, is a Chinese,
one out of four cars is a Fiat,

show a preference for ratios normed by "one out of . . . .".

A recipe "boeuf à trois moutardes" for four persons,

the meal unit is four, which in many cases will save conversions. (It is, however, not recommended that one trust ratio in cooking and baking.)
    The quality of drinking or swimming water is indicated by

this much salt in one litre

or

that many coli bacteria in one cc,

where the quantities that are actually drunk or bathed in are of quite another size, and the quantities actually analysed are again of a different size.

The production of refuse is measured by a vague unit such as

>inhabitant's equivalent,

which only serves to estimate and to tax the refuse production of families and industries.

The power of nuclear explosions is measured in

>kilotons of TNT,

a strange norming that serves to compare nuclear bombs with each other rather than with conventional explosives.

If the cost of living is put at 100 in 1965, it is 147 in 1975 — an example of the much used index figures, where by preference a basis 100 is chosen. In other cases an average is normed at 100, for instance, for the I.Q.:

>the average score in a certain population (at a certain age) is put at 100 in order to measure individual scores with it.

This number 100 links up with the decimal system, while on the other hand, decimal *fractions* are avoided as much as possible. In traditional instruction in arithmetic, percentages and interest were closely connected. This, however, is not an old tradition; interest was expressed rather by "one to . . . " (the "tithe" means one tenth). By the decimalisation of money, percentage interest arithmetic became effective. Today the most usual application of percentages is in "compositions":

>the whole is put at 100 in order to express the parts numerically.

The aim is

>to make different compositions comparable.

The comparison can be

>supported by visualisation,

for instance, by sector diagrams. The need to make composition data comparable is at present the strongest motivation for percentages; moreover, percentage is a device that presents itself most naturally as soon as, on behalf of comparability, totals must be normed uniformally. Thus not

>if the Netherlands were as large as the FRG

nor

>if the FRG were as large as the Netherlands

but

>put the area (number of inhabitants) of both of them at 100 (or perhaps 1000), then . . .

Resuming the preceding analysis, one can state a few levels

> with respect to making compositions and constructs more perspicuous by norming one component,

> with respect to making compositions comparable by norming the whole (in general on 100), while the absolute data and the scale factor play such a subordinate part that they are more or less disregarded:

>> understanding the norming,
>> understanding the rationale of the norming,
>> performing normings where they are required,
>> performing normings where they are useful,
>> understanding this activity operationally,
>> describing it,
>> and putting it into a larger frame.

6.10. A mistake related to norming is: forgetting about the unnormed data and the scale factor:

> absolute meaning is ascribed to data that depend on norming, in particular, data derived by different normings are compared without renorming;

> the number 100 plays an, as it were, magic part;

> percentages derived from different norming procedures are added and processed to — unweighted — averages;

> it causes surprise and is not understood if, for instance, a party in an election sees its percentages increase in all districts while the percentage over the whole decreases;

> double norming is applied as in the example taken from a newspaper: in 1972 the national product per capita of Bresil increased 5%, but this increase is in the greater part absorbed by the 4½% increase of the population in the same time.

6.11. A more subtle and more dangerous feature is forgetting about the unnormed data, for instance, in statistics, if this includes forgetting about

> the precision of the normed data:

"one out of two", or 50% can have been obtained from a total of two, or by a rough estimation from a total of a thousand or a million.

Problems of precision can be caused by measurements or by stochastics — as a matter of fact, the source of imprecision in measurements, whether exact or estimated, is also stochastic. Precision will be dealt with in another chapter, but meanwhile it makes sense to have touched upon this subject already in connection with norming relative data. Even in the present chapter we will touch on it once more.

6.12. It can happen that normings take place or are asked for where they do not matter or are even disturbing. Examples:

A string closing around the equator is lengthened one metre and again closed, loosely, around the equator. Can a man creep through under it?

The problem is often answered with a question regarding the diameter of the earth, which, in view of the linear relation between diameter and circumference of a circle, does not matter.

John and Pete live and work at the same address. By bike it takes John 30 minutes and Pete 40 minutes to go from home to work. John leaves 5 minutes after Pete. Where does he catch up?

The usual reaction is to ask for the distance between home and work, which again for reasons of linearity does not matter.

An even more drastic example: a student who must switch from the metric to the Anglo Saxon system of measures asks: how much is π here?

The preceding can be summarised as follows:

> Insight into the irrelevance of normings in the case of linear relations.

6.13. *Visualisations*

Understanding ratios can be steered and deepened by visualisations. One can illustrate

> expositions by histograms and pictorial statistics,
> compositions by sector diagrams and other planar divisions.

Example of visualised expositions: The EEC countries are represented, with respect to their areas, by

> rectangles with the same base and heights proportional to the areas

which are placed side by side as in a histogram; the numbers of population by

> a group of human figures (for instance, each representing a million),

where both representations can be combined by

> placing the human figures into the corresponding rectangles,

in order to visualise the different densities of population (ratio of number of inhabitants to area).

Example of visualised compositions: A circle divided into sectors corresponding to, and with respect to, area

> proportional to the use categories of the soil of a country,

for various countries side by side, in order to illustrate

> the differences with regard to the use of soil

(more or less agricultural, and so on).

Such visualisations are a kind of ratio-preserving mappings, with ratios other than those between distances of pairs of points considered – in the last example, on the one hand,

the ratios of areas, population numbers, use categories,

on the other hand,

areas of planar figures.

A sequence of levels could be:

understanding histograms, pictorial statistics, division of areas and similar visual representations as ratio-preserving mappings of expositions and compositions,

constructing such visual representations,

deciding conflicts in constructing them,

understanding the principles of such visual representations, and describing them;

recognising preservation of ratio as the common principle in the visual representation; and

describing it.

Furthermore, as regards comparing two or more expositions and compositions represented in this way:

deciding questions on "relatively more, as much, less" by means of those visual representations,

making such decisions possible by means of manipulating the material; understanding the principles of such decisions; and

describing them.

6.14. *Visualisations by Means of Constructs*

Constructs can serve to visualise not only ratios and proportions, but also entire linear connections. One can distinguish graphic and monographic methods:

the graph of the linear function (Figure 55),
the sun shadow (Figure 56),
the lamp shadow (Figure 57).

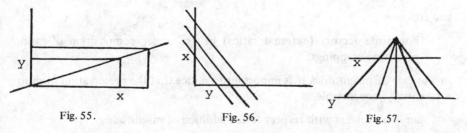

Fig. 55.   Fig. 56.   Fig. 57.

Though used too little, these visualisations are particularly effective didactically. They are models that fit quite well the ideas on the geometrisation of elementary instruction. They have a good chance of being seriously exploited.

> Internal and external ratios
> and their mutual relations

can be efficiently

> seen, understood, described

by these models.
When reading Section 6.15 one should remember this fact.

6.15. *Algorithmisations*

The counterpart of visualising is processing numerically. Verifying preservation of ratio of a mapping f is simplified by the remark that the validity of

$$w(A) : w(B) = w'(f(A)) : w'(f(B))$$

need not be verified for all all pairs $A, B \in \Omega$. Indeed the

> validity for A, B and B, C

implies the

> validity for A, C,

the transitivity of ratio preservation.

(In the case of constructs more simplification can be used which rests on geometrical facts; in the plane it suffices to check ratio preservation for the distances from two fixed points; the remainder is guaranteed by congruence theorems.)

It is less trivial to grasp that preservation of ratio can be described by the existence of a constant scale factor, that is, by an external ratio.

Another important insight is that the

> composition of ratio preserving mappings again yields ratio-preserving mappings

and to know

> how scale factors (external ratios) behave under composition of ratio-preserving mappings.

In the case of magnitudes it is important to notice that the preservation of ratio is essentially recognisable as

> an isomorphism with respect to the addition of magnitudes.

I am going to formulate a few levels:

Simplifying the verification of ratio preservation by means of
transitivity of ratio preservation,
geometrical congruence properties,
external ratio and scale factor,
isomorphism with respect to addition within magnitudes,
behaviour under composition of mappings;

simplifying the construction of ratio-preserving mappings by the same principles;

deciding conflicts in applying these principles;

understanding these principles operationally, and describing them;

understanding relations between these principles operationally and describing them.

In the course of algorithmisation this is complemented by

understanding ratios operationally in the context of the arithmetic of fractions; and

describing this relation;

understanding properties of ratio operationally as properties of fractions; and

describing this relation;

understanding ratio preservation of mappings of magnitudes operationally as linearity; and

describing it as such;

understanding their properties operationally as properties of linear mappings; and

describing them as such.

The converse, which properly belongs in the chapter on fractions, may explicitly be added:

understanding fractions operationally in the context of ratio; and describing this relation;

understanding properties of fractions operationally as properties of ratio, and

describing this relation.

understanding linear mappings in the number domain operationally as ratio preserving mappings; and

describing them as such;

understanding their properties operationally as properties of ratio preserving mappings; and

describing them as such.

Ratio-preserving mappings not only serve in visualisations, but also have their own cognitive function as models, as shown by our first example, the uniform motion as a ratio-preserving mapping of the magnitude time on the magnitude length.

The ratio-preserving mappings themselves are illustrated

> graphically (the straight line as an image of the linear function),
> nomographically,
> by means of the slide rule,

and algorithmised by

> proportionality tables (proportionality matrices),
> formulae for linear functions.

> Levels to be mentioned might be

> reading;
> constructing;
> understanding operationally the principles of the devices; and
> describing them;
>> isolated and in their mutual connection.

6.16. *Criteria for Ratio Preservation*

The principles by which one

> recognises and predicts

that a mapping preserves ratio are more profoundly rooted and less accessible. They can hardly be cleared up without a prior didactic phenomenology of particular magnitudes. The following discussion tries no more than to sketch how this can take place.

I start with an exemplary list of adjectives, whose meaning will soon become clear:

> many, big, long, wide, high, thick, much, full, long-lasting, heavy, fast;
> strong, old, sharp, blunt, soft, dense;
> bright, warm, red, loud, wet, high;
> sweet, beautiful, painful;
> clever, interesting, sleepy, difficult;
> valuable, expensive, rich.

Some of these words have several meanings (such as "bright"). The adjective "high" appears twice in this list, in the first place it may mean a property of mountains, in the second a property of sounds, but this does not matter here.

One can ask the questions:

> Which properties admit comparatives?
> Which properties admit doubling?

("Doubling" stands here as a paradigm; more general would be "multiplying", maybe also halving, dividing, finally also adding.)

How to check comparatives?
How to check doubling?
How to make comparatives?
How to make doubles?

These are questions on factualities, though with a considerable logical or linguistically analytical touch.

The central question is that of doubling. The process of doubling is that of combining two equals. This is how to transform a tower into one of twice the height, namely by putting an "equal" tower on its top. A weight of sugar is doubled by adding an equal one. Temperature shows that it is not always that easy; the temperature of a liquid is not doubled by adding a liquid of the same temperature; likewise the speed of a rolling ball is not doubled by uniting it with one of the same speed.

Parameters that, when things are combined, behave additively are called

extensive

— number, length, area, volume, weight, energy, brightness (of a light source), electrical charge, all have this property; others like temperature, colour, sweetness are called

intensive.

Yet even parameters like temperature, or rather temperature difference, can be interpreted as *extensive* parameters, though of a process rather than of a state. So what are combined are not the states but the processes. As to temperature, for instance, a difference of temperature which is obtained by means of heating with a source of heat W during a time t, is doubled if the "same" process is repeated (actually this holds only within certain limits). In the case of — vectorial — velocities this combining with the aim of doubling looks different again: if A with respect to B and B with respect to C have the same velocity, A has double the velocity with respect to C.

The principle by which the ratio preservation of mappings can be recognised and predicted can now be formulated as follows:

Two parameters which are extensive under the same operation of combining are in a ratio-preserving relation.

I do not claim that this digging has brought profound wisdom to the surface. The result is, in a wealthy wording, the criterion to which each able teacher will appeal, more or less consciously, if he wants to convince his pupils about where they may use the "rule of three" and where not. "He who works double the time gets double the money" he says for instance, and perhaps he puts twice the amount of money under two equal intervals of the time axis. Or "double the distance, in double the time" with a similar illustration.

It is clear why one cannot draw any inference from the number of wives of Henry VIII to that of Henry IV, since the rank number of kings of equal name can never be explained as an extensive parameter by any combination. The rule of three does not apply to the problem "if a man covers a distance in 3 hours and his son does so in 2 hours, how long do they need if they walk together?" because going together, for instance by people who are equally fast, does not change at all the time required. Yet also in the problem of the working men who do certain work first individually and then together, the central question is: does the required time double if two equals work together? No it halves, so the reciprocal time emerges as an extensive parameter. And so it emerges in the case of the man and his son, provided they do not walk together but to meet.

I note down the following levels:

deciding on the ratio-preserving property of mappings in factual contexts and problem situations;

recasting context and problems in such a way that ratio-preserving properties gain prominence;

deciding conflicts under these circumstances;

understanding principles of such decisions and constructions operationally; and

describing them.

Auxiliary activities may be required on the following levels:
In order to become oriented to ratio preservation

considering pairs of parameters that are extensive under the same combination; and

looking for such parameters;

grasping the importance of such parameters for ratio preservation; and

explaining it.

In these auxiliary activities the following levels can be distinguished:

deciding with respect to parameters of states and processes whether they are extensive according to a certain way of combining,

finding extensive parameters for given ways of combining,

finding ways of combining that make given parameters extensive;

finding parameters and ways of combining that fit with each other;

understanding what extensive parameters are operationally; and

describing them.

6.17. Non-linearity

A great variety of phenomena suggests that proportionality, ratio preservation, linearity are universal models; the faith in these models is reinforced by their frequent use. Approximately at least the linear relation looks appropriate in many cases as a phenomenal tool of description. We indicate cases where this primitive phenomenology fails for theoretical reasons:

the non-linear behaviour of areas and volumes under linear multiplication;

the non-linear variability of precision in measurements and stochastic data under multiplication of the sample size.

A historically remarkable example:

the bet on at least one six in 4 throws with a die was considered equivalent to that on at least one double-six in 24 throws with two dice.

Another historical example:

the idea to solve the "problème des partis"* by a linear procedure.

Faith, acquired by long practice, in linearity (or the rule of three), where fresh principles were at stake!

6.18. The Use of Ratios and Proportions

The general use is to predict a fourth term when three in

$$a : b = c : d$$

are given.

The ratios on both sides can be meant as

internal,

and related to

equal or different magnitudes,

for instance

two strips a, b and two money values c, d,

or

two paths on a map a, b and two lengths c, d.

* Mathematics as an Educational Task, p. 584.

Or they can be meant as

external,

for instance,

two strips a, c and two money values b, d,

or

two paths on a map a, c and two lengths b, d.

It can also happen that one of the ratio is

explicitly given as such,

which actually means two data only, for instance in the case of the external ratio

path to time,   weight to volume

explicitly as

velocity,   density.

Earlier in Section 4.19 we analysed a strategy of comparing approximately cardinals of what we then designated as

k-homogeneous sets.

In our present terminology the relation between the cardinal # and the character k of a set would be termed (approximately) ratio preserving. As announced there, it can be used to estimate cardinals and ratios of cardinals.

A practical use of proportions in general includes

changing the middle terms in order to profit from the relation between internal and external ratio,

processing the data, independently of the place of the known,

exploiting the definitory properties of internal and external ratio, using visualising models,

composing and splitting up proportions,

estimating parameters by means of approximately linear dependent ones.

6.19. *"Ratio" in Learning Processes*

The present section has been suggested by experiences in teacher training institutions, though they are rooted in principles which will be illustrated later on by more examples.

In Section 6.7 I stressed that in visual contexts children – even at the kindergarten age – can grasp the relative view and ratios (Figures 58 and 59).

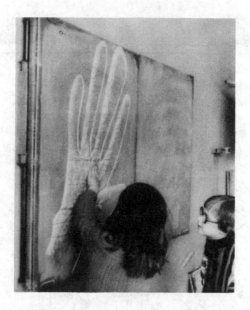

Fig. 58.

In the IOWO theme "The giant's greetings" children estimate the giant's size (and many other related sizes) by the trace of the giant's hand on the blackboard. This direct approach is possible because *no numerical data* are introduced by the text. The theme "Camping ground", however, is much less directly accessible because it introduces *explicitly numerical data*, and it addresses itself to children who have learned to identify measuring with the use of the ruler.

A similar case:

Monica (5;8) builds towers with congruent blocks. She is quite good at comparing towers of different height, even if they are placed on different bases. She has put 11 blocks on top of each other. I ask her to show me the height of a tower of 20. She shows a height 2–3 blocks higher. I let her continue building. At 13 I repeat the question; her answer is somewhat better. I ask her how many should be added. Her lips are moving. Obviously she is counting from 14 to 20 and every time is disappointed again because she does not know how many she should add. I teach her to raise the fingers while counting.

This story is told to unmask my incompetent didactical behaviour: the premature and unnecessary translation of a ratio into a numerical problem. This is quite characteristic of the traditional dominance of arithmetic over mathematics instruction.

To teachers in training whom I observed, "ratio" is either a vague relation which has not been made conscious, or an entirely algorithmised or automatised phenomenon — in the most favourable case expressed by proportion matrices.

Fig. 59.

The mental objects "relatively" and "ratio" have been blocked by numerical associations. The student teachers have great difficulty in creating models by which they can open to their pupils the entrance to the mental objects: they do not even grasp the relevance of such models. Obviously this is a consequence of their own process of learning ratio which has been directed straight on to algorithms.

By this I do not mean that these students have never gone through a period of insight with respect to "relatively" and "ratio". There is no need to suppose — and it is even not probable — that they have experienced these notions from the outset in an algorithmic way (in order to automatise them later on). It is more probable — and this is typical of many learning processes, especially in mathematics — that the original sources of insight have been clogged, and the way back to insight is blocked by the processes of algorithmising and automatising. Autonomy of algorithm and automatism is a strong inclination, which is understandable: too much insight can be a hindrance under certain circumstances. Anyway, we have to view critically the bad consequences of such blockages. What can we do against them?

I will answer this question at several opportunities. It is most often necessary but not sufficient that algorithms and automatisms are *acquired* by insight. The learning process must be steered in such a way that sources of insight are not clogged during the process of algorithmisation and automatisation. This can be achieved, in my view, by returning again and again during the process of algorithmisation and automatisation, and even afterwards where it fits, to the sources of insight. This process aims at an ever greater consciousness of what initially was subconscious, and an ever sharper verbalisation of what initially was not verbalised at all. With regard to "relatively" and "ratio", this means that the visual models are repeatedly recalled and abstracted into thought models. What is wrong in many methods is a satisfaction with the uniqueness of certain decisive steps in the learning process and with repeated exercises of the *consequences* of such steps, instead of repeating the steps themselves. A corrective measure: repeating the step if something goes wrong with the automatism. But more important is prevention: repeating the step from insight to automatism before things go wrong, in order to guarantee the ability to repeat the step.

CHAPTER 7

STRUCTURES: IN PARTICULAR,
GEOMETRICAL STRUCTURES

7.1. Chapter 7 was originally "Geometry", which later was changed to "Geo-metrical Contexts". I stepped straightly into a phenomenology, which was a bit didactically tainted. But I soon had to exchange the phenomenological thread for a methodological one. In order to make things clear I had to take so many side steps that the frame of the chapter was in danger of bursting. I started again. What follows now is simply mathematics or, as far as it might be valued as phenomenology, it is one with its object at a very high level, the phenom-enology of a quite advanced mathematics. I am afraid this will not be my last struggle with the revision of the whole idea.

7.2. Without much ado I used the word "structure" many times. I will explain it now more systematically.

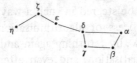

Fig. 60.

Look at Figure 60. It represents a structure, a graph consisting of seven nodes and seven connecting lines. The figure resembles the "Big Dipper", which was indeed the intention of the drawing. (The constellation, though more extended, is properly named Ursa Maior, but this does not matter.) I could have drawn the "same" graph differently (Figures 61 and 62), but then you would not have seen any constellation in it. The three figures — considered as graphs — are isomorphic or, in other words, *combinatorically* equivalent.

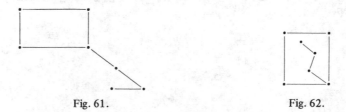

Fig. 61.   Fig. 62.

A *graph* is a set of "nodes" and "edges" with a relation "each edge joins two nodes". Visually the nodes and edges are rendered by points and — preferably straight — connections. *Isomorphism* for graphs means that the one can be

210

mapped one-to-one on the other such that nodes, edges, and joinings correspond to each other. A variant is the *directed graph*, where each edge is directed from one of its nodes to the other. Then isomorphism includes preservation of direction under the mapping.

Graphs as combinatoric frames are a frequent phenomenon:

A city plan with the corners of the streets as nodes and pieces of streets as edges.

The network of the Netherlands Railways with the stations as nodes and the direct junctions as edges.

A box of blocks or a jigsaw puzzle with the particular blocks or pieces as nodes and neighborhood as edge.

A cube with its corners or faces as nodes and its edges as edges.

7.3. Combinatoric structures are relatively poor. In general, physical or mathematical systems possess more structure. Take a trellis-work or wire-netting of squares or hexagons (Figures 63 and 64).

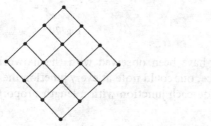

Fig. 63.   Fig. 64.

They can be structured purely combinatorically, and then they are equivalent to the graphs of Figures 65 and 66.

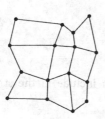
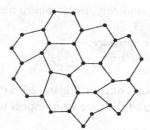

Fig. 65.   Fig. 66.

However, Figures 63 and 64 suggest more structure than Figures 65 and 66. First of all, that of a rigid body: Figures 63 and 67 are *congruent*, as are Figures 64 and 68, that is, they can be mapped on each other such that *all mutual distances are preserved*.

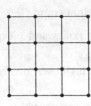

Fig. 67.

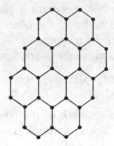

Fig. 68.

Figure 69 is *similar* to Figure 63, as is Figure 70 to 64; that is, they can be mapped on each other while all *ratios of distances are preserved*.

Fig. 69.

Fig. 70.

This kind of structure could have been observed with the network of the Netherlands Railways; for instance, one could note at every junction the distance in km or minutes or even provide each junction with a length proportional to the distance.

So the structure can include

> *distance* of pairs of nodes,
> *ratio of distance* of pairs of nodes.

Correspondingly, isomorphism means

> *congruence*,
> *similarity*.

If we consider Figure 63 as a graph we can include in the structure the relation of distance with respect to

> *all* pairs of nodes,
> *joined pairs* of nodes (edge lengths).

The second structure is weaker. In this sense the structures of Figures 63 and 71 are isomoprhic. They are equivalent under transformations we designated as *flexions*.

Fig. 71.

Vertical fences like Figures 63 and 64 are mapped by the sun into shadow images like Figures 72 and 73. The shadow mapping conserves *rectilinearity* and *parallelism*. Such a mapping is called an *affine mapping*.

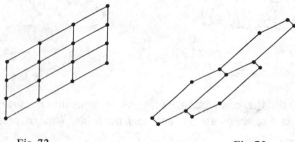

Fig. 72.   Fig. 73.

In general, structures are weakened if, rather than distance or ratio of distance, they include only

rectilinearity and parallelism.

Isomorphism then means

affinity.

If the fences of Figures 60 and 61 are projected as slides on a screen, we can get figures like Figures 69 and 70, at least if the slide and the screen are parallel. If they are inclined to each other, we get figures like Figures 74 and 75. This kind of mapping preserves *rectilinearity*.

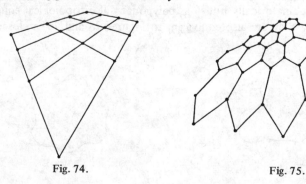

Fig. 74.   Fig. 75.

Structures like the preceding ones can be weakened so as to include only

  rectilinearity

as a structuring relation. Isomorphism of structure then means

  projectivity.

If the projection screen is not flat but bumpy, then the image of Figures 63 and 64 can look like Figures 76 and 77. What is preserved in this kind of

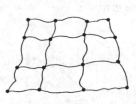

Fig. 76.

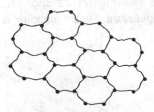

Fig. 77.

mapping? Only the connection of figures: a continuous curve passes into a continuous curve; there are no cuts and no folds. What is preserved in such mappings is

  neighborhood – the topological character.

Isomorphism here means

  topological equivalence.

Topological and combinatoric equivalence look in some aspects similar. A graph where two nodes are joined by at most one edge can also be defined as a set of nodes with a relation of

  being a neighbor.

It is a coarser kind of neighborhood than in the topological case. Figures 60–62 are not only combinatorically but – as polygons – also topological equivalent. But topologically they are also equivalent to Figures 78 and 79, a closed curve with a tail.

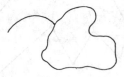

Fig. 78.

Fig. 79.

More examples of topological structures and their isomorphisms:

The surfaces of a sphere and a potato are topologically the same, but they differ from the surface of a ring, which as far as its topology is concerned can also be represented by a rectangle (Figure 80) where the opposite sides are imagined to be stuck together according to the arrow.

Fig. 80.

A cylinder is topologically equivalent to an annular domain bounded by two circles (Figures 81 and 82), where boundaries correspond to boundaries; the annular domain without its boundaries is topologically equivalent to the cylinder without its boundaries, as well as to a cylinder of infinite length, and finally also to a plane in which one point has been pricked out.

Fig. 81.   Fig. 82.

7.4. The examples of structures I gave were geometric or illustrated geometrically. I stressed this kind for didactical reasons but it would be a shortcoming if I were to leave it at that. As a matter of fact I already dealt with various structures in the *number system*: the *order* structure, the *additive* structure, the *multiplicative* structure. I can also put these kinds of structures on other sets, for instance, a multiplicative structure on a set of four elements e, a, b, c by the multiplication table

e	a	b	c
a	e	c	b
b	c	e	a
c	b	a	e

which is to be read in the usual way. This then is the so-called "group of four".

It is, however, unusual to define structures as explicitly as has been done here (and in Section 7.3). Most often it is done implicitly; that is, one introduces

a set

with

certain relations on it,

and requires these relations to observe

certain postulates.

For instance, a group is by definition a

set G

with

relations $ab = c$

and

as postulates,

associativity : $(ab)c = a(bc)$
an identity element e: $ea = ae = a$
for each element an inverse a^{-1}: $aa^{-1} = a^{-1}a = e$.

This does not define just one group but rather the group concept, which can be examplified by many (finite or infinite) models; and for each pair of groups one can ask whether they are isomorphic, that is, show the "same" structure.

This implicit approach is more fruitful than its explicit counterpart. In order to add one more geometric example, I take

metric space,

a

set R of "points"

with a

distance relation for pairs of points,

that is a function ρ, such that

$\rho(a, b)$ is a real number ≥ 0,

subject to the requirements that

$\rho(a, b) = 0 \longleftrightarrow a = b$,
$\rho(a, b) = \rho(b, a)$,
$\rho(a, b) + \rho(b, c) \geq \rho(a, c)$.

Metric spaces can again be compared with each other; isomorphic ones are also called

isometric.

A weaker structure is

topological space,

a

set R of "points"

with a

relation of being close to each other,

subject to requirements which I do not specify. The most usual topological spaces are in fact better handled if approached from metric spaces by weakening their structure. Then being close to each other can be defined technically via the metric:

V is called a neighborhood of p in R if there is an $\epsilon > 0$ such that all points at a distance $< \epsilon$ from p are lying in V.

This transforms the *quantitative distance* into a *qualitative closeness.*

7.5. Mappings are important not only between *different* structures. The combinatorial structure of Figure 60 admits a mapping onto itself which interchanges α and γ while all other nodes remain in their places. This is an

automorphism,

of course, in the combinatorial sense; that is, if I am satisfied with the combinatoric structure of the graph. If I consider Figure 60 as a picture of the Big Dipper with the correct distance ratio on the firmament, I am not allowed to interchange α and γ.

The one-to-one mappings of a system onto itself that preserve structure, are called *automorphisms.* They form a group.

The n-gon, combinatorially viewed, admits $2n$ automorphisms; that is, if I number the vertices subsequently by

$$0, 1, \ldots, n-1 \mod n,$$

I get the n automorphisms

$$x \rightarrow i + x,$$

and the n automorphisms

$$x \rightarrow i - x.$$

Together they form the so-called dihedral group.

An n-gon with more structure will in general admit no automorphism except the identity. Suppose the sides posess lengths which are considered part of the structure. Only if all these lengths are equal are the combinatorial automorphisms − flexions − also automorphisms in the sense of the richer structure. If the lengths of the sides and also the angles between the sides are both understood to belong to the structure, the above automorphisms exists as such only if all angles are equal.

The graph consisting of the vertices and edges of a *tetrahedron* admits all 4! permutations of the vertices as automorphisms; if the lengths of the edges are comprised in the structure, the group of automorphisms may shrink. The graph of vertices and edges of a *cube* has a group of automorphisms that is twice as large, under similar conditions as for the tetrahedron.

The *euclidean plane* and *space* are particular rich structures with the relations of

>   collinearity of three points,
>   coplanarity of four points,
>   order on the line, in the plane, in space,
>   congruence of line segments,
>   congruence of angles, and so on.

Distance does not belong *a priori* to the structure of euclidean plane and space, though it is a fact that assigning a length, say 1, to one single line segment bestows unequivocally lengths on all line segments. That is what is called gauging. Thus

>   gauging transforms the euclidean space (plane) into a metric — euclidean — space.

A posteriori it appears that

>   the euclidean structure is determined by this metric structure.

For instance, the order relation

>   *q* between *p* and *r*

can be brought back to the metric relation

$$\rho(p, q) + \rho(q, r) = \rho(p, r),$$

and collinearity of three points can in turn be brought back to betweenness.

The mappings of the (metric) euclidean plane or space on itself that preserve distance — the isometries or congruences — also preserve collinearity, coplanarity, order; they map lines on lines, planes on planes. So they are automorphisms of these structures.

The automorphisms of the metric euclidean plane are the

>   translations, rotations, slide reflections;

those of metric euclidean space are the

>   translations, rotations, screwings, slide reflections.

If, however, the metric is dropped and only the euclidean structure is left, the group of automorphisms increases; it becomes that of the *similarities*.

A weaker structure of plane or space than the euclidean is the *affine* structure, whole defining relations are

collinearity and parallelism.

The automorphisms of this structure are the affinities — a larger group than that of the similarities.

If one starts, as we did, from euclidean space, then to get the affine structure, one need not require preservation of parallelism separately. Indeed, mappings of euclidean space that preserve collinearity map straight lines onto straight lines, planes onto planes; parallel lines are by definition lines in one plane that do not meet — a property that is consequently preserved under such mappings.

In restricted parts of the plane or space, however, preservation of collinearity does not imply that of parallelism; there non-affine mappings exist that preserve collinearity, though they cannot be extended to the total plane or space unless the plane or space itself is extended, that is, by adding so-called points at infinity (lying on a line or plane at infinity). This produces the projective plane or space, a structure with the relation of

rectilinearity,

and as isomorphisms the

projectivities.

One more step is maintaining in one of these structures — euclidean plane or space, projective plane or space — only a

relation of closeness;

then a

topological structure

arises: euclidean plane or space, projective plane or space considered as

topological spaces,

with as automorphisms their

topological mappings onto themselves,

which form very large groups.

7.6. The group of automorphisms of any structure includes the identical mapping of the structure onto itself. It can happen that this exhausts the group. The examples of Section 7.5 showed large groups of automorphisms. For studying a structure its group of automorphisms may mean a great deal. Congruence theorems is an example: the fact that triangles are congruent can be the source of many aspects and properties they have in common. The midpoint of a line segment is an affine, yet not a projective, concept, which implies that with only the ruler one cannot halve a line segment, though it might become possible as soon as one has at one's disposal an instrument that produces parallelism.

The importance of group theory for geometry was chosen by Felix Klein as the theme of his *Erlanger* program. Felix Klein grasped and stressed that the geometries dominating his work together with their mutual relations were to a high degree determined by their groups of automorphisms and the mutual relations of these groups. Starting, for instance, from projective space and its group, one could pass to affine space and to the affine group as a subgroup of the projective group by fixing one plane — to be considered as the infinite plane. Further, by fixing in this plane a non-degenerate imaginary conic, one could pass from affine space to the heavier structured euclidean space, or by fixing a real or imaginary quadric, to non-euclidean spaces, and to their automorphism groups as subgroups of the projective group.

So the group of automorphisms of a geometry came to the fore — a slogan: geometry is the invariance theory of a group. It is a fruitful idea, which has been fully elaborated with new perspectives in E. Cartan's "homogeneous spaces". It is a sound principle, which applies to structures with such a degree of homogeneity that they can be defined by the expression of this homogeneity, that is by their groups of automorphisms.

But notwithstanding its importance, this is only one aspect of the study of structures, in particular, geometric ones. In our exposition on structures this aspect has been brought to the reader's notice as late as Section 7.5. In Sections 7.2–4 we were busy with various kinds of structures; we asked whether systems of the same kind of structure were *isomorphic*. Such knowledge too can be of great importance: discovering that two structures obtained in different ways are isomorphic, or managing by means of restructuring to make them isomorphic, may create opportunities to transplant concepts, properties, extra structures from the one to the other. In fact this use of

> *isomorphisms between structures*

is much more comprehensive and in general more fruitful than that of

> *automorphisms within a structure*,

which is tied to quite special, and particularly homogeneous, structures.

The automorphisms of a structure form a group. Finding out whether, and stating or ensuring that, certain structures are isomorphic has nothing to do with group theory. Nevertheless, these two things have often been confused. It is quite a serious confusion; it concerns more than the terminology of what to call a group and what not.

The germ of this confusion is already visible in Klein's *Erlanger* program. (The name under which that inaugural lecture has become known and is cited is also rather responsible for the confusion: it has been interpreted to mean *the* program for geometry, whereas it was an annex to the program of Klein's courses at Erlangen University.) Yet thoughtless copying is a danger that even mathematics has to face, unless it is hard mathematics, supported by formulae and logical reasoning. It is a sad story, but so far this misconception has not been extirpated. Even in mathematics research it shows up as soon as the *Erlanger* program is around. So we should not be surprised to meet it in the philosophical, psychological, and mathematical-didactical literature. A large part of Piaget's work, not only that regarding geometry, is dominated by this misconception, at least as far as it is concerned with mathematically tainted theory.

7.7. I will explain this more precisely.

The marvellous spaces with their beautiful automorphisms that we dealt with in 7.5 are not aims in themselves, and – to add immediately a psychological and didactical argument – neither are they starting situations. The spaces are to lodge figures, which can be both starting point and aim in itself, but if anything within these figures is worth being explored and is a challenge to exploration it has little to do with the automorphisms of the surrounding space.

If two triangles (or tetrahedra) have correspondingly equally long sides (or edges) they are congruent, which implies that this correspondence can be extended to a length-preserving mapping of the space. For quadrilaterals in the euclidean plane or space the analogue does not hold: the isomorphism of flexion isomorphic quadrilaterals cannot in general be extended to the plane or space in such a way that the flexion character of the data is reasonably accounted for. Flexions of curves and surfaces within the euclidean space are not susceptible to reasonable extensions into space; hence if more objects are concerned, such flexions cannot be brought together in a group, and in order to relate them to the group of automorphisms of euclidean space in a more profound way, highbrow devices such as the theory of sheaves are required.

The Figures 78 and 79 are topologically equivalent, but there is no topological mapping of the plane onto itself that realises this equivalence – indeed such a mapping cannot exchange the interior and exterior of the closed curves where the two tails are situated. It could be realised by moving into space, but there one can again find topologically equivalent figures that are not equivalent by means of a topological mapping of the whole space onto itself. Moreover, the topological structure of figures in space is much more important than the unfathomable group of topological mappings of space on itself.

In particular, prejudice arises against the combinatoric structures if automorphisms are too much stressed. The combinatoric structure of Figures 60 and 61 is not accounted for by any classical geometry of the plane: there is no reasonable mapping of the plane that continues the isomorphism of Figures 60 and 61, and in general there is no way to embed the isomorphisms of isomorphic graphs into a group.

A great deal of geometry is and occurs *within* space. It is true you can attach to a planet a copy of space in order to interpret the planet's motion as a motion of one euclidean space with respect to the other, that is, as a sequence of mappings of space upon itself, a sequence of automorphisms. Moreover this can be extremely useful if you want to study the centrifugal forces exercised by, say, the rotating earth on a free falling object or on air and water streams. It can perhaps even be useful to attach a whole moving space to a driving car in order to understand what happens if it goes into a curve. But a car also has wheels, and there is no need to attach separate spaces to the wheels in order to drag in automorphisms. Moreover, a car is flexible, its doors can be opened and closed, and this mapping of the open on the closed car does not extend in any reasonable way to space in order to be put into a group. Bodies of animals in movement are flexible systems according to certain combinatoric structures; the mappings expressed by their movements can be understood as isomorphisms, but no part is then played by automorphisms, either in the sense of some classical geometry or in any other more extended sense.

A drastic example of that confusion is found in almost all New Math textbooks: opening and closing a door as inverse elements of a group. What group, and on what set does it operate? On the two states of the door? But what does opening perform if applied to the open door and closing to the closed door?

This is the point where the prominence of the group of automorphisms, as an aspect of the *Erlanger* program, fails: in all that happens *within* the space, and which has not extension, or no relevant extension, to space as a whole.

CHAPTER 8

PUTTING INTO GEOMETRICAL CONTEXTS

8.1. *"La représentation de l'espace chez l'enfant"*

The title of this section is the same as that of a book by J. Piaget–Bärbel Inhelder* which, in spite of serious objections about its global approach and about many of its details, I rate highly. Even more than Piaget's other work that touches mathematics it should have deserved from mathematicians serious criticism rather than mere shrugs of the shoulders.

Initially I did not mean to discuss Piaget too much in this chapter, but while I was writing it, the need became more and more urgent. Nevertheless I will restrict my criticism of Piaget as much as is feasible because in spite of the variety of its aspects his work does not contain enough elements to take one's bearings on it.

I put the title of Piaget–Inhelder's book above the present section because my first considerations will center around two words contained in the title, "espace" and "représentation".

8.2. *Space*

"Space" is an expression that from the title to the last page occurs a thousand times in Piaget–Inhelder's book, often as an adjective. I did not put it into the title of the present chapter, and there were reasons why I did not. Space, whether as a mental object or as a concept, is rather the endpoint of a development, though not in the sense of a bearing — this would again be a wrong perspective. Not until a highly advanced mathematical context is reached, does "space" get a meaning. On long trajectories the word "space" can be dispensed with as a mere term, and even as a mental object it is not required. In no way does the constitution of the usual mental objects in geometry depend on that of the mental object "space", whatever this may be!

Greek geometry and philosophy do not possess an equivalent of our "space". The universe is finite and the fact that, according to one of Euclid's postulates, every straight line can indefinitely be extended does not imply anything about the medium in which this should be possible. Mentally such a medium may somehow exist, but no attention is paid to it, up to Cusanus and Newton, say. The etymological root of space is *spatium*, which means distance. Space and its

---

* Paris, 1948. English Translation: *The Child's Conception of Space*, Routledge & Kegan Paul, London, 1960. — For a few quotations this English translation has been used, though at some places it does not match the French original; at one place to be quoted, however, the translation is an improvement on the — incomprehensible — French text.

analogues in other languages originally mean a closed thing, and to indicate the very big space modern terminology was enriched by "outer space".

Before going further, I shall give examples of the often more technical use of the word "space", with the intention of showing the direction I wish to avoid.

The euclidean space — so called because the geometry studied in and according to Euclid's Elements implicitly presupposes at least something like this — includes points, connected by lines, contained in planes, forming circles as large as you wish, all over the world right angles are equal, and what happens far away with the geometrical objects can be predicted here, because in all triangles, however big they might be, the sum of the angles is two right angles. From the closest neighborhood one extrapolates to ever bigger distances, and this then is the mental object, called Euclidean space, for which Euclid himself had no name.

"Geometry" originally meant measuring the earth, as performed by surveyors, but this practical use was never stressed in Greek geometry; it was rather held in contempt. Eratosthenes managed to measure the whole Earth from a restricted piece of land, and Aristarchos did the same with the distances and sizes of the Sun and Moon. The proper domain of astronomers, however, was measuring angles. How this was done, is not told in the Elements though angles of a certain measure — right angles — occur in its theorems. Line segments of a definite length are of course not met with in the Elements, and only with an illiterate slave does Socrates speak about squares of so many feet.* What counts in euclidean space is the equality and the ratio of line segments.

The euclidean space with all its objects is a rich structure, although it is poor if compared with all I perceive around, its colours, polished and rough surfaces, sounds, smells, movements. But thanks to the impoverishment it furnishes a certain context, which for some reasons suits us extremely well — this is a point I still have to consider more closely. Anyway this context has been accepted as geometry for centuries, this mental object of euclidean space as if it were an objective datum, though efforts have been made to describe it more precisely and more efficiently than Euclid ever did. More *precisely* — this means axiomatics like Pasch—Hilbert's; more *efficiently* — this means the algebraic approach from Descartes to the modern version of metric linear space of three dimensions. Elsewhere** I have sketched this development.

This euclidean space has never been an aim in itself, but rather it has been the mental and mathematically conceptual substratum for what is done in it: for constructions with a pair of compasses and a ruler, or with only a pair of compasses, or with a ruler only, for constructions by means of algebraic equations or purely mechanical constructions, for deducing properties of such figures, for proving or refuting hypotheses about them.

In the more recent development of geometry it was an important discovery

---

* So does Theaitetos when he quotes Old Theodoros.
** *Mathematics as an Educational Task*, Chapter XVI.

that with the ruler alone one can go a long way, for instance in the theory of perspective and for all properties relevant in perspective. A methodological principle of mathematics, ever more systematically applied since antiquity, is purity of method. In the example I just cited it meant that as soon as one studies properties which depend on the ruler only, one must choose a substratum which is restricted to points, lines and planes in their mutual relation. This then is an even poorer structure, the projective one, but the relative poverty can be an advantage. Wealth, if dispensable, can be a hindrance. In the present case, in one respect, the poverty was too pinching; to embellish the structure, space had to be enriched by points, lines, and a plane at infinity, and this then was projective space. A variant on that principle: if parallelism is included in the fundamental concepts besides rectilinearity, one gets a structure — affine space — which is poorer than euclidean and richer than projective space.

Another evolution was that towards non-euclidean space. One started doubting whether the neighborhood, as one thought to perceive it, determined the remote depth of space as it had been assumed. If the sum of the angles in a triangle would systematically differ from the supposed value of two right angles, space would look different — curved, whatever this might mean.

A third evolution away from euclidean and the other spaces was into more dimensions, even to an infinite number of dimensions. Geometric language became a suggestive and creative device to organise quite different domains — analysis in this case.

And then the fences came down: structures are created according to one's needs, and if they are related to structures that had formerly been called space, or if they involve visual elements to be uncovered or stressed, they are called spaces: metric spaces, topological spaces, discrete spaces, and so on. There are good reasons why mathematicians did this: insight can be deepened and terminology can be simplified if various structures are brought together under one heading.

8.3. "Représentation" – The Mental Object

In the English version of Piaget—Inhelder représentation has been translated by "conception" which shifts the stress even more to concept attainment. I am not sure whether this is correct.* If one looks only at the titles, theoretical introductions and conclusions of the chapters, and sections of Piaget—Inhelder's work, one can indeed get convinced that the authors have tried to investigate the child's conceptual approach to space or rather to find out which features of

---

* Another work – J. Piaget, Bärbel Inhelder and Alina Szeminska, *La géométrie spontanée de l'enfant*, Paris, 1948 – has become in the English version *The Child's Conception of Geometry*. One may rightly doubt whether there is any spontaneous geometry in this work but anyway "conception of geometry" is nonsense. In the footnote to Section 8.1 differences between the versions were signalled as a general fact; actually they are rather frequent.

some adult conceptual approach to space can be traced in the children's mind. This underlying adult conceptual approach is then the one the authors knew from the literature, in particular the one about the *Erlanger* program.

However, I am not sure whether for Piaget—Inhelder's work this conceptual approach meant much more than an organisational pattern and some theoretical frill. The requirements are rather designed to observe the child's representations, *Vorstellungen* in Kant's sense, intuitions as some others say, or "mental objects", as I prefer to call them. They repeatedly distinguish the perceptive, the representative and — far away — the intelligible space, but the aspect they focus on is representation — sometimes mentally and most often mentally-graphically recorded.

8.4. *The Mental Object in Geometry*

In no part of mathematics do mental objects serve so long before, or even without, concept formation as in geometry. Images and imageries are more efficient if they represent figures and spatial constellations than if they represent numbers. *Small* numerical quantities can be supported efficiently by images, actual and imaginary ones, but this support does not reach far in the quantitative world and so is soon renounced. The numbers 3 and 5 are unsatisfactory paradigms of arbitrary natural numbers and their sum and product fail as paradigms of operations on arbitrary pairs of numbers. On the other hand each triangle that is drawn in a not too specific way is a good paradigm of *the* triangle, each pair of line-segments is a paradigm of *the* pair of line-segments if the aim is to show what the sum or the product is of two lengths. One can *show* other people what a parallelogram is, a rhombus, a square, what are diagonals and what it means to say that they halve each other, that they are perpendicular to each other, that they are equal. Without bothering oneself or the other person with concepts, one can introduce words to indicate them and restrict oneself to examples to explain what the words mean. One can explore widely the geometrical domain without forming concepts, so widely that finally over-ripe concepts drop in one's lap. One can even disregard the formalism characterising traditional geometry and for a long time be satisfied with demonstrative linguistic means and wait for relative and more symbolic linguistic tools to announce themselves. For these and many other reasons geometry is the field where one can fruitfully look for symptoms of learning processes, were it not that each investigator carries his own geometrical education as blinkers for himself.

By "geometrical education" I do not mean something that starts with the first traditional geometry lesson. Many geometrical objects and concepts have been formed early, most of them at the primary school age and some of them even earlier, though they do not yet bear verbal labels, or at least not those labels that we have learned to attach to them in our geometry lessons.

If one compares this with the content of Section 8.2, it becomes clear that my aim is not *the* geometry, nor a system of geometries; before I can arrive at space or spaces as mental objects, I must deal with mental objects which are understood as geometrical objects, are lodged within the space. As geometrical objects they will in a later stage be placed in a space, but as mental objects they are first of all in a *context*, namely a geometrical context. I have indicated earlier* the didactical significance of contexts, grasping a context as a necessary condition for more than mere algorithmic action.

8.5. *The Context of the Rigid, Congruently and Similarly Reproducible Bodies as an Example*

I have no doubt that geometric education starts very early, and that this early start has much to do with the fact that the geometric context suits and pleases us so well. Colour seems to be a more subtle case than geometric shape.

First of all, *rectilinearity*, in the natural environment of man exemplified by the straight posture, the stretched limbs — arms, legs, fingers — the stalks of plants and trunks of trees, and the straight way, which is the shortest, the most direct. Among the first tools, made by man, is the arrow, paragon of rectilinearity, and as civilisation progresses, so more frequently and forcefully man is confronted with objects and processes and elicited to actions that suggest or represent rectilinearity: sticks, pins, rims, edges, paths, folds, cuts, stretched strings.

Flatness is perhaps even more frequently and forcefully suggested, by paving stones, floors, walls, ceilings, tables, benches, roofs, sheds, boxes, lids.

He is confronted with *parallelism* as often as he is with rectilinearity, again of borders of objects, roads, gates, in planar divisions, in wire-nettings, palisades, rows of houses; *right angles* are suggested by perpendicularity but also by the angles of more or less carefully made objects. Even in the natural environment man has become acquainted with *mirror symmetry, polygonal symmetry*, and *axial symmetry*; so did stone age man try to imitate them and by this means educated others to see and appreciate them.

Objects that suggest *circles* are rare in the natural environment but they exist: Cross-sections of trees, sun and full moon, the horizon. After the wheel was invented, man was — already in the cradle — showered with round objects. Balls and rolling playthings suggest *spheres* and *cylinders*; to tell a child what is a *cone*, you say "a clown's hat". However regularly or irregularly something is formed, it is influenced by geometry and suggests geometrical shape. Natural production, craftmanship, manufacture, and industry have taught us *congruence* and *similarity*, in particular the similarity of playthings that imitate the world of adults.

I will leave it at that. I shall return to these examples to discuss details. Here they have served to make it clear how the geometrical context comes about —

---

* *Weeding and Sowing*, p. 242 sq.

I mean the context of euclidean geometry, or phenomenologically described, that of the

geometry of rigid, congruently and similarly reproducible bodies.

They are acquired early, all these mental objects, so early that it is difficult to distinguish that which perhaps is innate. Mental objects like rectilinearity, flatness, parallelism, rectangularity, rectangle, square, sphere, cube, symmetry, congruence, similarity are, as far as they are mental objects determined in a simpler, clearer, and sharper way than non-geometrical ones like plant, tree, animal, colour. If this is duly realised, we need not be astonished about the certainty of our recognition of geometric objects.

Notwithstanding the early start of our informal geometric education, the fact that the formal one starts so late is to be understood in the frame of the history of education. Intellectual education, which included geometry, was generally conceived of as concept formation. The mental objects, as an ideal material to work with, were most often neglected and so they are even today, in educational theory and psychology. Or rather, one disregards the distance between mental object and concept, identifies them, confuses them and by this means does not do justice to either.

There is a world of difference between our examples of geometrical context and the laboratory context, in which Piaget's subjects are placed, as well as with the theoretical frame of the *Erlanger* program. In the world as we view it the first and foremost things are bodies in space. Their congruence or similarity is of course related to certain mappings, which may be made explicit according to the needs that are felt. To be sure, these mappings can be extended to the whole space; they are restrictions of those global mappings which together form the group of similarities of space, but this is an idea, far away, at a remote distance where geometry has already been sufficiently mathematised. (I admit that even in Piaget—Inhelder's work it is not operational — group theory has been dragged in as a mathematical frame and as an organising element.)

One may object that in any case the group of similarities does not clash with that which we have in mind as the context of rigid bodies. This is correct. To say it in a terminology I have used in another connection,* the *Erlanger* program idea of the group of similarities is the apotheosis of this context. In the next section, however, we introduce other geometrical contexts which do not fit into the frame of *Erlanger* program.

8.6. *The World of the Boxes*

With much eloquence I have exerted myself to convince the reader of how the mental objects of euclidean geometry are forced upon us, from bow and arrow to baby spoon to television aerial. This demonstration, however, was a bit

* *Mathematics as an Educational Task*, p. 171.

simplistic. One might be led to believe that there were nothing else but this. If it has been understood this way, I have missed my aim. Yet I put into the title of the section the warning "as an example". There are other contexts, however, and the following is one of them.

I chose the term "boxes" for the geometric objects I am going to consider (right parallelepipeds), because this article is available in the richest variety — planks remind one of something with one aspect much larger than the other two, and bricks are represented by too small a number of models. But if I speak of boxes, one should think as well of chests, cupboards, rooms (of appropriate shape), books (with flat backs), and as many other things as one would like to invent. You may consider the box as a structure with 8 nodes and 12 edges, but then of well determined lengths, you may if you like add the side and space diagonals, you may number the vertices, orient the edges and concoct even more variations. They can be quite diverging structures but the conclusions I wish to reach will remain the same.

I said that mental objects like congruence and similarity are suggested to us by the world we live in. It is true, but sometimes other kinds of equivalence are even more strongly suggested as mental objects. In the world of the boxes we are told that a box is a box. Yes, a box is a box, but not in the way that a cube is a cube, or a sphere a sphere. Cubes are similar to each other, as are spheres. Boxes are not. What then do we say about boxes?

In order to map one box onto another in a gentle way, we have to do something with its edges, shrinking or expanding, but of course the same with parallel edges. What kind of mapping is it? Take an "origin" and a rectangular system of axes, put a box with one corner into the origin and with its edges along the "positive" axes. The mapping of one box on another is expressed in coordinates on this system of axes by

$$f: \ulcorner x_1, x_2, x_3 \urcorner \to \ulcorner \alpha_1 x_1, \alpha_2 x_2, \alpha_3 x_3 \urcorner,$$

a multiplication along the axial directions with factors α_1, α_2, α_3, respectively. It is a mapping extending to the whole space — line-segments parallel with the first, second, third axis are multiplied by α_1, α_2, α_3, respectively. Let me illustrate it by drawings in two dimensions, that is, a rectangle D instead of a box, two axes, and $\alpha_1 = 2, \alpha_2 = 3$ (Figure 83).

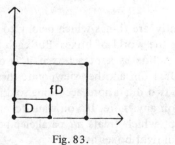

Fig. 83.

Such a mapping is called a *dilatation according to three orthogonal axes* —
let us briefly say: a *dilatation*. It is an affine mapping although not a very special
one, since *each affine mapping* can be split into a *rotation* and a *dilatation*.

Is this all we have to say about boxes and their mutual relations? No, I
can do more with boxes than change their edges. I can also displace them as
rigid bodies. By a translation such a box *D* gets a position *D'* which does not
any more lean against the axes but with edges still parallel to the edges of
D (Figure 84). *D'* too is a box and if I subject it to our dilatation *f*, I again get
a box; the two boxes *fD* and *fD'* differ only with respect to place; corresponding
edges of *D* and *D'* were multiplied by the same factor, so they are equal.

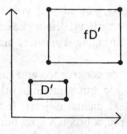

Fig. 84.

Rather than to translations, I can subject the box to rotations. By rotating
D I get it into a new position *D''* (Figure 85), with edges which are no large
parallel to the aforementioned axes, a fact which has momentous consequences.
If now *D''* is subjected to the dilatation *f*, the result is no larger what we have
called a box — rectangularity is lost.

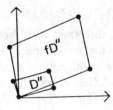

Fig. 85.

Dilatations apparently are things which belong to boxes. They are the mappings that characterise the world of boxes. But each box has as it were its own
stock of dilatations. A box is kept a box if it is diatated according to axes
parallel to its edges. Or, from another viewpoint: the dilatations do not form a
group, the product of two dilatations according to different triples of directions
need not be a dilatation any more. If you want to make a group of it, you get
the total affine group, which treats *all* parallelepipeds alike, and which does
not appreciate beautiful right boxes as such.

Well, there we are: the euclidean geometry with its group of similarities does not allow us to assert that "a box is a box", and the affine geometry does not know any boxes at all. What are we to do about it?

The mistake — if one may call it a mistake — is in extending to the whole space the mapping that transforms one box into another. I stressed this as early as Section 7.7. Here it is confirmed again. It is often meaningless or even obnoxious to extend mappings of figures in space onto each other to space as a whole — think of the parked and driving car, the closed and open car! Why then is it done in other cases and with much success? It is done in order to force the whole thing into a group, to interpret the various mappings of figures, say spheres (or cubes), onto each other as mappings of a larger structure, euclidean space, onto itself — extended mappings, which together form the group of automorphisms of euclidean space. Yet if one tries the same with boxes rather than spheres, one does not succeed as pleasantly: one gets embroiled with a group that does not respect "boxness".

On the other hand the suggestion that "a box is a box" is vigorous, as vigorous as the suggestion of supporting the assertion by a good mapping which makes from one box another one. These are strong visual stimuli in a stage where there is not the slightest need for extending such a mapping to space as a whole.

Mappings have still to be viewed phenomenologically more broadly, but this much can be said, that if mappings present themselves in any geometric context whatsoever, they are first of all mappings of restricted parts of space, which can be indicated or filled by bodies. Extending such a mapping to the whole of space, even though it be possible and meaningful, is not something that goes without saying — how long have geometers tinkered with congruence theorems where mappings would have made things easier? In history the step towards mappings of the whole space has been consciously taken as late as the 19th century — this is witnessed by the projective space, which has been created in order to be able to extend projective mappings to the whole space, and by the device of Moebius geometry where space is augmented by one point at infinity to account for inversions and make planes and spheres the same sort of things. Historically it was an extremely important step to view mappings more than locally, to embed them into a group, with momentous consequences, though also with that of a dogmatically interpreted *Erlanger* program — a bodice for the adult mathematician, and an oversize suit, bought for growth, for the young learner.

I could have put this section also in Chapter 7 — in or after Section 7.7 — as a counter example to the predominance of automorphism groups of structures. After some hesitation I put it here, in order to stress its positive meaning, as an example of a geometrical context.

I could give more examples, but I will wait; first I should say more about geometrical context in general. At this moment only a few variants of the world of boxes:

the world of rollers,

(or cylinders, if you prefer), a continuous sequence, stretching from flat discs to thin reeds,

the world of dunce's caps,

that is cones,

the world of pointed roofs,

square pyramids — and so on. Similar stories as those I told about the world of boxes can be told about these worlds.

8.7. *Intermezzo—Piaget*

Piaget's way, at least regarding the structure of his work mentioned in Section 8.1, has been dictated by what he had experienced as the gist of the *Erlanger* program, or to use our terminology of Section 8.6: he has bought a suit that looks like a bodice but in fact is dangling around his laboratory experiments, and as far as it is operational it functions as blinkers.

According to Piaget the development proceeds from the poor to the rich structures, from large groups of automorphisms to small ones, according to the sequence

topological, projective, affine, similar, congruent.

There is not any reason why this should be so, from poorer to richer. If there were a definite line, one would bet that it is the other way round: the richer structure presents itself with the greatest aplomb; impoverishing means abstracting, taking away. People and things are first singular objects, with proper names, and they finish with sorts and labels — from richer to poorer. Yet it is not that simple. Initially all that moves, is a car; then VW, Duck, Peugeot become more important concerns, but finally "car" wins. Initially each old man is grandpa; then the term is restricted to two specimens at most, with perhaps the grandpa of a friend added; finally almost every child appears to have a grandpa. Initially, Utrecht, say, is the close neighborhood, and the other side of the Amsterdam— Rhine canal is The Netherlands; then Utrecht becomes a label for a vast array of streets, squares, parks, which look like those of the neighborhood; later on it becomes the proper name of a geographical unit enclosing all such spots, and finally some political administrative unit.

It is neither a trend of ever more abstract, nor is it the contrary. It is rather a case of developing new contexts, which may overlap as they may be incomparable. In Section 8.5 I pleaded for the context of the rigid bodies with arguments which I could have extended tenfold or a hundredfold, arguments for its developmental primordiality, but if Piaget were right, this context would be the highest rung of the developmental ladder.

As a matter of fact levels have to be distinguished for forming contexts, for placing objects and operations into contexts — Piaget speaks of the perceptive, the representative, and finally the conceptual space. He admits that perceptively a child may be on a higher rung of the ladder than representatively. "Representative" is what I indicated as "mental object", though Piaget uses it also in the sense "being able to draw the object". In Part One "Topological space", for instance, "espace perspectif" is followed by "espace graphique" as a second chapter — the quality, or rather the lack of quality, of drawing becomes the criterion of the topological character. But even later mostly — almost always — the drawn record is the only criterion of the mental representation: in order to prove their existence, mental objects are to be drawn. I think this is wrong, but I will delay discussing this point.

Piaget's claimed priority of topological space is also a point to be discussed. In one respect, I anticipate the discussion: what is presented as topology by Piaget, is next to nothing, and partly it may even be asked whether it *is* topology. I tackle this here because it is essential for the question of how far the development of topological space must have proceeded before that of projective space can start — a question which must be asked again for other pairs of rungs of the *Erlanger* program ladder. If such questions are not answered, the claim that *this* space precedes *that* does not mean much, if anything.

Is it topology if the child does not distinguish a ball and a potato, or if the mental object "topological sphere" (closed surface of genus 0) has been formed? It seems that Piaget is satisfied with the first. With regard to the straight line (he says "the projective line" though it is the ordinary one), Piaget's requirements are much higher. Here he requires, indeed, the formation of the mental object "straight line", if projective space is to become manifest, since for Piaget "projective space" means the ability to look and draw according to perspective, from one's own or an imagined viewpoint. This projective space is the precondition of affine space, which again involves so little that one wonders why so much progress in perspective should be required as a precondition. Moreover — this too must be tackled later on — the ability of looking and drawing according to perspective is a totally different thing from the constitution of mental objects like straight line, plane, parallelism, congruence. Without laboratory experiments everybody knows that perspective as an ability is much more difficult and is acquired much later than the euclidean context surrounding the rigid body. But let us skip over this trivial fact.

For the acquisition of the mental object "straight line" Piaget makes exorbitant demands. In Section 8.5 I casually said how straight lines appear as

> arrows, trunks, sticks, pins, rims, edges, paths, folds, cuts, stretched strings,

but none of these can meet with Piaget's approval. Straight lines must be acquired as vision lines, and even stronger, the *global* constitution of the straight line must be preceded by the local one of the collinearity of three points.

Certainly, the vision line — as a light ray — is extremely important, and just as certainly we will draw to it the attention it deserves. It would be of interest to know when this appears in the development.* But the straight line as vision line is an advanced stage — there are even many adults who do not know how to use this property. Visual line is a property of the straight line but I refuse to make it a constituting property.

For Piaget it is just this. Why? Because of the constraint of the system. Projective precedes affine precedes similar precedes euclidean, projective means looking and drawing perspectively, perspective vision takes place along vision lines, thus it is required that the straight line is earlier as a vision line than as the mental object that has to do with rigid bodies and their edges, with flexible long objects in their preferential state, with movements directed to an aim. To satisfy the system the straight line must be constituted as vision line and in no weaker way.

Is this to be judged as a disastrous influence of pseudo-mathematics? Yes and no. No, because it may be appreciated that somebody has investigated, or tried to investigate, how the understanding of the straight line as a vision line comes about. Yes, because this pseudo-theory may have prevented investigators from looking for the true origin of the mental object "straight line".

I will now leave the so-called projective space and turn to Piaget's affine** space. The experiments about it were made with the so-called Nuremberg scissors or lazy tongs, in my terminology a context of flexibility rather than affinity. Well, it has to do with parallel lines, and this may be termed affine. One could even materialise a two-dimensional flexible lattice that as Nuremberg scissors demonstrates a special affinity of the plane — a very special one which is still dominated by the idea of flexions since it is the diagonals rather than the sides of the compartments that is variable.

Experiments with the Nuremberg scissors are undertaken in order to observe "conservation of parallelism" by the subjects. It is the only place in Piaget's work where the transformations are made explicit with respect to which conservation is meant: it is affine transformations, albeit of a very special kind and in a specialised materialisation. The subject must first predict what will happen when the scissors are opened, or more widely opened, in particular that rhombuses come into existence with "conservation of parallelism" of the sides. As far as I can judge there was no account taken of whether any of the subjects already knew this plaything. Anyway this — conservation of the parallelism structure under the movements of the Nuremberg scissors — was the original interpretation of conservation of parallelism, but while the experiments were going on the interpretation shifted; the experimenters became inclined to understand "conservation of parallelism" as the ability to copy parallel lines by drawing — in Sections 8.9–9 we will separate clearly this ability of reproduction from that of constitution of the mental object, which have been mixed up here.

* When reading the reports of experiments from Piaget's laboratory, I was astonished that it is never mentioned whether subjects closed one eye while aiming.
** In the translation one reads "affinitive" instead of "affine".

Yet the experiment with "conservation of parallelism" is wrong as such; it is produced by the same dogmatic interpretation of the *Erlanger* program which we identified earlier: parallelism is an affine concept; so as a mental object it is not well constituted until its invariance with respect to affine transformations is established. In the case of topology Piaget did not make such heavy demands — the topological mappings were not even touched. But the demand itself is by no means justified, like that of the constitution of the straight line as vision line. History at least is a proof to the contrary. Since affinities are quite a recent discovery, may one conclude that mathematicians before that time did not have a good conception of parallelism?

There is not the slightest doubt — and in Section 8.5 I have tried to convince the reader of it — that parallelism as a mental object starts early; it cannot be difficult to prove this by good "conservation" experiments. There is no reason why this conservation should be related to the affine group. Parallelism is perceived within the context of rigid bodies and the first and best way to observe it is in just this context. A subject is shown the parallel edges of a ruler, of a sheet, of a box and asked what happens to them if the object is moved. I do not know whether anybody ever made this experiment, but anyone who has studied children's behaviour will not have the slightest doubt that this conservation of parallelism is constituted early. And this decides the question. It is nothing but dogmatism to require more. The fact that parallelism is preserved by a larger group than euclidean motions — the affine group — is no argument to require invariance under this larger group as a criterion for the constitution of parallelism. With the same right Piaget could have required the invariance under the Moebius group for the mental constitution of the circle, which allows one to consider straight lines as a kind of circles; or to require for the incidence of point and line as mental objects that invariance is established with respect to all contact transformations. A mental object need not wait to be pronounced constituted until its invariance is established with respect to all mappings that can be contrived in all possible contexts.

In the preceding I have tried to do justice to Piaget's expositions on conservation of parallelism as far as they are intelligible and consistent. The experiments have little to do with these expositions; the experimenters observed other things than the conservation of parallelism. Moreover the theoretical text around the experiments is often unintelligible, possibly because of internal contradictions, which might be the consequence of double authorsnip. The experimenters said they had difficulties in explaining to children what "parallel" meant, or rather they were obliged to have these difficulties by the system (p. 316):

First of all, how is one to pose the question in order to make the idea of parallelism comprehensible? Presumably by asking whether the lines "slant the same way", since any mention of ideas as 'equidistance' introduces far more complicated notions and measurements which themselves depend on assumptions about parallelism (i.e. the parallelism of lines at right angles to those under consideration, thus yielding a circular definition of parallelism in terms of equidistance).

I stop here for a while to give the reader the opportunity to think about it. The definition of parallelism by equidistance involves no circle at all, but the reasoning between the parentheses contains a twist that deserves to be straightened out. I pass over the question whether one *should* define parallelism by equidistance, whether it is *didactically* the best way or even a good way. Anyway it can be done, and it has been done in the finest setting by Kuno Fladt. Distance from a point to a line can be taken in the sense of *shortest* distance or of *orthogonal* distance and there is not the slightest reason why measuring along fresh parallel lines should be required *a priori*. The experience that in euclidean geometry the lines along which the distances are measured, are again parallel, is an *a posteriori* fact, which at a certain level can be formulated axiomatically.

But let us continue reading:

Alternatively, to speak of two straight lines "sloping the same way" means introducing the concept of both straight line and space orientation [in the original: *inclinaison ou direction*]. Now we have already seen in Chapter VI how late the straight line comes to be visualized . . .

One would be inclined to say that if "straight line" comes so late, parallel lines come even later, thus why is there any further argument (indicated here by the dots, where I interrupted the quotation). But this should be understood as follows: parallelism of straight lines is discussed with ordinary drawn lines, whereas the lines that are said to come that late, are the vision lines of the so-called projective geometry. In order to comply with the system, where projective space presupposes affine space, the straight lines are forbidden developmentally, but somehow they must be allowed, because otherwise no experiments on parallel lines could be performed.

. . . and as for the concept of orientation [in the original: *inclinaison*] this is a matter of either measuring angles or else finding some other method of determining the inclination [in the original: *l'identité de direction*]. But the idea of parallelism appears at the same time as that of angles, and this is hardly surprising since a pair of straight lines intersect to form an angle wherever they are not parallel . . .

Similarly one could say: the concept of straight line occurs at the same time as that of radius of curvature because as soon as a line ceases to be straight, a finite curvature can be calculated. Or: the concept of length occurs at the same time as that of area, because as soon as a line ceases to be thin, it gets an area. Let us admit that any idea calls up its negation at the same time. The negation of being parallel may indeed be formulated as "forming some (positive) angle" but this is far away from the idea of angle itself, which includes knowing at least what equal angles are. As a matter of fact this whole story is flatly contradictory to what has been said a few pages earlier on the relation between affine and projective geometry:

. . . and in the next chapter we will see that these twin [original: *complémentaires*] concepts are psychologically interdependent. If this is the case, it necessarily follows that the concept of angle cannot precede that of parallelism, nor can it serve as measure of parallelity of a pair of oblique lines.

The French text is here incomprehensible probably by a clerical mistake (omitting a few words). The translator tried to make the best of it, though I would have preferred "serve to measure whether two lines are parallel". It continues:

This leaves us with the idea of identity in orientation [*direction*] but this is soon ruled out when it is realized that the concept of spatial orientation [*direction*] is the foundation [*point de départ*] of the co-ordinate systems themselves. And as will be seen in Chapter XIII, their development is an extremely complex and protracted affair . . .

A similar reasoning would be: Addition is difficult because multiplication rests on it. Or: the area of a rectangle is a difficult concept because it is needed for the definition of integral.

As to the whole story: anyone who has ever been busy with children and in education knows that angles and angular measures are much more difficult mental objects than parallelism. It is not a problem to explain, say, to a five-year old what parallel lines are: show him one example or two, and perhaps a counter example. With somewhat older children one can even successfully analyse the phenomenon "parallelism". To do this one need not be able to measure lengths, let alone angles.

Moreover the claimed dependence of parallelism on angles is inconsistent with the claim that developmentally affine space precedes euclidean space. But let us read further:

In short, it is no simpler to imagine the parallelism between two lines than between the sides of a closed and well-organized figure like the rhombus. But it may be asked, surely it would be simpler to perceive, even if not to imagine? Here the result of comparing perceptual estimates [*données*] is very much to the point, for actual study of the perception of parallelism leads to the conclusion that the idea of parallelism precedes their accurate perception rather than being a consequence of it as might have been thought.

Wursten (*op. cit.*) carried out the following experiment: twenty adults and twenty children aged between 5–6 and 12–13 were asked to compare the lengths of oblique lines drawn on cards. Alternatively they were invited to draw vertical, horizontal, and oblique lines, or else adjust pivoted metal rods in a parallel position. Wursten's findings were as follows: first, parallels are never perceived entirely without errors, even by the experienced adults. This is a further confirmation of the intellectual, logical character [*caractère rationnel*] of geometrical concepts which govern and influence [*informent et corrigent*] perception rather than being wholly dependent on it . . .

The last remark is correct, though not as a conclusion of the preceding. A certain concept is independent of perception not because perception is liable to errors, but because the concept enables us to establish the fact that the perception is wrong.

. . . Second, and most important, comparison of variations in thresholds and constant errors showed perception of tilt [*inclinaison*] and spatial orientation [*direction*] to be extremely poor below the age of 7–8. The reason why young children are better than adults at comparing the lengths of lines pointing in different directions is precisely because they remain indifferent to their relative orientation.

Here a particularly revealing foot-note is added:

Less difficulty is experienced with vertical and horizontal parallels. Hence in these two cases it would seem that perception of parallelity precedes the idea [notion] of it.

Take a long draught of it. "Vertical" and "horizontal" here mean, as in practically the whole book, even in interviews of children, directions not in space, but on the table between experimenter and subject. "Vertical" is the direction of the one to the other, "horizontal" that orthogonal to it, along one's chest. It is in this world that the coordinate systems of Chapter 13 of the book arise. Walking around the table, at least mentally, rotating the table, or the drawings on it, is forbidden. It is a world locked up in a rectangle. In the whole book *Représentation de l'espace chez l'enfant* space practically does not occur — the world is flat and is most often a table, direction means orientation with respect to the edges of the table. Only the little ones cannot be forced into this frame. They perform better in the meaningful experiments, which do not depend on the frame. By disregarding the meaningless frame they show more genuine mathematical insight than the experimenters allow the older and adult subjects to show.

I leave it here, but I cannot but ask myself: Is this really Piaget, or did he never see the proofs?

8.8. *Reproduction — Symbolic and Ikonic*

Researchers — piagetian and others — often show that they did not grasp the fact that lack of names for mental objects and actions — or lack of knowledge of the conventional names — does not prejudice anything with respect to the possession of the mental objects or actions themselves. But even grasping that fact does not protect one against serious misconceptions. Even the so-called non-verbal tests need not prove anything. One cannot measure whether or to what degree the mental objects (or the concepts) circle, square, straight line, and so on, are present by having the subject draw or in some other way reproduce that figure. Anyone who has observed children, is familiar with the technical difficulties they have in expressing their intentions with drawing instruments on paper or otherwise — intentions that can simply get lost in their failing attempts. I fail to understand how some researchers — Piaget included — will dare to interpret the failure to reproduce figures by congruence or similarity as a proof of priority of *topology* on *euclidicity*. Meanwhile more critical investigators have shown that young children are certainly able to distinguish a better copy of a circle from a poorer one and to appreciate them as such. There is a big gap between recognising two figures as congruent or similar and being able to copy them as such, not only for little children but also for adults unless they are gifted with extraordinary graphic talent. Yet there are still researchers who forget about such clear differences.

The misapprehension is, however, more deeply rooted, in the relation between reality, image, and language, or rather in the way that this relation is experienced. Everybody understands what a tree is, though there are border line cases where one can doubt whether something is still just a tree or is not a tree any longer. A picture of a tree — drawn or sculptured — will be recognised, at least in our cultural environment — an Indian from the Amazon or a Greenland eskimo may view it differently. Here too there are border line cases: how far can the artist go in the sketch or in the impressionist manner to get something accepted as being the image of a tree? But besides trees and images of trees there is the word 'tree', accepted all over the world by English speaking people to designate a tree, though misapprehensions are possible between people who pronounce it differently. And finally there is the graphic image t-r-e-e for the word 'tree' which in print or writing can look differently, known as such to people who have learned reading and writing.

In each particular case we know very well which 'tree' is meant. If the teacher pronounces the word tree, it depends on the situation whether the pupil points to a tree, or to the picture of a tree, whether he repeats the word tree or writes it down. Of course miscomprehension is possible, though in general it is not serious. It is more serious that most of the authors of set theory textbooks for primary and secondary education got into troubles or dragged users of their books into troubles with sets whose elements can be trees or pictures of trees or names or graphic pictures of names of trees and finally all of them in the same Venn diagram.

I tackle this here, not in order to identify this kind of misconception in experimental tests (they are being made, especially with logic blocks), but because I am afraid that misconceptions about reproduction might unfavourably influence the communication between experimenter and subjects.

The child gets acquainted early with two fundamentally different ways to reproduce objects and events — that is, fundamentally different in our view: the picture of a fire-engine in action on the one hand, and besides that the printed text, which according to the reader contains the word 'fire-engine' and a story about extinguishing a big fire. The child himself can interpret the pictures and he can check the authenticity of the story by having it read once more, by the same or another reader. How does the child experience this patent contrast — patent to us — between ikonic and symbolic means of reproduction? I cannot answer this question. Is the contrast really felt as such or is the one picture for the child just as much pictorial as the other? Is the adult *more* able to look at pictures, in the same way that he can take longer steps, climb higher, speak louder? 'Writing' and 'drawing' are often synonymously used by children, as are 'reading' and 'looking at pictures'.

One can certainly observe with children, as regards their internal and external means of expression, a development from the ikonic to the symbolic. Yet the question that puzzles me is whether and when a child draws a border line between ikonic and symbolic representation. If a 2—4-year old draws — as it seems at random — a few lines and asserts this is Uncle John or a doll-house or if he answers the proposition "draw a ..." with a seemingly disorganised system of scratches, is it then true that he is ikonically busy, that he has ikonic intentions, or does he act on the same legal grounds as the adult who claims to see a "tree" where there is nothing that looks like a tree? Would it not be more likely that a child who starts drawing or reproducing in some other way some reality, is moving in a sphere where the ikonic and the symbolic are not yet separated? I once observed such a separation becoming conscious:

Bastiaan (4; 8) who is encouraged to start early to make a list of suggested gifts for his birthday: "Writing by words or drawing by pictures?"

But even if the separation starts early, how long does the process last? I am sure there are people who never manage it completely, who remain convinced that a tree is called 'tree' because in some strange way the word 'tree' is similar to a tree, that the word graphically reproduces the tree — it is kind of magic belief. As a matter of fact set theory in textbooks is a proof of what difficulties even adults can have in this field.

Even if a borderline is drawn between ikonic and symbolic reproduction, it need not be the one we adults are used to. A drawn picture of a house, which we think is ikonic, can perhaps be meant symbolically, or symbolism may have dominated its production. Even school-children who are asked to draw their own house may produce a stereotype which does not resemble their own house, with stereotyped details such as corner curtains such as they can only have seen in standard houses in picture books — the symbol of a house.

How does a psychologist manage to have children producing or interpreting pictures without being sure that the child understands the assignments as they are meant, whether they conceive the ikonically meant ikonically, and the symbolically meant symbolically, whether they draw the borderlines as they are intended and whether they know such borderlines at all — at least operationally?

For the experimenter, a drawing of a circle with an inscribed equilateral triangle has a structure determined by his geometrical experiences; for a child who has not had much experience with geometrical figures, the figure can be meaningless, or ornamental, or a picture — ikonic or symbolic — of something, and the particular view that it has of the figure determines how it would react to the assignment to copy it. The child may have seen symbols of Fiat, PTT, VW, and recognises them by some structural resemblance, even though the various instances are not at all congruent or similar. But the circle with the inscribed triangle — what does it symbolise and which details do matter if it is to be copied? Must the circle be truly round, the triangle precisely inscribed and equilateral, or which deviations are admissible?

Adults who have not the slightest difficulty to recognise the symbol of the Netherlands Railways (Figure 86) do have the greatest difficulty to draw it from memory, and even when copying it, they repeatedly look back at the model.

Fig. 86.

Why? Because it is an arbitrary symbol with no clear context. The circle with the inscribed triangle, however, can be placed into a geometrical context, built up from geometrical objects.

Researchers have children copying models without checking whether they know what it means to copy something. Perhaps up until that moment these children have only drawn *objects*. In order to copy something, one must know what matters. Is the model to be copied an ikon or a symbol? For instance, the picture of a direction post where the angles of the various "hands" have to be respected can instead be the symbol of a railway crossing, with the symmetric disposition of the crossed arms? Learning to write letters and figures seems to be a struggle between ikonic and symbolic reproduction. Though ikons of one geometrical figure, the letters *b, d, p, q*, have different symbolic values.

Bastiaan (5; 6), who permits himself all liberties with the images of the figures 0, 1, . . . , 9, protested when — in a typewritten text — the figure 1 was indicated by the letter l.

Knowing what matters is a precondition of copying — experimenters are not in the habit of telling their subjects, perhaps because in their own world it goes without saying, or perhaps because if they explained it, the problem would be too easy. Whether Uncle Sam, if copied, gets the prescribed height of his hat, is different from whether the "stripes and stars" are forgotten. It is a different thing to draw some set of teeth, or a set of teeth as a symbol of Jimmy Carter. There are many shades between the ikonic and the symbolic. Caricature can show more resemblance than portrait, but then a resemblance with a person that has become a symbol.

8.9. *Reproduction From Geometrical Context*

In Section 8.5 I dealt with the context of the rigid, congruently or similarly reproducible bodies, in Section 8.6 the world of the boxes, also reproducible, though not that rigid; we will learn of other contexts, even less rigid or so individualised that no thought of reproducibility comes about. As a matter of fact, the objects in our contexts are not necessarily bodies, even if the term body does not include three-dimensionality. In any case for geometrical objects the possibility of reproduction is an important feature — reproduction by means available or created for this special aim, sometimes with great difficulties. We know wire models, plaster models, cardboard models of three-dimensional figures, but the most usual reproduction is: on the blackboard or on paper, in books and on sheets.

The prescription which the adult is expected to observe is to represent things visually in the way he sees them. Archaic and primitive art, however, shows that it is not that easy to see things "in the way one sees them". A geometrical theory — perspective — has been created in order to deduce how things are seen, and experimentally this theory can be confirmed by means of a camera, at least as far as it is allowed to identify the lenses of camera and eye with each other as well as the photo-sensitive layers of a film and a retina. Yet developing a film and processing a retina image in the brain do not seem to be isomorphic procedures. In any case the mental image of, say, a cube seems to differ considerably from the visual one prescribed by the theory of perspective. Seeing, interpreting, and producing perspective drawings is no naïve ability, but something that must be learned. By no means can I say what the mental image of a cube looks like — as a matter of fact it depends on a variety of circumstances. Certainly it involves more and other features than that which one sees or is expected to see. It involves as much as one needs to recognise, to make, to produce, and to reproduce cubes. It includes six faces, though one cannot see more than three at a time and may be unsure about the actual number, four, or six, or eight. About a man one is facing, one knows he has a back even though it is invisible; about a house, that it contains rooms and stairs behind its walls.

One has more sense-organs than two eyes, and a drawing may be used to communiate more than visual perception. The little child does not yet divide his knowledge about the world into compartments according to the so-called five senses. A 5-year old, ikonically precocious boy amplified his marvellous drawings of airplanes with images of the noise produced by the jet engines. A 6-year old reproduced the turning flashlight of an ambulance by a triple of three lights, one in forward direction, one right, and one left. The desire to depict things other than those which the eye perceives according to the theory, is not restricted to childhood. The great problem each painter has struggled with is to process his impressions in a way that they reproduce a more objective reality than his impressions do. The symbolic is entangled with the ikonic; viewpoints are chosen in order to have the ikonic doing justice to the symbolic — a rule even observed by the photographer.

5—6-year olds draw houses with a front and side facade. It looks like an attempt at perspective though it can be an imitation of a badly understood drawing method. It can also witness a reflection, although not yet governed by perspective: from certain viewpoints one can indeed see two sides of a house, but it does not yet matter how this happens in detail. It is difficult to ascertain what is achieved by posing questions as the child readily produces ad hoc answers:

Bastiaan (4; 3) has drawn a house with many rooms and appartments. There is only one bathroom. When asked about it, he answers: "All bathrooms are the same, aren't they?".

Traditional geometry instruction does not even face the problem of reproducing. The child is expected to have caught somehow and accepted the adult methods of reproducing. A cube is drawn on the blackboard, allegedly in perspective, though to avoid too strange a look, with parallel edges. Afterwards this can mathematically be justified by looking from an infinite distance, as they say, though a cube so far away would look infinitely small. Even all the contradictions are taken over from the adult methods of reproduction, even those which can only be justified by their usefulness and intuitivity, such as the usual representation (Figure 87) of the globe with equator and poles — counter to perspective but suggestive and convincing.

Fig. 87.

The perspective is *représentation de l'espace*, not in the sense of a mental object but of reproduction on a piece of paper, a method acquired by imitation, which is systematically exercised by teaching the pupil to see what he (the pupil) sees — lines, planes, light, shadow — and which is finally rationalised in a fully developed theory. But to stress it once more, primarily perspective is not a geometrical context but a kind of reproduction, side by side with others, and this remains unchanged for a long time.

This does not hold only for perspective. Initially, and to a high degree, reproduction is a matter of imitation, even before kindergarten. As soon as the child comes into contact with older ones, his method of reproduction is influenced. As a matter of fact adults too sometimes produce baby drawing methods alongside the more familiar baby language in order to be imitated.

Daphne (5; 1) gave me the drawing of a house with two chimneys which as usual were drawn orthogonal to the planes of the roof, rather than vertically. I took her to the window and showed her a roof with a chimney. Immediately she made a correct drawing. This spontaneous reaction is astonishing. Clearly the original drawing was nothing but imitation of what she had seen from another child.

A somewhat complex drawing of a child is a composition of more or less obligatory parts, combined in a somewhat functional way — the reproduction of a combinatorically flexible structure; what counts is the combination of the parts: the eyes are in the head, the ears and limbs, perhaps also the belly are systematically connected to it, the one at the right, the other at the left, the one besides the other, the one below the other. Parts can be meant ikonically, but their presence and location is symbolic. This is already a geometric context, not topological but rather combinatoric, with parts determined by the flexibility of the structure. The size relations are symbolically rather than ikonically reproduced — this is again not a matter of defective mental objects but of principles of reproduction (as in the case of perspective) where the symbolic is dominant or not yet separated from the ikonic.

However one looks at a cube or turns it, one cannot see more than three faces at a time. To make a cube, one needs six. The technique that accounts for it, is networks. On the network of the cube you can see the six faces at a time, and as far as it is not clear, one can indicate the edges that have to be identified, mentally or by adhesive tape. This too is a reproduction of the cube, not a perspective but a combinatoric one, from parts that respect the similarity of reproduction. Blue-prints are the same: each storey of the building is indicated separately to scale, where the combination of the parts is marked by special signs as well as by stairs and lifts; flexibilities, such as of doors and windows can also be indicated. Another method involves three projection planes; plan, front elevation and side elevation.

Let us call this kind of reproduction − that is combinatoric from ikonic parts and accounting for flexibilities − *compository*. It is more flexible than perspective; and in geometry it is at least as important for reproducing objects. The child's method is predominantly compository. If he wants to draw the interior of a house with two stories, after the frontroom of the groundfloor has been plotted, he has to solve the problem of the composition of the back-rooms of the groundfloor and the upper front room. Somehow he solves it − you could say in a primitive way, were it not for the fact that most adults do not know what to do either. There are techniques of reproduction required to solve it, such as the network of the cube, the blue-prints, descriptive geometry, or artistic sophistication.

But what I would stress here is that the compository method of reproduction, for instance, of a cylinder by means of a rectangle and two circles, or of a cone by a triangle and a circle, which somehow are attached to each other does not at all bear witness to defective mental objects. On the contrary, this way of reproducing can prove a better view of the mental objects than reproduction by means of perspective acquired by imitation.

Another method of reproduction is the *topographical* one, as used on geo-graphical maps, railway nets, motoring maps, most of them to scale but not ikonic, with cities, towns and villages indicated by too big spots, while rivers, roads, railways are shown by too thick lines; with airports symbolised by draw-ings of airplanes, ways out by circles, bridges and ferries by other symbols. Is it then a mad idea of a 5−6-year old who draws a network of streets to lay the stop and priority signs as it were on the ground? The symbolism in the adult topography is more subtle but it is symbolic and most often conventionally symbolic. If we do not draw from the adult topographic reproduction the conclusion that certain mental objects, such as perpendicularity, are lacking, we are not allowed to impute to the child such deficiencies.

8.10. *Grasping Of, and Putting Into, a Geometrical Context*

The examples of the context of rigid bodies in Section 8.5 showed how geomet-rical contexts come into being. Natural production, craftmanship, manufacture and industry have made us familiar with geometrical figures, their congruence and similarity, with rectilinearity, orthogonality, symmetry, parallelism. Tables, doors, sheets of paper, windows, beds are rectangularly produced and impose the rectangle as a mental object on us; we are being prepared to accept the *name* "rectangle" and to name each rectangle (even a square) a rectangle. Of course things are not always as easy. Is a diamond a rhombus or a square? Standing on a corner can be a more important property than having equal sides. A long rectangle looks different from a tall one, a cylinder lying down different from a standing one — typical consequences of a geometrical instruction guided, rather than by the objects, by their drawings. Rhombuses in jigsaw puzzles, cubes and cylinders from the construction box, fit better into the context of congruently and similarly reproducible rigid bodies.

Contexts should not be taken for granted, but once grasped, they can function reliably. This presupposes that the characteristics which matter in the context are paradigmatically clear.

A child understands early what things are to be classified as chairs, but in a certain context a chair can be appointed to be a locomotive or a ship. Words like triangle, square, rectangle can be meaningfully used by little children, even to recognise these geometrical structures where they are obscured by roughness, imprecision and rounded corners. A bench (without a back) made from three parallel planks with two interstices is seen as one rectangle or as three of them depending on your preference. The *gestalt* forming procedures that are active here, are not restricted to geometry. They do not differ at all from those by which we interpret a constellation as a dipper, or a cloud as an elephant.

The context required for recognising and reproducing figures can be determined more or less sharply by the data, and whoever wants to interpret the behaviour of others in such activities, should first analyse how the results are determined by the suggested context.

Suppose a person is given material, say plates, that differ with respect to

external shape — triangles, circles, squares, and so on,

finish — rounded or sharp corners, rough or polished surface, with grooves or prickles, and so on,

internal shape — with various numbers of holes of various sizes and shapes (triangular, round, squares) and in various arrangements,

thickness,

colour or colours,

matter — wood, plastic, metal.

Conclusions shall be drawn from the way the subject classifies the material. For instance one expects some subjects or age classes of subjects to classify primarily and by preference according to certain criteria, and one undertakes to test hypotheses on this behaviour. By varying the number of objects representing

a certain class, by stressing more or less some differences between class characteristics, by controlling the distribution of the characteristics over the classes, by
coupling characteristics more or less closely with each other, one can influence
the results in a decisive way. It can happen that thickness is the most striking
characteristic, because only two thicknesses occur or because there are ten of
them that are regularly ranked. By admitting only two or three, strongly different
colours, the stress can be shifted to colour. If the material is appropriately
chosen the most striking features can be small and big, or sharp and rounded,
or with round and not-round holes.

Geometrical criteria of classification can be

congruence,
similarity,
affinity,
combinatorial equivalence,
flexion-equivalence,
topological equivalence,

but the geometrical context in which such criteria shall be applied, is not at all
self-evident. One can tell a 5-year old to disregard thickness, colour, finish, but
if such instructions are lacking, even a 13-year old might be unable to put
the material in a geometrical context, and this will certainly happen if strong
enough distractors are built in. This kind of experiment, if undertaken in order
to investigate developments towards geometrical contexts or within geometrical
contexts is *a priori* useless.

Ethologists have experimented with more or less vague pictures of owls
shown to singing birds to illicit fright behaviour; they can tell you how far
they can go with dropping certain characteristics, for a male stickleback the
red belly colour of a putative rival is the signal to defend its territory. Man
— child, adolescent, adult — recognises places, things, persons, and identifies
classes in order to classify, by means of a small number of criteria which
rarely become conscious. Yet with regard to geometrical objects, the mental
development can lead to making criteria of recognition and classification
conscious.

At least so it looks. Without expressing it verbally, one can make absolutely
conscious to oneself and others what is a triangle, a circle; what are intersecting
lines, what is the structure of a cube. But it is much less clear why we ascribe
to an ivory die with rounded edges and vertices the shape of a cube, or more
poignantly said, the same shape as to a wooden die with sharp edges and corners.
What are the criteria? How is the die placed into the geometrical context where
it is judged to represent a cube — in fact it stands as well in the contexts of
gambling and of probability. How are we able to agree about how badly a
rectangle may be drawn to be accepted as such, where the tolerance terminates
and where sharper requirements are to be made?

Bastiaan (5; 6) describes the shape of a piece of wood he says he needs, as "like the front of a car". It appears that he means a rectangle. Though adults would not say so, it is correct that the front of many cars is roughly a rectangle.

Bastiaan (6; 0) says about an empty beer can which is somewhat compressed from two sides and shows roughly a square section: "This is a quadrilateral".

Why are they rectangles and squares (this is what he meant). One could have continued the conversation: "A rectangle has sharp corners, but this can is all curved." But I myself had admitted it was roughly a square, and he would have answered the same. Under certain circumstances "rectangle" and "square" can be excellent descriptions of things.

Bastiaan (6; 2) plays with a stick with a longitudinal groove and two bottle tops he found in the forest, as though it were a machine gun and two bullets. "What does a gun bullet look like?" he asks though actually he knows how to draw it. I add: "A cylinder with a cone on the top." I do not believe he knew the word cylinder. Anyway he only asked what is a cone. I said: "A clown's hat". Then I let him show cylinders: pieces of trunks of trees and trash baskets. I show him flat discs. He agrees that they are cylinders though with the reservation: "We shall rather call them discs." I ask him what you see if you cut a cylinder "this way". His answers have nothing to do with the geometric shape; they are related to the particular cases. I help him with the word "circle", which he apparently did not know. I show him examples like the section of a tree, the rim of a trash basket, a button, and I mention sun and moon (or he himself did so), and finally I show him the circular hole in the top of a beer can. He protested: "This is a bit long". Indeed it was elliptical – a difference of less than 10%. The next day he used the word "circle" correctly.

The sections of trees looked of course much less like circles than the hole in the metal. But then he did not protest; about the hole he did. Why? Clearly you cannot require so much geometrical shape from the section of a tree as you can from the sharply bounded symmetric hole formed by a smooth curve. The section of the tree did not pretend to be a circle but the almost circular hole did and consequently it had to be judged by sharper criteria – something like this must have been the background of his evaluation.

At the opportunity which I related in Section 1.28 I explained to Bastiaan what is a half – he did not know this word, at least not related to length or distance – by breaking a stick (not exactly) in half; he protested because one half was a bit longer. Here again we observe the presence of the mental object and the testing of the example – the first example – by the mental object.

Or do I abuse the term "mental object" and should I rather speak of visual imaginations? Well, it is visual imaginations but then different from those we have of animals, trees, stones. The context of geometry implies that they are normative imaginations, something like Plato's ideas, though I would not like to argue about the origin of these norms – it does not matter whether they are objective, genetically determined or developmentally acquired.

Am I allowed to name imaginations with this degree of exactness "mental objects"? It is a meaningless question. I would rather ask another one. What is the next step in the development? The concept circle, square, half? A definition like "a circle is the locus of . . .", or in a modern style "a circle with centre M

and radius *r* is . . ."? No! The next step is a *question* — the question is how to make a circle, how to produce a square, how to halve something. One can suggest the answer by handing our material, or letting the child choose from the material that is offered. One can also aim for a mental construction by sharpening the question to: "How can you make it more precisely?".

Bastiaan (6;4) asks: "Where is the centre of Netherland?" (Possibly he had heard about Utrecht as such.) I tell him it is not easy to determine, and then: "What is your centre?" He shows on his top. I argue it should rather be in his belly. Then I ask him about the centre of a tile of the pavement (Figures 88 and 89). First he denies its existence. Then he shows what is approximately the centre. I ask him to do it more precisely. He produces the groove between the next row of tiles and cuts it with an estimated mid-line between the other sides. I explain to him that it is easier with oblique lines. He draws the diagonals. I mention the word diagonal. At a bench I ask him to indicate the diagonal of its bottom. He draws a line that forms an angle of 45° with the sides of the rectangle. I show indignation. He corrects himself immediately.

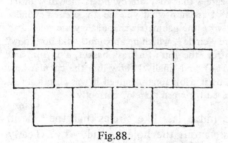

Fig. 88.   Fig. 89.

Is the context of geometry not grasped until the question of the precise construction arises in order to be answered? Anyway the question is characteristic of a certain context. Even then the answer can be different according to how precision is mentally measured — this too requires a context.

Summarising: Which symptoms indicate the ability to grasp a geometrical context and to put objects into it?

By showing knowledge of what matters in the context, by way of

    recognition,
    classification,
    material reproduction,
    naming,
    mental reproduction

of mental objects and processes and by

    making conscious to oneself and
    describing

these activities.

And how is "what matters in the context" determined? By

natural, craftsman, manufactured, industrial reproduction,
paradigmatic examples,
explicitation.

CHAPTER 9

TOPOLOGY AS A GEOMETRICAL CONTEXT

In the original version, I continued Chapter 8 by the question "What is topology?" The answer led me so deep into topology as a geometrical context that the frame of the chapter was in danger to explode. Finally I felt compelled to put topology as a geometrical context outside Chapter 8 as well as the planned "Topography as a geometrical context" in order to resume the thread of Chapter 8 in Chapter 11.

9.1. *What is Topology (Not)?*

As a joke topology is sometimes defined as the art of drawing badly. I will be glad to take up this joke, but before I do so, I must offer some serious mathematics in order not to interrupt the discussion at an inappropriate moment.

What is a curve? A point moves in the plane or in space, and a (planar or spatial) curve is the path described by the point. I mean of course a mathematical point, rather than that of a pencil.

"A point moves in the plane or space" — what does it mean? The mathematician is ready to answer such a question: the place of the point is a continuous function of a parameter t (the time).

Unfortunately, Peano gave an example of a continuous function that maps a line segment (a time interval) upon a square (or a cube). A point that moves continuously according to this function can behave madly enough to describe a whole square (or cube). No, it is no madness — Peano's examples are quite reasonable and intuitive.

But of course that is not what people mean if they speak about curves. Our definition was invalid as a description of the mental object that is present in people's minds if they speak about curves. They mean something like a thread, not a piece of surface, not a piece of space.

What is wrong? I mean: why is there a gulf between the mental object and the concept, between intuition and logic?

The villain is continuity. Continuity too is both a mental object and a concept. One has tried to define continuity in a way that justice is done to the mental object, but apparently one did not succeed as one should have done.

Can the wrong be redressed? Can the concept of curve be defined in order to exclude this kind of abnormality? Certainly, it can; for instance, by requiring that the function defining the curve is differentiable, or that the curve defined by the function has a well-determined tangent in each of its points — the Peano curves lack this property practically at each point. This prevents abnormalities but at the same time it excludes a whole host of legitimate curves, for instance

250

each broken line, which as it happens has no tangent at the corners. In order to save them, more compliance is recommendable: for instance by admitting piecewise differentiable functions, curves that lack a tangent in a finite number of points at most. Yes, this is an escape, but properly said, it is a loop-hole. It is not satisfactory. A concept like curve belongs to topology and should be defined in a topologically invariant manner, which means that each topological (one-to-one continuous) image should be of the same kind. Yet topological mappings do not respect differentiability.

"Curve" should be some thread-like figure. The problem of how to define such a thing reasonably, has been solved, but it would take us too far away to explain and to justify it; it would require too much theory.

The curves we tried to define were continuous images of a line-segment. This means that they may have multiple points, the moving point may cross its own path. Let us turn to a more handsome kind, simple curves, as it were. Let us define:

A *simple arc* or *Jordan arc* is the topological image (one-to-one continuous image) of a line-segment.

A *simple closed curve* or *Jordan curve* is a topological image of the circumference of a circle.

Jordan's name is attached to these objects because C. Jordan first proved the famous

> *Jordan's theorem*: A Jordan arc in the plane does not divide the plane; a Jordan curve in the plane divides the plane into precisely two parts.

The last assertion can even be strengthened: Let the plane Jordan curve K be the image of the circumference C of a circle by means of the one-to-one continuous mapping f. Then f can be extended to a topological mapping of the total plane, which in fact maps the interior of C upon that of K, and the exterior of C upon that of K.

One would not expect it otherwise. With the naked eye one can see that such a Jordan curve divides the plane in two parts, both of which look like the interior and exterior of a circle, topologically viewed.

Can you really see it with your naked eye? Figure 90 is such a Jordan curve, but Figure 91 too, with more bends and trunks, and you need your finger or a pencil to ascertain what is the interior and what the exterior.

Fig. 90.   Fig. 91.

Topology as the art of drawing badly — indeed in order to put a drawing of Jordan's theorem on the blackboard, one need not exert oneself to come out with a beautiful product. On the contrary one's effort may even be directed towards producing an ugly curve, neither convex nor smooth, with bends and trunks, as Figure 91. Or rather, one is advised to spare no pains to make it as wild as possible in order to show the learner how eccentrically a Jordan curve may behave. In fact, this is not superfluous if one wants to convince him that such an obvious looking theorem as Jordan's needs a proof — a proof that is not at all obvious and even not easy.

Once he has understood how complicated a Jordan curve may be, he can be satisfied with simpler models. Or can he? Is an extremely complicated drawing really enough to convey the full idea of a Jordan curve? I recall an experience from my own academic study, which gives me food for thought.

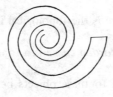

Fig. 92.

Take two similar logarithmical spirals (Figure 92), given in polar coordinates $\ulcorner r, \varphi \urcorner$ by

$$r = \alpha^\varphi \qquad \varphi \geqq 0$$
$$r = \alpha^{\varphi + \pi} \qquad \varphi \geqq 0$$

with fixed α ($0 < \alpha < 1$), which turn an infinite number of times around the origin, add the origin itself and join the other ends by a line-segment. It yields a Jordan curve K going through the origin. Sure, it is a Jordan curve, as nice as a circle, and the origin is for this curve a quite common point, though it does not look that way.

Now my own experience with this curve: I was decently familiar with topology, knew that Jordan's theorem requires a proof and knew proofs of it, knew all that was known at that time on mappings of manifolds and mapping degrees, and yet I was dumbfounded when I discovered that this was a Jordan curve like others. Since nothing of this kind had ever be dreamt of by me as a portrait of a Jordan curve, I got second thoughts about proofs of Jordan's theorem as I knew them. Possibly in such proofs appeals were made to — too restricted — visual images of a Jordan curve, rather than to its formal definition, — a serious mistake or a source of mistakes. My suspicion was unfounded, all was correct, and meanwhile I got accustomed to this kind of curve.

The logarithmic spiral

$$r = \alpha^\varphi, \qquad \varphi \geqq 0$$

is the topological image of the set of non-negative reals $\varphi \geqq 0$, and if $\alpha < 1$ this image approaches, for $\varphi \to \infty$, $r = 0$ legitimately, that is, the origin. The set of non-negative numbers completed with a point at infinity is topologically the same as a line-segment, even though in mapping the one on the other one has to run through it at an increasing speed. The spiral together with the origin is consequently the topological image of a line-segment, a Jordan arc. Two such arcs that have only the endpoints in common (presently the origin and $\varphi = 0$, $r = 1$) together with a connecting line-segment make up a neat Jordan curve — nobody can doubt it.

Okay! But this mad point, the origin, which pretends to be a well-behaved point of K, could it not behave badly with respect to the plane, that is, influence badly how K lies in the plane? No, it cannot. The strengthened version of Jordan's theorem asserts that a Jordan curve divides the plane, locally and globally, as a circle does, that is a given topological mapping of the curve K on the circumference C can be extended topologically to the whole plane.

How then does the impression arise that the Jordan curve K composed by the two spirals is lying differently in the plane than does a circle C? Well, ordinary circles with the centre at the origin intersect K infinitely many times whereas with C they would not do so more than twice. Yet this is not the way to look at K if one wishes to study the neighborhood of the origin. One has to distort the surrounding circles like spirals as one did with C in order to "straighten out" the image.

All this then is topology:

the mental object of a closed curve with no self-crossings,

accounting conceptually for it by the definition of a Jordan curve (the one-to-one topological image of a circle circumference),

the hesitation as to whether this includes the "pathological" K,

the confirmation that the definition includes K,

the question of what caused the hesitation,

the certitude that the hesitation was unjustified.

This is topology, and it is a non-trivial sequence of steps in topology. Non-trivial, because the course could have been different, which I will show by another example.

Let us mount one dimension higher, where even bad drawings serve no purpose, good or bad. Let us define:

A *Jordan disc* as the topological image (in space) of a circular disc.

A *Jordan sphere* as the topological image (in space) of an ordinary spherical surface.

Jordan—Brouwer's theorem: A Jordan disc does not divide space; a Jordan sphere does divide it into precisely two parts.

However, with the strengthening, as formulated for Jordan curves, it goes wrong: the interior of a Jordan sphere can now behave very badly. It need not look like the legitimate interior of an ordinary sphere, as appears from an example of Antoine—Alexander — the "horned sphere".

The mental object (or imagination) of a

> sphere-like surface

is imperfectly accounted for by

> the conceptual definition (topological image of an ordinary sphere).

So I can acquiesce in the fact that

> my mental object is less handsome than I thought,

or try

> a revision of my conceptual definition,

such as not to admit of the discovered pathology. This has indeed been tried by distinguishing

> tame Jordan spheres

which are lying as neatly in space as I intended with my mental object.

Properly said the difficulties started much earlier, with the concept of continuity (of functions and mappings), which I already indicated. The concept of continuity evolved in history from a mental object that for centuries had been clear enough in the minds of mathematicians that it needed no precise definition. The reason why they could do without it was not the clear mind and vision of these mathematicians, but the mathematics they cultivated. As long as discontinuity played no great part, there was no need for more clarity on continuity.

A continuous function f that is negative for a and positive for b, must vanish somewhere in between (Figure 93) — blind man can feel it, and also that the

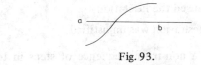

Fig. 93.

same would not be true for discontinuous functions. The drawing to illustrate it can be as bad as one likes it, or — by preference — even worse. It is a remarkable fact that Cauchy who gave the modern definition of continuity — almost simultaneously with Bolzano — initially did not care to prove this theorem; initially he appealed to the drawn image. On the other hand, for Bolzano the need for a proof of this theorem was just the starting point for his analysis of the mental object "continuity", which led him, too, to the modern definition. Yet Bolzano's interest and strength in mathematics was the analysis of mental objects rather than the creation of mathematical contents. Cauchy, however, continued working with the mental objects even where he had performed conceptual analyses — in limits, continuity, derivative, integral, and so on.

Once the definition of continuity was accepted, it appeared to admit functions as continuous which trespassed the limits of primitive imagination. It lasted for quite a while until the majority of mathematicians acquiesced in this unavoidable consequence. That is to say, difficulties arose where the step from the mental object to the concept had not been taken in an early stage. New generations soon got accustomed to this. Taught by the accepted definition of continuity, they revised the primitive mental object. But do not forget: the primitive mental object was indispensable — historically in the development of mathematics — and it remains indispensable, in the mathematical development of the learner, as an endpoint, or as a stage on the way to the more sophisticated mental object.

With objects that one habitually recognises as geometrical, it is different. As I stressed several times, one can advance very far in geometry without transforming the mental objects into concepts. If finally one lends them more precision by putting the mental objects as concepts into a logical system — algebraic or axiomatic — one will not be taken by surprise. Made even more precise they will show the expected properties.

Some restrictions should be made. The mental object "rectangle" might originally not have included the square, but as one progresses one gets convinced that it is better to include it, although in everyday language one would not call a rectangle something that is clearly a square. One would prefer to count the sign of a priority road among the rhombuses rather than the squares. It was probably not anticipated in one's mental object that a parallelogram has no axes of symmetry. But once drawn to one's attention, this is soon redressed.

A more serious thing is paradoxes, well-known in elementary geometry: by the wrong imagination of lines intersecting inside — rather than outside — a triangle, one can be led into contradictions. It is remarkable that the source of this mistake is somehow topological. It is a shortcoming of the visual insight into order properties — order, indeed, is a weak form of topology.

Nevertheless it may be asserted that ordinary geometry admits of a deep penetration before the mental objects must be sharpened into concepts, and even this can be restricted to a local procedure. It does not require a system, a global frame. Earlier on* I have shown how far this road can be followed even in analysis, how long continuity as a mental object suffices, as does the integral, as a geometrically viewed area.

In spite of all visuality in which topology can flourish, it is a different case. Since the watch-word Modern Mathematics has been heard, topology has more

---

* *Mathematics as an Educational Task*, Chapter XVII.

and more been propagated as part of mathematical instruction. The harvest has not been impressive: tesselations, graphs, the Euler polyhedron formula, the Moebius stip, perhaps even surfaces of higher genus, but in spite of all good intentions, it remains short runs and dead ends. With regard to our preceding exposition this is easily explained. Mental objects do not lead far in topology; concept formation is required to pass beyond the limits, and then concept formation which means more than local organisation. The concept former has mental objects in his mind, and ever new mental objects are formed on ever higher levels — spaces and varieties of arbitrary dimension and structure and their mappings, connected with algebraic structures — in order to be conceptualised again.

And how much is not required for the conceptual sharpening, for limit, boundary, continuity — not to speak of dimension and connection. It is concept formation on a high level, sophisticated alternations of quantifiers — for each ϵ there is a δ — and then not separated from the mental object but starting with it, keeping it in one's grasp, falling back on it, because unless much routine is acquired, it is a hard thing to manipulate the concept divorced from the mental object.

Or should we look for the conceptual sharpening of these mental objects elsewhere, not in topology?

9.2. *The Topological Context – Is This Topology?*

I started the preceding section with the pun of characterising topology as the art of drawing badly. Perhaps I had looked too much at Piaget* who deduced from the poor drawing techniques of the little ones that "représentation de l'espace" starts topologically. I explained meanwhile that the argument of the bad drawing techniques rests on false delimitations and the paralleling of ikonic and symbolic intentions of the experimenter who sets the task, and of the subject who performs it.

Meanwhile I myself have stuck to the argument of bad drawing although at an earlier stage I had asked with regard to representation that we should distinguish imagination and reproduction, and not identify the mental object with its picture. Of course this meant that my topology is left stranded in the plane because that is the domain where drawings are made. Properly said, I should rewrite the preceding section, but you cannot keep changing. Moreover it is a good opportunity to show the reader what was wrong. It is a habit I will never unlearn, in spite of efforts to change my own life and that of others. Again I started at the wrong end, at the topology which I know, to be sure at a low level but then from its upper rather than its lower side. "Topology" soon suggests "topological mappings". It is readily understood as identifying objects by one-to-one continuous mappings and forgetting about more primitive ideas preceding it. I took too seriously that topology to which Piaget had paid lipservice.

* J. Piaget and Bärbel Inhelder. Chapter 8.1.

Bad drawings — you can never draw so badly that it becomes topology. One can draw a rectilinear triangle so badly that it is neither right nor isosceles. (Unfortunately not so that it is neither acute nor obtuse and this then creates the chance to be misled by the mental object.) One cannot draw a continuous function in a manner that it is *only* continuous, even not approximately. Well, in the case of the Peano function one succeeds in suggesting by a few steps the whole sequence, and all the so-called pathological cases are being approximated systematically by non-pathological ones. But all this is far away: it presupposes too much sophistication.

I shall start at the other end, not with the concept, in order to uncover the mental objects where they are rooted, but in a more naive way. Yet how naive can you behave if you have learned a lot of topology?

I note down a few words:

> connection, arc, dimension, hole, border, tunnel, cave, path, circulation, braids, knots.

Where should I start? How systematically? How unsystematically?

9.3. *Connection*

Connection looks the most primitive. One's own body is connected, though disconnected from others. A tree is a connected whole from the roots to the top, separated from other trees, but you can cut it, into pieces. The network of streets of a city — a connected whole. A river from source to mouth with all its tributaries — connected. The pathways around a block of houses, but — mind! — do not cross! "Continent" means connected; an island is a thing detached. But how about a peninsula? Three sides water, the fourth land — I learned. How wide is the fourth allowed to be in order to leave it a peninsula? And how if you pierce the isthmus? And then build a bridge over the canal?

9.4. *Jordan Arc*

A string is connected until it is cut, and then pieces can be tied together — the break—make transformation. But the string can also be split, into connected threads, side by side. Was the string disconnected before it was split? As a twine of threads it was connected; if they are untwined, it is another object, which is disconnected.

The string suggests a mental object

> a curve, which may cross itself

if I throw it casually on the table, but if I avoid self-crossings it suggests a

> Jordan arc,
> the simplest connected figure.

The simplest because it can be

> split by a pointlike cut,

and because it can be seen as

> part of any connected figure;

the

> points of a connected figure are joinable arcwise.

Though the string is flexible, there is not much stretch in it. In this respect it is not examplary for the mental object arc. A rubber string is better, but also too special, because its stretching behaviour is a similarity.

> All arcs are the "same".

One can see it and understand what it means. One can materialise it

> by laying them upon each other, with stretching and shrinking.

9.5. *Continuity*

One more step? I hesitate. The word "same" I used above can be put into a broader context by defining for all that is connected, continuity of mapping:

> continuous is that which nowhere breaks the connection.

(Etymologically "continuous", indeed, means connected, this then was the mental object that preceded the modern concept of continuity.)
   Furthermore one defines for a mapping

> one-to-one-ness: it neither folds nor glues; what is different, stays different.

A mapping can be

> topological: one-to-one, continuous.

Two Jordan arcs are topologically the same.

9.6. *Linear Order*

If I cut a string, I get two strings, which I can again compose. Is the point of the cut broken into two points, which are afterwards amalgamated? Well, the concrete string is concretely cut; the cut is no mathematical point. Cutting the arc can be described as you want it, among others, by having a point divided in two points, which afterwards are identified. I may also demand that the point of the cut is attributed to one part only. Then I get an

> arc with an open end.

Equally well can I imagine an

    arc with both of its ends open,

such as

    the infinite straight line.

If this one is cut without doubling the point of cut, I get two arcs

    with one and two open ends,

respectively.

    On an arc there is a natural concept

    between;

b between a and c if by the cut at b the points a and c get into different pieces.
    As a consequence of betweenness an arc possesses

    two — opposite — orders.

Topological mappings

    preserve betweenness,
    an order passes into an order.

9.7. *Mathematical Comments*

How do the various objects depend on each other?

Connection looks like the most primitive concept though initially it is rather vague.

Jordan arc, illustrated by rope or rubber string looks more sharply determined.

Afterwards connection is more sharply described: the arc is an example of connection; moreover it is stated that two points of a connected figure are arcwise joined — as it were a criterion to test connection. It is, however, not the official definition; there are examples of (not too mad) sets which one would like to classify as connected though they are not so in the sense of arcwise connection.

Continuity in the sense of preservation of connection is not the official definition. One can make real functions that map intervals on intervals (accept between two values any intermediate value) but which are not continuous in any reasonable way. Continuity of real functions requires the originals (rather than the images) of intervals to again consist of intervals. However, for one-to-one mappings conservation of connection is a valid criterion of continuity in the usual sense. Consequently, with the suggested definition of continuity the definition of a mapping to be

    topological as one-to-one continuous

is valid. By this means

> *Jordan arc* as a topological image of a line-segment becomes more explicit, though the logical circle

> arc

> connected = arcwise joinable
> continuous = connection preserving
> topological = one-to-one continuous

> arc = topological image of a line-segment

is not broken.

9.8. *More About Connection*

A closed rubber string is connected, a Jordan arc the ends of which are tied together:

> a Jordan curve.

One cut suffices to transform a Jordan curve into a Jordan arc. A Jordan curve can be seen as the

> topological image of a circle circumference.

The Jordan theorem says that a Jordan curve in the plane breaks the connection of the plane into an interior and an exterior — a visual property which does not ask directly for a proof.

Is a chain connected? Roughly viewed it is. But the fine structure of a loosely held chain suggests a system of linked Jordan curves — if tightened it looks connected. If I am right to require that the points of a connected figure can be joined by arcs, the loose chain according to its fine structure may not be dubbed connected. But how to separate its parts? How to separate in general two linked Jordan curves in space? Concretely, I open a link in order to close it after the separation. A pair K_1, K_2 of linked Jordan curves can be mapped topological upon a pair K_1', K_2' of unlinked ones: K_1 upon K_1', and K_2 upon K_2'. This mapping, however, is restricted to the curves themselves; it cannot be extended to space as a whole. The K_1, K_2 and K_1', K_2' viewed as such are topologically the same but they are lying differently in space — a visually clear fact that does not ask directly for a proof.

The example of the loose chain shows that "connection" as a mental object is not as simple as it looks. The linked curves K_1, K_2 though separated, cannot be pulled from each other. The pair K_1, K_2 is topologically equivalent with the unlinked pair K_1', K_2'. We consider the figure consisting of K_1 and K_2 as not connected, because we mean connection as an intrinsically topological concept, independent of the situation of the figure in space.

9.9—9.16. DIMENSION

9.9. *Three Approaches*

Old geometry texts start with the somewhat classical definitions:

> a point is that which has no parts,
> a line is a length without width,
> a surface is a length and width without thickness.

Another set of definitions is the following:

> a line arises from a moving point,
> a surface arises from a moving line,
> a space arises from a moving surface.

Starting at the other end one gets the sequence of definitions:

> surface is the boundary of a body,
> line is the boundary of a surface,
> points are the boundaries of lines.

All these definitions interpret visual experiences, but they do it in a way that does not directly lead from mental objects to concepts, though for centuries philosophers have believed in this possibility.

9.10. *The First*

As to the first sequence:

> there are objects (spots) small enough to suggest something so small that it can be divided no further,
>
> there are objects (threads) the width of which pales in significance besides its length, suggesting in this way something with a length without a width,
>
> there are objects (sheets) whose thickness pales in significance besides the other extensions, suggesting in this way something with length and width and without thickness.

The suggested mental objects point, line, surface are conversely useful to describe certain properties and aspects of real objects (spots, threads, sheets).

In order to sharpen the mental objects conceptually one has to start at the visually clear "height, width, thickness" — three aspects of use of the same mental objects, which, however, are not interpreted as a measured magnitude. How long, wide, thick does not matter; rather the extensions are meant qualitatively. The simplest object of this kind is a Jordan arc — the simplest length without width; the simplest surface would be the cartesian product of two Jordan arcs. But one would certainly not be satisfied with these choices. A

Jordan curve and much more complicated thread-like figures shall be considered as lines, and spherical surfaces (which are not cartesian products of two factors) shall be so as surfaces. This can somehow be redressed by interpreting the definitions locally, but even then one is left with difficulties — how should I understand that an 8-like curve has a length but no width in the double point?

9.11. *The Second*

Let us pass to the second sequence of definitions. They are more precise than those of the first sequence, but we already know that as early as the first step the sequence goes wrong. The moving point — the continuous image of a line-segment viewed as a time interval — may cover a square and even a cube. This can be redressed by adding differentiability requirements, so that the line becomes a differentiable curve, the differentiably moving line a differentiable surface, the differentiably moving surface a differentiable piece of space. With the rise of Calculus lines, surfaces, spaces have, indeed, been interpreted in this way. But adding a new "motion" parameter, lines, surfaces, spaces were described, that is, by functions of one, two, three parameters, a "variable" point with co-ordinates x_1, x_2, x_3 was given by

$$x_i = f_i(s),$$
$$x_i = f_i(s, t),$$
$$x_i = f_i(s, t, u),$$

respectively. The functions f_i were supposed to be "continuous", which in fact included differentiability of any desired order. Moreover "independence" of the functions f_1, f_2, f_3 was assumed in order to exclude degenerations of spaces into surfaces, surfaces into lines, lines into points.

   With all these sophistications we are far away from the intended mental objects line, surface, space.

9.12. *The Third*

Let us now look at the third sequence of definitions. The striking distinction compared with the others is the start at the top, with the bodies, bounded by surfaces, which are defined by this capacity, that is as boundaries. Well, the definition yields closed surfaces only. This can be redressed by admitting extended pieces of surfaces again as surfaces. Their boundaries in turn yield lines — primarily closed lines, but afterwards also pieces of them. Lines in turn are bounded by points. It seems to work better than the first and second approach. It starts at the top, with bodies, in three dimensions. The descent to a lower dimension is systematic: the object is deprived of its fullness; the peel is left; the thickness of the body is lost, it is reduced to its boundary, the width of the surface is lost by leaving only its border, and the length of the line by leaving the endpoints.

The first two sequences are clearly inspired by the idea that as in arithmetic one starts at 1, so in geometry one has to start at the point. But as a cognitive development geometry certainly does not start with points. Earlier on I have put the rigid bodies first and foremost in the development, and if there were anything that I would allow to be detracted from bodies, it would be the solidity, rather than the bodylikeness.

9.13. *Surfaces*

Surfaces occur primarily — as the name says — as faces of something, as walls, tabletops, floors, waterlevels, peels, skins, clothes that wrap, bags that comprise, barrels that contain something. Primarily, I said, because in the long run we can detach the surfaces from the bodies of which they are boundaries, even while using the word surface.*

Sails and flags in the wind, leaves of trees, sheets of paper, curtains are objects that suggest surface without being surface of something. But unlike the surfaces of something they have two sides, right and left, or upper and lower, and if indeed they wrap something, inside and outside — sides that can differ by their look, but which primarily have to do with their situation in space.

Yet quite different physical objects can suggest surfaces: a fence, some wire-netting, a railing. They delimit space though not in the strict sense of inaccessibility; in spite of the holes they mark boundaries. The filled net is a particularly striking example, a surface pervious to water, but not to fish. Even one closed curve in space can suggest the surface it spans, and this holds to an even higher degree for nets of closed curves as found in wire-netting.

One step further: Independently of any embedding in space a cut-out or an atlas may suggest a surface, even surfaces which are not without deformations or not at all realisable in space, such as the well-known rectangle model of a torus (Figure 94) or the circle model of the projective plane (Figure 95).

Fig. 94.   Fig. 95.

9.14. *Lines*

After the efforts to show the phenomenological origin of "surface" as "surface of something" one may expect, in order to get a well-shaped closed system, that

---

* The German language knows *Fläche* and *Oberfläche*. English and French are restricted to one term, surface.

a similar origin of "line" is postulated, that is as boundary — or piece of a boundary — of a surface. Systematics is excellent in an *a posteriori* synthesis, but not as an analytical principle. Systems are artful and artificial and for these reasons sometimes useful, but let us not yield to this seduction.

There are, indeed, numerous examples of lines suggested by borders or pieces of borders; in particular circles first appear as rims of cups, dishes, bowls, buckets, wheels, of the sun and the full moon, and as the horizon. If we use the term circle, but also triangle, square, rectangle, we often do not know whether we mean the surface or its boundary. The child is early taught to reproduce an object by drawing what appears as its circumference — perhaps against a natural inclination to draw a whole surface, which attempt is interpreted by the adult as scribbling. On the other hand we know that for other lines, in particular straight ones, a quite different phenomenological origin can be indicated. In Sections 8.5 and 8.7 I have argued this forcefully for the straight lines. As objects and processes that suggest straight lines, or at least rectilinearity, I mentioned

> arrows, trunks, sticks, pins, rims, edges, paths, folds, cuts, stretched strings.

Some of them can, depending on the actual situation, also suggest curvilinearity, for instance strings if not stretched. Rims, edges and folds derive from circumferences; cuts too, and certainly so if something is cut out. Nevertheless there are enough examples left of another kind of origin of the mental object "line" than as a border. In order to systematise this wealth of examples, I put on record four roots of the mental object line:

> arrow,
> string,
> path, cut,
> border,

and their mathematisations

> arrow: line segment;
>
> string: continuous image of a line-segment with a surveyable number of self-crossings;*
>
> path as well as cut (viewed as covered in time): continuous image of a time interval with a surveyable number of self-crossings;*
>
> border: boundary of a piece of surface.

Whereas in the case of surface the phenomenological analysis led to a primary "surface of something" and a secondary abstraction from the spacelike substratum, we recognise in the case of "line" a fourfold root the components of which are phenomenologically equivalent or at least almost equivalent.

---

* To avoid in an informal way pathologies like the Peano curves, which fill whole squares and cubes.

9.15. *Points*

Here we can put it briefly. Two possible and equivalent aspects can be distinguished: the point and the spot. The point as the end of a line-like object (the point of a pin) or as the end of a surface- or space-like object (the peak of a clown's hat or a pinnacle). The spot as the "smallest" piece of surface or space, perhaps produced by the point of a pen, a pencil, or a pin.

9.16. *Mathematical Comments*

Mathematically two aspects of dimension must be distinguished, the differential-topological and the purely topological aspect.

In differential topology the objects to be studied are created by mapping line-segments or products of line-segments, where more stringent requirements are imposed on the mappings than continuity only: differentiability of some order and independence of mapping functions. The dimension of the object is then defined by the number of independent parameters used to describe the object.

It is unsatisfactory that a seemingly primitive mental object has to be sharpened into a concept so far away, in analysis. This has first been made conscious by H. Poincaré, who at the same time proposed a new approach, afterwards realised by L. E. J. Brouwer, P. Urysohn and K. Menger, in so-called dimension theory. The approach is closest to the third sequence of definitions, in Section 9.9. In order to stress what is visually essential, I avoid exaggerate precision in my exposition:

> A pointset is by definition n-dimensional if it admits everywhere cutting out arbitrarily small pieces by cuts that are at most $(n-1)$-dimensional (but not already at most $(n-2)$-dimensional).

This is a so-called inductive definition. One has to start somewhere:

> The empty set (and no other) is (-1)-dimensional.

Thus a set is 0-dimensional if everywhere arbitrary small pieces can fall off without cutting. So is for instance the set of rational numbers, and Cantor's discontinuum. A line-segment, a Jordan curve — even tne union of a countable set of Jordan curves is one-dimensional, and so it gets on to higher dimensions. It is, however, a non-trivial fact and hard to prove that the cartesian product of n line-segments (the "n-dimensional" cube) is n-dimensional in this inductive sense. This again creates an enormous distance between the concept of n-dimensionality and the mental object of which it should be the conceptual sharpening, though this distance is not yet felt in the definition itself but in essential applications.

"Cutting out" and "arbitrarily small" as meant above require more precision

— the first by means of the concept of connection, the second by means of that of topological space.

9.17. *Dimension and Measure*

It looks out of proportion paying so much attention to dimension and it would indeed be exaggerate were it not that dimension is an indispensable tool if magnitudes and their mutual relations are at stake.

The first sequence of definitions, of Section 9.9, is the traditional approach to formulae like

> area equals length times width,
> volume equals length times width times height,

definitions which are of course restricted to rectangles and planks, and which then are supplemented by formulae for

> circumference (of the rectangle)
> surface (of the plank),

and perhaps more formulae for elementary figures.

Didactically this algorithmic treatment can of course only be the end of a development which runs through various stages.

From magnitudes one can form new magnitudes by mutual multiplication (and division). By putting twice or thrice "length" as a factor one gets the magnitudes

> area

and

> volume.

If measuring these magnitudes one chooses the units again as products of the length-unit, for instance,

$$cm^2 = cm \cdot cm$$
$$cm^3 = cm \cdot cm \cdot cm$$

as a unit of area and volume, respectively.

If magnitudes are to be measured, the first thing that matters is to know what kind of magnitude it is — this is a general requirement — length, area, volume, time, velocity, weight, work, and so on. At this point dimension comes in, at least as regards the first three: What dimension does the object to be measured have? Or rather: Which dimensions of the object are relevant? For instance a road can mean for me

> a length if it is leading me from here to there,
> a surface if I must pay asphalting,
> a space if a groove must be made for it,

and correspondingly what is measured, are

> lengths, for instance in km,
> surfaces, for instance in m^2,
> volumes, for instance in m^3.

There are more primitive aspects concerned, which should be stressed for didactical reasons,

> the behaviour of geometric measures under geometrical multiplication, depending on the dimension.

Linearity is such a suggestive property of relations that one readily yields to the seduction to deal with each numerical relation as though it were linear. Understanding that multiplication of

> length by d,
> areas by d^2,
> volumes by d^3,

go together with the geometrical multiplication by d, is mathematically so fundamental, that, phenomenologically and didactically it should be put first and foremost. This fact rather than formulae for circumferences, areas, volumes, should be primary. The behaviour of various numerical characteristics of the same object, depending on the dimensions they are related to, plays a part not only in physical enlargements and reductions but also in reproducing and reading reproductions of such objects. We may conclude this exposition on dimension with the somewhat paradoxical remark that dimension, though conceptually located by us within topology, has its significance as a mental object by its relation to measures — lengths, areas, volumes — rather than topologically.

9.18. Border and Boundary

The mental object "border (or boundary)" has been mathematised in two divergent ways: *set theory* topologically and *algebraically* topologically. The first version is about the *boundary of a set S in a topological space R*: a point p of R is called

> *interior* to S if there is a neighborhood of p entirely inside S,
> *exterior* to S if there is a neighborhood of p entirely outside S,

and the remainder is the *boundary* of S with respect to R, thus p is called a

> *boundary* point of S if any neighborhood of p possesses points in and not in S.

Boundary points of S with respect to R can or cannot belong to S itself.

An essential feature of this concept of boundary is its relativity with respect to the embedding space. A circular disc, considered as part of its plane has the ordinary circumference as its boundary, the same disc as part of the (ordinary) space is entirely boundary. A hemisphere as part of a spherical surface has a circle as its boundary, but as a part of a full sphere it is all boundary.

In contradistinction to this *relativity* the "natural" mental object border has an *absolute* meaning: the border of a sheet, of a footpath, of a bucket, of a city depend only on the object of which they are considered as a boundary, independently of any embedding in a larger something. At least, so it looks. However, one can view it also in another way:

> if we speak of the border of an object we relate it to a "natural" extension of this object, we view it as naturally embedded into a larger whole.

Such an embedding is most natural for flat objects in space: the border (or boundary) is determined with respect to the plane in which it is included. The endpoints of a line-segment form its boundary with respect to the straight line that extends the line-segment.

This then is the viewpoint of the algebraic version of the border concept. One considers points, line-segments, triangles, tetrahedra and higher dimensional analogues and defines their border in a "natural" way, independently of any embedding, albeit with a certain algebraic sophistication: the line-segments, triangles, tetrahedra, and so on, are assumed "oriented"; they are linearly combined, and borders are defined as linear combinations (Figure 96):

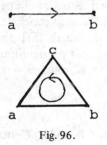

Fig. 96.

$$bd[a, b] = [b] - [a],$$

$$bd[a, b, c] = [b, c] + [c, a] + [a, b],$$

$$bd[a, b, c, d] = [b, c, d] - [a, c, d] + [a, b, d] - [a, b, c].$$

In the case of flat or rectilinear objects a natural extension is uniquely determined to wit as a plane or a straight line. With a certain latitude the same can be said of many objects which can be given, even if not precisely, yet with a certain clarity, and delimited in a larger whole: parcels in a landscape, countries on the globe, rooms in a house, fences, landmarks, and walls can represent or suggest borders either isolated or in relation with each other. *Territories*, as we shall call them, can be delimited against the world around or mutually by concrete, marked, or symbolic frontiers, restricted accessibility and defense conditions. Romulus kills his twin brother Remus who steps over the not yet *concretely* operational wall of Rome, in order to stress its *symbolic* function.

Bastiaan (2; 5) in the park; I drew a circle around him with a stick. "You are locked up, you cannot go out." He accepted it. Only after I wiped out a piece as a door, he stepped out.

A barrier across a road determines a frontier, though concretely I can pass under, above, and around it. By his mere presence a policeman closes a street, with his stretched arms a child marks a blockade of the footway. The law requires a white line on the street before a traffic-light; a no-entry sign suffices to forbid entrance. According to the circumstances the two halves of a street are one territory or two with a continuous or interrupted line as a border. To decide whether a person is indoors it does not matter whether some window is open or closed, unless he leans far outside, but even with closed windows the frontier can be violated if due to too much attention from outside, or if with a draught one feels as if one is "sitting on the street".

The kerb of a side-walk has a certain width; the frontier between two countries is actually a strip of no man's land. On the other hand land territories extend to an undetermined height and depth into the soil and the air. It depends on the functionality of the border how strongly the bounded territory is of a higher dimensionality than the boundary.

The words border and boundary from the title of this section have been used synonymously, sometimes the one preferred to the other — a third word I used was frontier. "Border" accentuates the separation of territories or of inside and outside; "boundary" is rather the end of an object. The extension of a territory beyond its border is concretely given — again a territory. Extending a cup, or a vase, a dish, a cap beyond its border or boundary is a matter of imagination. In order to extend such objects so unambiguously that a boundary can be claimed with respect to the extended object, another latitude is required than in the former examples. It now matters how the cup, or vase, or dish, or cap would continue; rather than how far or how thick, it matters how curved, how vaulted. Or rather, even this does not matter because finally what is the border does not depend on the way of extending. One can see and feel where the objects terminate. Two-dimensionally viewed these are surfaces with a boundary; at ordinary points one can move in all directions, there are disc-like surroundings; at the boundary the freedom of movement is restricted, in some directions one would drop off. But also three-dimensionally viewed the border of a cup, vase, dish, cap behaves differently at the boundary: the surface of this body is sharper curved in the points one would call boundary than in their neighborhood.

Let us summarise: The set theory topological concept of boundary rests on a relation with the embedding topological space. The mental object "border"

can be sharpened to this concept, if as embedding topological space a natural
extension is presumed which is either concretely present or derived from the
structure of the object by imagination.

In the following, border and boundary will be viewed from the aspect of their
coming into being.

9.19. *Prick, Cut, Slit, Passage, Hole, Tunnel, Cavity*

The words in the title of this section aim at phenomena which can be concretely
present in some object or — concretely or in the imagination — caused by acts
which are described by (similar) verbs

  pricking, cutting, slitting, drilling, digging.

Inverse acts are

  fusing, stitching, gluing, shutting, filling.

The prick in a one-dimensional object makes it fall apart — at least locally;
a new one-dimensional object comes into being, or two or even more. A two-
dimensional object, say a sheet or a disk is pricked in order to attach it somewhere or to have it turning around the prick axis. Sewing is stitching together
by joining prick holes by threads — how is this done by the sewing machine?
A three-dimensional object can be pricked in or through — a candle pricked in
the pin of a candlestick, chestnuts pricked through by means of a darning needle
— but curved pricks, too, are admitted. The prick suggests a damage of the
lowest possible dimension — 0-dimensional in 1- and 2-dimensional objects,
1-dimensional in 3-dimensional ones. The damage caused by a cut in a sheet
or in a body by a slit is one dimension higher. Cutting can be cutting in as
well as cutting through, slitting can be sawing in or through. Cutting in a sheet
at the border does not change the structure too much; if the sheet is elastic one
can stretch the cut-in sheet to look like the original one. However cutting in
the interior of the sheet creates a slit within the sheet, a stretching of the prick.
Cutting out creates a hole, which can be viewed as originated from stretching
the prick or the slit. The more holes I make in a 2-dimensional object, the more
complicated its structure becomes; there are 2-dimensional objects with systems
of holes: sieves, nets, curtains.

This is the starting point for a full-scale mathematisation: the function theory
plane (or spherical surface), from which a closed set is taken away — a system
of punch-holes, point-like ones, line-like ones and true holes — it does not
matter — but such that the surface does not fall apart. If the number of punches
is finite, say n, the surface is *n-fold connected*, which means that n cuts from
border to border cause the surface to fall apart: one cut for the once punched
surface — it is simply connected; two cuts for the twice punched surface — it is

doubly connected; for the thrice punched surface (the swimming trunks) three cuts — it is triply connected. These are visually obvious facts, but a proof that p cuts from border to border certainly split the spherical surface with p holes, is not easy.

Likewise 3-dimensional objects can be split superficially or thoroughly. Superficial splitting does not change them essentially; if cut through, they fall apart. Splitting is a 2-dimensional damage. A 3-dimensional damage is suggested by drilling, which again can be superficial or piercing. A pit does not change the terrestrial surface essentially nor does a blind alley; a tunnel, however, which emerges elsewhere, does. Tunnels can appear in various ways: a tunnel is produced by the handle of a cup or bin; with the legs spread or the hands clasped a tunnel is formed; children when playing form an archway passed by others; "front door in, back door out" is as it were a tunnel through the house. The power line crossing the road between two poles looks one-dimensional, but the radio black-out caused by it makes it a true tunnel. Underpasses for pedestrians below city squares form a branched system of tunnels; the table with its four legs on the floor creates a similar tunnel system, with well distinguished passages. Or, even more involved: the tunnel system produced by the edges of a cube — the passages through it are determined by the sequences of faces of the cube which are being pierced.

The surface of the once or multiply pierced ball deserves to be considered: the surface of the once pierced ball is essentially a ring surface, or in other words, a torus. The twice pierced ball has a surface which is known as that of a pretzel; it may also be represented by a cup with two handles. A ladder with $p + 1$ rungs is equivalent to the p times pierced ball; its surface is of "genus p".

This kind of surface can also be subjected to a complete mathematisation. They are, apart from the punched sphere, closed surfaces, though of another kind than the spherical surface. By a "return cut" the torus can be changed into a cylinder, which in turn gets simply connected by a cut from border to border, and by one more cut falls apart. In other words by two return cuts the torus can be made simply connected. For the pretzel it can be reached by four return cuts. In general: in order to make a surface of genus p simply connected I need $2p$ return cuts; it is $(2p + 1)$-fold connected.

Two return cuts make the torus simply connected (Figure 97).

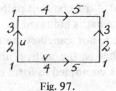

Fig. 97.

Four return cuts make the pretzel surface simply connected (Figure 98).

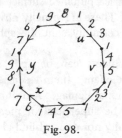

Fig. 98.

Moreover if such a surface is q times punched, it becomes a bordered surface, which is $(2p + q)$-fold connected.

Again the mathematisation requires profound and difficult reasoning.

Among the words in the title of this section there is one left that requires some comment. It is the word cavity, by which I mean a place you cannot get in if you are outside, nor out if you are inside: the cavity within a soap bubble, in a closed tent, in a closed room. It need not be spherical as is the interior of a spherical shell; the cavity may be a tunnel within a globe, or within a tunnel. It may be a multiplicity of cavities as those in foam.

9.20. *Multiple Connection*

"Connection" was the starting point of our sequence of topological pictures. In the course of our exposition the concept of "connection" underwent a sharp differentiation.

A sheet of paper and a spherical surface are called simply connected because they fall apart by any cut from border to border or by any return cut, respectively. A sheet with n holes requires n cuts to make it simply connected, a surface of genus p requires $2p$ return cuts in order to be spread as a simply connected domain on the plane. Conversely, these surfaces — closed or bordered — can be built from plane networks by sticking the cuts together (Figures 97 and 98).

This "homological" approach to connectedness can be confronted with the "homotopical" one: On the surface one considers *closed paths* starting and finishing in a fixed point, which are allowed to be deformed while the start and finish remain fixed. In the plane and on the spherical surface each closed path can be contracted into a point, "the constant path". On a torus (Figure 97) this is obviously no longer true: the paths indicated by 1231 and 1451, briefly called u and v, cannot be contracted, nor can they be deformed into each other, or into their inverses u^{-1}, v^{-1}, that is, the paths u and v completed in the inverse sense. The paths u and v can be combined to form new paths, like

$$uv^2 u,$$

which, read from the right to the left, means completing first u, then twice v, then again u. The paths obtained in this way, however, are not all different in the sense of deformational equivalence. For instance uv and vu can be deformed into each other. So u and v can be considered as commuting; thus by deformation all paths can be reduced to

$$u^m v^n \text{ with integers } m, n.$$

Similarly Figure 98 of the pretzel surface shows paths u, v, x, y, from which all other paths can be combined while

$$(*) \qquad x^{-1} y^{-1} x y u^{-1} v^{-1} u v$$

is a contractible path. From x, y, u, v one can form all "words" — not in a commutative way but as though (*) represents the unit.

The same kind of concepts apply to bordered surfaces. Take for instance the swimming trunk, that is the twice punched plane (Figure 99). The paths

Fig. 99.

considered start and finish at 0; the path u circulates in a certain sense around the one hole, the path v around the other one, u^{-1} and v^{-1} are their respective inverses. They can be combined to form new ones, while paths that can be deformed into each other are considered as the same. All closed paths starting and finishing at 0 can be combined from u, v and their inverses, while uv and vu are now to be taken as different.

The paths considered on such — closed or bordered — surfaces constitute what is called the fundamental or first homotopy group of the surface. I have briefly tackled this subject in order to show another variant of the mathematisation of multiple connection: besides the homological one, by cuts, we also have the homotopical one, by paths.

9.21. *Circulation*

If I travel along a closed path in the plane, there are points that are orbited, and others that are not; I can even distinguish whether the path turns left or right around some point. The paths considered are allowed to cross themselves — see the examples Figures 100–105.

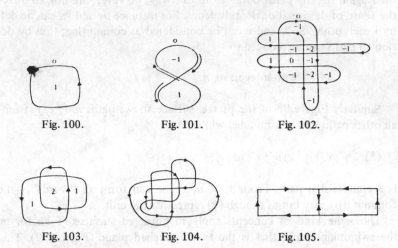

Fig. 100.   Fig. 101.   Fig. 102.

Fig. 103.   Fig. 104.   Fig. 105.

In order to provide more substance for the present section, I am going to tackle a mathematical problem, which I will solve in all rigour.

There is a celebrated topological problem, the so-called four colour problem: how many colours are required to colour a map — that is, a geographical map picturing countries — such that each country gets a colour and countries with a common border are differently painted? For instance, the map of Figure 106 requires four colours. It is an old problem as to whether four colours suffice in all cases: the four-colour problem. After numerous fruitless attempts the problem has recently been solved, as it seems: the answer is positive.

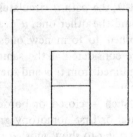

Fig. 106.

The problem I am going to tackle is much easier. There are maps that can be coloured with less than four colours. I will indicate a kind where two colours suffice, say black and white (if these are rightly called colours).

A closed path divides the plane in a number of domains. If the path shows a *finite number of self-crossings*, two colours suffice. This is clearly seen in Figures 100—104, whereas Figure 105 requires three colours.

In Figures 100—103 you may notice certain numbers the meaning of which I am going to explain.

If C is a closed path and p a point not lying on C, I can ask the question how many times does C revolve around p? I draw a ray from p to the point x that travels along C and I count how many full turns (of 360° each) the ray px has completed if x is back to its starting position. As usual in mathematics a clockwise turn is counted negative, an anti-clockwise turn positive. It can happen that the ray px runs back and forth, but what matters is the final result.

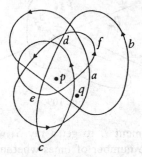

Fig. 107.

In Figure 107 the path C circulates 4 times around p; if C is replaced by its inverse, it becomes -4 times. Around the point q, however, the path C circulates 2 times: starting with x at a the first revolution is completed if x arrives at b, the ray qx starts running back at c, from d to e it resumes its forward motion, from e to f it runs again backwards, and from f to a forwards to complete its second revolution. (Look now at Figures 100–103 and verify the numbers in these figures.)

An easier way to determine the *circulation number* is the following: Draw a fixed ray S from p. The path C pierces the ray S several times, sometimes in the positive sense, sometimes in the negative one. Add these numbers while taking the sign into account. The result is the circulation number. For instance, if in Figure 107 the ray S is drawn from p horizontally to the right, it is just 4 times. If S is drawn from q horizontally to the left, it is three times positive and once negative, which makes it 2.

The number I get by this prescription does not depend on which ray I have drawn from p. If I change the ray S, I can lose or gain piercings by C but this happens always pairwise, a positive and a negative together — look at the six positions of S in Figure 108! (Look also at Figures 100–103!)

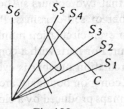

Fig. 108.

But what happens if I move point p? I take two points p_1 and p_2 and look how many times C circulates around each of them. I join p_1 and p_2 by a line-segment L. Suppose L is not touched by C (Figure 109).

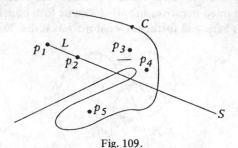

Fig. 109.

I produce the line-segment L to get a ray S, which as a matter of fact is intersected by C the same number of times whether it is seen from p_1 or from p_2. This means that C circulates the same number of times around both of them. The same holds with respect to p_2 and p_3, which can be joined by a line-segment that is not met by C. It again holds for p_3 and p_4, for p_4 and p_5. All of them are circulated around by C the same number of times. This shows:

Points of the same part of the plane, determined by C, are circulated around by C the same number of times.

What about neighboring parts? Let p and q be two points of this kind (Figure 110). Draw the ray from p that contains q. The numbers of piercings of this ray, considered from p and from q just differ by a positive or negative unit. Thus:

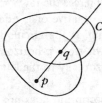

Fig. 110.

The circulation numbers for points of neighboring parts of the plane differ by ± 1. (Consider also Figures 100–103.)

We undertook to prove that two colours suffice to paint maps created by a curve C with a finite number of self-crossings. Paint those parts of the plane white where points show an even circulation number, and black where it is odd. Then each part has one colour. For parts with a common boundary the difference is ± 1. If in one part the circulation number is even, it is odd in the other one, and conversely. Parts with a common boundary are differently coloured.

This, indeed, proves that maps produced by a closed path with a finite number of self-crossings can be coloured with two colours.

9.22. *Weaving, Braiding, Knotting, Linking*

Warp and weft — the warp is a system of parallel threads, which are divided into two layers between which the weft is shot. The result is a fabric which can combinatorically be described as follows: if the warp threads are numbered 1 to *n*, the way of each weft thread is described by indicating whether the warp thread is passed above or below, thus by a sequence of minus and plus (for below and above). This sequence of signs changes from one weft thread to the next, according to a periodic pattern if the fabric is regular. The simplest pattern is the chessboard distribution of plus and minus, which is also used to darn socks. Plaiting mats — a kindergarten activity of former times, perhaps not yet entirely abolished — is a primitive example of weaving: the warp is a coloured sheet with parallel slits, the weft consists of paper strips of various colours which are interlaced by means of a darning needle.

From weaving mats and baskets one is led to braiding pigtails. The most primitive braid consists of two strands (Figure 111); strand 2 passes alternatively above and below strand 1. A more solid braid is made up of three strands (Figure 112): strand 3 passes first above strand 2 and then below strand 1, which is

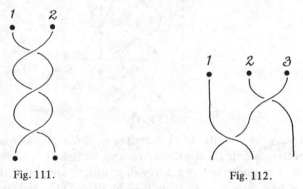

Fig. 111.   Fig. 112.

continued with a cyclic permutation of the strands: strand 2 passes first above strand 1 and then below strand 3; strand 1 passes above strand 3 and below strand 2; and so on. The number of strands and the pattern can be variated ad lib. This combinatoric description can be built into a whole mathematical theory of braids.

Knots can be described in a similar way. Mathematically a knot means a closed curve in space. To conform with this terminology the ends of the knotted string are joined to meet. The simplest non-trivial knot is the so-called clover-leaf knot, which exists in a righthand and a lefthand version, depicted from the right to the left in Figure 113. It seems to be a matter of taste which one is called right and which left, but it is not — such questions of orientation will be dealt with later on. It is remarkable that right-handed people prefer to tie righthand knots, that is, knots where the end of the string in the right hand is led above that in the lefthand.

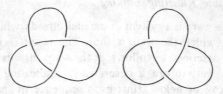

Fig. 113.

Knots can be contrived arbitrarily. If the lower and upper ends of the braid of Figure 111 are joined, one gets the knot of Figure 114. Knots can be described combinatorically by the pattern of crossings. Whether two knots are the same in the sense of deforming the one into the other, is a question that in general cannot be decided at sight. For instance a profound proof is required to show that the right and left clover leaf knot are different. The theory of knots has been developed to a high degree of sophistication. There is, however, no general method to decide whether two knots are equivalent.

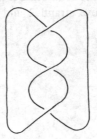

Fig. 114.

The last term in the title of this section is "linking". As one closed curve may be knotted, a pair of closed curves may be linked. So are, for instance, the links of a chain, a key and a key-ring, locked arms. But there are more sophisticated linkings, such as exemplified by Figure 115. Again it requires a lot of mathematical theory to prove that closed curves that are obviously linked cannot be deformed into an unlinked state without crossing each other.

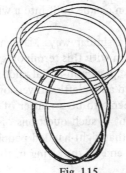

Fig. 115.

9.23. *Conclusion*

I have been reviewing quite a lot of topology — mental objects, processes, relations — of a forceful visuality, which however do not demand the birth of
topological concepts. Though mathematics had for centuries been pregnant
with topological objects, neither philosophical speculations on dimension,
point, line, surface, nor more mathematically tainted ones on knots and links
gave birth to concepts; labour, if there was any, ended with miscarriage. The
topological concepts require difficult procedures of constitution, whereas on
the other hand there is no urgent need for such concepts unless we move far
away, into a theoretical sphere. The mental objects, processes, and relations
which have been discussed here, are certainly important. We are accustomed to
them, but we are not accustomed to realise this fact for ourselves and even less
in a didactical context.

Notwithstanding the wealth of phenomenology, displayed in this chapter,
there was hardly any didactical phenomenology involved, or as far as it was
didactical phenomenology, it were so in a negative sense. What I have sketched,
has scarcely been tested on an elementary level. Everybody knows what "inside"
and "outside" means, what is a line, a surface, connection, border, knot, link.
It turns up spontaneously in one's mind, and since there is little need for conceptual precision, it is not easily seen from the point of view of teaching matter.
Topological subject matter such as that offered in more advanced expositions
as teaching matter, lacks the character of necessity while the mathematics
developed from it with the sole aim of doing some topology in mathematical
instruction, lacks sufficient motivation and is leading to a dead end. A popular
subject — combinatorics rather than topology — is travelling through a graph
such that each edge is used once and only once — a nicety known as the
Königsberg bridges. It does not take much trouble to find a necessary condition
for graphs to be travelled in this manner, but proving that this condition is
sufficient is much more than can be expected from a non-professional mathematician. Moreover, it is an unmotivated and isolated subject.

If I try to survey the examples of this chapter, the only one that promises
more in a mathematical didactical respect, is the circulation number, which is
motivated by a true problem and can be placed into a larger mathematical
context. I do not discuss at which age or on which level it could be realised.

This does not mean that the other examples of topology would be didactically
irrelevant. They should be appreciated as what they really are — mental objects,
processes, relations — and not wantonly subjected to badly understood mathematisations. The context in which they may play a part — among others — is
the theme of our next chapter.

The time has come to discuss once more the developmental priority of
topology, compared with classical aspects of geometry, such as is suggested by
Piaget and stressed by his followers. Earlier on I explained that Piaget's claim
rests on confronting quite diverging levels with each other: in topology he is

satisfied with the constitution of quite primitive mental objects, yet with respect to what he calls projective, affine and metric geometry, he makes high demands of concept formation. It has been shown in the present chapter that even with more sophisticated mental objects in topology the need for, and accessibility of, concept formation is far below that which we know in the didactics of classical forms of geometry.

CHAPTER 10

THE TOPOGRAPHICAL CONTEXT

10.1. *Order of Coexistence – Order of Succession*

The pair of terms in the title of this section stems from Leibniz' correspondence with a newtonian. Leibniz attacks the absolute space and absolute time of Newton's *Principia mathematica philosophiae naturalis*. For Leibniz space is an order of coexistence, time one of succession, motion is displacement of an object with respect to other objects that may be supposed fixed. I am not going to tackle things philosophically, I am rather dealing with Leibniz' exposition as an expression of common sense and as starting point to describe what I mean by topographical context.

Time, one-dimensional, looks simpler than space: the whole order of succession is described by "before and after", "earlier and later". Roughly described, that is: refining the relation is not that easy. Gradually in the development of the individual the past and the future become differentiated. Philosophically time appears even more difficult – a famous pronouncement of Augustine witnesses this uneasiness. That which was is gone, that which shall be is not – this expression of helplessness can be mitigated by saying: that which was is *already* gone, that which shall be, is not *yet*.

But let us stay away from philosophy. The phenomenon "time" can be spatially caught by clock, calendar, and time axis for practical and didactical aims, and the only thing to be careful of is not to disturb the spatial "catching" of time by the spatiality of space.

Space, with its three dimensions, is much more complicated, a complicated order of coexistence.

Bastiaan (6;4) after a talk at a three-forked road, says, about three- and four-forked roads in general: "There is much more than right and left, front and back". "How much?" "Certainly twenty." (I do not remember whether he also mentioned "above and below".)

Some of the coexistences can be stated at a glance, but what I mean by the topographical context includes

the coexistences that mentally must be constituted,
the mental system of coexistences,
coexistence as a mental possibility,
the mental system of coexistence as a mental object,

and perhaps some more variants of these subtle distinctions.

(To avoid misunderstandings I stress that if I say "space" I mean space. The topographical context is not restricted to two dimensions nor to two flat dimensions decorated by light bulges in the height and depth. The context I mean may include the layer structure of a building, the aerials on the roof, the airplanes, sun, moon, and stars if need be, even the world as seen from a frog's perspective, or from a bird's nest, or from the moon, it can extend to the interior of Earth, Sun and planets.)

In a few words I shall continue with Leibniz' order of coexistence. Leibniz shows how "place" is defined: Two things A and B are at the same place provided they possess the same "references of coexistence" with respect to things C, E, F, G, and so on, which have not had any reason to change their "references of coexistence". Then he continues: One must here restrict oneself to the references of coexistence rather than admitting all references since otherwise A and B would be the same according to the principle of indistinguishability.

Leibniz knows that his definition explains equality of place rather than place itself, like Euclid's (properly said, Eudoxos') definition of ratio, which aims at equality of ratio rather than ratio itself — a way of concept formation, which mathematical methodologists have become used to in the meantime.

With the definition of place the coexistence is detached from the succession. The order of coexistence must also be established, at least in principle, for objects that are not simultaneous or not simultaneously perceived or perceivable or even imaginable, and this happens, as is usual in the physical world, and is systematised in physics, in an indirect way, by intervening objects and sequences of objects, which can be placed pairwise coexistently in the order of coexistence. In order to illustrate it by a quite concrete example, I mention the procedure of the surveyor, who establishes the mutual position of two far distant points on the terrestrial surface by linking them through a chain of "small" triangles, with vertices in each other's neighborhood. Well, at present such things are done by photogrammetry from airplanes, but finally the pictures from such a film must be knitted together to make up a total picture — an order of co-existence constructed indirectly, built from direct orders of coexistence. Direct ones? No, as soon as details are considered, a pair of points of any particular aerial photograph is not coexistent in the sense of visual perception of per-ceivability, but they are so only in the mental relation.

This, then, is the topographical context, briefly summarised:

the catching of space (or of the space) as a mental coexistence of places, that is, of

places of objects,
places of objects and perceivers,
places of perceivers

in their mutual physical and mental relations.

Topography literally means "place description". Is this translation of "topo-graphy" not all we need? Or is this section a somewhat highbrow philosophical, or philosophically looking, introduction to a chapter with a potentially rich content that will be imperfectly realised, by means of examples rather than by a long list which anyway would be unsystematic because of the lack of any classification, and boring to the reader? As mentioned before, this chapter has come as a separate birth out of the womb of geometry. Its content is geometry, though of a kind that does not lead to concept formation in the sense of geometry — the geometry of rigid and flexible bodies, of motions and other transformations, of topology and combinatorics.

10.2. *Means of Expression*

Little if anything seems to be known about the development of the topographical context in general. Obviously its development will heavily depend on the world of the developing individual, whether it is restricted to one room, to an apartment, to a house with many stories, with or without a garden, whether this world is a plane or a mountain, whether the way from home to kindergarten is travelled on foot, or an mother's bike or by car, whether the environment of the home is explored at a young age, whether the outside world is imitated by picture books and television and by construction kits. However it develops, the mental context which I have called topographic will not show in the end much dispersion among members of the same cultural community. The verbal and other means of expression, however, by which this context is understood and described, as well as the variety of related concepts will sooner or later diverge greatly, depending on the general development of the individual.

A few simple examples may show the tension between the possession of a topographical context and the means to express it:

Monica (4;4) wants to indicate some place by means of the way leading to it. The index finger lifted but motionless, she says: "... and then you go so, and then you go so, and then you go so ...". Obviously she sees the way clearly with her mind's eye but she lacks verbal and even mimic means to describe it.

I want to go with Bastiaan (6;4) to a certain place. He agrees that the shortest way is "over the locks". "And then you go so, that is also a shortcut". I understand that he means "after the locks", and if I had asked it he would have been able to elaborate it mimically.

Even adults can experience difficulties if they have to show another, say a motorist, a way, but then the reason can be gaps in the mental topography — how many corners, how many traffic lights, a one-way street, striking characteristics of spots where the direction is to be changed; for instance, traffic signs leading to a goal beyond the desired one.

Modern manuals, from kindergarten onwards, show an increasing appreciation of the topographic context. The stress is on teaching passive and active mastery of verbal means of expression. This is welcome but not enough. The context itself deserves more attention. Moreover more means of expression deserve to be paid attention to in dealing with this context. On a lower level than, and as a preparation for, the verbal means of expression, there are

ostensive means — the way, the size of an object is shown —

accompanied or not by

>a verbal component — "there", "that large"

and

>a mimic component — the gesture "this way around the corner", "that far
>>away",

possibly

>accompanied by these words.

More sophisticated visual means of expression in the topographic context
show a great variety:

>diagrammatic sketches,
>views according to various principles,
>ground-plans,
>blue-prints,
>maps,
>atlases,
>globes,
>models,

supported by symbolic means such as

>coordinates,

and verbal means, such as

>legends.

Finally there are

>exclusively or preponderantly verbal means of expression
>>in colloquial language,
>>in a language created for topographical aims,
>>in mathematical language.

It is impracticable to illustrate this by examples detached from the applications,
which will be more closely viewed.

10.3. *Polarities*

The naive space shows the polarity

>*above — below*,

which I can transfer from one place to another, thanks to my balancing-organ
and the perpendicular fall of heavy bodies. Geography and cosmography have

taught us otherwise, but the gulf between naive and cosmic space can be bridged only by theoretically educated intuitions, supported by topographic means.

Things display polarities, which can be determined by their normal situation in space as well as be independent of space. Man can stand upright in space, from head to toe, but he can also lie down or stand upside down. Houses, trees, mountains stick fast from cellar to attic, from root to crown, from foot to top; cars move such that normally their above–below coincides with that of space. Tables, chairs, cups, and bottles possess an above–below, which keeps its value independent of their situation in space, even though among these situations there is a preference for those that match the space polarity.

The polarity "above–below" determines a gradient and an order. A thing is above or below another – *directly* above or below, or higher or lower. What is the common basis from which this "higher" and "lower" is measured? What part is played in this by the oblique lines running from high to low, whether they are concretely realised, realisable, only mentally present, or abstract mental or theoretical conceptual constructions?

How are the outside visible vertical lines of a building related to the layer structure of the stores, to the staircases within the building, to the leaping numbers in the elevator?

Notwithstanding the – necessary – stress on space, I must not neglect a usual restriction of the polarity "above–below" to the plane – I mean not the plane as such but the drawing and writing plane, the plane on which communications are written and objects pictured. How has this plane acquired this above–below polarity? Obviously it has been transferred from walls and other standing planes as communication and picture planes with their natural "above–below" to more or less horizontal planes, and with some arbitrariness the side near the viewer has got to play the part of the below. As a matter of fact, in perspective representations this can produce a conflict between the polarities "front–back" (foreground – background) and "above–below". Young children experience difficulties with this overlap, even the convention of identifying "near the viewer" and "below" is only slowly accepted in their own production.

Another polarity is that of

head – tail

in living creatures, transferred to other objects such as arrows, pencils, roads. A sausage or a rod has two ends which can arbitrarily be assigned the part of head and tail. Roads can be neutral, or one direction can be stressed: to town, out of town, up hill, down hill, one-way traffic.

"Above–below" and "head–tail" are

polarities of direction,

mental objects, which can be sharpened to the mathematical vector concept. Another kind is the

polarities of side and opposite side,

of a street, a river, a wall, a coin, a leaf. With respect to one's body this polarity is sharpened into

front — back,

and in a more neutral form to

in front of — behind.

With a curved or closed surface the polarity of side becomes

inside — outside,

of a room, a house, a cup, a bend in a cycling-track, a river, a garment, a bag, a fence. This — static — polarity determines a movement

from one side to the other,
from inside to outside —

crossing the street, uptown. These look again like polarities of direction, but it is a more vague direction: one can cross the street in various ways, there are many directions across the street, to the reverse of the medal, in and out of the forest. The distinction, however, is not sharp: there are also many ways from the head to the tail of an animal. It depends on the stress whether some polarity is understood as one of direction or of sides.

Things that show a polarity of direction can be symmetric around this direction (vases, bottles, pots, arrows) or indifferent (mountains, bushes, or they can possess a second polarity. Two polarities of direction — mathematically two vectors — determine a plane, and consequently a polarity of side and opposite side. It is a well-known feature of our own body. Besides the polarity of "head-to-toe" it possesses that of front and back, which is transferred as such to our environment. They become conscious and are verbalised at an early age. Because of the near-symmetry of our own body the polarity right—left takes longer, and quite a few people experience it as a difficulty, though mathematically the polarity right—left is determined by the two former ones, "above—below" and "front—back". Mathematically, indeed, though what is one's right and one's left side, has to be learned empirically. As regards our own body two phases can be distinguished,

the operational and
the verbal

distinction of right and left, similar to the distinction of colours: distinguishing colours and knowing their names. Even if the child is used to writing with the right hand and shaking the right hand, the question "which is your right hand?" or "which hand is this one?" can cause hesitations, and a correct answer to this question does not imply that the child can indicate his right foot, eye, ear.

New difficulties can arise as soon as the relation to one's own body is severed, or the distinction between right and left must be coordinated with motor acts or experiences of one's own body. I used to remember the fact that as late as the age of 9–10 I had trouble to distinguish right and left (although I am righthanded). If now I analyse this recollection, I think the case was different. My recollection is tied to gymnastics lessons: I had to imagine myself standing in the drill-hall in a certain direction in order to decide what is right and left. Yet this relation to gymnastics indicates something else: understanding the commands right turn, left turn. Neither operationally nor conceptually, indeed, does the distinction right—left imply that of right turn — left turn. It is a mere definition that right turn means that the front vector, that is, the sight or the stretched arm, moves on the shortest way to a position at the right — a conceptually acceptable definition that need not be operational. Certain tools (keys, taps, corkscrews) have to be turned right to carry out certain operations, and likewise the clock hands turn right, but that is only metaphorically related to the right turn of one's body — a point we shall deal with later on.

The distinction righthand—léfthand can be transferred from one's own body to other bodies and objects by means of displacements or other continuously acting transformations. As soon as the polarities above—below and front—back are given for any object, the polarity right—left is also established. Somebody who faces me eye-to-eye has a front—back vector that is opposite to mine, while his above—below coincides with mine, and this means that his right—left is the opposite of mine, his right hand touches my left hand. If I am lying on my left side and somebody else at my side but with his head at my feet and looking the same direction as I do, then his above—below is opposite to mine, his front-back the same as mine, thus his right—left opposite to mine, which means that he is lying on his right side.

With non-living objects we behave less consequentially. Of a cupboard in front of me, the side that faces me is considered as its front. But I would not call that side of the cupboard right or left which I would do if the cupboard were a man. On the contrary the side of the cupboard that is on my right is called right, and that on my left is called left. With a portrait group in a newspaper one may doubt what is meant by "to his/her right" if it is not somehow explained. Books on a shelf are placed with their backs in front of us, with the consequence that the pages succeed each other from the right to the left and in the usual arrangement of volumes from the left to the right the first page of Volume 1 is close to the last page of Volume 2.

Inconsistencies like this one are also found in the use of in front of — behind. Somebody sitting in a room, seen from the street — is he sitting in front of or behind the window? Which one of two objects, seen in a straight line, is behind the other? (The English language knows the subtlety of "in front of", lacking in other languages.) If in some respect the objects form a procession or a queue, the front—back is uniquely determined. A house has a front and a rear. What does "in front of the house" mean for somebody standing at its rear? Does "in front of the station" mean the main entrance? Does somebody who lives "across the river" say at home that he lives across the river? Does a citizen of Transvaal live on "the other side" of the Vaal?

The background of these difficulties is a polarity, which I confess I had almost forgotten; clearly because I was too strongly preoccupied with the three-dimensional structure of space and bodies in space. I mean the polarity

here – there

– if related to objects: this – that. We know nothing about the sequence in which the various polarities which I enumerated, are experienced by our children, mentally seized upon, and conceptually structured, in which order they are refined, entangled, isolated. Piaget claims that the topological polarity of inside–outside precedes the others, and in certain respects he may be right. If I should assign priority to one of them, albeit on shaky grounds, it would be that of the "here–there". One of my arguments would be that in my first approach I had forgotten about making it explicit, probably because it is more deeply rooted than the others.

The polarity "here–there" can be weakened to "here–not here" or refined in the way that the "there" means a certain direction or a more or less precise region, or quantitatively be refined to "nearby" and "far away". The "here" in "here–there" can be meant mentally rather than actually, which results in confronting two theres with each other, one of which is more here than the other. A variant of it is: this–that direction, place, region.

The "here–there" is the means to relativise the other polarities. Misinterpretations of some polarities can sometimes be explained by implicit assumptions involving "here–there" decisions. In the antipodes the "above–below" differs from what it is here. The "in front of and behind the table" depends on which "here" is stressed. For the child constructing or viewing a departing procession the "behind" might be ambiguous. The right–left side of the cupboard is determined by the "here" of the user. The man who says he lives across the river, locates the here in the centre of the city.

I am going to reconsider the polarity right–left, in order to answer a well-known paradoxical looking question: why does the mirror interchange right and left, though not above and below? I recall that the polarity right–left derives from our body. Other living and dead objects can be assigned a right–left by comparison, as it were by mapping my own body on them; that is by assigning to my own above–below and front–back polarities corresponding polarities in those objects. In some cases such an assignment can be defined in a natural way: if the other object is a man I have my "head-to-toe" and my "front–back" to correspond to his, and then his right–left is uniquely established. In other cases the assignment may be more arbitrary, but once I have settled, say, about a box, what is the above–below and the in front of–behind, then its right–left, as seen from the box, is unambiguously determined.

What then does it mean that the mirror interchanges right and left? Well, I suppose that in the image that the mirror reflects of my body, right and left are interchanged, that is, not determined according the rule I am accustomed to. However, at an object like "my mirror image" I can distinguish right and left only on the basis of some presupposed above—below and front—back. On the other hand, it is quite natural to understand the above—below of the mirror image as the direction mirror top to mirror bottom, and the front—back of the mirror image as the direction mirror man front to mirror man back, and then for the mirror man what he has to call his righthand and lefthand is established, provided I apply my own rules. Then since the above—below of the mirror man equals mine and his front—back is the opposite of mine, it follows that the mirror man, if he lived in my world would call his right hand the one that mirrors my left hand. In this sense the mirror interchanges right and left: The mirror image of my left hand is that hand of the mirror man he would call right if it is supposed that he lives in my world. It is not an isolated right—left that is interchanged, but the right—left of a body, coupled to an above—below and a front—back which remain unchanged as it were by definition.

With an object where the above—below and in front of—behind are not as naturally determined one will reach different conclusions. If a column is placed between a mirror and myself, I would probably call the sides of the column and the mirror column in front of me their front sides — then the mirror would interchange front and backside and leave right—left unchanged. One could also suspend a box above a horizontal mirror between one's eye and the mirror and now call the visible sides of the box and the mirror box upside — then the mirror would interchange above and below and preserve front—back and right—left.

This then is the answer to a paradoxical question. We will resume it as soon as we have dealt with the screw sense in space.

10.4. *Connections*

Connections are as it were syntheses of polarities but they are more than this. They can remove polarities but they can also create new ones, by their existence or lack of existence. The country is structured by land and water ways, railways and highways, the city by streets, lanes, boulevards, alleys, which are meant as connections. Less rigid are the unbeaten paths across meadows and through woods, the furrows and trenches, the airways — sometimes marked by white condensation trails. Crossing a parking lot, between two cars, then diagonally and again between two cars — using gaps and spacious fields — is another mode. Low walls and curbs, meant as borders, are interpreted by little children as connections to walk on. Ladders, stairs, elevators are connections between below and above; bridges over rivers and viaducts over roads, zebra crossings connect this side with that, but holes can do likewise, the hole in the fence where one must bend down, the longer one is the deeper, to creep through, the service-hatch, the ticket-holes, the window, the key-hole for spying, tunnels and subways.

Connections may be composed of pieces which are mentally, or by carto-graphic means, stuck together, a road visible at a distance, marked by driving cars, which appear and disappear behind bushes, houses, hills; the Champs

Elysées at l'Etoile, at Rond Point, at Place de la Concorde, the highway E8 at the Hague and at Hengelo, the Rhine at Basel and Cologne which, when travelling by train, one looses sight of and meets again and again. Underground sewer canals are witnessed by arrays of manhole covers, mole-tracks by hills in the meadow. Game tracks and traces of a fugitive can stop and continue again.

Connections can form networks like that of the railways, of a river with its tributaries, the underground systems of sewerage pipes, waterpipes, gas, electricity, telephone, which can cross each other in a chaotic arrangement, or above the ground the roads at a large roundabout, but also the indoor system of passages — inside and outside doors, corridors, staircases, possibly even the windows (to climb on the roof), chimneys and ventilation channels — can be a display of connections.

The detour structure, as I called it, is the child's first experience in this field: two paths between two spots, the normal one mother goes, and the other one, diverging, a child's adventure, safe-guarded by the knowledge of coexistence, visible coexistence if there are no obstructions between the two paths, partly visible and invisible coexistence if the paths are separated by bushes or walls. A similar experience — the closed path, pieces of which are simultaneously visible and invisible but mentally coexistent; in a suite of rooms, one room in and the other out; around the block where one is living. Decisions which one of two ways is longer can be difficult for children of the age of 6–7 even in spite of a lot of experiences; this holds for adults, too, in involved connection structures if no cartographic support is available.

10.5. Standpoints

I do not mean opinions — points for a disputant to stand on. I really mean spots where one can stand, from the concrete "here" from which I look at the world, up to the mental one at which Archimedes said he could move the Earth by a lever. Well, here, on this spot, I can turn objects to view them from all sides, I can touch one side while I view the other. Gradually I learned to coordinate sense-organ and tool, eye and hand, and if need be, to isolate them from each other, to check with the eye what I did by touch, and to have my perceptions elicit actions, the — wanted and unwanted — consequences of which can be approved or disproved by my eye. I have learned to point my ears to the sound and my nose to the smell, to constitute objects as far as they can be constituted from the "here".

My subject is not the constitution of spatial objects in general, and the geometrical objects are only a special case which will be dealt with later on. The constitution of most objects requires a variety of standpoints. For the present I am not interested in the objects but in the standpoints in their mutual relatedness — the space as order system of standpoints experienced or posited as coexistent.

Standpoints are connected with each other, primarily by the path which leads from the one to the other, physically or mentally. The continuity of the path is certainly an important factor by which images — not only visual ones — perceived at different standpoints are mentally related as derived from the same object: the constitution of an object in space and on its place in space. Yet this continuity is not a strict requirement: as early as the age of four if not earlier a child is able to identify a familiar spot at, say, one side of a canal from the other side and to experience this identification as a meaningful activity.

Experience and knowledge regarding the interrelatedness of standpoints is required to anticipate from a standpoint here a phenomenon such as it would appear from another standpoint there. In order to better see, hear, feel, smell something one changes one's standpoint.

Bastiaan (5;4) and Monica with me on the way to the playing ground. "I wonder whether it will be open" I say at a distance of 50–100 m. We were somewhat obliquely crossing a meadow in the direction of the door of the playing ground. The fence of the playing ground is hidden by hedges and bushes, which also hide the entrance unless one approaches it perpendicularly. After my words Bastiaan ran 10–20 m parallel to the fence until he could look straight into the entrance and state that the door was open (Figure 116).

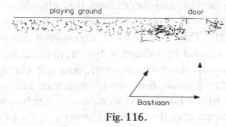

Fig. 116.

One runs to the window, to the front or rear, upstairs, in order to locate the origin of a sound or a light effect, down the street, around the corner, on the roof, in order to remove obstructions to perception, which may be effective at one place though not at an other. The child asks to be lifted in order to perceive something beyond a fence, or wall or the ridge of a roof. Obstructions and the way the foreground is delineated on the background, are also used to locate some object for another's eye: "Stand here, then you can see the star exactly above the aerial."

Misinterpretations of these "standpoint procedures" are worth mentioning as well:

Monica (almost 4;8) asserts, while it is still quite clear twilight, she sees Venus near the crescent; she is probably right, whereas my visual acuity fails. She asks me to sit on my heels in order to see the planet from her standpoint.

Something is approached to be better perceived, or viewed from a larger distance for a better global sight. A six-year old knows that at a larger distance objects look smaller, and in a way he knows how to draw conclusions from this fact, in order to explain phenomena. However, that a line-segment viewed

obliquely looks shorter than viewed straight — this correction of the mental
by the visual reality, which is manifest in perspective drawing, seems not to be
made by six-year olds. They know too well that the rim of a cup is a circle to
be able to see it — from various standpoints — as an ellipse. The primitive realism
where object and perception are identified, is still waiting for a process of
sophistication: disentangling object and perception. Mirror image and shadow
— familiar phenomena at an early age — do not yet disturb the harmony of
object and perception, since they belong to the objective world, but with the
knowledge that objects look smaller at a larger distance, with the phenomenon
of the apparent size, a step — the first? — is being made on the road of a — still
primitive realistic — distinction of essence and appearance. At the same time
the child learns to distinguish between true stories and fiction. It would be
worthwhile investigating the development of the distinction between essence and
appearance, truth and fiction — I do not remember having ever read anything
about it that drew my attention.

Whereas it can cost time and trouble to have a child grasp the fact that from
another standpoint things may look different, it can cost as much time and
trouble to have him accept and understand that elsewhere — in the holiday
resort — the sky is the same as at home — the sky, that is, not the clouds but
the starred firmament. The 6–7-year-old knows and grasps that with each
movement the world around him changes, that is, the near world. The cathedral
tower at a few kilometers distance, however, does not change if I move a bit.
And qualitatively proportional: Sun, moon, and stars are so far away that a
long journey like that to the holiday resort counts as much or as little as winking
one's eyes does with respect to the cathedral tower — again an application of the
polarity "nearby—far away".

The most surprising event I came across when observing the mental change of
standpoint with young children, again concerns Bastiaan (7; 4).

Back from a long walk we cross a slightly ascending bridge — Monica and myself on the side
walk, and Bastiaan on a small wall along the side walk, ascending discontinuously by steps
with in between horizontal pieces of about 5 meter (Figure 117). At a certain moment
Bastiaan says: "Now I am higher, but then you will be higher, it amounts to the same". In
fact we were always lower. He meant the difference between the continuous and the discontinuous ascent. His verbal means of expression were still imperfect, but it is clear what he
meant: we are ascending gradually, whereas he lags in order to catch up at the next step.

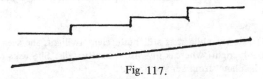

Fig. 117.

10.6. *Reciprocity of Change With Respect to Place of Object and Standpoint*

One can hang a picture upside down but the same effect is attained by oneself's
standing upside down. There are situations where one inclines one's head to the

right or to the left in order to get a straight impression of an object that is somehow inclined. One can turn a sculpture to consider its rear but one can as well change one's standpoint and walk around the object. More generally:

> In order to attain an intended relation between an object and its perceiver or manipulator, there is a choice between
>> change of place of object, and opposite change of standpoint,
> and such a relation can be maintained by
>> change of place of one of them and a corresponding change of the other.

This is something that in concrete cases can be conscious to little children. A more difficult case is the apparent motion: sitting in the train one gets the impression that things in the foreground move across a fixed background; ditches perpendicular to the railtrack seem to turn.

A general principle:

> In order to modify the relation between an object in the foreground and its background one has the choice between change of place of one object and the opposite change of standpoint.

This is also conscious to young children. A different thing is:

> In order to explain a change of relation between an object at the foreground and its background, one has the choice — or feels the dilemma — of interpreting it as a change of place of the object or as a change of standpoint.

With six-year olds I could not observe any symptom of awareness of this choice or dilemma. They are able to repeat that rather than the Sun that rises and sets, it is the Earth that turns, but this is done with no insight, although they are able to understand the monthly course of the Moon along the planets as its rotation around the Earth.

10.7. *Obstructions*

By obstructions I mean obstacles for ear and eye, for walking and acting. In Section 9.26, Border and boundary, I already gave examples. Obstructions are not necessarily absolute limits, they are rather obstacles that thwart the penetration into another territory, the transition from here to there.

A wall, fence, street can be an obstruction to sensory perception and motor activity, but there is a chance left that one can go around it, across it, through it. A mountain is an obstruction in the landscape, passable by crooked paths, pierced by tunnels. A river is an obstruction to direct passage from one bank to the other; the traffic across the river is restricted to bridges and ferries, but by swimming and flying, people and animals can reach the other bank. Swamps and deserts are moderate obstructions to any traffic, a highway is such for pedestrians who wish to cross. "No entry", a series of poles on short distances, a chain obstructs motor vehicles and yields to pedestrians. A red light is an obstruction to traffic that respects it. Some roads obstruct only heavy traffic. A barrier blocks traffic, but the blockade can be lifted.

A water or sewer pipe can be obstructed. A crooked pipe obstructs putting in straight bars, but a flexible long object can conquer the obstruction; the more crooked the pipe is, the more flexible — transversally and stiffer longitudinally — the object must be to go through.

A window and a door is passable for small objects rather than for big ones. Things that cannot pass transversely may succeed upright. An object may not exceed a certain length to be carried around a corner in a corridor; bends and corners can form obstructions.

Chains rest on the principle that each link obstructs the removal of the next. Threads twisted in a rope obstruct each other's stretch. By means of ropes and chains persons and objects are obstructed to move. Springs and spirals grant a greater margin, a lock between the spoke of a wheel obstructs turning. A plug in the floor obstructs the swing of a door, a bolt or hook its opening. A door, once opened, can obstruct opening another, or the passage through a corridor.

Aiming is a procedure by which one attempts to just or just not obstruct the view on an object. Obstruction and non-obstruction may be mutual or not. If A can see B, so B can A. Or can he? The peeper at the keyhole or behind the curtain sees but it is not seen himself. Viewing eye to eye, the line from the other is not obstructed, but the remainder can be partly so. One-sided transparent windows and mirrors, as well as periscopes interrupt or obscure the reciprocity of viewpoint and object, eavesdropping instruments that of the path of sound. The way from above to below can more easily be passable than the converse.

10.8. *Combinatorics*

Line, plane, and space are gifted with combinatoric structure: that of the immediate coexistence, the vicinity. Vicinity should be understood with a pinch of salt: beads at a string that touch each other, tiles in the paving or countries with a common frontier, bricks in a wall separated only by a layer of mortar, faces of a polyhedron with a common edge and their networks, bones of a skeleton in a common joint, vertebra joined by a spine, but also neighbors in a street, possibly across the street, towns along a road, stations and airports communicating by direct traffic.

In Sections 7.2–3 I paid attention to a mathematical tool to describe such structures: graphs. Mathematically viewed there is not much to be added, since no profound properties of graphs matter in this context. It should, however, be mentioned that the geometrical image of the graph also serves to visualise non-spatial relations of vicinity. Genealogy is visualised by pedigrees, processes by flow diagrams, electrical circuits by networks, which in no way reproduce the spatial situations of parts but "electrical vicinity", a social system with its components as nodes and its influences as paths.

10.9. *Conclusion*

In which way does this chapter fit into a didactical phenomenology? In the chapter Topology the didactical component was almost lacking. Its aim was to signal the tension between mental objects and concepts, which is not — probably cannot be — bridged by didactical means at school level. Those mental objects have again emerged — together with others — in the present chapter, purely phenomenologically, unrelated to mathematical concept formation, but also without tying more than a few didactical connections. This does not mean that they do not exist. On the contrary, they do exist, more implicitly, more globally than if numbers in the arithmetical and figures in the geometrical sense are being discussed. Arithmetic and geometry can boast explicit didactics, elaborated in detail. I can deal in detail with didactical aspects of counting, of arithmetical operations, of perspective. I have illustrated what I called the topographical context by examples but I would not be able to divide it according to details of didactic action. To say it more concretely: what is above and below, front and back, how to come from here to there, how to look around the corner, what I can see through a key-hole, what is a bridge, a viaduct — and so on and so on — is not being learned, let alone taught, point by point but in a rich context, which presents itself, which if presented is modified, which is recreated from the bottom upwards, and in which — consciously and unconsciously — the wanted elements are processed, in an integrated way. Such a context can serve to form the mental objects, relations, and operations, to become aware of them, to verbalise them, to learn using and understanding them as tools of understanding and acting and finally to remodelling them conceptually.

This chapter has been *didactical* phenomenology in the sense that it aims at making conscious the topographical context as a global didactic medium.

CHAPTER 11

FIGURES AND CONFIGURATIONS

11.1. *Abstracting*

How many properties must be abstracted to make a thing a figure? Colour, material, weight, taste, smell, irregularities, roughness, the place of the thing, the place of the perceiver, which determine its appearance, use and emotional relations. What is finally left? Shape and size. Or should the size also be wiped out by abstraction in order to get the bare figure? Yes and no. The child copies the same figure as drawn on the blackboard on a smaller scale. But stepping through Madurodam* and a true city are different things.

Classification tests bear witness to the fact that six-year olds are able to perform this kind of abstraction. They classify objects primarily according to shape and size characteristics (length, width, thickness, height); they understand explicit instructions to classify according to shape and size (but also according to other criteria) and they understand the words "shape" and "size". Only if they are asked for *new* classifications, can it happen that the place of the objects is introduced as a new aspect of classification, but even this is easily averted.

The objects from the topographical context are tied to their places: the ground-plan of a room, the façade of a house, the net of streets, the solar system, are mapped in the plane or in space according to their more or less pronounced individuality. A line, a circle, a cube is not tied to any place; models of these figures, even if specified by one specimen, are also models of mental objects. Of course, there are intermediate cases: the pattern of a crossroad or a round-about with right-of-way indications and traffic lights, is not as closely attached to one place as are many other topographical representations.

I already explained how geometrical contexts arise, in particular that of the rigid congruently and similarly reproducible bodies, how accidental features and, if need be, thickness and width are wiped out by abstraction — aspects which with the greatest ease can again be included, as can colour, material, weight, taste, smell. Place is more difficult; its influence has been eliminated so early and in a so strongly implicit way, that it costs a lot of trouble to make it conscious again, for instance in the use of perspective.

11.2. *Production, Reproduction, Occupation, Manipulation, Transport*

Several times I stressed that at a young age and in many ways children are confronted with things that suggest geometrical mental objects. As products

---

* The midget city near The Hague.

they are the result of production and invite reproduction, occupation, manipulation, transport — concrete actions which can mentally be repeated with corresponding mental objects, and this mental element is the one that counts in the geometrical context. It is a rich variety of actions which I shall display in the sequel — probably in a quite unsatisfactory way. We are satisfied too soon if we have introduced a mental object concretely in one way. On the other hand we ask the child to be receptive for a multiplicity of approaches whereas in the didactical situation we prefer to narrow our own view.

The didactical feature of the present phenomenology is the multiplicity of concrete approaches. Among the numerous terms that describe the concrete actions none is redundant; on the contrary I am sure there are not enough of them.

11.3. *Planes*

I start with planes, or rather pieces of what one imagines to be infinitely extended planes. There are reasons why I do not bestow priority on lines — planes come earlier. First of all, in the topographical context, horizontal and vertical planes, floors, ceilings, walls, bottoms, covers. Among the oblique planes the most striking are roofs, covers of chests and slides. Objects with faces can be bounded by oblique planes, depending on their position. Water in a vessel does not behave as a rigid body; its surface remains horizontal even if the vessels are inclined. (A glass with powder, beads, or peas behaves as though it were halfway between liquid and solid matter. Contrary to what Piaget claims it has nothing to do with logic but all to do with physics whether such a surface is horizontal or inclined and how much it is inclined.)

A remarkable feature of planar pieces is their unique continuation — as it is of straight lines. In fact, it is the criterion of flatness of a planar piece whether it continues its own sub-pieces in an virtually unambiguous way, and the same property plays a part in the mental constitution of planes. According to my experience 6–7-year olds are able mentally to extend planar pieces. Some evidence will be displayed later on.

Thanks to unambiguous continuation, planes can be laid down and slide on other planes.

If a surface is everywhere horizontal, it is certainly flat — "level" as the carpenter and bricklayer call it after the instrument by which they check it. Whether something is vertical is checked with the plumb: a vertical wall is as it were a combination of plumb-lines along a level-line.

Knives and saws make planar cuts because they are flat themselves and continue unambiguously flat indentations. The shaving tool, called a plane, transfers its planar bottom to the shaved material. Scrubbing is removing roughness — in order to glide, a thing must be flat. The imprint of a plane is again a plane — this is in general the simplest method of checking and reproducing. Smoothing upon a plane creates again a plane; a table cloth on a table is a plane as is the table.

Among the surfaces the planes are characterised by the property to comprise with a pair of points the connecting line-segment — planes are "closed" with respect to rectilinear connecting. Through two intersecting straight lines a plane can be laid by pointwise connecting the lines. Three points, not lying on one line, determine a plane by means of their connecting lines. The straight lines drawn from a point to a line-segment or line, determine a planar piece or plane.

The rectilinear connectivity of the plane, is the reason why rolling a soft matter (dough) on a plane produces a plane: A roller is a cylinder, described by straight lines, which are imprinted in the soft matter; rolling repeatedly in different directions assures rectilinear connectivity. Cylinders and cones can be rolled upon planes such that at every instant a generating line is lying in the plane. (Surfaces with this property are called torsi; apart from cylinders and cones these are the surfaces composed by the tangents of a curve in space.)

Tests of rectilinearity — which will be discussed later — can also serve to test the flatness of a plane: applying the ruler, stretching a rope, aiming. Aiming along a plane gives a straight line.

Planes can be suggested by apertures: a gate, an arch, a football goal, a sequence of lampposts. Planes can be piecewise given, by housefronts, interrupted by sidestreets, or by networks of lines such as wire-netting. A linear object together with its shadow determines a plane.

A plane can be suggested by mirror symmetry, that is, as the plane producing the symmetry. Such a plane can be explicitly marked within the symmetric object or the object can invite the marking of it mentally, as happens if an object is divided into congruent parts.

Planes can be flexible; they may be rolled into cylinders, cones, and more generally, torsi; by this process soft planes acquire a certain stiffness, which they did not have before.

Planar pieces divide the space locally as do the total infinitely extended planes globally. If — in a topographical context — they are horizontal, the parts may be called above and below — even though the planar pieces are somewhat inclined. For vertical or almost vertical planar pieces the predicates right—left or front—back may fit, according to whether the observer imagines himself placed within the plane or in one of the spatial parts.

Planar pieces can be lying as though one produces the other, that is, as parts of the same extended plane. They can be parallel — parallelism will be dealt with later on — or they can meet. Depending on how far they extend, this can happen in various ways: in a single point, in a number of points, in a number of line-segments. In the truly mathematical context one means the infinitely extended planes, which if not parallel meet in an entire straight line. It is a well-known fact that this meeting along whole lines is obscured by drawings where planes are symbolised by parallelograms. I experienced, however, with 6—7-year olds that if a pair of planar pieces are concretely given in space, they can indicate where the extended pieces would meet each other and devise methods — use of the ruler — to carry out the construction more precisely.

(This led me to assume that children can perform the construction of planes from planar pieces mentally.)

11.4. *Polyhedra*

Polyhedra with which the child gets acquainted at an early age, are blocks, prisms, pyramids. The context is often topographic: blocks of houses, roofs, pointed roofs. Toy building bricks provide an opportunity to detach such figures from their topographical context and to place them in the geometrical one: to relativise the phenomenon of base, to have blocks balancing on edges and corners, to view roofs as lying prisms, to have pointed roofs resting on lateral faces. It is well-known that older pupils, after many years of planar geometry, can have difficulties with these changes of perspective as soon as they are supposed to learn solid geometry.

It does not take much trouble, even at a young age, to recognise vertices, edges, and faces and to name them as such; there is little reason to stick to other terms that are believed to be more adapted to children's language. Likewise the relations between vertices, edges, faces (lying on, passing through) are accessible at an early age. The same holds for networks, building polyhedra from, and splitting into, networks, understanding a network as a combinatoric pattern.

In spite of the didactical value that the network derives from its manageability, there are reasons to warn against the "flight into the network". Polyhedra are primarily surfaces of solids and must be mentally constituted and conceptually approached as such. This is the way to start; if this has successfully been undertaken, the analysis of the surface into a network may follow. (This does not exclude confronting the learner incidentally with a network and the indications to build from it a — perhaps surprising — polyhedron, but it should not be recommended as the normal approach.)

Let us take the case of a — not necessarily triangular — prism. Twelve-year olds, and even older ones, do not lack the vocabulary but rather the mathematical ability to describe this class of surfaces. Supplying the child with a description bears witness to a lack of didactical understanding, but the flight into the network is no more justifiable. One should rather exploit the fact that they are surface structures of solids, and this is done most efficiently by constructing the solids themselves, from clay or potatoes. This then is the way towards a conceptual analysis of the prism as a class of surfaces. It starts with modelling from clay, or cutting out of a potato, a disc, which can be irregularly bounded at its sides. In order to arrive at a prism, one remodels the sides: cutting away pieces, perpendicularly through the disc, in order to get a right prism. This construction implies a conceptual description: congruent base and top polygons connected by rectangular walls. (The parallelism of the edges — in the usual approach the primary element — is now a consequence.) Piling up prisms of the same kind or sawing parallel to the base and top side produces new prisms. These parallel cross-sections are congruent polygons related to each other by right or skew translations — a relation that leads to a new definition of prism: a polygon moved in space sweeps out a prism — a right one if it is moved perpendicularly to its own plane. At a later stage this conceptual analysis leads to the definition of the solid prism as the cartesian product of a planar polygon and a line-segment, or even an infinite line.

Similarly one can analyse and define pyramids and truncated pyramids. The cross sections parallel to the base are homothetically situated. So a pyramid arises by homothetic displacement of a planar polygon towards one point, the top. This leads to a unified description of prisms and pyramids, finite and infinite ones.

These solids are an appropriate start for simple, and thus particularly effective, logical-geometrical analysis. Other polyhedra are by their characteristics didactically important to learn structuring: in order to count the vertices, edges, faces, for instance of a cube, the sets are structured:

> four vertices below, four above,
> or four vertices at the front-side, four at the backside,
> or four vertices at the right, four at the left,
> four edges below, four above, four upright,
> or four edges in the width, four in the length, four in the height,
> a ground face, a top face, four around,
> or two faces, front and back, two right and left, two above and below.

The set of the twelve face diagonals is structured

> two in each face,
> two sets of six forming two tetrahedra,

that of the spatial diagonals,

> one by each pair of opposite vertices.

A more difficult problem is the structure paradox:

> Six faces with four edges in each of them, is 6×4 edges, rather than 12.

Young children do not experience it as a problem but from the age of ten onwards they may be susceptible to it.

By their great diversity polyhedra offer a rich context. The cube is a good start but other polyhedra are as well worth considering. That is, not only closed ones (with no boundary) but also polyhedra with a boundary and non-orientable ones (the Moebius strip), and polyhedra given combinatorically by networks which are not embeddable in space (the projective plane).

Polyhedra can be concretely represented and constructed: as

> closed surfaces of solids,
> surface structures (cardboard models),
> edge structures (wire models).

Structures within a polyhedron (space diagonals in the cube, face diagonal tetrahedra in the cube) are patently visible in wire models.

Polyhedra, in particular cubes, can be piled upon each other to get buildings (Figures 118–120). Structuring such buildings is a significant activity. Describing them can be an inducment to create linguistic means. Strangely enough, does the toy industry know better than the educators do how to profit from the children's building activity?

Tilting polyhedra over edges on a plane is also a fruitful subject. In fact it is a predecessor of unrolling certain curved surfaces on the plane.

Particularities of convex, regular, and semi-regular solids will be dealt with later on.

11.5. *Direction*

In my original design "plane" was followed by "straight line", but while I wrote that section, I became once more critical of my preconceived judgment: caught within logical geometrical structures, which phenomenologically are

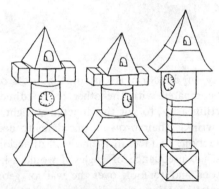

Fig. 118.

Fig. 119.

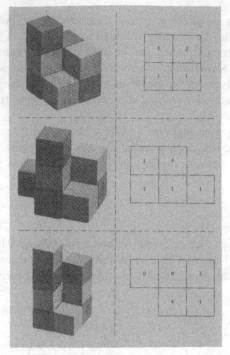

Fig. 120.

prejudices. Constancy of direction is one of the roots of the mental object
"straight line" but compared with the other roots "direction" sprouts much
more than mere rectilinearity, to wit the directed straight line and parallelism.
Axiomaticians may wrinkle their brows if one dares to define the straight line
by constancy of direction. In fact, in their world direction comes much later,
after congruence, as a consequence of angles. I would ask those virtual critics
to stand on their toes and to look over the wall of geometrical axiomatics.
In differential geometry, for instance, direction and transport of direction are
presupposed and straight lines are derived afterwards — let us call them *straight
ahead* lines — under the term "geodesics". I feel that this approach of space
from the infinitesimal nucleus matches more closely the natural local approach
required by phenomenology.

Straight on running is a natural activity of people and animals and thanks to
inertia even a habit in the non-living nature as long as there are no forces to
disturb this motion.

Straight on towards the goal means

globally: that at every moment the fixed goal is invariably kept in view,
locally: that at every moment the once chosen direction is continued.

This unique continuation recalls that which characterised the plane, but while
the planar continuation was all-sided, this one is one-sided. One goes on, or

something goes on or is moved forwards, yielding neither to the right nor to the left, neither up nor down; something rises straight or descends straight, and the continuation is determined in each point not by an artificial track but by the geometry of space.

Well, one can retrace one's steps, backwards in one's own footsteps, or after a turn interchanging start and finish, at each step take the opposite direction, the opposites of the "same" direction are again mutually the "same"; the opposite of a directed straight line is again a directed straight line — for people, animals and in the non-living nature — the straight ahead line is also a straight back line.

The things moving or moved in that way are points or something that is considered as points, though it extends to rigid bodies. The body moves straight ahead, without yielding or wobbling: all its points move in the same direction.

Equality of direction generates not only directed straight lines, but also parallelism, the phenomenon of being parallel. Earlier I cited parallelism among the mental relations which from early childhood onwards are suggested by natural objects, and even more frequently by the products of human technique. "Direction" also belongs to this list, and even "vector" — the arrow that possesses length as well as direction — and "parallelogram" — the figure that claims that vector $\vec{a}$ displaced along vector $\vec{b}$ yields the same result as vector $\vec{b}$ displaced along vector $\vec{a}$.

Let us review these mental objects later on, since at present the stress is on how much "direction" contributes to the mental object "straight line".

The order "plane — direction" methinks is phenomenologically the right one. Plane — the static element, the resisting wall; direction — the straight ahead element, against the wall, through the wall. But thanks to its comparability everywhere the direction is not bound to its carrier line. The same direction is everywhere recognisable, and its carrier line that pierces the wall or screen, can be restored behind them.

The comparability everywhere of direction is suggested by the parallelograms in our technical environment. Actually we are living on a sphere (and perhaps in a curved space), which does not allow for a global parallelism, but this is a secret, not betrayed by the small parallelograms which we manipulate — again a point to be elaborated on.

11.6. *Straight Lines*

Earlier on I made a choice among the great variety of objects and situations suggesting straight lines. I am going to enumerate the ways straight lines originate. Origin of the straight line:

by copying (drawing by ruler),
as intersection of planes,
as cut line,
as fold line,

as straight-ahead line,
as shortest line,
as stretched string,
as vision line,
as symmetry axis (in the plane),
as rotation axis (in space).

These modes of origin are not independent of each other. The sharp edge of the ruler is something like the intersection of two planes. The cut line is as it were a cutting copy of the sharp edge of the scissors. The fold line arises when on a piece of paper is imposed the shape of two intersecting planar pieces, and the same explanation seems to hold for the rectilinearity of the symmetry axis.

Like the straight ahead line the shortest line (shortest between each pair of its points) is a familiar mode of the straight line, in particular in inorganic nature. The rectilinearity of the stretched string can be explained by physics as well as a mode of straight ahead line, and of shortest line.

The most subtle in this list is the straight line as vision line or light ray, that is the rectilinearity of the propagation of light — it would require much more comment as would the symmetry and the rotation axis. As in the case of the plane as the symmetry plane in space, one can imagine a great many situations where symmetric planar figures suggest or explicitly display straight axes. Likewise rotation axes are a frequent phenomenon though few people will be aware of the fact that any rotation in space takes place around a fixed axis. As a matter of fact it takes a great deal of didactic trouble to make it conscious — again a point worth being discussed.

The first four entries of our list are as it were mechanical, whereas the other six are more theoretically related to the physics of our common sense space, though they certainly do not characterise it. In more general metric spaces — in spaces with a Riemann metric — it still holds that straight ahead lines and shortest lines are the same; the stretched string on the globe (and on other curved surfaces and in curved spaces) is also the skipper's straight forward course and the path of shortest distance. The rectilinear propagation of light can be explained by similar principles as the straight ahead line or the shortest line. Straight line as symmetry and rotation axes exist also in non-euclidean spaces.

Up to now I did not distinguish between lines and line-segments. The unambiguous continuation of the line-segment — involved in the idea of straight ahead line — is a fact experienced early. It is a mere technical shortcoming if even in the higher grades of primary education pupils who measure distances do not pay attention to the rectilinear continuation as soon as the measuring instrument is repeatedly laid down. It is a striking feature that they proceed more carefully if the measuring instrument is to be laid down parallel to the preceding situation than if it takes place in its extension.

Between each pair of points there is a rectilinear connection, which is unique.

This uniqueness is suggested by the "straight ahead towards the goal" as well as by the "shortest path" if compared with detours. I would guess that this uniqueness exists mentally at an early age.

Logically equivalent with the uniqueness is the statement that two straight lines have at most one point in common.

With respect to lines that do not meet, it is usual in solid geometry to distinguish between parallel and skew lines according to whether they are comprised in a common plane or not. This is a typical case of didactical inversion: a sophisticated logical approach is preferred to a phenomenological one.

Bastiaan (7; 4) — spontaneously though perhaps taught at school: "Equidistant lines* never meet." I ask him whether lines that never meet, are always equidistant. After some fumbling with two forks he exclaims: "Two highways above each other, they never meet, but go far away from each other." The same question discussed for planes.

An important aspect of the straight line is the direction, as explained in Section 11.5. Constancy of direction was mentioned as one of the origins of the straight line. Equality of direction of directed straight lines is an intuitively primary phenomenon. The discrepancy of direction of intersecting as well as skew line pairs is a striking phenomenon. The fact that there is a plane through parallel, though not through skew, lines is comparatively secondary and not immediately obvious. There may be reasons to choose this property to define parallelism of lines in a *logical* system of geometry, but they are not at all compelling. One can equally well imagine a system in which direction or equality of direction is one of the fundamental concepts.

As I pointed out at another opportunity, Piaget designed his experiments as though the traditional logical structure of geometry reflected the development of the geometrical mental objects. Whoever is familiar with the mathematical method and methodology, knows what part is played by the inversion: the final result of the developmental process is chosen as the starting point for the logical structure in order to finish deductively at the start of the development. This genetic-logical inversion expresses itself as a didactical — or rather antididactical — inversion.

The traditional definition of parallel lines is one among a lot of examples of this inversion. Phenomenological analysis is a means to uncover them. No doubt direction plays an important part in the constitution of the mental object "straight line" — the straight ahead line; straight towards a goal, walking along a — most often — straight line are early activities in the development, which are verbalised just as early and instrumentally imitated, though perhaps insufficiently. Compared with these symptoms the vision line is so late in the development that it is a preposterous attempt to have the constitution of the straight line conditioned by that of the vision line. Earlier on I criticised the fake arguments by which Piaget lets the vision line precede direction as a mental object —

* This is the literal translation of the Dutch term for "parallel".

arguments taken from a traditional structure of geometry which is considered as though it were compulsory. There is, however, sufficient reason to pay attention to the vision line — as well as to the symmetry axis and the rotation axis — in a special section.

11.7. *Vision Lines*

Light propagates in straight lines — a fundamental statement in physics, which immediately requires amendments. First of all in geometric optics: light does not propagate straight ahead but, according to Fermat's principle, along the fastest way, which in media with varying speeds of light and in the presence of reflecting surfaces, means deflection from the straight line by refraction and reflection. In wave optics interference phenomena cause the light to bend around corners. For many observations, however, and certainly in the naive view, the rectilinear propagation of light is the zeroeth order effect compared with which the others are negligible, or accounted for as first order perturbations.

Between the experience that an object can be made invisible by an intermediate screen and again visible by an appropriate hole in the screen (the hand befor the eye with slits between the fingers) and the statement of the rectilinear propagation of light and its applications, is a long journey, aided by a long sequence of discoveries. First of all I will formulate the

vision line principle:

If the eye O sees the object B covered by the object A, then O, A, B are rectilinearly connected.

If the eye sees the objects $A, B, C, \ldots$ covering each other, then $A, B, C, \ldots$ are rectilinearly connected.

If a long object is seen as a point, it is rectilinear.

These principles, resting on the rectilinearity of light propagation, can serve

to establish whether objects

are lying on a straight line with the eye,
are mutually lying on a straight line,
are rectilinear,

to manage, by displacing one object or some among them, or all of them, or the eye, or by deformation, that objects

are lying on a straight line with the eye,
are mutually lying on a straight line,
are rectilinear.

Here it makes a difference whether

the eye and the objects are

concretely realised,
or experimentally simulated,
or partly or totally imagined,
or symbolised in drawings or otherwise,

to occupy certain places.

I would not be able to tell how and under what circumstances and in which order these principles are being

discovered,
operationally applied,
refined,
transferred,
made conscious,
verbalised.

In vain I have looked for any research on such questions. We observed, however, that learning processes for such activities,

are possible with seven-year olds,
can be necessary for eighteen-year olds.

At all levels of the primary school (1st–6th grade) and in the two lower grades of the secondary school the IOWO people paid attention in experimental and in elaborated instruction situations to the geometry of rectilinearity under the aspect of the vision line.

The picture of an island with all that is on it, is hanging in the class room:

What do I see if I stand here?

At which point am I standing if I see this or that?

How does one object move with respect to the other if I move this way?

How shall I move in order to have one thing displace itself with respect to the other?

Pictures of the school with a tall building behind it, of apartment buildings, supermarkets, public gardens, bridges, churches, sculptures, monuments, and so on, in the city quarter around the school pass in review:

Where was the photographer standing?
Was he closer when taking this picture or that?
Was he more to the right or to the left?
Was he on the street or on this or that floor?
Where should he have stood to get a certain picture?

Photographs of block buildings (piled cubes) are analysed:

From which side was it taken?
Can there be more cubes behind this one?
How should it look from the backside, from the right, from the left?

Photographs of the coast and the harbour taken from a ship travelling along, which must be arranged in order:

> From which place can you see the lighthouse, the mill and the church in this mutual situation?

Aerial photographs of a landscape:

> Where was the plane?
> In which direction did it fly?
> How should it fly in order to see this or that?

Photographs with a camera turning around a horizontal or vertical axis:

> How to arrange these photos?
> How to interpolate?
> How to extrapolate?

The sky is lightly cloudy:

> Why does it look heavily cloudy at the horizon?

An analogue-model, an ape behind bars (Figure 121):

Fig. 121.

What from the exterior world does the ape see?
A hole, a window, a mirror:

> What can you see if standing here?
> Where should I stand to get a certain picture?
> Is there a place or not where you can see a certain object?

A tree, screen, wall:

> Where can you hide?
> Where shall I stand to see somebody hidden?

The preceding examples are activities embodying the principles proposed earlier.

> concretely,
> imagined,
> symbolised in drawings or otherwise.

There is, however, more involved in them than the vision line principles such as formulated, to wit:

the mutual relations of vision lines,

right and left,
above and below,
obliquely right above, and so on,

of each other. What I formulated in Section 10.5 under standpoints and in Section 10.6 as reciprocity under change of place of object and standpoint, is under the vigour of the vision line principles sharpened to

viewpoint

and

reciprocity under change of place of object and viewpoint.

There is even more to it, resulting from the principle of the straight line,

X, Y, Z are on a straight line

can also be pronounced as

X is lying on a straight line with Y and Z,

or

Y is lying on a straight line with X and Z,

or

Z is lying on a straight line with X and Y,

statements that are equivalent, which for the vision line means:

if the eye O sees the object B covered by the object A,
then the eye at the place of B sees an object at the place of O covered by A,

or with the object B interpreted as another's eye:

if the eye O sees the eye O' covered by the object A,
then the eye O' sees the eye O covered by the object A,

or more symmetrically:

A is between O and O'.

Of course, it is not as simple as that, and this is early understood. O can spy on O' around the corner A while O', looking in another direction, does not perceive O. Or O can see around the corner A parts of the body belonging to O' though not O' and by this way shield itself against O''s look.

11.8. *Lightrays*

I could not devise a better title — shadow line would be misleading though
actually I mean precisely the fact that light source, object, and its shadow
are in a straight line. In the case of what I called the vision line the object seen
— or to be seen — is a *secondary* light source lying on a straight line with the
observing eye and the screening object. Now I mean the line between a *primary*
light source, such as the sun or a lamp, the object, and its shadow, which is
cast because the object screens the light source. In both cases we are concerned
with light rays and their rectilinearity, and in both of them I could have spoken
of the light ray as a straight line. In the first case the term "vision line" was
available, whereas I could not find as good a term for the second case.

I have good reasons to deal with "light ray" separately and after "vision
line". Whereas the vision line is operational early and becomes more or less
conscious at the age of about 7 years, understanding shadows requires more
time. Shadows, as well as mirror images are perceived early and their origin is
qualitatively understood early, but I could not observe any understanding of the
geometry of shadow production at the age of 7–8-years. Even less than in the
case of the vision line, would I be able to indicate when this phenomenon is

   discovered,
   operationally applied,
   made conscious,
   verbalised.

What matters here is the

   shadow principle

with its consequences:

   light source, object, shadow are lying on one straight line,

   from two of them conclusions can be drawn regarding the third,

   displacement of light source or object correspondingly induces displacement of the shadow, and change of shape and size,

   deformation of the light and shadow-receiving-surface correspondingly
   induces deformation of shadow images.

The adverb "correspondingly" is a bit vague, but I used it in order not to be
drowned by details, and nobody will doubt how the correspondences are
meant —

   shadows get longer if the sun is setting or objects farther away from the
   street lamp — what is the common element in these phenomena?,

   the elliptic shadow of a circle, the perspective image of a circle, the trunca-
   tion of the cone — what is the common element?,

the solar image of a hole at the opposite wall, the track of dust in a sun beam, the sunlight circles under foliage — how are they connected?,

the increasing vagueness of the shadow on the ground with increasing height of the object,

the clouds in the sky and the changing illumination of the landscape — how are they related?,

the phases of the moon — where is the sun?,

at which side does a lunar eclipse start?,

silhouettes and shadow images — how are they connected?

— a large number of queries, as meaningful for 18-year olds as for 8-year olds — meaningful pieces of geometry, even for adults.

11.9. *Straight Lines in Views*

The three-dimensional world projected on the two-dimensional retina is in the central nervous system reinterpreted as a three-dimensional pattern, thanks to sources of experience other than visual. We even succeed in suggesting this three-dimensional world by means of two-dimensional pictures of this world, thanks to an involved system of experiences and conventions, which is in no way watertight, as appears from experiments on optical illusions.

How do we find out whether two lines in the optical field are skew to each other? An eye movement can inform us about the intersection, whether it is genuine or apparent. Homogeneities in these lines can inform us about which passes before which, but there may be quite other experiences that contribute to this mode of structuring. Conversely, such experiences can also be exploited to provoke wrong judgments.

I do not intend to analyse here the theory of perspective and the methods of descriptive geometry. Both subjects have in common the fact that they try to do justice in two dimensions to the three-dimensionality of space and that these attempts are astonishingly successful. The parallel-perspective image of a cube, the central projection of a street is accepted as the representation of a piece of geometric reality and reproduced in submission to the same principles — principles that may, or may not, have been made explicit.

A third means to get a grip on space in the plane, is "folding down" planes, which when performed successively leads to networks of polyhedra, which can be rebuilt from them. Shortest lines on polyhedra are recognisable in the network as shortest, that is, straight lines, and as such they can be reconstructed on the polyhedra themselves.

As an exam of the freshmen year of a teacher training college students with a quite rudimentary mathematical education were asked the following question:

This is the frame for a tent (Figure 122). The canvas is stretched straight on from the oblique roof to the ground with no kinks or folds. Construct the line in the drawing where it reaches the ground.

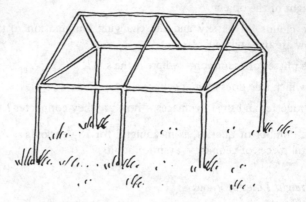

Fig. 122.

Terms like plane have been avoided because they suggest too much technical mathematics. The problem was thought to be difficult, but in my view it was even too difficult. I observed a student perform the required constructions without any hesitation. Though it was clear that he fully understood why the construction was correct, he was not able to justify it. He did understand, however, an *a posteriori* explanation. Likewise other students who succeeded in solving the problem, acted by mere insight, though because of lacking a geometric terminology they were not able to prove the construction formally.

The two lines asked for are found by connecting points which themselves are intersections of two lines: a line *g* on the ground which connects the feet of two poles behind each other and another line *t* that runs obliquely from the roof. The two lines *g* and *t* intersect on the paper, but is there anything in reality that corresponds to the intersection point on the paper? The students saw that it was the case but were not able to motivate it. With some elementary knowledge of solid geometry it is a simple trick: imagine a plane through the two poles. Indeed, if this is possible, *g* and *t* cannot be skew. But why is there a plane through the two poles? Well, the plane is clearly visible, it is determined by the poles and it extends to the triangle in the roof, containing *g* as well as *t*.

When solid geometry was still taught in the upper grades of our *gymnasia*, paper and pencil constructions of plane sections of all kind of bodies was a popular subject — popular because it was an opportunity even for poor pupils to achieve something by viewing and doing. Descriptive geometry was a similar case at Modern secondary schools. These activities rest on recognising, verbalising and processing such simple principles as:

in order to find the intersection of two planes, look for lines in the planes that intersect each other,

in order to find lines that intersect, look for a plane comprising both of them,

in order to find planes, look for lines that intersect or are parallel and imagine a plane comprising them.

The didactical folly of this subject was that it was offered to 17-year olds rather than to, say, 10—12-year olds, who in turn were tormented with a deductive system of planar geometry. Why? Possibly because in Euclid plane geometry precedes space.

11.10. *Polygons*

We already met polygons as a combination of line-segments under the term of graphs. One need not restrict oneself to the plane if discussing polygons. The edges of a polyhedron form a polygon, which can be studied with regard to combinatorics as well as flexibility. Even a closed skew quadrilateral can be a rich source of experience. With its diagonals it forms the edge polygon of a tetrahedron.

A *line track* is a directed graph, with vertices $a_0, a_1, \ldots, a_n$ and segments $a_i a_{i+1}$; it is closed if by chance $a_0 = a_n$. Self-crossings need not be forbidden, though with what is usually called *polygon*, they are not admitted; figures like the pentagram (Figure 123) are called star polygons.

Plane closed polygons with no self-crossings divide the plane into an *interior* and an *exterior* domain, as do Jordan curves; the name polygon is also bestowed on the interior; then the line track is called its circumference.

Plane closed polygons may possess what is called *re-entering angles* (Figure 124), but most often if one speaks about *plane* polygons, one means convex figures. Convexity will be dealt with in the next section.

Fig. 123.

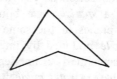

Fig. 124.

11.11. *Convexity*

"Convex" and "concave" first prompt associations with lenses. Both terms also apply to arcs, and then it matters from which side they are viewed. If it is the graph of a function it is understood that this happens from below to above.

In Figure 125 arc 1 is convex and 2 is concave. A closed plane curve such as
Figure 126 behaves convexly at 1 and 3 and concavely at 2 and 4. Similarly with
a dented ball one can distinguish the dents as concavities from the convexity.

Fig. 125.   Fig. 126.

Convexity such as is defined in general for plane or spatial figures is a characteristic example of an unusually successful mathematisation, close to visual
reality, of a mental object. One defines:

> a set *S* is convex if with any pair of points it contains the connecting line-
> segment.

(So the line of Figure 126 does not enclose a convex set since the line-segment
pq is not contained in it.)

A fundamental fact:

> the intersection of convex sets is convex.

A plane in space determines two parts,

> *half spaces* both of which are convex.

A line in the plane determines two parts,

> *half planes* both of which are convex.

A point on a line determines two parts,

> *half lines* both of which are convex.

By forming intersections of suchlike sets one gets new convex sets:

The intersection of two half spaces — an infinite *disc* bounded by parallel
planes or a *spatial angle* bounded by intersecting planes; the intersection of
two half planes in the same plane — a *strip* bounded by parallel lines, or a
planar angle bounded by intersecting lines; the intersection of two half lines
on the same line — a line-segment. Three half spaces, if not too particularly
situated, have a *solid angle* in common; a *tetrahedron* as intersection of four
half spaces is convex, as is the *triangle surface* as intersection of three half
planes in the same plane.

At a glance as it were one can see whether a planar or spatial figure is convex.
If, however, such a figure is given mathematically, the problem arises as how
to prove convexity. This is not at all easy if the figure is bounded by a curve or
curved surface — say, a circle, ellipse, parabola, sphere, and so on. For instance
in the case of the circle one has to prove:

if p and q have a distance $< \alpha$ from m, all points of the line-segment pq have the same property.

This in turn follows from:

for x on the line-segment L the distance (m, x) assumes its maximum at an endpoint of L.

In other words:

for x on the straight line L the distance (m, x) assumes nowhere a (relative) maximum.

11.12. *Circles*

Acquaintance with and mental constitution of the mental object "circle" is an early developmental phenomenon. The term "circle" for this mental object appears relatively late, because parents and educators — as it seems in all languages — believe that "round thing" or suchlike words are easier — propagating baby language is an educational phenomenon worth being investigated.

Even the construction of circles can be a relatively late phenomenon, drawing round a circular object can precede the construction by means of a pair of compasses. Incidental observations showed me that 8-year olds had much trouble in using a pair of compasses but in a third grade where the children were familiar with the use of the instrument, nobody had any trouble. The precise construction of circles is of course preceded by free-hand drawing (or foot drawing) of circles on paper (or in sand).

Attempts at provoking definitions are useless even with children that can handle the compasses, or do not yield the result that one would expect. One gets statements like

circles are round,
circles are everywhere as wide (or thick),
circles are everywhere as round.

If one asks what "as wide" means, parallel tangents (or support lines) are drawn and their distances are indicated as equal. Of course a figure with this property need not be a circle: there are a great many other figures of constant width (Figure 127).

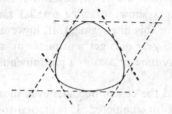

Fig. 127.

"Everywhere as round" is a good local characterisation of the circle — a curve
of constant curvature, where curvature is an intuitively visual datum. S. Papert
has stressed this local definition; in his approach it is the computer that draws
a circle on the screen:

    the "turtle" gets the instruction (Figure 128)

Fig. 128.

    a little step of fixed length forward,
    a turn over a little fixed angle.

The result is a polygon with equal sides and angles — indeed, a circle in the limit.
    Curvature of a curve is defined by differential geometry as the limit ratio of
tangent turn and arc length; Papert's program expresses the constancy of ratio
"turn of direction to path covered" which indeed in the limit gives the constancy
of curvature. As a matter of fact among all plane curves the circles are the only
ones with a constant curvature. "Everywhere equally round" is a good *local*
definition of the circle.
    A circle cut out fits in many ways into the hole it left; I can *turn* the circular
disc transitively in the hole, that is, such that each border point passes into
each other point. I can attain the same by turning the cut out circle around a
diameter (a *reflection* at the diameter).
    This is a more global definition of the circle, which needs profound mathe-
matical reasoning to be derived from the former local definition. Visually the
connection can be made via Papert's approximating polygons.
    The usual global definition of the circle as the set of points x that from a
fixed point p (the *centre*) have a fixed distance α (the *radius length*) is not at
all obvious. The vast majority of circles in the environment do not betray their
centre.
    How to discover the centre? From people who have not been trained in
geometry, one may not expect the construction that is taught in the geometry
lesson. The usual attempts show: drawing parallel tangents, connecting the
contact points and halving this line-segment. If, however, the circle is given or
suggested to be cut out, 8-year olds get a diameter by symmetric folding. This
is the straight line as a symmetry axis — a phenomenon which will occupy us
later on.
    The query how to find the centre becomes even more tricky if the circle is
given by an arc that must be completed. It is natural to seek the centre on the
axis of symmetry, that is on the orthogonal bisector of the largest chord of the

arc. But where on the bisector? A change of perspective to a smaller partial arc is of help: the orthogonal bisector of a smaller arc as a second locus for the centre. Finding the intersection of the two bisectors boils down to the classical construction of the centre of the given circle — a solution not parachuted but suggested by the seemingly more difficult problem of completing a circular arc to a whole circle.

From here it is a little step to the circumcircle of a triangle and the insight that a quadrilateral in general does not possess a circumcircle.

With this exposition we have run far ahead on a didactically warranted approach to circles. The approach by means of a circle drawn on the backboard is much too narrow. One can have children experimenting with concrete circles (round discs) of various sizes, paving the plane with a large number of congruent circles, discovering that each circle is touched by its neighbors in a regular hexagon, that opposite contact points are connected by a diameter, and that these diameters intersect in the centre. One can also have the children stand in a circle and order one of them to stand in the centre — where is it and how to find it precisely? One can have circles drawn with a string, with a pinned up strip, or by enlarging and reducing a given circle, before handing them out the pair of compasses.

With the pair of compasses one can stake out and transfer distances. The shortest distance of a point p to a set S has something to do with circles: the compasses pinned in p are being opened until meeting S. The set S can be a straight line, of course the shortest distance is: straight ahead on the line. The open circle just touches the line; the line S is the tangent of the circle, the shortest distance is the radius, tangent and radius are orthogonal — a principle that extends to functional analysis: the orthogonality of the line of shortest distance. Let S itself be a circle; finding the shortest distance from p to S produces a circle contacting S. Both contact each other, both radii drawn to the point of contact are orthogonal to the common tangent, so the radii are extensions of each other. And so it goes on — there are a great many approaches to the geometry of the circle besides the one that is prescribed by school geometry.

Measuring of arc length was discussed in Chapter 1. One can measure the circumference of the circle by unreeling a string or rolling it along a straight line. Rolling circles on straight lines or on circles is itself a source of many phenomena which are worth being considered.

A well-known puzzle asks: If I roll a moveable florin around a fixed florin, how many "turns" will it have performed when it has returned to its original place? The surprising answer is: two. How to explain it? Though the answer does not fit in the present context, let me anticipate it. One can observe the same phenomenon with, for instance, congruent regular 3-, 4-, 6-gons, which makes it more perspicuous. Localising is even more efficient. I roll off the angle C on the angle C' (Figure 129) — how much is the turn? It is the sum of their exterior angles. A simple closed polygon has the sum 360° of its exterior angles. If they are such that the one can be one-to-one rolled around the other, the total of turns is 360° + 360°.

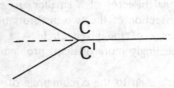

Fig. 129.

I said "one-to-one". I did it intentionally. Take a triangle and a hexagon, both regular and with same side length. I roll the triangle around the hexagon. How many turns does the triangle perform, and why?

Among all plane figures the circle shows the highest degree of symmetry. It is easy to disturb the symmetry and it is an art to profit, as least partly, from this fact by drawing regular polygons and stars in the circle and to surround one circle by a wreath of circles which touch and intersect each other — phenomena to be considered elsewhere.

11.13. *"Venn-diagrams"*

This is no section that got lost and properly belongs to Chapter 3. I once introduced the term "natural Venn diagrams"* in order to distinguish Venn diagrams as used as a tool in serious mathematics from the artificial ones, which in school mathematics is an aim in itself. A natural Venn diagram appears if one draws a simple closed curve and pays attention to its interior as a point-set. This means that the set A is the point-set within a curve which is labelled A rather than a set of objects or letters placed or indicated within the curve. Only these natural Venn diagrams occur in normal mathematics although the plane sets within the curves can be models for other kinds of sets whose unions, intersections, and complements are to be illustrated.

I put Venn-diagrams between quotation marks because I mean that I have the figures in mind, primarily as geometrical figures, while the opportunity to exercise set theory operations on them is a bonus. "How shall I teach sets in the first form of secondary education?", a teacher sighed after my well-known criticism of what usually happens with sets. There is a vast choice for everybody who does not suffer from what I called the constraint of the system. The system, that is, straight lines, half planes, circles, circular discs, and so on, at the far end of a long development which starts with sets at its first step. I need not argue that system constraint is contrary to didactics based on phenomenology.

The intersections, unions, complements of geometrical figures which I am going to exemplify are again meant as geometrical figures. Terms like

* *Mathematics as an Educational Task*, p. 342.

set of the ... with the property that ... ,
inclusion,
intersection,
union,
complement

will serve as linguistic means — as is the case almost everywhere in mathematics — to

describe given figures by means of others,
formulate processes of construction,
explain connections between figures.

Training these linguistic means can be useful by transfer to other subjects where they apply, the figures and the operations performed on them can be models for vaster abstractions.

Our "elementary" sets are

line-segments and straight lines,
half planes (it does not matter much whether the boundary is included),
circles,
circular discs,

where the last are explicitly defined as

set of the points at a distance α of p,
set of the points at a distance $\leq \alpha$ of p.

Starting there

angular parts of the plane, polygons, polygonal pieces,
circle segments, circle sectors,
circle annuli,
types of lunules and their complements,
circular triangles and polygons and their complements

are

constructed by means of set theory operations,
analysed with respect to their mutual set theory relations,
described by set theory means (see Figure 130).

11.14. *Spheres*

In the development of geometrical mental objects the sphere — as a ball — precedes perhaps even the circle. In traditional geometrical instruction the sphere was a latecomer, dealt with if at all only through formulae for surface and volume, which were learned, confounded and forgotten.

A local definition of the sphere, as of the circle, would be that it is everywhere

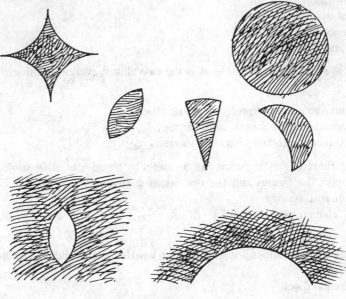

Fig. 130.

as round. There are, however, with respect to the sphere experiences which are more original than constant roundness. A ball can roll — and differently from cylinder and cone — as easily in all directions. Or does this hold of all round convex bodies — ellipsoids, eggs, acorns? No, the sphere — at least the substantially homogeneous ball — does not know any prefered position; at which ever point it touches the ground plane, it is in — indifferent — equilibrium, whereas other convex round bodies are distinguished by certain positions of stable equilibrium. Mechanically this can be expressed as follows: the sphere is the only round convex homogeneous body, the centre of gravity of which is as far from all boundary points — and this point then is also the centre of the sphere itself.

But let us review the characteristic of being everywhere as round. In the case of the circle the degree of roundness was mathematised by the concept of curvature — the curvature at a certain point was the limit ratio of the turn of the tangent and the arc length, where instead of the tangent vector one could as well use the normal vector. In differential geometry the curvature of surfaces is defined in a similar way as that of curves: one imagines in each point of the surface a normal vector of unit length; by assigning to each point of the surface its normal vector, the surface is mapped on the unit sphere; the inverse limit ratio at a point p of the areas of a piece of surface around p and its image on the unit sphere is by definition the *Gauss curvature* of the surface of p. For a sphere with the radius r it is everywhere r^{-2}.

I just mentioned this kind of curvature though it does not reflect what we mean by everywhere as round. The *Gauss curvature* accounts only for the intrinsic properties of a surface, not for its appearance in space. A cylinder and a cone have everywhere vanishing Gauss curvature as has the plane, and they can be unreeled upon the plane without stretching or shrinking — at least locally. The sphere has a constant Gauss curvature but there are more surfaces with the same constant curvature, which can be unreeled upon each other.

The "everywhere as round" means the appearance in space. The sphere shows locally everywhere the same image. As a matter of fact, this holds also for the cylinder, but for the sphere it does so at every point in each direction. A mechanical production of spheres rests on this "being everywhere as round". If two stones or pieces of metal are ground upon each other, they will turn out to display finally two congruent spherical surfaces — by chance they might be planes.

Once the circle has been identified by means of its centre and radius length as the set of radius distance from the centre, the same characterisation of the sphere is natural. Of course the centre of the sphere is less accessible than that of the circle, but a pair of antipodes can be found on a concrete sphere, and so can a diameter.

Cuts of spheres or the water level in spherical bottles show that the plane intersections of spheres are circles, but this fact is also mentally understood as soon as it is clear how to find the centre of this circle: the foot of the perpendicular from the centre of the sphere on the plane. Or, more perspicuously: viewing the sphere as a solid of revolution with as its axis the diameter perpendicular to the plane. Likewise the intersection of two spherical surfaces is recognised as a circle; then the axis of revolution is the line connecting both centres.

Shortest paths on any convex surface can be obtained experimentally, by held strings. It looks natural to consider such a path also as a straight ahead line. There is, however, also an independent, intuitively acceptable, definition of straight ahead line on a curved surface. Straight ahead line is a local concept; if two surfaces are being rolled upon each other a straight ahead line is copied as a straight ahead line; if a curved surface is being rolled upon a plane such that the contact points form a straight line, the corresponding line on the surface is a straight ahead line — a geodesic.

Applied to spherical surfaces, this shows that the great circles are the straight ahead lines. It requires more profound arguments to show that arcs of great circles are also shortest lines.

It follows from the foregoing that spheres cannot be unwound to become planes, even locally. A spherical triangle unwound upon the plane would represent itself as a straight triangle with the same lengths of sides and angular measures; this would result in a sum of angles of 180° whereas in a spherical triangle the sum of angles is certainly larger. The fact that the sphere cannot be unwound upon the plane is responsible for the variety of projection methods by which cartographers try to map the globe.

Nice partitions of the sphere most often use arcs of great circles — not because they are the straight ahead, and the shortest, lines but because the planes of the great circles are symmetry planes of the sphere and the circles themselves symmetry lines on the surface.

Well-known partitionings of the sphere are the so-called dihedra — great circles through the poles together with the equator (Figure 131). The regular

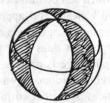

Fig. 131.

bodies inscribed into the spheres project themselves from the centre onto the surface as regular partitionings of the spherical surface. Similar effects are obtained with semi-regular solids; the best-known is the "football", a partitioning by 20 regular pentagons and 12 regular hexagons. There are other opportunities to deal with this subject.

11.15. *Revolution Surfaces*

Revolution surfaces arise if a plane curve — the profile — is rotated around an axis lying in the same plane. The spherical surface is an example, with a diameter as axis and a great circle as profile.

If, however, a circular arc turns around a chord, which is not a diameter of the circle, one gets a lens or a vault. A decent surface does not arise unless the axis is lying outside the circle; it is a ring, also called torus, next to the sphere one of the most familiar surfaces. Revolution surfaces are produced in a rich variety on the potter's wheel; the rotating clay is being profiled into the wanted shape. There is a striking difference between points where the profile curve is convex or concave with respect to the axis. In differential geometry these points are distinguished by the sign of the Gauss curvature: positive and negative, respectively.

If the profile curve is a straight line-segment, the surface arising is a — truncated — cone or a cylinder. It has been earlier mentioned that these latter surfaces can be unwound on the plane, and on each other.

11.16. *Angles*

Elsewhere* I have extensively analysed the various concepts of angle. I started with the mental objects in order to reach the mathematised concepts as fast as

* *Mathematics as an Educational Task*, pp. 476—494.

possible. Though the present approach is a bit different, duplications would be unavoidable if I were not allowed to call on the former exposition.

In agreement with the title of the present chapter, angles are dealt with here as figures or configurations. Just as squares differ from oblongs and the "same" square or oblong is possible at various places, so angles will be distinguished as different or equal before *measuring* angles is discussed.

Angles present themselves in a variety of ways — early or late — in the development. If these presentations shall be classified phenomenologically, it is quite natural to look at the components that contribute to the angle concept, and at the degree in which they are.

> concretised,
> made explicit,
> indicated,
> suggested,
> imagined.

By the components I mean

> the sides of the angles,
> the planar or spatial part enclosed by them

if the angle is a *static* datum, and

> the turn

if *kinematics* comes in.

I introduced angle concepts in the plural, because there are indeed several ones; various phenomenological approaches lead to various concepts though they may be closely connected. Even in practice one single version does not suffice.* This is witnessed as early as Euclid's Elements; in Book I after the definition VIII of the angle as the mutual inclination of two lines, definition IX speaks of the angle enclosed by two lines, and though by definition angles are less than 180° this does not prevent Euclid from adding angles beyond 180°.

The most concrete though also the most misleading phenomenon of the angle is

> the concrete part of the plane or space;

one is expected to restrict one's attention to the neighborhood of one vertex or one edge of an angular sector or an edged body, while this local view is disturbed by the actual limitation as to extension and shape of the sides and of the figures as a whole — two angles are equal if they

> fit each other locally.

that is,

> if the one fits locally in the hole of the other.

---

* *Mathematics as an Educational Task*, pp. 476–494.

The local character implies that one is allowed to anyhow break off an angular corner in order to seize the angle (Figure 132); the

nondescriptly broken off corner

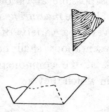

Fig. 132.

then is a means to suggest what is meant by an angle. Whoever has observed children working with jigsaw puzzles of polygonal pieces will have noticed that they are familiar with this mental object of angle, though there are 7–8-year olds who fail in that matter for lack of former experience.

The stress laid on the enclosed planar or spatial part is weakened if the enclosure is removed as much as possible, as happens with instruments like

drawing triangles,

where the inner border suggests the same angle, and suchlike moulds. Even leaner concretisations are possible – the sides only concretised,

as legs of a pair of compasses,
as two hands on a clock dial,
as sides of an angle folded from a sheet of paper,
as main and side branch of a tree,
as two bones or staffs in a joint,
as sheath and blade of a clasp-knife,
as two canvas walls of a tent,
as a telephone wire that changes direction at a pole.

The enclosed planar or spatial part can still be suggested by the sides as a skeleton:

the steel wires of a fence, or wire netting

form angles of squares or hexagons,

the wings of a folding screen

hide a part of space, the

edges of a wire cube

form angles and at the same time squares. The leanest representation of an angle is a drawn pair of

> line-segments from one point,

which are supposed to represent line rays, with possibly an arc or a system of arcs between the sides added, which however does not represent the enclosed part of the plane, but rather indicates that the figure means an angle, and well the "same" angle if the same arc or system of arcs is repeated.

But it is even possible that the two sides are not — or are not both — concretised, made explicit, indicated but at most suggested. This happens with the angle as

> deviation from a rectangular or flat continuation,
> branching off, breaking

when one of the sides is not concretised but imagined or imaginary as the continuation of a concrete line or plane — the dotted feature in Figure 133 — that is an

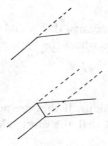

Fig. 133.

> adjacent angle of a concretely given angle.

A polygonal disc or prism shows concretely a number of interior angles; if they are walked around, the change of direction is expressed by the exterior angles — a subject usually neglected in instruction.

Subjects in Piaget's experiments describe the reflection of a ball at a wall by means of the angles between the wall and the direction of the movement, which is quite natural — the deviation of concrete legs. Piaget, however, accepts as a proof of full understanding only the formulation by means of the so-called angles of incidence and reflection (that is with the perpendicular on the wall), which is the way it is taught in textbooks.

Of the pair of sides of an angle there can be

> one explicit and the other implicit,

for instance in a topographical context: in the terms

> upright, oblique, lopsided, leaning.

A line or a plane deviates from something that is considered as normal:

> an oblique roof,
> an upright wall,
> an oblique ladder,
> a painting hanging askew
> a steep road, ladder, mountain,
> an inclined plane,
> a smoothly sloping landscape

aim at an angle with respect to the horizontal plane, whereas

> crooked growth, out of plumb, slanting in the curve

may rather mean an angle with a vertical line or plane. In

> slanting letters, sloping line, oblique square

the margin line of the sheet, the lines or squares on the sheet may be the norm with respect to which the deviation is stated.

> An oblique nail

forms an angle with an imaginary nail that would be right in the plank.

So far we mentioned cases where the "other side" of the angle is apparent. In the case of the angle between

> line and plane

the other "side" is not as obvious. If they are not orthogonal, it requires a profound analysis to understand that the other side is obtained by projecting the line on the plane.

The angle is even more stripped of its concrete elements if both sides are imaginary, as is the case with the

> vision angle

under which an object or an (as imaginary) line appears. The sides of the angle are vision lines, imaginary lines that connect more or less concretised points. In the case of the

> elevation of the pole and the angular distance of stars

vision lines provide the sides of the angle, which afterwards are interpreted as radii of a sphere, the celestial globe. In the case of angular distance on the Earth, such as

> geographic longitude and latitude

the sides of the angle are entirely imaginary lines, drawn from the centre of the Earth, and extended to the zeniths if the measurement is performed by astronomical means.

If an

    arc is interpreted as an angle,

as happens in geometry and trigonometry, the sides of the angle have entirely disappeared, though this angle can be interpreted as

    change of direction

(Figure 134) — continuous rather than instant change.

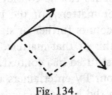

Fig. 134.

We have now arrived at the

    angle as turn,

that is, neither the static pair of sides nor the enclosed planar or spatial part, but

    the process of change of direction,

the transition from side to side, where the enclosed piece is swept out. It is my experience that nobody except the teacher, who exerts himself to explain it, has any trouble with this aspect of angle.

    The key in the key-hole,
    the tap,
    the hands of the clock,
    the knob of the watch,
    the handle of the door,
    the screw cap,
    screw and nut,
    the cork-screw —

in all cases one has quite precise kinesthetic ideas about the meaning and consequences of a certain turn, how far one can and may turn: a quarter or a third turn of the key if the door is simply closed, and one of two turns more if it is bolted, a quarter of a turn of the tap for washing one's hands, and many turns if a bath is to be filled; angles at clocks tell you the time even if there are no figures on the dial, and each of the familiar turning instruments possesses characteristics that tell you how far you can and may go.

Benjamin (7; 5) is shooting with a dart folded from paper — called a Concorde — at his sister. Twice the dart lands at the same spot at her right. The third time he aims while taking the deviation angle into account, and hits.

When starting this subject, I said that only people who try to explain it to others, have trouble with it. The reason why conceptualisation of this mental object requires so much effort may be the fact that it is the most natural, the most instinctive aspect of angle.

I refer to the earlier quoted exposition in order to stress that the mental object "angle" is conceptualised in various ways, depending on the approach, and the possibilities and needs of application. Euclid's angle finishes at 180°, as does the usual protractor. The angle of trigonometry is represented on the full circle protractor. The pair of sides can be considered as ordered or it is understood that the order does not matter. For the turn angle it can be essential whether after a full turn one counts further or starts anew, and whether it is a right or a left turn. Whether this or that matters, this depends on the context. The mental object is being constituted in a context.

The famous joke of one of our TV entertainers a few years ago — in a political context — that three turns to the left is the same as one to the right, rests on intermixing two contexts. In the gymnastics lesson both of them give the same result, but if a quarter of a key turned to the right opens a door, three quarters to the left will not do — on the contrary the door gets even more closed. Whether a gymnastics lesson or opening a door is a better model for politics, fortunately need not be discussed here.

But we may ask why what is true in the gymnasium does not hold for the door lock. The action of the key in the door is: bolting and unbolting. A rough sketch (Figure 135): a toothed wheel moving a toothed bar. When the wheel has

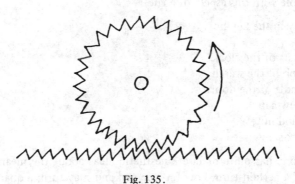

Fig. 135.

made a full left turn, the bar has moved a full circumference of the wheel to the right. If the bar were long enough, one could move as far as you like it by means of the wheel, which itself stays on its place — as far as the wheel is concerned, it may look as though nothing has happened. But the converse can also be realised, that is with a rack-and-pinion railway, where the moving toothed wheel of the train steps into the fixed toothed bar between the rails. Or, without teeth: a circle wheeled on a straight line.

In all these cases turn angles that count unlimitedly, are tied to angles where a full turn is counted as though it were nothing — two mental objects of angle, the relation between which is clear enough as long as one stays away from conceptualising. In order to pass from the turn angle to the angle determined by two sides, one must abstract from — consider as inessential — that what happens in the meantime when one side is being turned into the other. Making this explicit can be difficult, indeed. It can obscure the insight while its strength may be spared if it is left implicit. This then may be the reason why teachers meet difficulties when they try to explain the turn angle. I did not investigate this problem but it would be worthwhile trying it.

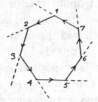

Fig. 136.

A heptagon (Figure 136) has the sum of angles 5 · 180°; 5 times that of a triangle. What kind of angle? If I walk around the heptagon according to the arrows, the change of direction in each vertex corresponds to the exterior angle, which is after a whole walk a full left turn (360°). This then is the sum of the exterior angles. Interior and exterior angles together are 7 times 180°, which leaves the interior ones together with (7 − 2) times 180°. How does this angle arise? I am turning the line 71 around 1 clockwise until it is lying along 21, then 12 around 2 clockwise until lying along 32, and so on, until finally by a turn of 67 around 7 I come back to 17. That is 5 times 180° turned clockwise.

CHAPTER 12

GEOMETRICAL MAPPINGS

12.1. *From Figures to Mappings*

In the original version I slipped at this point straight from the figures into the mappings. I slipped though geometrical mappings as cultivated now are phenomenologically all but selfevident. Historically they were a late phenomenon – this will be discussed later. I doubt whether in the individual development there might be something like a spontaneous genesis of geometrical mappings as mental objects; didactically viewed the most urgent problems of geometrical mappings have not yet been reconnoitred.

For the many people, who have not yet conquered Aristotelism, concept formation, even in developmental psychology, is synonymous with classification. In our last chapter the titles of the sections may evoke classificatory associations; those of the present chapter will rather conjure up ideas of relating objects with each other, which indeed is the germ of mapping. A mapping requires three components: the thing to be mapped, the image, and the process of mapping. In the course of the present chapter stress will shift from figures as the objects to be related with each other via the mapping as mental operation to the mapping as mental object. To be sure, from the beginning onwards mappings will be considered, though within a restricted context: as a means to compare figures, to recognise them as the same, to distinguish them, to structure them, according to varying criteria.

12.2. *Comparing Figures*

Let us start with the last examples of the preceding chapter on angles.

I face a planar angle. I fix one side and move the other – I map it from one position into another. The angle shrinks or swells – I mean the part of the plane enclosed. Or I follow with the second side the movement of the first such that they move as fast. The part of the plane enclosed changes its place though not its shape. I can move an angle over the whole plane, pass with it into space, the sides rigidly coupled to preserve the angle or hinging in a joint such that the object remains an angle.

Mathematicians might take offence at this exposition. What remains the same, if the angle is rigidly displaced, is the angular measure *rather* than the angle. It is the same as with the line-segment, where displacing preserves the length only. We learned, indeed, to avoid sloppy terminology. We indicate two line-segments, two angles, two triangles, the one a reproduction of the other at another place, congruent rather than equal.

Right, we act this way as soon as we have put objects into a geometrical context, or rather if we have done it so consciously that it is a *logical* geometrical context.

On my writing desk I am facing two cylindrical plastic little barrels, exactly the same, that is so exactly that at a glance I can be mistaken about which is which. One contains pills, the other is empty. If I lift and possibly shake them, I know which is which, I can distinguish them. No, I can distinguish them right now: the left one and the right one. If I interchange them, I will have interchanged *these* attributes. But the other thing, their substance, has not been interchanged. I can move, turn each of the barrels, put them in my pocket, and they remain the same. Moreover, they remain the same as geometric figures, perpetually congruent. The one Peugeot 404 (1972) is congruent with the other, and each with itself, but with itself it entertains a closer relation than congruence only.

A talk with a group of 10–11-year olds: a sum 3 with two dice — how is it possible? The dice are lying on the table. They manipulate the left one to show 2, the right one 1. But it can be done differently: to show it they have both dice change places. A desperate discussion. The aim to have the left die manipulated in 1 and the right one in 2 is unfeasible, because it is much easier to have them change places. With the best of intentions they do not succeed because the concreteness of the material dice prevents them from forming the mental objects "die 1" and "die 2". Dice of different colour would have been more useful.

What is wrong here? Of two somehow equivalent objects

substantial and local exchange

are confused. Elsewhere* I have explained how much trouble textbook authors cause themselves and their pupils with this confusion; though I hardly scored any success with this exposition, I am weary of repeating it once more. It is clear what is wrong here. In order to have two dice function as an ordered pair, I must be able to distinguish them as actualisations of one die, independently of their localisation. Speaking about figures I neglect these elements. At a glance both barrels are somehow cylinders. This is their figure. There are different kinds of cylinders, thick ones and long ones, but both of these ones have the same figure. It is one and the same figure, twice actualised. I do not say: embodied. There is one natural number 5. I can embody it by the fingers of my left hand, of my right hand, by the five continents, or by any imaginary set of five things. Likewise I can embody a cylinder in many ways, and this happened here by the pair of barrels. But I can also take one of them and place it differently in the space. Both barrels are not only different embodiments but they are also together in an "order of coexistence".

"The same" means various things in colloquial speech. Fortunately! A profusion of words can be annoying. But a profusion of meanings of the same word

---

* *Mathematics as an Educational Task*, pp. 377–387.

is no evil as long as the meaning is determined by the context. Look at this pair of cubes, and say what is different. Colour — no, we do not want to pay attention to colour. Plastic or wood — no, the material does not matter. There is a dot on one of them — it does not count. A scratch — as little. The place? They are at different places. Does it make a difference? If the one faces you with a side and the other with an edge, it might be considered as a difference, which, however, might be reasoned away. It is a pair of cubes, which can produce a new figure by putting one upon the other, as one can add up the same two fives.

How long can this terminology be maintained? When does the need emerge for the word "congruent"? I would say, in such a context as congruence theorems. Two cubes are the same if their edges are the same — this can still pass. But: two triangles are the same if their corresponding sides are the same — this walks with a limp. Is it that much better to say "two triangles are congruent if corresponding sides are congruent"? Well, let us say: "Triangles ABC and A'B'C' are congruent if AB, BC, CA are congruent with A'B', B'C', C'A' respectively. But be cautious! A triangle is more than a triple of points and a triple of line-segments. Triangle then includes also an order of the vertices (a linear, not a cyclic one), because arranged in another way they may no longer fulfill this condition. Well, I could also say: "Two triangles are congruent, if I can assign their vertices in such a way, A' to A, B' to B, C' to C that the corresponding sides are congruent."

It appears from this formulation how far we are advanced in a logical geometrical context. "Being congruent" means here the existence of a congruence-mapping, which as such has been eliminated by the existential quantor. Are these superfluous sophistications? Yes and no. One can come into situations where a pair of figures is congruent in various ways and the way in which, that is the particular congruence, does count for this or that reason. An isosceles triangle is congruent with itself as is each figure, but it is so in a non-trivial way, to wit by interchanging the vertices at the basis; an equilateral triangle is self-congruent in as many as six ways, a square in eight ways — I will deal with symmetries later on.

The phenomenon signalled in the last paragraph is quite frequent in mathematics. Two mathematical objects are "the same" thanks to a certain equivalence but this equivalence may have been drawn from a stock of equivalences, and at a certain moment it may matter under which one the objects are the same:*

Third graders are building four cubes houses. All of them should be different. Houses that can be turned into each other, are the same, but mirror images are being considered as different, $\boxed{2}\boxed{1}\boxed{1}$ and $\boxed{1}\boxed{1}\boxed{2}$ are mirror images, but at the same time they are turn images of each other. Thus, mirror images will be considered as different if they are not at the same time turn images.

The cubes of the quoted experiment were differently coloured but nobody paid attention to this fact. They were of the same size, thus congruent; they

* Wiskobas Bulletin 6, nr. 2.

must touch each other along full faces. It was tacitly accepted that they were all the same cubes, interchangeable. But another stock of mutually congruent cubes, of another size, would as tacitly have been accepted as "the same" material. Comparing figures as meant in the title of this section need not be restricted to stating congruence.

If some class of figures is recognised as squares, rectangles, cubes, planks, prisms, pyramids, cylinders, cones, spheres, a comparison takes place in order to establish agreements and differences. Comparing can have the character of point-by-point mapping, but it can also be of a more global, combinatoric kind. If two rectangles are recognised as such and compared, the attention can be restricted as far as the common features are concerned to the fact that all angles are right and opposite sides are equal, and the sides of both of them can be compared with each other, but this comparison can also be extended to a — not necessarily congruent — mapping of the one upon the other, possibly embodied by the deformation of elastic material. In the case of parallelograms, which as such are recognisable by the parallelism of opposite sides, more characteristics can diverge: an angle and the sides around it. In this case, too, the mapping — an affine one — can be extended by a systematic construction of parallelograms (Figure 137): in the case of rhombuses with constant sides and variable angle this can be embodied by a frame of threads. If a pair of triangles is given it is natural to extend the assignments of vertices to each other as an affine (barycentric) mapping. Prisms, and likewise pyramids, are distinguished by base and height; as to the base the kind of polygon matters; a quadrangular prism can be divided into two triangular ones or two standing faces can be turned with respect to each other in order to produce a triangular one.

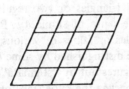 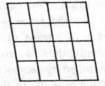

Fig. 137.

I close the present section with these examples in order to stress that primarily, comparing figures does not mean pointwise mapping though comparisons can be unified under the aspect of mapping.

12.3. *Distinguishing Figures*

Comparing and distinguishing are two aspects of the same activity, differently shaded. Equality does not mean much without inequality; it matters what is stressed. The means to distinguish figures, specified in the preceding section, were place and size. In elementary geometrical material, but also in drawings on the blackboard, colour can be used as a means to distinguish: the red and the green triangle, the red lines and the green lines — of course I could have distinguished them also by letters or numbers. The faces of a die are distinguished by primitive number symbols. The crystal lattice of household salt shows more structure than the ordinary cubic lattice: by chemistry the spot of the Na atom is distinguished from that of the Cl atom. Does this still belong to the geometrical context? Yes, and no. We can pick it up, as the colour of triangles and straight lines, in order to refine a structure, to compare more sharply, and in this sense it is still geometry though not expressed in purely geometrical terms.

As a means of distinguishing, new kinds of characteristics are added to the purely geometrical ones, not only in order to stress certain geometrical differences by material and colour factors, as assumed in the beginning of this section, but to create new structures. Triangular pyramids that are congruent when viewed as tetrahedra, are distinguished as pyramids by extending to one of their vertices — for each of them another one — the predicate of top (and to the opposite face, the predicate of base). Are these new characteristics of a geometrical nature? Base and top seem somehow related to the terrestrial horizon and the plumb, but this is not the intention; pyramids with 4, 5, 6 side faces are recognised as such however they are situated in space; tops may show downwards and side faces lie horizontally. If, however, a tetrahedron is to be considered as a triangular pyramid the top is not determined by internal data but by a *stigma*: this shall be the top. If now I am going to compare triangular pyramids with each other as was done in the preceding section — whether they are equal or different according to this or that characteristic, mappings play a part: assigning vertices to vertices, and of course top to top. This can lead to distinguishing figures by extra structure.

Two planar partitions into equilateral triangles or two regular partitions of space into cubes are the same — in the sense of similarity. But by painting the plane or space by means of triangles or cubes of various colours, I can create diverging configurations, which are distinguished by some extra structure, which is not determined by geometrical figures but by "stigmata".

The most striking example in this respect is the cube transformed into a die. As a matter of fact, I could interpret the familiar number symbols of the die or the domino as geometric spot figures, but this would be a preposterous way to save the geometrical context. Indeed, I do not make a difference between say ∶∶ and ∶∶∶ , which shows that the pictures on the faces of the die are arithmetical symbols rather than geometrical figures.

The faces of a cube are numbered to transform it into a die. It is not done arbitrarily. It is a convention that opposite faces add up to 7. In front of me on my desk there are two dice. I manipulate them to get the 1 at the top. Then I turn them to get the 2 in front of me. Where is the 3? For the left die it is at the right, and for the right die at the left. Thus they are not the same, they are a different manufacture, printed in different moulds, mirror images of each other. Well, as cubes they are the same, and not until they were marked with different stigmata, did they differ from each other; as dice they are different. As cubes they could be mapped on each other but by the stigmata the possibilities of mapping were restricted.

This kind of example can be diversified *ad lib*: A cube is different from the open box in which it fits; one of the faces has been provided with the structuring stigma of "lacking". A cylinder differs from a tube which is open at one end or at both of them. A rectangular sheet can be folded in many ways, accordeon-like, that is alternating to the right and the left, or always in the same sense, or first threefold in one direction and accordeon-like or otherwise in the other direction — the inventiveness of map and folder producers is unlimited. All these objects are the same if viewed as rectangles, but as folded figures they differ by the extra structure, which looks accidental though it can reasonably be put into a geometric context. Which among those structures are the same and which are different, is again determined by the existence or non-existence of mappings respecting the extra structure. It may happen that I am in no way concerned about the length and width of the rectangle and that all that matters is the fashion of folding; then it depends on the existence of a mapping that respects the folding stigmata whether two folding structures are the same or not. This too is geometry, albeit not the traditional brand, and of a heavily combinatoric kind.

12.4. *Structuring Figures*

In our badly formalised — and for just this reason practical — colloquial language, plurals of nouns may mean many things. "Comparing figures" and "Distinguishing figures", which was discussed in the preceding sections, means comparing the one with the other (perhaps within a given stock of figures). "Structuring figures" means a plural of "structuring a single figure", once this one, and another time that one — maybe in order to state that the first admits other structures than the second (which is again a way of distinguishing) or that certain structures are common to both (in which sense they would appear as the same). Yet as a principle structuring as meant here will be a way of comparing a figure as it were with itself. Comparing and distinguishing as discussed in the preceding sections involved mapping. Since this plays also a role in structuring, the present section is a counterpart to the preceding one.

I cast another glance at the little barrels on my desk. As we know they are the same. In fact I can dispense with one of them. I move the one that is left, not by displacing it, but by turning it around its axis. It remains the same on its place though not in all details. Rather than with another figure it is compared with itself. In fact to get this done I need not turn it, I can rather view it from the other side to compare it with itself. Or, transferring the ikonic and enactive to the mental mode, I can imagine the barrel mapped upon itself by a rotation or a reflection, while it remains the same. "The same" because I have neglected the labels and the print on the labels, which do not belong to the geometrical context, in which my barrels are cylinder-like figures. How about the screwcap? Does it belong to the geometric context? It prevents me from turning the barrel upside down — a stigma as I called it. But I have got on my desk also a long plastic tube closed at both sides with a screwcap, which I can turn upside down on its place without changing it. Of course instead of this I can look at it from above and below to get the same result and finally I can imagine it done, say by reflecting the tube at a transverse plane cutting it in halves, in order to state that the mirror image is the same.

Structuring figures as it happened here, is done by mapping the figure on itself — properly said, we just look for structure phenomena expressed by mappings. A line can be articulated to become a garland of beads, a closed curve to become a chain of beads — the look jumping from the one bead to the next produces something like a mapping. The more striking structuring feature of a figure in mirror symmetry, often coarse and only global, sometimes restricted to parts of a figure or configuration, sometimes disturbed by stigmata of geometric or non-geometric character. Leaves, simple or composite ones, often display a plan of symmetry, with sometimes systematic deviations, but the assignment to complete half a leaf is by young children well understood as a construction of symmetry. Star-fish display a five-fold symmetry, flowers according to their species a three-, four-, five-, six-fold symmetry and a circular one, which is not as precise as that of crystals or technical products, though its plan cannot be mistaken. Rotational symmetry is approached by trunks of trees, pupils, all kind of fruit, though as far as precision is concerned, the products of the potter's wheel and the lathe are more perfect: a vase, a cup (the symmetry of which can be disturbed by a handle), a tube, a drill hole. The screw symmetry of winding plants is surpassed by that of the screw-thread and the winding stairs. For all that, we have the habit of putting the less perfect realisations of symmetry into a geometrical context. Why? Because these sometimes global, sometimes coarse, sometimes only partial mappings of the figure on itself make it easier to grasp the figure structurally.

In front of me on my desk a button is lying, invariant under rotations, though as becomes a button, pierced by four holes — that is stigmata reducing the rotation group of the button to a four-cyclic group. My desk-lamp is mirror symmetric, with rotation invariant parts, such as the switch and the bulb — no, it is a bulb with a screw fitting, which in itself suggests screw invariance.

Suggests — that is what I said. Let us be cautious. It is only a finite piece of fitting. At some place the screw motion stops or the lamp drops out of the fitting. The mappings I discussed so far, are lacking something. They serve to compare, to distinguish, to structure figures and they are infected by the imperfections of these figures.

12.5. *Comparing by Mapping*

The stress is going to be shifted, from the objects that are being compared, to the act of comparing, in particular by means of mapping. "Mapping" indicates more sharply what is meant than "map" would do — the act rather than the result. "Mapping" means a process while "map" suggests the result of the process.

Mapping geometrically is kind of — concretely or mentally — superimposing. It is easier with one- and two-dimensional figures than with three-dimensional ones, where the act will in general be restricted to the mental reality. As far as the objects to be compared are concrete, the possibility of comparing is often guaranteed by the knowledge about the origin of the objects — having arisen from the same mould, a mass product, a mechanical or optical copy, DIN format.

For rectangles only one term is available, if the limit case of the square and vague expressions such as "big", "small", "narrow", "long" are disregarded. Triangles, however, exist and are named in a great variety: equilateral, isosceles, right, acute, obtuse — and some of these properties combined. The size does not matter in this case; all of them are characteristics of similarity. Several times I claimed that congruence and similarity of figures are mental objects formed at an early age. Mathematically similarity can be defined by means of the invariance of ratios of distances, but this is an *a posteriori* statement, the result of a logical analysis. It is a hard thing to establish similarity of two rectangles at a glance — the ratio of the sides is not perceived unless the divergence is big, as long as one's attention is not drawn to it or one has to copy something. Triangles are quite different: equilateral ones differ strikingly from right ones, obtuse triangles from acute ones. If similarities are perceived and established, factors other than distance play a part, in particular the angles are an important factor. Unfortunately a rectangle has only right angles, which makes rectangles strongly resemble each other. As soon as the diagonals are added, the image changes; more visual structure arises, in particular because of the new, non-right angles, which makes the recognition of similarity at a glance easier. The same effect can be obtained by other means. Rectangles as pictures representing something are with greater ease tested on similarity or established as similar than mere frames (Figures 138—140), and according to the increase of detail the impression of similarity and dissimilarity gets stronger.

Fig. 138.   Fig. 139.   Fig. 140.

12.6. *Mental Constitution of Mappings*

However, what matters now is not comparing figures but constituting mappings mentally, a process that again precedes that of concept formation. The child becomes acquainted earlier with images of objects and knows how to relate

> the object to its image,
> two images of the same object to each other.

They are

> global relations,

"head—head", "neck—neck", "belly—belly", "foot—foot", "door—door", "roof—roof", which are refined to

> point—point-wise relations —

connections

> between different objects

or

> between different images of an object

or

> symmetries within an object or an image.

The connections can be given

> to be discovered

or proposed

> to be constructed,

such as if an object or an image is to be copied; or they can partially be given or proposed in order

> to be completed.

The connections can be fixed

> in all detail,

or indicated or suggested

> by some characters,

which is the usual procedure in geometric representations. It will depend on the actual situation, how and how far these characters, indications, and suggestions in fact determine the connection. Figure 141 explicitly relates the six vertices of two planar hexagons to each other, and at the same time suggests an extension

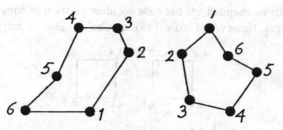

Fig. 141.

to the joining sides, which by preference would be linear, or rather affine, for instance by starting with relating midpoints of line-segments to midpoints, an

affine interpolation.

Extending this mapping to the interior of the hexagons is less obvious. (In the case of convex figures it could be done in a piecewise affine way.) As regards the two circles of Figure 142, first of all the equally numbered points are related

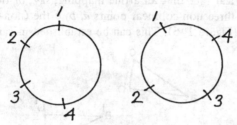

Fig. 142.

to each other; the drawing suggests extension to the arcs, by preference linearly with respect to the arc length. Two less regular arcs (Figure 143) can suggest

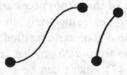

Fig. 143.

many things beyond the relation between the endpoints, linear extension according to the length, or somehow topological. The net of the cube (Figure 144)

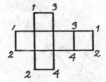

Fig. 144.

asks of course to be mapped on the cube by identification of equally numbered
vertices and edges. However (Figure 145), a cube can also be mapped on three

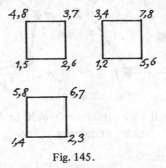

Fig. 145.

projection planes while the numbers at the projections correspond to vertices
of the cube and relations between the projections are created by means of
the number identification.

It is a mathematical fact that an affine mapping, say, of the plane on itself
is determined if of three non-collinear points a, b, c the (non-collinear) images
a', b', c' are given (Figure 146). This can be established in various ways — the

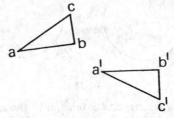

Fig. 146.

easiest being a barycentric construction: each point of the plane can be understood as the centre of gravity of three (not necessarily positive) masses at a, b, c,
which are transferred to a', b', c', respectively, in order there to yield another
centre of gravity. A more visual geometric method is that of constructing a net
from a, b, c and a similar one from a', b', c' that covers the whole plane, and to
refine it gradually (Figure 147). Space can be dealt with in a similar way by
starting with four non-coplanar points.

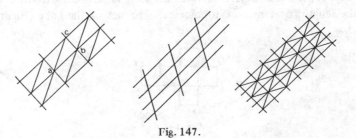

Fig. 147.

If the original triangles (tetrahedra) were mapped by congruence or similarity, the mapping affinely extended to the whole plane (or space) is again a congruence or similarity.

From Euclid onwards up to school geometry as taught until about the middle of our century, statements about congruence and similarity were restricted to figures in the plane or space — often even to triangles, since statements about other figures could be reduced to the corresponding ones about triangles. Extending such congruences or similarities to the whole plane or space is an idea that came rather late in history, properly said not until the importance of other mappings — affine and projective ones — had been discovered. Projecting entire pictures already bears witness to a broader view of the object to be mapped than restricting it to a single figure. Regular planar pictures and in particular symmetries as encountered from olden times in ornaments, suggest something like mappings of the whole plane onto itself by means of the congruent displacement of the picture or by means of the interpretation of symmetry as reflection.

It looks as though it requires a long step from comparing figures by mapping the one on the other to mappings of the whole plane (or space), and indeed a long step it was and still is for people educated in traditional geometry, as appears from numerous mathematical and didactical mistakes in modern textbooks.

The habit of reducing the congruence of figures to congruence of triangles was a historically determined method to guarantee the validity of statements and proofs. The quadrilaterals of Figure 148 are flexibly equivalent, since

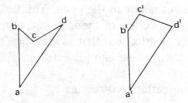

Fig. 148.

corresponding sides are congruent. However they are not congruent as quadrilaterals are understood by tradition. According to the traditional view, though not at the heart of the matter, it is the difference of angles that disturbs the congruence. It is more to the point that the congruent mapping of the union of the line-segments does not extend as a congruence to the plane. An even more drastic example is Figure 149, a parallelogram, which strongly suggests

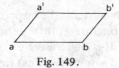

Fig. 149.

a congruence $a \leftrightarrow a'$, $b \leftrightarrow b'$, that is a mirror symmetry — a frequent mistake, which is easily understood. This figure betrays a visually structural symmetry which can be extended affinely, but not congruently, to the plane. Traditionally one would explain this phenomenon by the acute and obtuse angles at a and a', respectively, though it is more to the point to compare the diagonals, the short and the long one.

The classical method of discovering congruences is to extend the figures to nets of triangles — an extension of the mapping from a given figure to a larger one. Why has the ultimate step of extending the mapping to the largest possible figure — plane or space — been taken so late in history? Though a phenomenological approach is not tied to any historical track, it should include an explanation of the — possibly diverging — historical track. This analysis will occupy us now.

Undoubtedly Greek geometers were familiar with symmetries; they did not spurn symmetry as a heuristic tool, and even — as appears from pre-euclidean traditions — as a deductive argument. As late as the euclidean revision of the Elements, symmetries disappeared and for a theorem such as the equality of the basis angles of the isosceles triangle, a complicated proof by means of congruence properties was contrived. This artificial system of congruent triangles has been canonised in the traditional school geometry. We cannot tell for sure what moved Euclid to reorganise geometry that artificially, and his successors to stick to this form of organisation, but we can guess that one argument, if not the decisive one, against mappings might have been that of purity of method: motions is a subject of mechanics rather than of geometry.

Of course this argument is not to the point. This insight, however, that it is not to the point, requires a course of thought that, historically viewed, was not at all obvious. I already pointed out that as mappings the projections precede the motions, and that even affine and projective mappings drew the attention before motions did so.

Motion is first of all something that occurs

> to an object,
> within space (or plane),
> within time.

In order to view motions the way mathematicians are now used to, three steps must be taken:

> from the limited object to the total space (plane),
> from "within" space (plane) to "on" space (plane),
> from "within time" to "at one blow".

In particular the last step is important: the mapping must be viewed as a relation between the initial and the final state; there is no "in between", or as far as it exists, it must be neglected. It is quite usual to visualise a translation by a field of equal and equally directed arrows (vectors) (see Figure 150); the misleading feature in this representation is the suggestion that the points move along these arrows. A system of "curved arrows", with the same heads and tails (Figure 151) is as well to the point or is not.

Fig. 150.   Fig. 151.

(Of course a movement taking place in time can also be represented mathematically, though not as a mapping of the space R on itself: one has f to depend on a "parameter" t such that f_t is a mapping of the cartesian product of R and a time interval T,

$$\ulcorner R, T \urcorner \frown R,$$

but this does not matter now.)

The didactical consequences are obvious:

> either an approach to mappings where the dependence on time is *a priori* lacking, or is not to the point, or can easily be eliminated,

or with any other approach,

> stressing the irrelevance of the "in between".

There is a danger lurking in the second approach. Stressing the irrelevance of something includes drawing the attention to it in order to warn against it. In the long run mappings that invite embedding into the course of time, such as translations and rotations, cannot be avoided in any reasonable way. What one can do about it, is to watch that they do enter in a way and at a moment where the irrelevance of the "in between" need not be stressed any more, because it has already become obvious.

The mental object "mapping" is constituted along the way of constituting special mappings as mental objects. If this occurs by means of mappings that are by their very nature in one blow, one may expect that the principle of one blow becomes itself one of the constituting features of the mental object "mapping", so that any warning against the misleading "in between" can be dispensed with, or can at least be pronounced with less stress.

12.7. *Mappings at One Blow*

The most natural examples are

> projections, that is mappings by means of light sources at a finite or infinite distance,

which produce

light and shadow images.

They include a wealth of geometric information, which may not be side-stepped. It is a disadvantage that they are mappings of the space on a plane, or of one plane on another, rather than of space on itself, or a plane on itself. Another disadvantage is that the connection between image and original is too concrete — lightray, thread — which may block mentalising.

A second kind of mapping at one blow

endomorphisms, isomorphisms, and automorphisms of discrete structures,

such as

grids and other patterns in 1, 2, 3 dimensions,
the ordered set of natural numbers,
the ordered set of integers,
a square lattice in the plane,
a cubic lattice in space,
a hexagonal lattice in the plane,
ornaments,
crystal structures,
mosaics,

thus structures that structure also the straight line, the plane, the space around them, and determine structure conserving mappings, which extend to the surrounding medium. An example is shown by the rectangular pattern of Figure 152, which possesses a set of automorphisms, each determined by one lattice point and its image — a mapping of the plane performed at one blow.

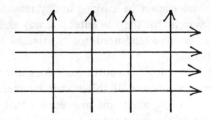

Fig. 152.

Without relying on an extra discrete structure one gets mappings of the straight line, the plane, the space on itself, performed at one blow, by means of

reflections,

in particular if the relation between point and image is performed by means of a mirror. Original and image are simultaneously present, without any lag, at one blow. The use of the mirror, however, as a producer of a mapping, should be fully exploited. One is easily seduced to place the mirror only vertically, which is an unnecessary restriction. On the other hand the actual use of the mirror, if maintained too long and too rigidly, can block the development of the mental operation and the mental object "reflection".

Reflection is first of all an

operation in space,

which relates each point to its mirror image with respect to a planar mirror. By reduction to a table or the drawing plane one gets the reflection as an

operation in the plane,

with a straight line in the plane as its axis. Besides these, one should not neglect

point reflections in the plane or space

— the first realisable by a half turn, but nevertheless a mapping at one blow — and the

spatial reflection at a line

— again realisable as a half turn, around this same line as its axis.

All these reflections are involutions, that is mappings relating two points mutually to each other as original and image, which reinforces their one blow character.

Less pronounced, though still present is this one blow character in the case of the so-called enlargements, and this is certainly so if the enlargement factor is negative: enlargement with centre O and factor ρ maps the point P on P' such that (in vector notation)

$$\overrightarrow{OP'} = \rho \overrightarrow{OP}.$$

In traditional school mathematics this was the only admitted mapping — again an argument in favour of the necessary one blow character if mappings are to be constituted as mental objects. Unfortunately in that instruction enlargements were only a tool to study properties of figures and configurations rather than an object of study as such. An example (Figure 153): The enlargement by the factor $-\frac{1}{2}$ from the barycentre of the triangle ABC into the triangle $A'B'C'$, which served to reveal a large number of beautiful properties.

Even by the mappings at one blow of space or plane onto itself, it is first of all figures and configurations which are being related to each other; the bare plane and bare space are totally inappropriate opportunities to step in. But even in this case reflections are privileged. A translation, a rotation, an affine mapping requires in general, in order to be visualised, the production of a figure and its image. For a point, line, or plane reflection it suffices to give the point, line, or plane, which serve as mirrors — at least if at the same time the mapping is explicitly supposed to be a reflection. The converse problem, however, requires a change of perspective: given a figure and its reflection image to find out the

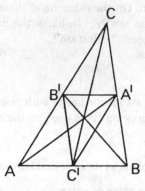

Fig. 153.

mirror that causes the mapping. This change of perspective can be hampered by misleading cues. A well-known error: in an oblique parallelogram the lines joining the midpoints of opposite sides are being considered as symmetry axes.

12.8–9. *From a Mental Operation to a Mental Object and to Concept Formation*

12.8. Of a figure or configuration I can make a visual or haptic image; of a mapping I cannot, at least not directly. Except for the graph of a function, there is in general no visual or haptic means to represent a mapping in its totality. All that can be done is to relate a number of points or marks to each other and then to extrapolate. The case of reflection is particularly favorable: one datum, the quite simple set of invariant points suffices. In order to extrapolate from such data, certain general rules are required. It is a situation quite similar to that of acquiring the arithmetical operations: at each stage the learner has certain ready data at his disposal, from which new relations are formed by means of certain rules which initially are handled by insight in order to be algorithmised in the course of learning.

Rules, as meant here, can initially be implicit in order to gradually be made explicit, and as far as the available linguistic means allow, be formalised. The learner is in particular confronted with the need to recognise a mapping as a mental object and to describe it if wrong constructions ask for clearing up and removing misunderstandings — such as is the case if "right" and "oblique" reflections are to be distinguished. Then explanations are required like

the image is as far behind the mirror as the object is in front of it,

or

the image is found by dropping a perpendicular on the mirror and setting out behind the mirror a piece as long as that in front,

or

object and image are lying on opposite sides of the mirror at the same distance and straight across each other,

or

the line-segment joining a point and its image is perpendicularly halved by the mirror,

or, with a change of perspective,

the mirror is the perpendicular bisector of the line-segment joining points related to each other

or

Σ is a reflection at the line (plane) S if and only if S is the perpendicular bisector (plane) of $\overline{p\ Sp}$ for each $p \neq Sp$.

The change of perspective also delivers as a byproduct an indication to answer the question

what is the place of the mirror mapping p on q?

if the change of perspective and the reformulation of the description of reflection has not even been provoked by this question.

According to the progress of awareness about the characteristic properties (of a reflection), concept formation progresses. But even without verbalising what is becoming conscious about the properties of reflection, the character of mental operation or mental object can be maintained, not only when mappings (reflections in the present case)

are being constructed from data,

or

are being recognised within data,

but also

when inverting them,

when composing them,

when stating that composition in a different order can yield different results.

Starting from reflections (on lines in the plane or planes in space) and composing them, one gets

translations and rotations

on the level of mental operations, or if subjected to closer analysis, of mental objects. Such as reflections by a mirror, so are translations given by

one translation arrow as a model of all of them (original-image),

and the model rule

to move all points the same way,

sharpened to the prescription of

parallel moves of the translation arrow,

and refined by a prescription as to how to act with points lying in the extension of the translation arrow.

In order to constitute rotations as mental objects two data are required,

the rotation centre (axis)

and a second, which may be

a model out of all pairs of points "object-image"

or (Figure 154)

a model out of all pairs of rays (half-planes) "object-image"

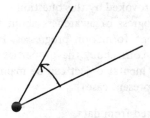

Fig. 154.

with the prescription that under a rotation

all points turn as strongly.

The precision required for "as strongly" cannot be obtained in the case of rotations as readily as it can for translations. The tool needed here — the substitute for parallel displacement of the translation arrow — is the

turning angle,

that is the oriented one: all turns through the same angle.

In order to progress from here to the group of congruence transformations, complete concept formation is required. In order to understand that the product of two rotations is again a rotation (or a translation), and why it is necessary to admit slide reflections and screw transformations, sophisticated reasonings are required, which presuppose concept formation at a high level.

Composing reflections (at a line in plane, or at a plane in space) is an activity where mappings function at least as mental objects if not as conceptual constructs. It requires insight at a high level to pay attention to the fact that

invariance of distance

is inherited from the reflections by their composition results in general. It is a change of perspective, characterising the activity of axiomatising, to ask for

all selfmappings of the plane (space) that leave distance invariant,

and a sophisticated mathematical organisation is required for the activity of

exhausting this sort of mapping by classification.

Similar considerations could be made for affine mappings (of plane or space). As we mentioned earlier they can be fixed by and constructed from a model triple of three non-collinear (quadruple of non-planar) points and their images; it is quite a simple construction by extrapolation to any point. This constructive process of taking possession of the affine mappings can evolve entirely on the level of mental operation. For the conceptualisation, however, an axiomatic change of perspective is required: an affine mapping defined by the preservation of rectilinearity (and parallelism).

12.9. As regards less structured mappings concept formation is even further away from mentalisation. If two simple closed plane curves are visually given, a topological mapping of the one on the other intrudes itself with the same visual force upon the viewer, and even a topological mapping of the interiors upon each other. Without much ado one would replace the word "visually" in this statement by "mentally", and even by "conceptually". However, in order to pass from the mental to the conceptual closed curve, one has to prepare a more or less explicit definition, which in turn presupposes the concept "continuity of a mapping", thus certainly presupposing the concept of mapping, of mapping a line-segment or a circle into the plane, and such a mapping as the object of investigation. If, however, such conceptual definition of a simple closed curve is available, the topological equivalence of two of them is a conceptual triviality and the existence of a topological mapping of the one on the other is an object of an entirely conceptual status. Then one is staying on a conceptual level where topological mappings (or continuous mappings in general) can be concepts. Only in this state can the question of whether the interiors of two simple, closed, planar curves really are topologically equivalent, as they look visually, be asked on the conceptual level.

The gulf between topological and geometrical mappings — congruences, similarities, affinities, projectivities — is obviously caused by the phenomenon of continuity, with its unusually big distance between mental object and concept. The elementary mappings, too, are continuous, but this continuity, if it ever draws attention to itself, does not ask to be made explicit and certainly not to be conceptualised. It is as it were, accidental, whereas for continuous mappings it is the essence.

Nevertheless there is a non-negligible bridge between elementary geometric and continuous mappings – I mean the piecewise affine ones. Graphs can be distinguished according to their combinatoric structure as equivalent or non-equivalent; if the graphs are built with rectilinear edges, then a combinatoric equivalence can be embodied piecewise affinely. Two simple closed polygons can be recognised visually and mentally as topologically equivalent, and this equivalence can be established even constructively, by mapping constituting line-segments on each other, which provides the mental topological equivalence with a primitive operational status, that excedes the purely visual one. This procedure can be extended from the polygons to their interior. By affine means it can be understood that the interiors of two triangles are topologically equivalent, and this can happen on a visual, mental, and conceptual level. A quadrilateral can be split into triangles, and the one of them can be pulled to form a larger triangle together with the first (Figure 155) – a procedure used to transform polygons into triangles of the same area.

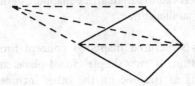

Fig. 155.

If one agrees to restrict oneself to simple closed plane curves, built from line-segments (simple closed polygons) it is not hard to pass from the visual via the mental to the conceptual mode. It can be realised while still avoiding continuity, which means however that the one element is lacking which is required for a genuinely topological context. So as nice as it might be, this procedure does not help to bring the concept formation of continuous and topological mappings closer.

CHAPTER 13

MEASURING BY MEANS OF GEOMETRY

13.1. *Geometry*

Etymologically geometry means measuring the earth, geodesy. By the need, after each inundation of the Nile, for restituting everybody the land he was entitled to, Herodot explains the allegedly Egyptian origin of geometry. In every somewhat advanced society the need for measuring land is felt: as a basis for levying taxes and rent, for delimiting and dividing land, distributing water and seed. In order to plan buildings and to establish their capacity, mensuration and geometry is required, as it is for the construction of roads, canals, tunnels, temples, pyramids, fortifications, which as geometrical figures are designed according to geometrical principles. Measuring distances by day's journeys is the basis of early cartography, measuring angles that of the measuring survey of the sky. This is not the place to explain how these methods were ever refined to trigonometry, spherical trigonometry, and differential geometry, in theories which were to explain the structure and shape of the universe.

As I stressed before, there exists also non-measuring geometry. In affine geometry line-segments must be parallel in order to be compared metrically. Projective geometry works without measuring though the cross ratio can serve as a substitute. In topology and combinatorics there is nothing left that reminds one of measuring.

13.2. *Measuring Length Along Straight Lines*

Measuring length was already discussed at the end of Chapter I though there the geometrical context was poor. In fact it was announced: "Geometrical insight leads to refined methods of measuring distance. Some of them are possible early. We will come back to this point."

Not only was the geometric context of measuring length insufficient. "Length" was a function of long objects, invariant under displacement (geometrically: congruence mappings), flexions, break—make transformations. Long objects could be broken into pieces and again composed to form new long objects, with an additive behaviour of length. Distances emerged as length of — possibly mental — long objects: roads that could be straight or curved. The long objects serving in primitive measuring — rulers, straight ropes — suggest straight lines, but straight lines and line-segments were hardly discussed in Chapter 1. Meanwhile much attention has been paid to the straight line (see Section 11.6). The geometric context in Chapter 1 is, however, insufficiently explicit; a closer analysis has meanwhile taken place, the essentials of which are resumed here, and which is continued now.

Lengths are being measured rectilinearly; as far as curved objects are concerned, measuring is reduced to that of rectilinear ones. Distances are understood by bird's eye view, unless it is otherwise explained, or clear from the context. In order to determine the distance along a "broken" path, the lengths of the "pieces" are added, a curve is approached by broken paths or otherwise straightened out. This rectilinear measuring first of all seems instrumentally conditioned: the measuring instruments — rulers, straightened ropes — are embodiments of straight line-segments, indeed. I said "seems instrumentally conditioned" because I am not sure whether this is also true genetically. The first spontaneous measuring acts of children I observed, were not instrumental, but pacing and spanning (between thumb and forefinger), or by means of the palm, or with parallel fingers in sand, or parallel hands at breast height, thereby not using any instrument that would suggest rectilinearity. Nevertheless rectilinearity plays a part even here: the steps are taken straight forwards, the spans are prolongations of each other — at least this is the conscious or half conscious intention. So the straight line is mentally rather than instrumentally present in such measuring acts.

In Section 11.6 we met the straight line in a large variety of phenomenological contexts. In each of them the straight line gives occasion to measuring procedures. A particularly interesting situation is that where the straight line along which a distance is to be measured, is not present in the data but has still to be constructed mentally, for instance as a straight ahead or as a vision line. Indeed, if the measuring staff (step, span, ruler, string) is too short and has to be applied several times in succession, the rectilinearity of the measured mental object has somehow to be guaranteed by the rectilinearity of extension,

> by overlapping
> by aiming along a vision line.

In Section 11.6 the straight line also occurred as the shortest line. Is it meaningful to posit that distances are being measured according to shortest lines, or is this a vicious circle? Indeed, before deciding whether something is a shortest line, I must know what "shorter", hence what "distance" means. No, it is no vicious circle. In order to be able to discuss shortest paths, I need not measure lengths; I only have to compare them with respect to the

> order relation

of "longer and shorter" and this length comparison according to order is first of all — and at an early age — actualised by the insight that

> the detour is longer,

there is more rope or rail needed to connect along a detour. Or to state it in terms of the rectilinearity of the ruler: the rectilinear path from *a* to *b* (Figure 156) is shorter than the polygonal one composed of line-segments. Mathematically this fact is known as the inequality of the triangle,

$$\text{dist}(a, b) \leqq \text{dist}(a, c) + \text{dist}(c, b)$$

(Figure 157), which purely logically extends to the

inequality of the polygon.

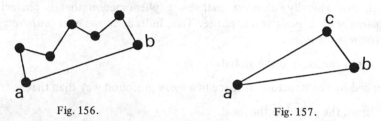

Fig. 156.   Fig. 157.

There is, however, even more to say about measuring distances along shortest paths. The — possibly mental — tools to measure distance first of all assure rectilinearity locally. Going straight on, cutting straight through, is primordially a local procedure; the global line arises from piecing together local segments; special arrangements are needed to globalise the procedure — overlapping or aiming. The vision line looks like a more global tool, suggested by the imagination, the mental substratum, of the light ray. One notices a certain transitivity: if for the eye point a covers point b, and point b covers point c, then a does so with c. The origin of the straight line as a shortest path, however, is

a priori global.

It can be conceived for points which are arbitrarily far away from each other, and realising it can progress from the local to the global domain with or without the stretched rope as its dynamic embodiment. The shortest path is shortest in all its parts, as a tight rope is tight in all its parts, and conversely. These are two perspectives of shortestness:

from the global to the local,
from the local to the global.

They are of a quite different character. It is trivial that a globally shortest path is also locally shortest; it simply follows from the additivity of length: if I can shorten the path somewhere, I can apply this shortening to the whole.

But the converse is not at all trivial. It is not caused by general metric properties but by the specific structure of our space, as can be seen by comparing its geometry with that of the spherical surface. On the sphere going straight ahead and tightening strings produced arcs of great circles. The great circles and their arcs are the straight-ahead-lines of the spherical surface. Small pieces of great circles are also shortest lines, but this ceases to be true as soon as the arc gets longer — longer than the spherical distance between antipodes. In the plane and in space the notions of straight-ahead-line and shortest line coincide; on the sphere (and on other surfaces and in spaces with positive curvature) they fall apart: going straight ahead you reach a point that you could have reached from the starting point on a shorter path – it even happens that going straight ahead one returns to one's starting point.

In the *plane* and in *space* an everywhere locally shortest line is also "globally" the shortest line. The great circles on the *surface of the sphere* are locally, though not globally, shortest paths – a phenomenon that is characteristic of spaces with a positive curvature. This, indeed, shows that with respect to shortestness the inference

> from the local to the global

is implied by the structure of space in a more profound way than that

> from the global to the local.

13.3. *Comparison of Length by Special Congruence*

If length is measured along straight lines the factual rectilinearity of the line to be measured is guaranteed by linear production, actualised by means of overlapping or aiming. Often such procedures do not suffice: the path to be measured may be inaccessible to these procedures; obstructions can block the path between the pair of points the distance of which has to be measured. Then the path to be measured must, at least partially, be replaced by another one that on geometrical grounds may be supposed to be of the same length. In the simplest way this happens if – for instance when measuring a room – a piece of the path is being replaced by a parallel piece. The geometrical insight behind this procedure,

> translation preserves length,

can be formulated less globally,

> in a rectangle, opposite sides are equal,

or more generally,

> in a parallelogram, opposite sides are equal.

"Geometrical insight" in this context means knowledge about such mental operations or mental objects as translation, rectangle, parallelogram – a knowledge that does not necessarily include geometrical deductions, nor the conceptuality of these mental objects, not even the knowledge of their conventional names. By numerous things that suggest the shape of a rectangle the

> existence of rectangles of arbitrary size

and their

> most important properties,

among which

parallelism and equality of opposite sides

are visually and mentally being suggested. Though parallelograms are less frequent among these concrete things, they can profit from those properties of the rectangle that characterise them; rather than as concrete things they play an important part as representatives of rectangles in parallel perspective drawings.

A change of perspective: the construction of rectangles and parallelograms by means of long things which may be supposed to be equal because they are of the same manufacture:

parallelograms made from straw, sticks and suchlike things,
rectangles by vertically planted poles,
larger parallelograms built from smaller ones of the same manufacture.

Rectangles and parallelograms can only serve for translating line-segments; so they can guarantee length equality only for equally directed line-segments. The most primitive means to ensure length equality of unequally directed line-segments actually or mentally is symmetry. By numerous symmetric things

symmetry as a mental operation

is suggested at an early age, and so are

the most important properties of symmetric figures,

in particular, the

length equality of corresponding line-segments.

Likewise

length equality of different directed line-segments

is being suggested by

isosceles, in particular equilateral, triangles,
squares, and more generally, rhombuses,

and, even more generally, by

regular figures.

A more global means to suggest equal lengths, is

patterns, such as mosaics and wall paper,

where

transitivity of length comparison

is an extra mental factor.

Comparing length by means of translation and symmetry is of course not restricted to line-segments. Curved arcs can be recognised as equally long by the same procedures.

The figure that produces at one stroke line-segments of the same length in all directions, is

the circle.

By means of a ruler or any other long object that is provided with marks, line-segments can be transferred to other places while length is preserved. The same can be done with a span of fingers, hand, arms, or with the points of a pair of compasses, although the compasses can perform more than this: producing the whole set of points at the same distance of a given point, and that for all distances within their scope.

13.4. *Measuring Lengths by Congruence Properties*

It is told about Thales that he had measured inaccessible distances (the distance of a ship off the coast, the height of pyramids) by geometrical arguments. For this reason it is the habit in some countries to call a certain theorem on similar triangles (Figure 158) Thales' theorem. This terminology cannot, historically,

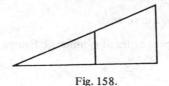

Fig. 158.

be justified. In the most trustworthy source Thales' method is linked to the congruence theorem "one side and two angles"; according to this method, which has explicitly been recorded in ancient sources, the inaccessible distance AB (coast to ship) is replaced by the accessible distance CE (on the land) (Figure 159). The height of a pyramid is said to have been measured by Thales by means

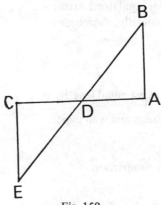

Fig. 159.

of its shadow at the moment of the day when a verticle pole and its shadow were equally long, thus with no appeal to similarity.

Both indirect procedures to measure length can also be understood by symmetry arguments, and the same holds if the equality of the diagonals of a rectangle is to be recognised. In many cases like this congruence theorems are too heavy guns.

A congruence mapping f of the plane is almost uniquely determined if the images a', b' of two points a, b, respectively, are prescribed (of course such that $\text{dist}(a', b') = \text{dist}(a, b)$). The image $c' = f(c)$ of an arbitrary point c then is subjected to the equations

$$\text{dist}(a', c') = \text{dist}(a, c)$$
$$\text{dist}(b', c') = \text{dist}(b, c),$$

which have (at most) two solutions. The triangles abc and $a'b'c'$ are congruent by means of f (congruence theorem, three sides). Congruence, however, also extends to the angles, which means that in the pair of triangles abc, $a'b'c'$ the corresponding angles are equal. After a change of perspective one can localise the point c with respect to a and b (and correspondingly c' with respect to a' and b') also by means of the angles at the side ab (and $a'b'$). If the line-segment ab is given together with the angles under which the point c is seen from a and b with respect to the line ab, the position of c is determined, but for a symmetry, and in particular the distance of c from a and b.

For the *practice* of distance measuring the congruence properties are only significant if related to similarity — the didactical consequence will be dealt with later on. It is, however, a *matter of principle* to know that a triangle is determined by

> three sides,
> two sides and the enclosed angle
> one side and two angles

>> as to shape and size,

and if two vertices are given by position, also

>> as to its position,

albeit but for a symmetry.

With this principle one can at an early age answer the question of how inaccessible distances — the height of a tower, the distance of Moon and Sun (with a terrestrial base line), the distance of a star (with a diameter of the terrestrial orbit as a base line) can be measured.

A particularly important case is that of the right triangle, thanks to various applications of the Pythagorean theorem.

13.5. *Measuring Lengths by Similarity Properties*

In an explication of how distances in the universe are measured, a seven-year old accepts without any hesitation a model drawing (Figure 160): "Suppose this is 1 km and one measures these two angles, then one knows how far away that point is lying." Unfortunately the explainer was at a loss to answer the next question: "Yet how can one *figure out* the distance?"

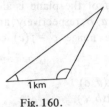

1 km

Fig. 160.

Several times I drew attention to the fact that from the picture books onwards representation by similarity is accepted as a most natural thing, and for this reason the representation by similarity is the *practical* realisation of what I called a matter of *principle* in the preceding section. For the *complete* realisation more is required: trigonometrical functions and tables in order to perform calculations.

Similarity is readily accepted as a means of faithful representation if maps and ground-plans are to be read. The scale explains and a drawing beside the scale shows how many metres or kilometres in the reality correspond to a centimetre on the map. It requires some insight into decimal fractions and the metric system to draw the desired conclusions from these data. Or, the other way round: insight into decimal fractions and the metric system can be acquired by interpreting ground-plans and maps. Besides lengths the map also provides information about angles, which in the open field is a way — possibly aided by a compass — to take one's bearings.

Didactically instructive cases of measuring inaccessible or less accessible lengths and distances is for instance

> determining the height of a tower from certain data,
> determining the spatial diagonal of a rectangular prism,
> determining the height of a pyramid from its network.

Insight — with no intervention of geometrical theory — suffices to deal with such problems by means of drawings in scale; from a didactical point of view they are entitled to be solved in this way at a stage where no algebraic or trigonometrical means are available. Surveyors have been satisfied with such methods for quite a time. The trigonometrical functions were invented by the ancient astronomers to solve spherical triangles; not until Snellius were they systematically used to solve terrestrial triangles.

What is said in the preceding sections applies also to measuring polygonal and curved distances and perimeters of figures and lengths of curves, which again can be determined from reproductions on scale. Applying similarity gets to be a matter of principle if figures are compared with each other:

> all equilateral triangles,
> all squares,
> all regular *n*-gons,
> all circles

are mutually similar. Similar figures with subdivisions that match each other, produce

> patterns

that are again similar. By similarity

> division of length (of a path) in a certain ratio

is carried over to its image.

13.6. *Measuring Lengths by Euclidicity*

As a summary of Sections 13.3–5 I state a number of geometrical insights and abilities that serve measuring of lengths:

> the insight that certain lengths and angles are determined directly or indirectly by others,
>
> the ability to use them in a productive way,
>
> the faithful reproduction of lengths from reality on ground-plans, models, networks and so on,
>
> the interpretation of such reproductions,
>
> the insight that the faithful reproduction of the internal ratios is equivalent to the reproduction in scale (the constant scale factor),
>
> the interpretation of the reproduction in scale as a similarity transformation,
>
> the interpretation of similarity transformations as reproductions in scale,
>
> the insight into the behaviour of lengths in reproductions in scale,
>
> the insight into the behaviour of angles in reproductions in scale.

13.7–13.22. MEASURING ANGLES

13.7. The Greek equivalent of "measuring angles" is goniometry. This, however, is not what is meant here. We rather mean it in an elementary way, with or without an instrument made to measure angles. One would like to say that such knowledge should precede the use of trigonometrical functions. Nevertheless it is a fact in teaching practice that the mastery of trigonometrical functions and the ability to use their tables does not necessarily include any insight into what is an angle and how it is measured.

Earlier on we have signalled various aspects of the angle (inclination of directions of straight lines; the part of the plane enclosed by half-lines, oriented and non-oriented angles). Here we will stress measuring angles, and then the primordial thing is to make clear to the learner what is measured and according to which criteria he has to construct something of a prescribed measure.

13.8. The fundamental distinguishing feature in measuring lengths and angles is the availability of a natural unit in the case of angles: the "full" angle, or if you like it better, the "stretched" angle, or the right angle – a full turn, a half turn, a quarter of a turn. This natural unit is even the most natural access to the mental object: the angle subjected to being measured – at the same time a not unusual approach to fractions. The most natural access – this does not mean the most familiar, the didactically most appropriate, and certainly not a unique access.

First of all, it is a didactical misunderstanding to believe that measuring anything should start with constituting, or even bringing about, a measure – anyway the existence of a logically natural unit in the case of angles, in contradistinction to that of lengths, is no argument to excuse it. There is a stage preceding the use of measures, natural or conventional ones, in which the need for measures can be developed. Acquaintance with length measures and instruments to measure lengths is preceded by a developmental stage, starting in the early years, where the child becomes familiar with length and comparing lengths. Didactics of angles starts differently, as a consequence of the fact that angles, in contradistinction to lengths, are being introduced and made explicit in an already heavily mathematised context. The angles of a room are without delay used as models for geometrical figures, to wit for the angles of planar figures in pictures and drawings, and very soon or even immediately for the bare angle, represented by two sides and a little arc between them. This bare angle can be attractively clothed by tart and clock dial divisions – an approach that is didactically sound and easily elaborated, though too narrow to be restricted to these concretisations.

At an early age children start showing comparing and even measuring lengths by means of the spaces between palms or fingertips or using strings or sticks as measuring tools. Almost no didactic attention has been paid to the corresponding activity in the case of angles, probably because angles are too late in instruction for such "childish" work. Yet it may be doubted whether it would really be that childish, in view of the lack of understanding which even adults may show with respect to magnitude of angles.

In the same way one can have children indicate with their hands how large, how wide, how high a thing is, one can ask them to use their limbs to show or to imitate how big is the inclination of a roof or a shed, how steep a ladder or a slope is, how a ball is scattered, how acute or obtuse a crossing is, how far a door is opened, that is to show or to imitate it by spreading their fingers, arms or legs or by flexing the elbow.

In order to compare angles, one can use materials and tools, cardboard angles, pen-knives, scissors, compasses and certainly also the clock-hands and wedges of cake. Activities in or towards the open air can prevent a premature mathematisation of measuring angles. Whoever points to the window of a building, to the top of a tower, to a bird in the air, to a star in the sky, defines an angle with respect to the horizontal plane; at a certain moment the need will be felt to view not only the shown object but also the slope of the indicating line and to make its magnitude explicit. As one turns the head of a child that looks the wrong way, into the right direction, as at a more advanced stage one has one's finger pointing to one object describe an arc to point to another one, there may be many opportunities to make the change of direction and its magnitude explicit.

The examples I just mentioned are predominantly found in what I called the topographical context. It is not that strange, or is it? In an advanced, more technical stage of topography (think of surveying, cartography, cosmography), angle is the most important object for measuring, much more comprehensive than measuring lengths. Would it not be recommendable to pay much more attention in the primitive topographical context we are didactically concerned with, to measuring angles albeit in a primitive, more qualitative way?

At this stage measuring angles is perhaps a less *practical* concern than measuring lengths. In the practice measuring angles is rather an auxiliary means to determine lengths. This possibility as such is created by the geometrical insight discussed in Sections 13.4–5 – a theoretical insight that can be present or made conscious early.

The 7-year old about whom we reported in the beginning of Section 13.5 had not the slightest difficulty with the term "angle". He even described – though indistinctly – an instrument to measure angles and finally brought a book to show the picture of a sextant.

13.9. Quite the opposite of the view I explained in the last section on measuring angles as a developmental phenomenon is Piaget's as it appears in an earlier quoted work.* The chapter on measuring angles starts with – and is almost exhausted by the following test.

The subject is shown a drawing (see Figure 161) of two supplementary angles ADC, CDB, and is asked to make another drawing exactly similar. He is not permitted to look at the model while he is drawing, but he may study and measure it as often as he wishes while not actually engaged on his own drawing. This requirement is met quite simply by having the model behind the subject. The latter is provided with rulers, strips of paper, string, cardboard triangles, compasses, etc., all of which may be used in measuring.

* J. Piaget, B. Inhelder, and A. Szeminska, *La géométrie spontanée de l'enfant*, Paris, 1948, Chapter VII. Our quotation follows the English translation, *The Child's Conception of Geometry*, London, 1960.

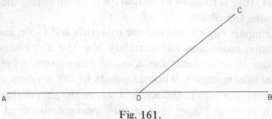

Fig. 161.

The first superficial impression of the reader might be: "a clever idea". But after studying the behaviour of the subject, one notices that up to the highest levels the problem is not at all interpreted as one of measuring angles, and indeed this result had been intended by the authors, who confessed later on*

indeed the figure was shown deliberately so as to permit of its being considered as a problem of angular measurement while not imposing that solution; we therefore rejected a figure consisting of one angle in favour of two supplementary angles.

The first impression of "how clever" is very likely to be followed by a second "too clever". Only the oldest subjects (10–11-year olds) can somehow bring it off, essentially by doing something with the perpendicular distance of C from AB. It has nothing to do with insight into angles nor can it be interpreted as such in spite of what is claimed by the authors. Copying such a drawing is a geometrical construction that requires a large amount of geometrical experience and instruction. Relating the task to angles and measuring angles would bear witness to such a degree of geometric insight and skill that can hardly be expected at the ages under consideration and should not matter at all in such an investigation.

If one analyses the formulation of the text more closely, the first question that arises is whether the letters A, B, C, D belong to the test data or have been added later on. The next question is what the words "and so on" mean in the list of the material made available to the subject. Probably nothing. In any case no material from the phenomenological and didactical context of the concept of angle – no scissors to cut out something, no hinging pair of sticks, no paper to be folded or to be used to copy. The pair of compasses the authors do mention sounds like an acrimonious irony – the great majority of subjects will never before have used this instrument or be able to use it, and in fact nobody tries it. The experimenters have succeeded excellently in concealing what they had the intention to test, but at the same time they blocked the path to any knowledge whatever about the mental object of angle and the mental measuring of angles. In Section 13.8 I sketched the kind of experiments with which to begin such investigations.

I have dealt with this test circumstantially because it reveals in a particularly

---

* The English is more explicit at this point than the original.

drastic way a certain weakness of Piaget's method: in order to test insight in a certain idea, the test itself is chosen at a rather high level and uniformly administered to all ages, a procedure which guarantees the required spectrum of levels: performing not at all, insufficiently, to some extent sufficiently, almost sufficiently, sufficiently. In the present case the level was unfortunately chosen so high that nobody succeeded. It looks like an investigation of how much weight various people can lift, with weights that only professional weight lifters can raise.

13.10. Of course the preceding criticism is not made against the extremely clever test as such. Recognising angles in this problem and attacking it by angle measurement represents a high level of geometric skill, which deserves maximum attention.

In the sections on length measurement we repeatedly mentioned angles — angles as a means to measure lengths, thanks to congruence and similarity theorems, which involve angles as well as lengths and length ratios. Measurement of angles consequently plays a part as a means of comparing angles when

> reproducing according to ratio ("on scale"),
> checking reproductions,
> reinterpreting the reproduction in the reality,

in general when

> identifying and checking directions with respect to some basic direction or basic plane.

13.11. The classical instrument to measure drawn angles and to draw angles of a given measure is the protractor — essentially half a circular ring, subdivided by ray segments into 180 degrees. For reasons I was unable to find out, this instrument has recently been superseded by an isosceles right triangle — called geo-triangle, solid, transparant, made of plastic — with an angular division radiating from the midpoint of the hypotenuse to the other sides. Well, inside the triangle half a circle with the midpoint of the hypotenuse as its centre is indicated, and from the position of the degree numbers it becomes clear that it is the semicircle that really matters.

One is inclined to say "an outrageously misleading instrument". An angular division should suitably be constructed on a circle or its periphery, rather than on the rim of an isosceles right triangle. Well, it is misleading but usefully misleading, a trap the learner should have walked into once in order to avoid it in the future.

In order to measure angles, one has to subdivide angles. Subdividing angles can be confused with subdividing lengths or areas. This source of misunderstanding should be uncovered and scotched as early as possible.

When a cake is divided into five equal wedges, equality of angle, area, and arc coincide, at least if it is a circular cake.

If it is a square or otherwise regular, division into equal wedges still means a coincidence of equality of area and border segment.

If, however, the cake is rectangular, or otherwise irregular, the equality of area and border arc cease to coincide.

Dividing, comparing and measuring angles is required in many figurative contexts. If from the start onwards it is performed exclusively or one-sidedly in the circular context, the learner is insufficiently prepared for other contexts, which in fact are more frequent than the circular one. Among these "other" contexts there are a great many where the equivalence of equality of angle, area, and border arc holds approximately, such as triangles with a pronouncedly acute angle where bisector and median are identified with each other by sight, or rectangles that look like squares and where the diagonals are understood as bisectors.

13.12. Perhaps it is a surprise that under the title of angle measurement so much fuss is made of angle division. However, this is a most natural thing. For angles there is a natural unit, and each angular measure is related as a part to this unit. It is not farfetched to require that the learning process of measuring angles properly starts with manufacturing a measuring scale, and there is hardly any other way to do this than by subdividing.

As stressed earlier the most natural substratum for angular subdivision is the circle (or the circular cake), which in fact are already instrumental in introducing fractions. But as in the case of fractions the cake, so in that of measuring angles by subdividing the circular disc is too poor an approach.

The simplest operation of dividing angles is halving and the most elementary and concrete way to perform it is folding; dividing into three equal parts by folding is more difficult to perform though it can reasonably be done; five equal parts by folding is almost impossible. Continued halving, which means dividing by a power of 2 is quite easy. Pleating an arbitrary sheet of paper produces an angle of 180° at each point of the fold line as a vertex, that is half a full angle. Folding once more such that the first fold is doubled produces a quarter of a full angle, that is right angle; if the sheet of paper is opened, one notices four right angles produced by the fold lines. Continuing the same way one can halve all four right angles at the same time, and so one can go on. It is the way to produce a compass card: the full angle divided into, say, 32 equal parts, *points* in the nautical terminology:

1 point = 1/32 of the full angle = 1/8 of the right angle.

To my astonishment a seven-year old knows the four cardinal points and can indicate them in the open field, he even knows the meaning of NE, SE, SW, NW, and immediately grasps the meaning of NNE, and so on.

I started with halving angles because with appropriate material it is the simplest operation that leads to measuring; as a point of departure I chose the full angle. I can, however, apply the operation of halving to every angle, given as a part of the plane bounded by two rectilinear sides. It does not matter what the remainder of the boundary looks like, but the fact that it does not matter is important enough to be illustrated by an explicit variety of examples, in particular by folding sectors that are not *a priori* symmetric with respect to the virtual fold line, in order to show clearly that

> halving the angle does not imply halving the area,
> halving the angle does not imply halving the border arc,

and conversely the even more important fact that

> halving the area or the border arc does not lead to halving the angle.

13.13. Adding and subtracting angles occurs in this course as operations of — temporarily — second order, where during the process of measuring the additivity of the angular measure is tacitly assumed — as was the case with the length measure.

First of all, the addition of equal angles — an operation so natural with a view to measuring as long as there are no prestructured scales, that it is unbelievable that didactically it has been neglected, if not overlooked.

A ten-year old boy tells me that in the afternoon he had "measured the shadow"; it was 3 and 9 (probably decimetres). I ask him which was 3 and which was 9, and though a bit surprised or annoyed by this query, he gives the correct answer. I ask him, if the stick had been 1, 2, 5, how long the shadow would have been, and he answers correctly. He tells me he would have liked to measure the angle too. I had neither a protractor nor tangent tables at my disposal; I did not continue the talk satisfactorily — a serious didactical failure. I should have had the boy draw a triangle of 3 to 9, cut it out and multiply the angle by turning it around — five times — to get with an incredible precision a right angle, and finally as an unexpected crowning of the work, divide 90° by 5 to get the angle of 18°, which was the aim of the boy's experiment.

If an angle is given, on paper or by means of show or vision lines, the first step to estimate its size is

> comparing it with a known angle, or
> multiplying it a number of times to get approximately a full angle.

The first can be done

> by superposition, the one on the other, concretely or mentally,

the second

> by reproducing the angle a number of times cyclically — cut out, by means of turning, folding, or mentally.

If the sequence does not fit into the full angle

> it is continued to two, three or more full angles.

If

> n copies of the angle complete — approximately — m full angles, the given angle equals — approximately — $\dfrac{m}{n}$ full angles.

13.14. One should fully realise that by this measuring procedure the *concept* of angle behind the *mental object*, if not the mental object "angle" itself, has changed. Angles are not somehow added anymore but adding angles has progressed to

> adding in a certain orientation,

and what is added in order to be measured, and arises as the result of the addition, is

> rather than the angle as a part of the plane or
> as inclination of two rays:
> the turn.

Euclid did not have any trouble with this transition, as little as did the Babylonian and Greek astronomers, who from the number of years and months elapsed between two eclipses at the same place of the sky calculated revolution times and angular velocities of celestial bodies. Earlier on* I explained how much is required to attain concept formation in the broad field around angles — I need not repeat it now. I only summarise step by step what happens here in the sphere of the mental objects:

> adding in a certain sense,
> meeting the natural unit, the full angle,
>> initially as an upper bound,
>> then to be trespassed by counting further,
> exhausting (or approximately exhausting) a number of units,
> interpreting a quotient as an angular measure.

13.15. Traditional didactics takes it easier. Nobody thinks about measuring angles by means of one's limbs or primitive instruments nor about making conscious the natural unit, the full angle, nor about measuring as exhausting this unit or multiples of it. Angular measures are offered prefabricated on the protractor. Miscomprehension, created in lower grades and not eradicated, can persist up to the highest grades. It can start quite early.

A teacher who tries to have children find out what is a circle, gets as one of the definitions: 360°.

* *Mathematics as an Educational Task*, pp. 476–494.

To be sure, the prefabricated subdivision should come up some time, but it matters whether this happens after the children have themselves learned to make subdivisions — in 32 points or in some other way.

Must the division in 360 parts, called degrees, be pressed upon the children in order to be accepted straight away? There are deviations from the decimal system pupils are familiar with, and have been so for quite a time: the division of the day into 2 × 12 hours, the hour into 60 minutes. (Anglophone pupils are acquainted with more deviations, for instance in the length measures.) The origin of these deviations in the Babylonian sexagesimal arithmetics, and the — likewise Babylonian — duodecimal zodiac corresponding to the twelve months of the year, is worth being discerned as — at least — an approach to a discussion about the difference between necessity and convention in mathematics — I would have liked to dedicate a whole chapter to this subject, and it is a pity I cannot tackle it here.

13.16. One of the characteristic insights of what is traditionally geometry is: transferring operations such as congruent displacements and subdivisions of line-segments from the context of measuring into the less instrumental and more theoretical-looking context of congruence theorems. This insight also extends to angles. Angles are being transferred by means of a pair of compasses, thanks to the congruence theorem on three pairs of equal sides, and halved, thanks to a symmetry property, which can be explained by the same congruence theorem.

Congruence, however, is only an *a posteriori* justification for a procedure of transfer of angles, which is phenomenologically more easily described. First of all the pair of compasses is used to delimit an unlimited angle (Figure 162).

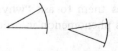

Fig. 162.

This extra structure means a change of perspective from

the angle

to

the arc or/and the circle sector.

More strongly than with the use of the protractor, this construction stresses the idea that

the angle is measured by the arc or the sector.

The object that is mentally transferred, is this extended figure. The triangle from which the congruence property is to be read is only present explicitly by two of its sides; the third side, the chord, can be added mentally, but the arc impresses itself more forcefully as the thing that is caught between the compasses and transferred. It is, in fact, tacitly assumed that the result

does not depend on the radius of the auxiliary circle (Figure 163).

Fig. 163.

It is not at all obvious that the angle can be transferred by means of the compasses, and the discovery of this fact looks like a surprise. It looks as though the chord measures the angle; at least when transferring an angle one is allowed to act as if this were true. But why, then, are chords inappropriate to measure angles? The objection that the chord is shorter than the arc is not to the point. Indeed one can measure the distance Amsterdam—Utrecht on a map, where it is — albeit to scale — unfathomably smaller than in reality. What really matters is understanding that the relation between angle and chord is not ratio preserving; the double angle is not matched by the double chord, nor half the angle by half the chord. This should be made explicit, not only in its geometrical context, but also as a paradigm of a non-linear relation.

The transfer of angles by means of compasses also serves to double, triple, ... angles. Pacing with the radius the circumference of the circle in order to return after six times to the starting point, is a construction many children learn to perform — the construction of the regular hexagon — long before this property of closure motivates them to ask "why?". I do not know whether anybody ever tried to use this phenomenon in order to test in the development of children stages such as

readiness to
need for
ability to

perform deductions. It is a particularly appropriate example because whatever one tries one cannot manage it by means of symmetries. Some kind of knowledge about the sum of angles in a triangle is indispensable. I will shortly return to the figure of the regular hexagon.

13.17. I started displacement of angles with the classical compasses construction. A simpler, though less general one is sliding along a ruler, to say it phenomenologically. In the traditional geometric context this construction is justified by theorems on pairs of parallel lines intersected by a third, and on the related angles.

As a matter of fact,

by sliding the triangle along an appropriately chosen third line, a quite special angle is being transferred,

the variable side of the triangle of the varying angle is considered to remain parallel to itself while gliding,

the concrete triangle with a given angle is replaced by a mental one in order to be mentally transferred,

which mentally produces the same system of parallel lines, from which by change of perspective equality of angles is inferred —

at least this would be a possible phenomenological course.

Yet similar effects can be reached by paving the plane

concretely, by drawing, mentally

with congruent triangles (Figure 164).

  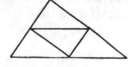

Fig. 164.

Two congruent triangles are juxtaposed to produce a parallelogram;

the figure is extended by a third triangle, which makes two sides extending each other;

a fourth triangle completes the figure to a triangle, similar to the original one;

and so it goes on,

producing three systems of parallel lines.

Why do the line-segments continue to produce lines? It can be a primitive experience, not subjected to further analysis, but it can also be derived from the displacement of triangles along rulers by change of perspective, which was discussed earlier.

13.18. If the plane is paved as in Section 13.17, in any vertex the angles of the triangle succeed each other cyclically, each twice. Thus the sum of the angles of a triangle is half a full angle. The angles viewed here are the interior ones. Earlier on we dealt with the exterior angles (even of an arbitrary polygon) as turn angles in the trip around the figure, with the full angle as their sum. Of course there is a close connection between both of these treatments. In a larger context the statement that the turn angles of a simply closed path add up to a full turn is equivalent to that of the sum of the angles in a triangle being half a turn and to that of equality of angles along parallel lines (the postulate on parallel lines).

In this connection I take up again the figure of the regular hexagon, which I started discussing in Section 13.16. The construction by pacing the circle with its radius can be interpreted as paving with *equilateral* triangles as soon as the lacking sides are actually or mentally completed. By doing this one understands why the construction closes, thanks to a change of perspective: the view

> from the circle to the inscribed regular hexagon

is changed into that

> from the regular hexagon to the circumscribed circle.

A more explicit deduction is required if the angular *measure* is called upon: each angle of the equilateral triangle is

> one third of half a full angle,

thus six of them add up around a vertex cyclically to produce a full angle (in other words, three in succession, half a full angle, to produce the extension of a side) (Figure 165).

Fig. 165.

13.19. The proportionality of

> angular measure, arc length, sector area

of corresponding

> angles, arcs, sectors

of a given circle is an intuitive fact and as such is easily accessible to conceptual analysis as soon as the need for such an analysis is felt. It should be preceded by an analysis of arc and area in general, obviously by means of the polygonal approximation of circumferences and areas. If the possibility of such definitions is accepted or understood, the additive invariance of the viewed relations and consequently the proportionality of the measures becomes obvious.

If the full angle is agreed upon as the unit of angular measure, the proportionality factor is

> for the arc: $2\pi r$
> for the area: πr^2

where r is the radius of the circle. In these expressions the factors r and r^2, respectively, are nothing but scale factors of the similarity which reduce the circle to the unit circle. The construction of the regular hexagon proves that $\pi > 3$. It is well-known how by continued halving of angles and doubling of numbers of sides this approximation can be improved. The step from 2π at the arc length to π at the sector area is made by considering the sector (as regards its area) as a triangle with the arc as its base, and the radius as its height.

13.20. Though it is not my intention to deal here with trigonometric functions, it is worthwhile indicating the part played in practice by the tangent of an angle as a measure of inclination, especially in the topographical context. The inclination is usually given in percentages. Most often the angles to which this is applied are so small that the angle and the inclination are approximately proportional.

13.21. In spite of the more general first approach, I restricted myself later on to angles of lines or rays in the plane.

I just said "the plane" – a typical formulation – as though there were one unique plane, the drawing plane where all things occur, horizontal or vertical (the blackboard in the classroom), possibly a bit out of level, but then solidly supported and certainly not floating in space. The straight lines and rays, the angles of which were measured, were tightly bound to such a plane, and it requires quite an effort to detach them from it.

The angle between the direction towards two arbitrary points in space (for instance two stars), swept out by the stretched arm, moving from the one to the other, can in the topographical context be a first opportunity to detach angle measurement from horizontal and vertical planes. Models of solids, in particular pyramids, but also cubes with their face and space diagonals are a means in the geometrical context to abolish the confinement of angle measurement to the plane.

Though the angle between planes and that between a plane and a straight line are, at least in the topographical context, as natural as that between lines, their measurement is practically reduced to that of the mutual inclination of two lines within a plane, to wit the plane that intersects the figure perpendicularly. Here too it is important that the plane in which the angle is measured has been experienced in many positions, not only horizontal and vertical ones.

Angles of skew lines are less farfetched than one would believe. A look out of the window is an opportunity to discover such angles in a meaningful context, for instance, the protuberances of TV aerials. There are many more examples: roads, tubes, conduits that cross each other, though not on the level. They define angles, measured after parallel displacement to one point or by turning the one into the other.* Again it is important to detach the situation of skew lines from the topographical context of horizontal and vertical (the TV aerials moved out of their vertical position on the roof).

* *Mathematics as an Educational Task*, pp. 479–480.

13.22. Solid angles, delimited by more than two planes (or parts of planes) have not been discussed here, though they are quite concretely suggested by the corners of a box, a drawer, a room. Obviously three mutually perpendicular planes delimit an "octant", the eighth part of the space around the vertex. Yet how to measure such angles of three or more planes in general?

Parts of the plane bounded by the sides of angles are firmly delimited by circles, and then measured thanks to the proportionality of angular measure, arc length, sector area on the circumference or inside the circle. This definition extends to solid angles: A sphere is considered with its centre in the point where the side planes meet, and the solid angle is defined as proportional to the part cut out on the surface or within the sphere. In other words, with the "full" angle as a unit corresponding to the total sphere area or volume: the solid angle equals the part cut out from the surface or the volume of the sphere.

13.23–35. *Measuring Areas and Volumes*

13.23. So close to reality as a mental object and so profound as a mathematical concept — I do not know where to start with area (and volume) as a subject and where to finish, and nevertheless to aspire to completeness. Nowhere is the variety of levels so patent, the restriction to one level of understanding and exactness so much in contradiction with the demand of mental growth and development as it is in the case of area (volume) as a didactical subject. Shall I first expose, or at least sketch, the mathematical theory, then take the phenomenological course, in order to finally put this phenomenology within a didactical context? I tried this disposition with subjects that lent themselves to it, whereas I dealt with others in a rather unstructured way. In the present situation the most urgent seems to be the phenomenological approach, albeit not towards area itself, but towards the current instruction of area.

The striking feature of both area and volume is their wealthy context — in nature, culture, and society — on the one hand, and the extreme poverty of the related instruction on the other hand. In education — at least in primary education, where attitudes are acquired and fixed — area is emaciated to "length times width", replenished with a formula for the circle, which has neither of them (or does it?); for volumes there are a few more formulae in primary and secondary education; and Calculus boasts a machinery to compute the areas and volumes of ad hoc fabricated figures. This is — from the lowest to the highest level — a degree of emaciation as I think has not been the fate of any other mathematisable subject in instruction.

There is even more to it: no other subject looks as little problematic with mountains of problems hidden to the unskilled observer; none is afflicted by a more rampant battle between the mathematical and the didactical conscience than is area (and volume).

13.24. Area is a magnitude to measure objects of a variety that looks more variegated than for any other magnitude:

a sheet or a roll of paper or cardboard,
a slab of wood or hardboard,
a roll or a cutting of fabric,
a hide of leather,
a football field or a swimming pool,
a plot of farmland or forest,
premises,
the territory of a state or parts of it according to their kind or use,
the tributary domain of a river,
the surface of a lake or sea,
a wall to be painted or plastered,
a floor to be covered with mats or parquet,
the floor or window surface of an office to be cleaned,
a street to be paved or asphalted,
a roof to be tiled,
a meadow to be mowed,
a field to be plowed, sown, planted,
a skin surface to judge its heat exchange,
a skin surface, to be clothed tightly or airily,
the leaved surface of a forest to judge its evaporation and gas exchange,

and then

   areas of geometric figures
      plane, developable, non-developable,
      expressed, or not, by formulae,
      interesting as such,
      or as a means of expressing lengths as functions of other lengths
         (as in the theorem of Pythagoras).

13.25. As a magnitude we already discussed length, a

   function of long objects,

phenomenologically. In order to form a magnitude in a category of things

   an equivalence relation,
   an order relation,
   an operation of composing

are required. In all these aspects area is more varied and more complex than length. An object to which an area is assigned, must have both length, and width, that is, in the phenomenological sense as meant by us it must be a

   two-dimensional object.

In the case of length the constituting equivalence was generated by comparatively simple operations, such as displacements, flexions, break-and-make transformations;

in the case of area the character and scope of these operations is much less perspicuous — this will be dealt with in Section 13.26. The operation of composing had a unique result in the case of long objects, that is, up to permutations of parts; whereas the objects to which an area is assigned can be composed in a multifarious way.

Comparing objects with respect to length without measuring them explicitly is in principle a simple task. Indeed, "longer" and "shorter" are easily reduced to "containing" and "being contained". For surfaces, measuring looks in most cases the most natural way of comparing.

To measure lengths one chooses some unit and determines how often the unit fits into the object; if a possible remainder is worth being accounted for, one switches to a smaller unit, from metres to decimetres, from decimetres to centimetres.

As a matter of fact, this method of

    measuring by exhausting with units

ultimately works for all magnitudes. In the case of

    lengths, weights, times, quantities of liquid

one can leave it at that, but for

    surfaces and solid contents

it is — at least as a unique method — unsatisfactory. A square as a unit for areas (a cube for solid volumes) is a natural figure to exhaust with congruent (possibly similar) copies

    plane figures bounded by irregular curves,
    spatial figures bounded by irregular curved surfaces,

but for

    figures bounded by straight lines or regular plane figures,
    figures bounded by planes or regular spatial figures,

this can be a perverse procedure. To show this by two extreme examples:

    A square divided into two congruent parts, produces a partial figure, which is directly (not by exhausting with little squares) recognisable as one half half of the original figure (Figure 166). A circular disc is obviously better exhausted by regular polygons than by little squares.

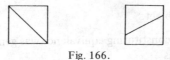

Fig. 166.

For all magnitudes

fair sharing

can be reduced to *numerically* dividing by measuring the thing to be shared. It is quite natural to do it this way if measuring is easy as is the case with

lengths, weights, times, quantities of liquid.

Measuring areas and solid volumes, however, is more complicated if not extremely difficult. Under such circumstances it is most natural to benefit from a geometrical structure of the thing to be divided, if there is any. There are many ways to divide a square, a circular disc, a cylindrical surface in three congruent parts, and each of them produces a fair division of the area (Figure 167). Even if the thing that is to be divided, shows less or no geometric structure, fair sharing by estimate will mostly be preferred to involved measuring procedures.

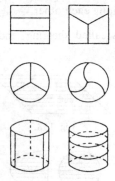

Fig. 167.

13.26. Let us put this confrontation also in the context of mappings! Which mappings leave length, and which leave area (volume) invariant? As regards length, this could be discontinuous mappings (cutting), continuous mappings of single figures (flexions), and continuous mappings of the whole plane, leaving each length invariant, that is, congruences.

Congruences of the whole plane (space) leave of course area (volume) invariant but there are many more mappings that do so. An affine mapping of the plane (space) transforms parallelograms (parallelepipeds) into the same kind and a definition of area (volume) for this kind of figure suffices to understand the behaviour of area (volume) under affine mappings. It appears that

an affine mapping of the plane (space) multiplies all areas (volumes) by the same factor.

(This factor, a property of the mapping, is called its determinant.)

Then there are

special affine mappings

(that is with determinant 1) which leave invariant the areas of all parallelograms (the volumes of all parallelepipeds) — and for that matter area (volume) at all.

There are still many more. A kind of area- (volume-) preserving mapping that has drawn too little attention is

>shearing.

Let us consider a plane figure as constituted of

>linear layers

in order to have them

>glide along each other

(a spatial figure as constituted of

>plane layers
>gliding (also twisting) along each other.

In mechanics, indeed, this is called shearing, and there I took the term from (Figures 168—170).

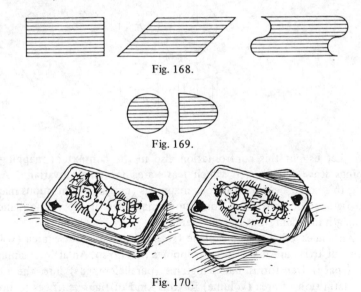

Fig. 168.

Fig. 169.

Fig. 170.

Shearing preserves areas (volumes). This is intuitively obvious: One ascribes a thickness to the linear (plane) layers, which means considering them as having been built from rectangular (plate shaped) layers, which glide on each other. There the invariance of area (volume) is an elementary fact, which with the thickness approaching zero is being taken along to the limit.

The invariance of area (volume) under shearing also follows from

>Cavalieri's principle:

Two plane (spatial) figures the sections of which at each height are correspondingly equal as to length (area), have the same area (volume). By this principle one proves for instance that

> pyramids with equal basis and height

have the same volume, whereas the equality of area of

> triangles with equal basis and height

is ascertained by simpler methods.

(Cavalieri's principle is also the basis of Archimedes' best known proof for the volume of the sphere. As far as volumes are concerned, the hemisphere is considered as the difference between a cylinder and a cone (Figure 171); indeed at any height the plane section of the hemisphere equals that of the difference between cylinder and cone.)

Fig. 171.

Even shearings do not exhaust the area (volume) preserving mappings, which of course form a group: two of them in succession or the inverse of one of them have the same invariance property. Yet the product of two shearings according to differently directed layers is in general not a shearing any more; it is just a fresh area-preserving mapping. (The product of two spatial shearings according to differently directed layers is a shearing along parallel lines; the product of three spatial shearings according to planes that do not meet in a line, is again a new volume-preserving mapping.)

The plane (spatial) shearings according to arbitrary layers generate a group of area- (volume-) preserving mappings. I do not know whether all area- (volume-) preserving mappings are obtained or at least approximated in this way.

13.27. It would be a serious shortcoming if I did not put area and volume in the frame of the mathematics of the *measure* concept. Measure, as understood in measure theory owes its existence to the mathematical need for an extension of such ideas as length, area, volume — originally ascribed only to geometrical figures — to sets in general. Axiomatic theories make clear how far this is possible.

What is a measure ? One is concerned with

> a substratum set E,
> a set $\mathscr{B}$ of subsets of E,

a function μ, the measure, which to every element of $\mathscr{B}$ ascribes a non-negative real number, possibly also ∞.

For instance E can be the straight line, the plane, the space. Anyhow $\mathscr{B}$ and μ will be subjected to certain postulates. The most important one is:

Additivity: The measure of a union of mutually disjoint sets equals the sum of the measures of the particular summands.

However, in this form the postulate is untenable. We wish to assign to the unit line-segment the (linear) measure 1. The line-segment is the union of all its points (considered as one-point-sets), which should have all the same measure, and 0 by preference. The postulate is defeated by this incompatibility. We shall be more modest. One choice is:

Finite additivity: The measure of the union of a finite number of mutually disjoint sets equals the sum of the measures of the particular summands.

Instead of "a finite number of" I may say "two"; step by step this extends to any finite number. Another choice is

σ-additivity: The measure of the union of countably many mutually disjoint sets equals the sum of the measure of the particular summands.

There are mathematical reasons why σ-additivity should be preferred to finite additivity. This is what actually happens in Lebesgue measure theory.

It would be nice if we could leave it at that. However, for a measure theory with the line, the plane, or the space as E we would also require the

invariance of the measure under congruence transformations,

which means that

congruent sets have the same measure.

Unfortunately this postulate is defeated by σ-additivity:

Let E be the circumference of the circle with radius 1, in other words the set of complex numbers of absolute value 1 (or the set of the angles). E is gifted with a group operation, the multiplication of complex numbers (or addition of angles). Let α be chosen $\in E$ such that for no integer n does $\alpha^n = 1$ hold ($\frac{\alpha}{2\pi}$ is irrational). α generates an infinite cyclic group F. Form a set V by taking exactly one element from each coset of F in E. Then V is fully divided into the sets

$$\ldots, \alpha^{-2} V, \alpha^{-1} V, V, \alpha V, \alpha^2 V, \ldots,$$

which are mutually disjoint. However, they are also pairwise congruent. If congruent sets

shall be assigned the same measure – which is fair – and E shall be assigned a positive measure, say 2π, we run into trouble. The only way out is to forbid the set V to have a measure.

Rather than E I could have chosen, with a slight variation, the straight line, the plane or the space.

In the space even finite additivity runs aground:

The spherical surface can be divided into *three* sets T_1, T_2, T_3 as well as into *two* sets U_1, U_2 all of which are pairwise congruent.

In order to arrive at a decent measure theory, one is obliged to restrict the set $\mathscr{B}$ of sets gifted with a measure; "pathological" sets should be avoided. It would take us too far afield to go into details. Our aim is not a measure theory for *sets* but the connections between the various phenomenological approaches to area (and volume) of *figures*. Even this will be difficult enough.

13.28. Intentionally we abstain from establishing precisely what a *figure* is. Phenomenologically viewed even attempts at precision would bring us into the same kind of difficulties as we encountered in the discussion of topological concepts. It would be a relief if we could restrict ourselves to planar (spatial) figures delimited by straight lines (planes), while neglecting circles, circle, sectors, circle segments, ellipses, parabolic segments, cylindric, conic, spheric surfaces, and parts of them (and similarly for the volumes). We cannot but admit curved boundaries, yet as soon as we do so we must require a certain smoothness, if we want to avoid trouble. What degree of smoothness and how to define it? By means of differentiability conditions, and then of what order? Piecewise differentiability — and should we allow an infinite number of pieces? Even this can cause pathologies.

Shall we in a planar (spatial) figure include its boundary, or are we allowed to neglect it, totally or partially, if it is the area (volume) that matters? This is the very question that goes to the heart of the matter. It is clear what the phenomenological answer will be: The boundary may not matter if areas (volumes) are to be compared or measured.

This then is the principal reason why the quest for area (volume) of figures cannot be satisfied by a measure theory for sets. As mathematical concepts dimension and measure are estranged from each other by phenomenologically unexpected mathematical consequences.

As an example I choose a —well-known — 0-dimensional set (on the line) with a positive linear measure:

From the line-segment $[0, 1]$ the open mid-third $]\frac{1}{3}, \frac{2}{3}[$ is deleted. From the remaining pieces $[0, \frac{1}{3}]$ and $[\frac{2}{3}, 1]$ again the open mid-thirds ard deleted, and so it continues indefinitely. The remainder in the limit is a closed set C (Cantor's discontinuum), which contains no line-segment — a 0-dimensional set. What is finally left from the length? At every step $\frac{2}{3}$ of the previous total length is left, that is, $\left(\frac{2}{3}\right)^n$ after the nth step, which fortunately converges to 0 with $n \to \infty$.

But let us now do it in a topologically identical, though metrically different, way:

At the first step the open mid-third is deleted, at the second step the open mid-sevenths, then the open mid-fifteenths, and so on. The remainder is, as regards its total length,

at the 1st step $\frac{2}{3}$ of the previous length,

at the 2nd step $\frac{6}{7}$ of the previous length,

at the 3rd step $\frac{14}{15}$ of the previous length,

and so on, and finally at the limit

$$\frac{2}{3} \cdot \frac{6}{7} \cdot \frac{14}{15} \cdot \frac{30}{31} \cdots$$

of the original — an infinite product with the limit $\frac{1}{2}$.

The modification leads to something which topologically is again a Cantor discontinuum, though according to our reasoning it should be assigned a positive linear measure $(\frac{1}{2})$.

A similar construction produces 0-dimensional sets in the plane (space) with a positive planar (spatial) measure.

On the line the set C is all boundary. The boundary of a plane (spatial) set, though of lower dimension, may possess a positive measure; a closed set with no interior point can nevertheless have a positive measure.

On the other hand it seems to be an intuitive requirement that when comparing and measuring areas (volumes) the boundaries do not count — a figure to be measured is exhausted by squares or other measuring units, which touch each other, and it does not hurt (or may not hurt) that some points are counted twice or even more often. This is a new example of mental objects which, sharpened to concepts, seem to show paradoxical features. Of course such a paradox can always been scotched by *ad hoc* definitions — in the present case by restricting the definition of area (volume) to sets, the boundary of which is smooth enough that its measure necessarily vanishes.

For our aims this would be a useless sophistication. In the sequel we will uncover more profoundly rooted impediments than paradoxes of seemingly pathological origin on the way from the mental object to the concept of area (volume).

Hence I will not look for more precision with respect to figures and continue to speak of their area (volumes) in a naive style.

13.29. The proper difficulty is the large variety of approaches to the mental object area (volume) of which it is not *a priori* clear whether they form a consistent whole, though each of them separately and all together are being used whenever they are needed.

Let us summarise them once more:

Planar figures can be compared with respect to their areas
> directly, if one is a part of another,
> indirectly, after
>> break-and-make transformations,
>> congruences and other area preserving mappings,
>> measuring.

Plane figures can be measured
> by exhausting with a unit of area,
> by interior or exterior approximation.

In this, use is made of
> the additivity of area under the composition of plane figures
>> that up to the (one-dimensional) boundary are mutually disjoint,
> or of convergence of areas under approximation.

Similarly surfaces in space can be evaluated with regard to area, as far as they are composed of plane figures or can be approximated by such. (A corresponding statement can be made with respect to spatial figures.)

13.30. It is not at all obvious that all these approaches lead to the same result. On the contrary, the proofs require efforts that would surpass anything that can be asked or realised, say, at the highest secondary grades, even if it would be possible to excite any interest in this complex problematic. Nevertheless I feel obliged to sketch this problematic, with the accompanying reasonings, at least in order to explain how much is required to raise the perspicuous looking mental object area (volume) to a proper mathematical concept.

13.31. The universal means of area measurement is exhaustion by units of area. This means can be applied efficiently by

> covering figure F by a square grid

The united squares

> contained in F approximate F from the interior,
> containing points of F approximate F from the exterior.

Their united area, which can be obtained by simply counting, is by definition, respectively,

> smaller than,   larger than

the area of F that is to be defined.
> Refinement of the grid leads to better approximation, respectively,

> from the interior,   from the exterior

by unions of grid squares.
> The limit under unlimited refinement is, by definition, respectively

> the inner area,   the outer area of F.

Area of F has a meaning as soon as both of them are equal. *A priori* this equality is not obvious – for arbitrary F it would not even be true. Equality of inner and outer area of F boils down to the property of F that

its boundary has an outer area 0.

It is rather easy to understand that the outer area of a line-segment of a union of line-segments, and thus the boundary of a polygonal figure vanishes; it is less clear which requirements a curve should fulfill to have a vanishing outer area. We shall suppose implicitly that our mental object "geometrical figure" satisfies them.

We required additivity of area. It is not obvious, though it is not difficult to prove, that the area, such as defined a moment ago, does possess this property, at least as far as finite additivity is meant.

Moreover invariance of area under congruence mappings was postulated. This invariance is not at all a trivial concern. Suppose we take a square grid R with its refinements fixed. Does it, with a view to area, matter how I put the figure F upon the grid? Of course, it does not matter, but how to prove it?

The question can be formulated otherwise. Suppose the figure F in a fixed position, and put the grid R with its refinements arbitrarily on F. Does it matter how this is done?

Let us start with a rectangle F with *sides parallel to the grid*. By simply counting and approximating from the interior and the exterior, I can verify the formula

area = length times width,

where as a unit of area one has chosen the grid square with side 1. What happens if I subject F to a congruence transformation?

Under a translation the rectangle F keeps its sides parallel to the grid. So we can apply the formula anew, which means invariance of its area under translations. This statement can immediately be extended to the case of an arbitrary figure F, instead of a rectangle.

Reflections leaving the grid invariant do not cause any trouble either. We are still left with rotations. Under a rotation D the rectangle F gets in a position with respect to the grid R which seems to resist the *direct* evaluation of its area. We will succeed by a trick (Figure 172).

Fig. 172.

Let us take as F a square with sides of length a, parallel to the grid. Whatever F may be, DF is certainly contained in a square parallel to the grid with a side length equal that of the diagonal of F, thus with an area $(a\sqrt{2})^2 = 2a^2$, and on the other side DF certainly contains a square parallel to the grid with a side length at least half that of F, thus with an area $(\frac{1}{2}a\sqrt{2})^2 = \frac{1}{2}a^2$.

It has been shown that under a rotation D the area of a square parallel with the grid is multiplied by a factor r depending on D such that

$$\frac{1}{2} \leqq r \leqq 2.$$

It is not difficult to understand that this factor is the same for *each* grid square. The area of an arbitrary figure, however, is defined by means of square grids. Consequently under the rotation D the area of *each* figure is multiplied by the *same* factor r.

This holds in particular for the figure DF, where F is again a square parallel to the grid R. Thus

$$\text{area } DDF = r \text{ area } DF = r^2 \text{ area } F.$$

One can continue applying D:

$$\text{area } D^n F = r^n \text{ area } F.$$

D^n, however, is again a rotation, with now the multiplication factor r^n. Thus

$$\frac{1}{2} \leqq r^n \leqq 2$$

as previously proved, and this should hold for arbitrarily large n, which it is impossible unless

$$r = 1.$$

Hence the area of a square parallel to the grid R is invariant under rotation, and thus under all congruence mappings, from which the same is easily derived for the area of any figure.

I could have proved this — more elementarily but also more cumbrously — by breaking and making. However, the proof I gave runs almost the same way in space (and even in arbitrary dimensions). My proper intention was to make clear how much is required to make sure a seemingly obvious mathematical fact.

13.32. The same holds if we choose a more elementary approach to area. Let us now restrict ourselves to rectilinearly bounded figures, and let us start with rectangles. The formula

$$\text{area} = \text{length times width}$$

can be obtained without lattices and approximations, by means of

break and make operations,
and similar triangles.

Indeed (Figure 173) by subtracting from the large congruent triangles PQC and

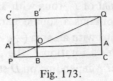

Fig. 173.

PQC′ the pairs of small congruent triangles POA′, POB, QOB′, QOA, I get for the remaining rectangles

$$\text{area OACB} = \text{area OA}'\text{C}'\text{B}'.$$

On the other hand, by similarity

$$\text{OA} : \text{OA}' = \text{OB}' : \text{OB}$$

thus

$$\text{OA} \cdot \text{OB} = \text{OA}' \cdot \text{OB}'.$$

Conversely, starting with this formula, one can reconstruct Figure 173 and the rectangles, from which it follows:

> Two rectangles have the same area if and only if they agree in the products of their side lengths.

Thus the product of the side lengths characterises the area of rectangles. It is additive if rectangles are stuck together to produce another rectangle.

One might remark that a similar argument holds if rectangles are replaced by parallelograms, whereas it seems that the final conclusion ceases to hold. In fact:

> Two parallelograms with correspondingly equal angles have equal areas if and only if they agree in the product of adjacent sides (Figure 174).

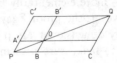

Fig. 174.

Thus the product of adjacent sides of parallelograms with the same angle characterises their area.

One prefers rectangles and in particular assigns the unit area to the unit square. One could as well choose as such the area of the rhombus with side 1 and angle 60°, or any other parallelogram. The whole difference is a fixed factor, depending on this choice. It should be stressed that area (volume) measurements presuppose the choice of a unit as does length measurement. In principle these choices are independent of each other. Even in practical metrology before the metric system the various units (also non-geometrical ones) were not connected to each other.

By break and make transformations arbitrary polygons can be reduced to rectangles as far as their area is concerned.

First of all, the parallelograms: Figure 175 shows how the parallelogram ABCD is transformed into the rectangle A'B'C'D' by sticking on and cutting off congruent triangles.

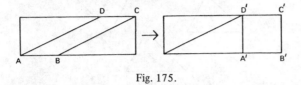

Fig. 175.

Then the triangles (Figure 176): ABC transformed into parallelogram ABMN by cutting off and sticking on a triangle. Moreover it appears that any triangle can be transformed in any other with the same base and height by a break and make procedure.

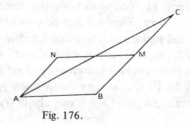

Fig. 176.

An n-gon (ABCDE, Figure 177) can be reduced to an $(n-1)$-gon (ABCE') in order finally to arrive at a triangle.

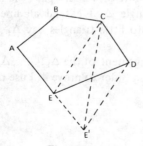

Fig. 177.

The most noteworthy outcome of this discussion is that for polygons equality and comparison of area can be defined by break and make procedures with no intervention of approximations nor use of algebraic formulae. By break and make procedures each polygon can be transformed into a rectangle with a prescribed base and, as to their area, polygons can be compared by comparing the heights of the corresponding rectangles.

A nicety I may not pass over: Actually one can restrict oneself to cutting off and glueing on *triangles*, and concerning the transport of such pieces translations and point reflections suffice, thus (Figure 178) a triangle I displaced in positions like 2 and 3 — more general congruences are not required.

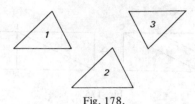

Fig. 178.

Yet new problems arise. If I insist on this elementary definition of area by means of breaking and making and avoid any appeal to the former definition by approximation grids, am I sure that I get a uniquely determined result? Would it not be possible by means of different break and make procedures to transform a polygon in rectangles with the same base and different height? Or to say it another way: Would it not be possible to transform a rectangle by means of a break and make procedure into a smaller one?

It is not possible, but this is not a trivial fact. A direct proof — independent of approximation procedures — requires non-obvious tricks. Only after a number of editions did Hilbert's *Grundlagen der Geometrie* manage to produce a reasonable proof, and even this proof is not easy. I will try to give the reader an idea of the proof.

Suppose a figure F is transformed by break and make procedures into a smaller one, F'. The pieces that are cut off and glued on are polygons, which I may suppose are each divided into triangles. So the pieces may be supposed to be triangular from the start onwards. I pack the figure F and all the auxiliary triangles into a large triangle Δ. I extend all line-segments in this figure to produce a division of Δ into little triangles, say $\Delta_1, \ldots, \Delta_n$. The problem can now be reworded:

Is it possible by rearrangement of the $\Delta_1, \ldots, \Delta_n$ to cover only a part of Δ?

Of course it is not, and in order to prove it, I use the — defining — expressions for the area of a triangle

$\frac{1}{2}$ base times height.

I shall show

(1)   area Δ = area $\Delta_1 + \ldots +$ area Δ_n

— the area of the whole triangle equals the sum of the area of the partial triangles. This, if proven, means that by rearrangement I cannot manage to cover only a part of Δ.

So we shall prove (1). This formula is certainly correct for a subdivision of a triangle in two, thus subsequently in the cases of Figures 179–182.

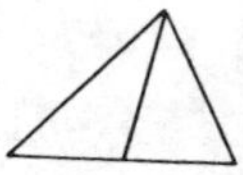

Fig. 179. 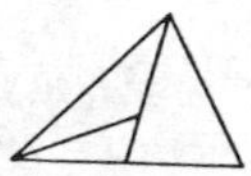 Fig. 180.

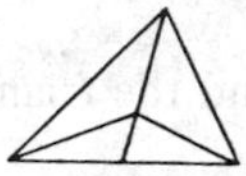

Fig. 181. 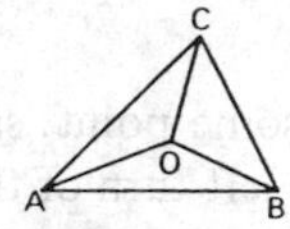 Fig. 182.

Thus

(2) area ABC = area OAB + area OBC + area OCA

This formula still holds if O is outside the triangle ABC, if it is supposed that I allow negative areas and for instance in the case of Figure 183 I am counting the area OAB negative. The general rule runs as follows:

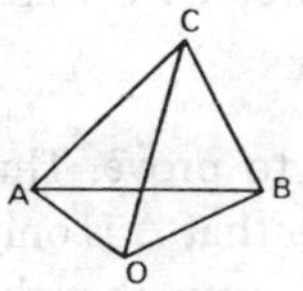

Fig. 183.

Suppose the plane is oriented, that is provided with a sense of left turn,* consider the triangle XYZ having walked around in the sense XYZX and count its area positive or negative according to whether this circulation matches the left turn of the plane or not.

If interpreted this way, (2) holds independently of the position of O.

Another remark: Suppose that the triangles ABC and ABC' are on different sides of AB (Figure 184) thereby having no common points other than those of the segment AB. Then the triangles ABC and ABC' have opposite circulations

Fig. 184.

Now back to the big triangle Δ, subdivided into $\Delta_1, \ldots, \Delta_n$. We may suppose, if need be, after supplementary subdivisions that partial triangles meet pairwise either in a common side, or in a vertex, or not at all (Figure 185).

* Orientation will be dealt with later on.

Fig. 185.

Let O be some point, say, outside Δ. I walk around the triangle $\Delta_i = A_iB_iC_i$ according the left turn of the plane, thus

$$\text{area } \Delta_i = \text{area } OA_iB_i + \text{area } OB_iC_i + \text{area } OC_iA_i.$$

If I add all these equations ($i = 1, \ldots, n$) then on the right hand side of the equality all contributions "area OPQ" arising from inner line-segments PQ cancel, whereas the little line-segments on the boundary of A just produce together the sides of Δ. Thus

$$\text{area } \Delta_1 + \ldots + \text{area } \Delta_n =$$
$$= \text{area } OAB + \text{area } OBC + \text{area } OCA =$$
$$= \text{area } \Delta,$$

which is the formula we wished to prove. Thus it is impossible to rearrange the partial triangles $\Delta_1, \ldots, \Delta_n$ such that Δ is only partially covered.

It is a tough job, this proof, even though I touched on some points (such as orientation) only superficially.

13.33. The area of rectilinearly bounded planar figures can be defined elementarily — that is by using break and make transformations rather than approximations — though justifying this procedure is not easy. The volume of planarly bounded spatial figures is a different thing. The equality of two pyramids with equal basis and height cannot be obtained by break and make procedures. To obtain this equality approximations such as Cavalieri's principle are required. So it is less elementary than the case of triangles.

13.34. The preceding exposition shows clearly the gulf between the mental object and the concept as far as areas (volumes) are concerned: obvious looking properties and relations require sophisticated arguments to corroborate related statements. It is a matter of course that at school, even at the highest level, no more than a few little steps can be taken toward concept formation. Didacticians of mathematics who consider concept attainment as the very objective of mathematics education, are very likely to pass over the subject area (volume) or to admit it only in the state of extreme emaciation. Whoever is convinced of the importance of mental objects for mathematical thought, will not shun it.

As early as with 8-year olds one can notice a well-founded visual idea of area. With no hesitation they perform such tasks as

   colouring half (a third of) an area red

as soon as the figure shows some suitable regularity, such as a rectangle, square, circle, circle sector; they even can produce various solutions of such problems. Even with irregular figures they succeed in solving them by reasonable estimates. With ten-year olds

   comparing and measuring areas by grid covers

proves that there exists a more profound insight into area. The observations bear witness to the presence of a rather sophisticated mental object. Not until the formula

   length times width

appears, does the pauperisation start, which characterises the upper grades of the primary school and where secondary instruction rarely or never adds any enrichment. If then mathematics of area (volume) is taught at all, approach, method and subject matter are estranged from the corresponding mental objects by which the phenomenon area (volume) is understood in everyday life and mathematical applications.

   This didactical situation is not at all unique; it is only more involved and for this reason looks less perspicuous than similar mathematical didactical situations do. In "Mathematics as an Educational Task"* (where I did not yet use terms like "mental object"), at the opportunity presented by the system of real numbers, I asked the question "describing or creating concepts — analysis or synthesis?" In my present terminology I would adapt the same idea as follows: the number line is a mental object, which is gradually and step by step learned to be seized upon, that is by localising on it the natural numbers, their negatives, the rational numbers, certain irrational ones and finally all the real numbers. The number line is being described by the real numbers but not created by some definition of real number — this is a didactics which fits the needs and possibilities at school and at the start of university studies.

   With areas (volumes) the situation is similar insofar as they are constituted and accepted as mental objects, and *a posteriori*, if need be, analysed in order to arrive at gradual concept attainment. The situation differs by the greater phenomenological wealth of the mental object area (volume), the variety of approaches, the greater profundity and sophistication required for concept formation as well as the built-in deceptive features, which may not be under-estimated.

* Pp. 212–214.

I summarise the approaches:

fair sharing
    by profiting from regularities,
    by estimating,
    by measuring;
comparing and reproducing (in another shape)
    by inclusion,
    by break and make transformations,
    by estimating,
    by measuring,
    by means of mappings,
        that is, congruences, affinities, shearings;
measuring
    by exhausting with units
        with even finer subunits,
    by approximations from inside and outside
        with fixed grids
        with adapted figures,
    by converting break and make transformations,
    by means of general geometric relations,
    by means of general formulae,
    by means of principles, as Cavalieri's,
    by means of mappings,
        that is, congruences, affinities, shearings.

Didactically all of them are acceptable, albeit with various weights. Being restricted to one of them in order to pursue purity of method, is bad didactics. On the contrary one is advised to profit from this wealth — notwithstanding the didactical impossibility of

a more than local organisation of this field of concepts,

that is, the didactical impossibility of

the question of the internal and external consistency,
the delimitation of domains of validity of the approaches,
the logical justification of the approaches in their mutual connection.

13.35. This does not imply an entirely naïve and uncritical activity in this field, which is a ground of traps and pitfalls. As soon as magnitudes are concerned, confusions can arise: distances, times, speeds, metres per second, and seconds per metre, gasoline per 100 km or the distance per litre are examples. For the constitution of magnitudes as mental objects it is also required that they are confronted with others. In the case of area (volume) the big misleader is perimeter (surface area). At the same time as children give evidence to be able to divide an area fairly, they succumb by appropriately misleading examples to judge the area of a figure according to its linear dimensions. Figure 186 shows a situation where some assign a larger area to the righthand figure because of its larger perimeter whereas others are misled by the narrowness of the figure — measured by the distance between the oblique parallels — to draw the opposite conclusion. Plato thought it worth stressing that the area of the Peloponnesos cannot be measured by sailing around it. Galileo's sack, sown up at different sides is another historical example that might be cited here.

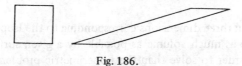

Fig. 186.

For the constitution of the mental object area (volume) one needs also

> examples of figures that in spite of misleading deviations in the linear dimensions have the same area,

such as the parallelograms with equal base and height, as well as

> examples of figures that in spite of misleading agreements in the linear dimensions, have different areas,

such as the rhombuses arising from a square by flexion (Figure 187).

Fig. 187.

Extreme cases are especially instructive: in Figure 186 where the parallelogram gets ever narrower and nevertheless keeps its area; in Figure 187 where the rhombus finally collapses to include no area at all.

With hindsight it is no problem to enclose *arbitrarily little* areas by a given perimeter. In the constitution of the mental object area the impossibility of enclosing *arbitrarily much* area by a given perimeter obtrudes itself. There is an upper bound to the area enclosed by a given perimeter, which depends on the kind of competing figures. This leads to the so-called

> isoperimetric problem:

to find within a class of figures with a given perimeter the one with the largest area. It is rather easy to prove that among the rectangles with the same perimeter the square has the largest area. It looks obvious that among the planar *n*-gons with the same perimeter the regular is the biggest, but except for the case $n = 3$ it is not that easy to prove. It looks just as evident that among the planar figures the circle solves the isoperimetric problem but proving it requires quite a lot of mathematics.

Perimeter and area are to a certain degree, yet not totally, independent: the perimeter imposes an upper bound to the area. The analogous feature one dimension higher in the relation between surface area and volume has as regards both aspects a great importance as a principle of explanation and control in nature and technique. In order to aid the exchange of heat or matter with the environment, nature and man, whenever it is-needed, create excessively big surfaces in a small space; in other cases in order to minimalise exchange, shapes are assumed or created which show as little surface as is compatible with the other data.

The problem in three dimensions corresponding to the isoperimetric problem, that is to enclose as much volume as possible by a given surface area, is mathematically even harder to solve than the isoperimetric problem, though here too it looks obvious that more regularity, in particular that of the sphere, guarantees more volume to the same superficies.

13.36—46. *Measuring Volumes*

13.36. So far we have dealt with area and volume together — the latter within parentheses. There was one exception: for the "elementary" theory of area (Section 13.32) there is no mathematical analogue in higher dimensions. On the other hand it will appear that volume can phenomenologically and didactically be approached along ways and dealt with by methods which have no analogues in the approach to area.

Compared with area, volume shows — besides the third spatial dimension — a large number of extra "dimensions", which so far could not be done justice. Concrete manipulations with volumes show a broader spectrum, moreover volumes are less exposed to phenomenological impoverishment than areas. There are two reasons why:

    the double aspect of content and volume,
    the concrete character of the volume preserving operations.

13.37. There are good reasons why the metric system in three dimensions knows two kinds of measures:

    cubic centimetres, decimetres, metres, and so on,

and

    litres, hectolitres.

In principle one of them could be dispensed with. In fact they are closely connected:

    1 litre = 1 dm^3.

In traditional measure systems this was different: stacks of wood and big luggage on passenger steamers were measured with different measures from oil, beer, and wine. The term "content" reminds one of a barrel to put things in, whereas "volume" rather suggests a thing that claims space. A container has a content, which tells you how much it can hold, but it has also a volume — a bit larger than the content — which matters if a number of containers of a certain kind are to be placed in a ship. So

> content as creating space,
> volume as claiming space
>> shall be understood
>> identified, and if need be
>> distinguished.

The relation between content and volume is complicated by what I will call the

> pack and stow effect,

which I take the liberty to introduce by means of the paradoxical inclusion syllogism

> I fit into my suit,
> my suit fits into my bag,
> so I fit into my bag.

Or, more serious situations, which everybody will be able to locate:

> it is a whopper but you cannot put anything in it,
> the trunk packs poorly,
> it is a piece of nothing but one cannot store it,
> if it has in the length what it lacks in the height, it would be possible.

Content measures serve to measure liquids, powder-like grained matter (provided it is not too coarse) — things which can be stored in barrels. But the content terminology reaches farther: to packing and stowing bags, containers, carriages, vessels, though ultimately less standardised measures are used to describe contents:

> 28 standing and 12 sitting places,
> 80 beds,
> 200 chairs,

to indicate "contents" of buses, hotels, theatres (or barber shops).
Later on I will deal with the need

> to understand, and
> to recognise the pack and stow effect,
> and to eliminate it in the relation between content and volume.

Meanwhile, after now having stressed the phenomenological relation and difference between content and volume, I will, in general, in order not to complicate the language, use one term: volume. The reader may add the word "content", where it is needed, in parentheses.

13.38. In the course of constitution of the mental objects length, area, volume an important part is meted out to procedures preserving these magnitudes: discontinuous ones such as break and make procedures, and continuous ones, the geometric correlate of which are mappings. It looks like a vicious circle to claim in the constitution of magnitudes an essential role for procedures leaving their magnitudes invariant, but as we know, it is not.

Indeed, in Sections 1.16 and 1.18–20 I linked the constitution of

> rigid body,
> flexible body,

to processes of

> displacing in a gentle way (not badly belabouring),
> reversible plying and bending with negligible effort,
> rolling off with no skidding

as physical correlates of

> length-preserving transformations.

The physical context "gentle treatment", negligible force, and non-skidding breaks the circle, which would be a vicious one, if I would have length constituted by the principle of length-preserving transformations.

The length-preserving mappings of the plane (and space), that is the congruence mappings, play also a part in constituting, comparing, and measuring area (volume) because they preserve area (volume) too; together with the break and make transformation they would even logically suffice for the constitution of the concepts area and volume. I recall furthermore the

> shearings,

a geometrical rather than physical kind of transformation. As regards area I could add the

> flexions,

a rather restricted means to compare curved surfaces with each other and with parts of the plane as far as area is concerned. As regards volume, however, I have more powerful transformations of a physical nature at my command, the

> pour and mould transformations,

which in the course of constituting the mental object "volume" certainly account for an important contribution. A liquid, poured forth and back, assumes, if it is wanted, the same, that is a congruent, shape, although the mapping, as far as the single particles are concerned, need not at all be a congruence. Plastic material can after moulding be brought back into its initial state to fill a congruent part of space but the pointwise mapping from initial to final state is not a congruence. The excavated soil fills again the pit (unless it is too forcefully stamped), but the soil grains will have been displaced in an unaccountable way.

This preservation of filling space then is interpolated to get a preservation of volume: in between the initial and the final state the volume has remained the same, albeit in another space. Of course this interpolating conclusion is only accepted if the deformed substance has not been too badly belaboured. A rigid body, molten or even gasified, can when resolidified take the same shape, although meanwhile it has expanded its volume. A gas can at equal temperature and pressure take place in the same container though if these parameters are changed it can expand or contract. But for all these volume changing transformations there are still kinds of

   pour and mould transformations

which by their physical appearance suggest an invariance that no doubt plays a part in the constitution of the mental object "volume".

Pour transformations are a well-known method of practically measuring volumes of liquids or liquid-like matter. The object to be measured is brought into a special shape, most often a cylinder with prescribed base such that the volume is proportional to the height. This is how gauged litre measures, calibrated measuring-glasses, bushels (to measure volumes of flour, grain, potatoes) are used. Gauged conic funnels are also not unusual.

Measuring glasses are also used to determine the volume of solids: the substance is immersed into a liquid and the rise of the liquid column is observed. The solid is as it were re-created into a liquid cylinder — a congruent displacement procedure followed by a deforming pour transformation. Let us call it by a specific name:

   immersion transformation.

Altogether this is a rich variety of transformations that are relevant to volume.

Let us add the remark that for areas there is nothing like this. The area preserving mappings are geometric constructs rather than correlates of physical operations. That is why "conservation of volume" has drawn more of the attention of psychologists than has conservation of area — unfortunately in a way that other invariances, in particular break and make transformations, have been disregarded.

Pour and mould transformations are as it were break and make transformations made continuous. This continuation of the break and make procedure need not be advantageous for the constitution of the mental object volume; one can even imagine that it would cause troubles. The fact that the same blocks fill again the same box can, because of the discreteness of the packing activity, be a more important contribution to the constitution of the mental object volume than the stream of pouring, gentle moulding, or immersing, which cannot be followed in detail. Experiences with malleable and liquid substances are less easily interpreted than those with solids.

Anyway, if "conservation of volume" is discussed, it is useful to specify the transformations with respect to which this conservation is meant. In Section 4.15 I formulated such general invariance principles as invariance

> in the course of time,
> under change of standpoint;

or

> if something is added, it becomes more,
> if something is taken away, it becomes less,
> if nothing happens, it remains the same.

With regard to volume I can now add

> shake transformations,
> shearings,
> break and make transformations,
> pour and mould transformations,
> immersion transformations

as physical transformations, and of course as geometrical physical transformations

> displacement and other geometrically defined mappings.

13.39. In Section 4.14 I stated:

> Up to the present day nobody can say whether the stress Piaget put on conservation was justified, whether indeed certain invariances characterise the constitution of certain mathematical objects, whether this might be true of all mathematical objects for which Piaget developed conservation criteria, for some of them, or for none. I guess that in principle Piaget chose the right way, but I believe he deserved to be followed more critically on this way in every detail than he has been in fact.

If I reread this and try to apply it to the mental object volume, I feel undecided about whether to take a stand about it: I am hesitating between a decided "yes" and as decided a "no". One of the reasons is the lack of phenomenological clearness in Piaget's work and even more in the research of others that continued it.

In those circles people generally distinguish between invariance of

> quantity (also called substance),
> weight,
> content (or volume).

As a matter of fact, for the *same substance* under the *same conditions* these three magnitudes are *physically identical* — I mean as seen from the viewpoint of adult physical concepts. The researchers are right not to accept this identification, but they are wrong to deal with these three mental objects as separate.

Conservation of volume is attributed by them to an even higher developmental age than that of quantity and weight, at ages that even exceed the high age stated by Piaget.

The difficulty is how to distinguish the three of them. By separate tests, or if you do it by a single one, how to formulate the test questions? Weight looks to be no problem — it is explained by the adjective "heavy". For quantity and volume the questions are of the kind "is it the same quantity?" or "does it take the same space?" — in particular the second question presupposes much of the linguistic development of the subject and of the linguistic ability of the interviewer.

13.40. Let me first of all analyse the concepts at issue from our own viewpoint. "Quantity" obviously means something by which only specimens of the same matter can be compared; it does not make sense to ask which of two pieces, one of clay and the other of iron, is more. Quantity is truly the thing that can be said to remain the same as long as nothing is added or taken away.

The same quantity of something also has the same weight — I disregard the hardly relevant difference between mass and weight (in the sense of force). But by weight *different* matter can be compared. Weight applies more universally.

The *same quantity* of a matter can under varying circumstances occupy varying parts of space in the sense of volumes. Metals if heated expand; many materials shrink under pressure. These are facts of physical experience. One can imagine that a subject with a clear view on volume reacts the wrong way on tests because he is not acquainted with certain physical facts or nurtures questionable theories about them.

13.41. Conservation of content and volume presupposes gentle transformations, not too hard nor too hot. In this respect pour transformations look the "kindest". Strangely enough nowhere in the literature I studied is pouring a liquid used as a test of conservation of volume. The reason is probably historical: when Piaget made his famous experiments on pouring liquids, he was interested in conservation of quantity and thus formulated his questions in this sense. This has slavishly been imitated. To my view, besides break and make transformations, pouring liquids is a particularly appropriate means to observe phenomena of conservation of volume, at least qualitatively. The most neutral experiment, which to the best of my knowledge does not occur in the literature, is the following:

> Cylindric vessels of different width and height, one of which is filled with water (powder, gravel) — the subject is asked to predict the level after pouring the content into another vessel.

Children from the age of four onwards with whom I tried it — unsystematically — answered it in a reasonably satisfactory way.

Obviously this really tests something with respect to *volume*, whereas the usual pouring tests are connected with questions like *"is it the same?"*

13.42. In contradistinction to this, conservation of volume is usually tested by mould transformations, with clay or plasticine as material: if a ball is deformed into a sausage, there is a large percentage of subjects conserving quantity and weight, but denying that of volume, among which are even 16-year olds, which runs counter to Piaget's general theory which attributes conservation of volume to the period of formal operations.

First of all we should ask the question as to what our certainty is based on that clay and plasticine do not change their volume when kneaded. The immersion experiment is verbally adduced as an argument but it is never performed. I would not stake my head on the statement that kneading is really such a gentle operation, and I can imagine that denying it has nothing to do with misgivings about conservation of volume. Kneading and moulding need not be that gentle. For rubber, if stretched, no conservation of volume holds, if I am not mistaken. Bread-crumbs can be compressed to small sizes, and this happens irreversibly; a ball of wool yields, though the operation is to a certain degree reversible. A full suit-case can be filled even more. Ironed linen takes less space than unironed. It is a rich variety of experiences in which clay and plasticine are singular cases. Then, what does clay and plasticine put in that singular position? This cannot be answered by geometry.

13.43. Difference of weight is unmistakable. One can feel it with one's muscles. It is easily understood why a child answers questions about weight reasonably though these answers need not prove much, and certainly not what is wanted, about weight. For quite a time children believe that they can make themselves heavier.

Children also know rather early that a bagful of paper is lighter than the same filled with iron. A bagful — what does it mean? I think something like a volume measure, though a bag can be packed more or less tightly.

13.44. I am inclined to think that the development of the mental object volume should be rather viewed within the context of packing and filling. In order to test it, one should design experiments of quite a different kind from those which are usually performed.

Take two bags of different shape, which can be filled with the same number of marbles (of the same size): a spherical bag, and an oblong narrow one. (Obviously the second has a larger content, though it is badly packed.)

Then ask what would happen if the two bags were filled with smaller marbles, or sand, or flour.

From the answers one could draw true information about how far a child is advanced in separating geometrical and physical properties.

I am sure it is no fabrication of mine to put it that an envelope is judged to be smaller of content if "it packs more poorly". A "non-conserving" subject of the age of 16;5 motivates non-conservation of volume if a plasticine ball is kneaded into a sausage in a way* that fits quite nicely my hypothesis:

The molecules may be more compressed in the ball than in the sausage; although the sausage has the same number of molecules and the same weight, its volume is not the same.

13.45 I am now coming to the immersion transformations. The question as to whether they matter for the constitution of the mental object volume is even more critical than it is for the moulding transformations. Numerous experiments show that subjects relate the rise of liquid to weight rather than to volume. This is not surprising. The most striking behaviour of solid bodies in a liquid is floating or sinking, and whether it is the one or the other, the child knows quite early, depends on weight, whether they are lighter or heavier than the liquid. Compared with this striking phenomenon the rise of a liquid as a consequence of displacement is an unspectacular phenomenon, never observed before by many among the subjects in this kind of experiment. It is not at all a matter of course to put immersion into a geometric context. To be sure, it is quite an achievement if a child does perform it — and even more if it happens spontaneously — but I would hesitate to admit that it is essential for the constitution of the mental object volume. It is one application among many others, which is not at all obvious. Adult researchers can be excused if they think that it is a trifle; they do not realise that it is an ingenious method, allegedly invented by Archimedes. In their mind it is *the* method, which they learned in the physics lesson, to determine volumes of solid bodies. They simply forget what they owe to physics instruction.

13.46. Piaget — and most of his followers — set out to study spontaneous development. What, heavens, is spontaneity? Does the child's environment consist only of objects (unconcerned adults included) or aren't they taught, the one more, the other less? And what do the tests reveal, is it really spontaneous or elicited by the tests? All right, let us forget about these questions. As a pedagogue one is fascinated by other things than a psychologist is. Not by stories about Caspar Hausers and wolf children, but by instruction and education.

The psychological approach can, and sometimes must, be adidactical. The educational experimenter wants to observe learning processes. Even if he renounces the steering of them consciously, his design will anticipate a desired, perhaps spontaneous looking, learning process. Piaget's and his followers' experiments are valuable for the educator who wants to learn how it should not be done from an educational point of view. (It is a pity that nevertheless some educators slavishly imitate the experiments.) Certainly it is worth a learning process to understand that kneading clay or plasticine preserves the volume (at least if it is true). Certainly a child should learn to detach the water displacement by an immersed body from the context of its weight. Certainly he should

* Quoted by J. G. Wallace, 'Concept growth and the education of the child', *Nat. Found. Ed. Res. In England and Wales, Publ.* **13** (1965), p. 111.

learn to eliminate disturbing *physical* factors to arrive at the *geometrical* volume. For all these, learning processes are required, which no doubt need to be made conscious to teachers and learners, but in the didactical context they are located at another place, far away from the first steps towards the constitution of the mental object volume.

It should start with

> break and make transformations,
>
> building, converting, rebuilding, pulling down of buildings consisting of, say, congruent cubes or simple parts of them,
>
> their volume and its additivity,

(which includes subtractivity) with a stress on

> the difference between volume and surface area.

It could continue with

> the equivalence of content of open barrels and volume of solid bodies.

Further with

> practical pour transformations to compare contents.

Beside this mathematically structured material one should deal with

> less structured material that with regard to volume is compared by estimate.

Moreover one should

> connect the packing effect to, and if need be, eliminate from, the "content".

One should tackle

> volume-preserving and non-preserving transformations,

though not in the petty context of clay and plasticine. The

> volume determination by immersion

could be in a sense the coping stone yet not until many other things, such as weight and force, have been dealt with in the course of instruction.

In the foregoing list I skipped a few items, which I kept for the next sections.

13.47–50. *Measuring in One, Two, Three Dimensions*

13.47. It looks so simple, the relation between measuring in one, two, three dimensions. A length unit determines an area unit, distinguished by the exponent 2 at the length unit, and a volume unit, with exponent 3. Area is length times width, volume is length times width times height, where width and height are only synonyms of length. Length and width are measured horizontally, and height vertically, though to make things worse, there is a horizontal and a vertical direction in the plane, too.

As far as mappings are concerned, the stress has been up to now on those that preserve length, area, volume. When I discussed the mental constitution of rigid body, I ventured to claim a larger part played by similarities than is usually admitted. I take up this thread to look at what it means for length, area, volume. Congruence mappings preserve each of them, similarities multiply their values by a certain factor, which depends on the magnitude but not on the particular figure. Each similarity can be composed from a congruence, and a geometric multiplication, which is solely responsible for the special behaviour of the magnitudes at issue.

A geometric multiplication by the factor ρ multiplies

> lengths by ρ
> areas by ρ^2
> volumes by ρ^3.

There is hardly any insight that is as fundamental as this one for the constitution of these mathematical objects and their measurement. It can be acquired

> paradigmatically

for

> line-segments, squares, cubes

with simple factors like

$$\rho = 2, 3, \frac{1}{2}, \frac{1}{3}, \frac{2}{3}, 10, 1000, \frac{1}{10}, \frac{1}{100}$$

and is transferred to other figures

> initially straight away,

and subsequently argued by means of

> break and make procedures,
> grid coverings,
> approximations.

This principle deserves, as far as the moment of constitution and the stress are concerned, priority above algorithmic computations and applications of formulae because it deepens the insight and the rich context in the naive, scientific, and social reality where it operates.

Formulae for perimeter and area of the circle, for area and volume of the sphere and other solids are didactically and practically overshadowed by the knowledge about their behaviour under enlargement and reduction, which applies in a large field, not covered by formulae. The world literature from Plato onwards is full of misapprehensions and warnings against misapprehensions regarding the different behaviour of one-, two- and three-dimensional figures under enlarging and reducing operations. The warnings are not superfluous. Reason cannot be dispensed with to correct the eye, if the eye does not agree to accept that of two vessels which at sight do not differ that much, the one contains 1 cc and the other 10 cc.

What is even more important is what I called the rich context. Both Lilliput and Brobdingnag are impossible for physical and biological reasons. Similarities can physically be realised only within a narrow range and only macroscopically. If miniature models of the tides in the North Sea are to be constructed, sophisticated measures must be taken to account for the fact that gravitation and size of sand grains cannot be miniaturised. Even more fundamental are the limits put on physical similarities by the physical constants, charge and mass of elementary particles, speed of light, and Planck's quantum, which are not susceptible to change. The size of cells is not as constant, but still not variable enough to allow similarities on a large scale in biology. A more mathematical feature is the restrictions imposed on similarities by the geometrical dimensions: forces and interactions, determined by one, two, three dimensions, which must be equilibrated, stress and strain and speed of flux in one dimension, exchange of matter and heat through surfaces in two dimensions, inertia and gravity in three dimensions. How can an elephant twice the normal size get its food and get rid of its heat? How much should a mouse half the normal size eat to protect itself against the loss of heat? How should the bones and muscles of a double size horse be built to prevent its knees giving way and to allow it to move its limbs? In Brobdingnag and Lilliput, how long, big, thick are the telegraph wires, the water conduits, the antennae, the rain drops and the hail stones?

13.48. We already know that

area = length times width
volume = length times width times height

are too simplistic formulae. There are, however, shades of simplicity and complexity. First of all, the formulae

area = base times height
volume = base times height

for figures which measure the same at each height. Then for triangles and pyramids, cones,

area = $\frac{1}{2}$ base times height,

volume = $\frac{1}{3}$ base times height.

Where do these factors $\frac{1}{2}$, $\frac{1}{3}$ come from? This question can be answered on various levels.

The simplest — paradigmatic:

a square with side a, thus area a^2, divided from the centre into four triangles with base a and height $\frac{1}{2}a$ (Figure 188), which consequently have each an area $\frac{1}{4}a^2 = \frac{1}{2} \cdot a \cdot \frac{1}{2}a$;

Fig. 188.

a cube with side a, thus volume a^3, divided from the centre into six right pyramids with base a^2 and height $\frac{1}{2}a$ (Figure 189), which consequently have each a volume $\frac{1}{6}a^3 = \frac{1}{3} \cdot a^2 \cdot \frac{1}{2}a$.

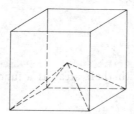

Fig. 189.

A high level — almost calculus:

a triangle with base b and height h is approximated by the unions of n rectangular strips (Figure 190) with

$$\text{base } \frac{1}{n}b, \frac{2}{n}b, \dots, \frac{n}{n}b \quad \text{and height } \frac{1}{n}h,$$

Fig. 190.

together an area

$$\frac{1}{n^2}(1 + 2 + \dots + n)bh = \frac{1}{n^2}\frac{(n+1)n}{2}bh,$$

which with $n \to \infty$ converges to

$$\frac{1}{2}bh;$$

a pyramid-cone with base area B and height h approximated by a union of n prismatic discs (Figure 191 with base areas

$$\frac{1}{n^2} B, \frac{4}{n^2} B, \frac{9}{n^2} B, \ldots, \frac{n^2}{n^2} B \quad \text{and height } \frac{1}{n} h,$$

Fig. 191.

together a volume

$$\frac{1}{n^3} (1^2 + 2^2 + \ldots + n^2) Bh = \frac{1}{n^3} \frac{(n+1)(n+\frac{1}{2})n}{3} Bh,$$

which for $n \to \infty$ converges to

$$\frac{1}{3} Bh.$$

Finally in Calculus:

Take similar triangles with their areas as a function f of their — variable — height h,

$$f(h) = \alpha h^2 \quad \text{with fixed } \alpha;$$

the difference

$$f(h + \delta) - f(h)$$

is the area of a strip of height δ on the base, the difference quotient

$$\frac{f(h + \delta) - f(h)}{\delta}$$

approximately equals the base, and the differential quotient

$$\frac{df}{dh} = 2\alpha h$$

equals the base,

$$b = 2\alpha h.$$

Thus indeed

$$f(h) = \frac{1}{2} bh.$$

Take similar pyramids-cones with their volumes F as a function of their — variable — height h. Thus

$$F(h) = \alpha h^3 \quad \text{with fixed } \alpha.$$
$$F(h + \delta) - F(h)$$

is the volume of a disc of height δ on the base, the difference quotient

$$\frac{F(h + \delta) - F(h)}{h}$$

approximately equals the base, and the differential quotient

$$\frac{dF}{dh} = 3\alpha h^2$$

equals the base

$$B = 3\alpha h^2.$$

Thus, indeed

$$F = \frac{1}{3}Bh.$$

13.49. The foregoing allows for a number of generalisations.

Consider a polygonal track L (closed or not) consisting of tangential segments of a fixed circle with midpoint M and radius h, and the planar piece spanned by M and L. It is a union of triangles, which can be considered as one figure with base L and height h (Figure 192). So it again fulfills

$$\text{area} = \frac{1}{2} \text{ base times height.}$$

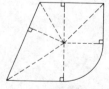

Fig. 192.

Let now all tangential segments or some of them shrink such that the base becomes a circular arc or composed of tangential segments and circular arcs, or the whole circumference. Still

$$\text{area} = \frac{1}{2} \text{ base times height,}$$

with the special cases

$$\text{circular sector} = \frac{1}{2} \text{ arc times radius}$$
$$\text{circular area} = \frac{1}{2} \text{ perimeter times radius}$$
$$\pi r^2 = \frac{1}{2} \cdot 2\pi r \cdot r.$$

Consider a polyhedron P (closed or not) consisting of tangential planar pieces of a fixed sphere with midpoint M and radius h, and the space spanned by M and P. It is a union of pyramids, which can be considered as one pyramid with base P and height h. So it again fulfills

$$\text{volume} = \frac{1}{3} \text{ base times height}.$$

Let now all tangential planar pieces or some of them shrink, such that the base becomes a spherical cap or composed of tangential planar pieces and spherical cap or composed of tangential planar pieces and spherical caps or the whole surface of the sphere. Still

$$\text{volume} = \frac{1}{3} \text{ base times height},$$

with the special cases

$$\text{spherical sector} = \frac{1}{3} \text{ spherical cap times radius},$$
$$\text{spherical volume} = \frac{1}{3} \text{ spherical area times radius},$$
$$\frac{4}{3}\pi r^3 = \frac{1}{3} \cdot 4\pi r^2 \cdot r.$$

These results can also be derived by the method exposed at the end of Section 13.48, that is, by considering similar figures as functions of the variable radius.

CHAPTER 14

TOPOGRAPHY WITH GEOMETRY

14.1–2. *Localisation*

14.1. The primordial parameter system to localise a point, consists of

standpoint
distance (of the point to be localised)
direction.

The standpoint need not be an

actual standpoint.

Indeed, if somebody tells somebody else how to go from P to Q, the point P is an

imagined standpoint.

The distance can be defined by true

length measures

or for instance by required

travelling times

— travels on foot, horseback, bicycle, carriage, car, train, plane, or by sound or light. The direction can be a

looking or showing direction,
sound direction,

— yah, there the sounds come from —

following a street, road, river, edge of a wood, and so on,

or more objectively determined by the

point of the compass

or by the

angle with a fixed line,
angle with a fixed plane.

Another quite usual localisation of a point is as a

crossroads,

that the crossing of two paths, rectilinear or otherwise conventionally determined.
Less obvious and less usual as a parameter system for localisation is

two or more standpoints,
their distances (from the point to be localised),

where the choice of only a pair (triple) of points in the plane (space) causes an ambiguity.
Much more usual is the choice of a

standing line or plane through two or more standpoints, the distance of which is known,

the directions (towards the point to be localised) defined by the angles with the standpoint line or plane,

that is, as intersection of two rectilinear paths, which in turn are given by standpoint and direction (Figure 193).

Fig. 193.

Iterative application of this kind of localisation is the principle of surveying, where starting with a certain base line the procedure continues with a long chain of triangles and angle measurements.
Localising of a global trend takes place

on the terrestrial globe,
on the celestial globe,

where in order to do justice to history I should rather interchange the first two items. A definition such as sometimes found in newspapers,

300 km South West of Gibraltar,

or in popular astronomy

3° South West of Aldebaran

are localisations as considered in the beginning, where on the celestial globe distance is measured in degrees.

14.2. So far we considered local means of description albeit in a global context. Localisation in a global frame is done by

coordinate systems.

The first example in history was the celestial globe, where a point is localised by

two angular coordinates,

each of which is the angle of the vision line with a plane. The two planes can be

horizontal and meridional plane

of the observer; the relevant angles are

height and azimuth.

Or

the equatorial plane and the plane through celestial poles, and vernal point;

the relevant angles are

declination and rectascension.

Or with the equatorial plane replaced by the ecliptic, the angles are

ecliptic latitude and longitude.

The second system has its match on the terrestrial globe: the

equatorial plane

is a natural datum; the angle of a radius from the centre to the point P with this plane is the

geographical latitude of P.

The other plane is conventional: the

meridional plane of Greenwich,

with respect to which the

geographical longitude of P

is measured (Figure 194). Whereas the geographical latitude is easily derived from the elevation of the pole, the geographical longitude caused navigators a lot of trouble in the past. The most natural way to find it was by means of the difference between local time — obtained by astronomic observations — and Greenwich time, mediated by a trustworthy clock. Meanwhile these methods of localisation have been outdated and replaced with

radar bearings,

Fig. 194.

which reminds one of the parameter system explained in Section 14.1. Big
planes, however, fly on a modern version of what seafarers called

> dead reckoning

– direction observed and speed estimated – where thanks to modern instruments
and computer processing, estimation has yielded to calculation.
    All these coordinatisations on spheres have in common

> structuring by means of
> > two antipodic points,
> > the bundle of circles through them,
> > and their orthogonal trajectories.

Localising in the universe requires besides that on the celestial globe,

> distance measurement in the universe,

which for "short" distances can now be done by radar bearings, and for long
distances thanks to the cooperation of all kind of phenomena in fundamental
theories – a fascinating subject, which is here out of issue.

14.3–5. *Cartesian Coordinatisation*

The run-of-the-mill rectangular coordinate systems have not yet been mentioned.
Coordinatisation in geometry is a recent acquisition. Though already in antiquity
Apollonius described conics by a "symptom", that is, an equation between
variable geometrical magnitudes related to the conic at issue, and Descartes
extended this method to more general curves, coordinatisation of the plane
(space) itself, that is, localising points by pairs (triples) of numbers, is an idea
that gradually gained ground only after Descartes, mainly to satisfy the needs
of analysis and analytical mechanics.
    The

> "cartesian" coordinatisation

of plane and space is now applied as a means of localising

> on geographical maps,
> on city maps,

on building plans,
on plans of buildings,
in the planning of building,
in city planning,
on a chess board.

Moreover it has become

a model for visual organisation

of material and information according to two or three characteristics:

Tables with two entries (rows and columns),
perhaps even stacked in a third dimension

is as it were the geometric image in the coordinatised plane (space) of the

pairs (triples) structure

introduced in the material.
The cartesian coordinatisation of the plane (space) implies

structuring by means of two (three) systems of parallel (perhaps mutually orthogonal) lines (lines and/or planes).

14.4. Once the cartesian coordinatisation of plane and space had been invented, as the aftermath of the functions of two and three variables in analysis and analytical mechanics, functions of an arbitrary number of variables came into being, for instance, a system of k particles, "localised" by the $3k$ coordinates of its members, soon complemented by its $3k$ momentum coordinates. The *term*

n-dimensional cartesian space

for the system of ordered n-tuples is of a more recent date than the application of geometric methods in this structure.
In our century

geometric terminology

and

geometric visualisation

have become a habit wherever from n systems

$$S_1, \ldots, S_n$$

the

"cartesian product"

is formed, that is the system of ordered n-tuples

$$\ulcorner x_1, \ldots, x_n \urcorner \qquad (x_i \in S_i).$$

Sometimes it is not much more than terminology, sometimes an illustrating model, for instance around the

time—path graph,

when the cartesian product is formed from

time and path axis

and

uniformity of motion and rectilinearity of the graph

go together. In a more general way the cartesian product serves as a model to combine

magnitudes in pairs, triples, and so on, and to represent connections between them graphically,

among others

proportionality by linearity.

14.5. According to Piaget* the plane is mentally being constituted as a cartesian product — a view that is the consequence of a logical systematisation of which the "product of relations" is an essential element. This view is supported by experiments where the subject is not allowed to do anything but structure the plane as a cartesian product. In another series of experiments** the authors, in spite of themselves, deliver all the evidence one can think of to refute Piaget's thesis.

In these experiments the subject has to reproduce a point P given on a rectangular piece of paper $ABCD$, by its image P' on a congruent rectangular paper $A'B'C'D'$. It appears that young children already understand the essence of this task and can contrive and sometimes even realise, intelligent methods to perform it. Unfortunately they cannot meet with Piaget's approval, because no solution bears the characteristic of cartesian coordinatisation. Almost all of them localise the point P (and accordingly P') by means of the first parameter system exposed in Section 14.1. The subjects choose one of the corner, say C, as "standpoint", measure the distance CP and give evidence of the idea to carry over the direction of the line CP. The experimenters, however, deny them all aids and appliances to perform this, because according to Piaget's theory directions and angles are only constituted after this has happened to the plane, and then by cartesian coordinatisation. One is astonished at the inventiveness displayed by some subjects, for instance, Col (5; 8) whose attempt at solving virtually boils down

* J. Piaget and Bärbel Inhelder, *La représentation de l'espace chez l'enfant*, Paris, 1948.
** J. Piaget, Bärbel Inhelder and Alina Szeminska, *La géométrie spontanée de l'enfant*. Paris, 1948. In particular Chapter VII.

to producing *CP* until it meets another side of the rectangle and to measure and carry over the line-segment cut off on this side — an entirely acceptable method, though not based on the "multiplication of relations" and therefore rejected. For the same reason solutions are rejected where the subject besides measuring the distance *CP* measures that of *P* to one of the sides of the rectangle. The only solutions evaluated as correct are those where *P* is localised by means of its distances from two orthogonal sides of the rectangle. Older children succeed in doing this.

What is wrong with these experiments from the start onwards is the reduction of the plane, which is, or is to be, constituted, to a rectangle. In both works I cited this is a bodice created by Piaget's "logical multiplication of relations". Why does Piaget not supply the subjects with a sheet in the shape of a circle, ellipse, triangle or some irregular figure to locate a point on it? It is clear that cartesian coordinatisation then is doomed to fail, whereas coordinatisation by other means would succeed. But this is not allowed because it would be contrary to Piaget's dogma on constitution of the plane as a coordinate plane.

In the case of the rectangle as a reduced image of the plane the cartesian coordination is indeed the natural one — that is, for subjects who have experienced rectangles sufficiently, and in particular more strongly structured ones than the bare rectangles administered by Piaget. It is quite probable that the older subjects who succeeded were accustomed to ruled or squared paper and consequently in possession of a stronger mental structure of the rectangle, which might explain their cartesian coordinatisation which just works with a rectangular sheet. Anyhow it is remarkable how well the younger subjects, though deprived of useful aids, do perform. If anything these experiments prove that the plane is not constituted the way Piaget asserts.

Piaget also — quite briefly — investigated localisation in space, which in a way appears to function better than in the plane. Piaget's explanation is not worth anything. The material of this experiment is a little box where a point is defined by a pearl on a vertical stick. Because of this vertical stick the right angled parallelepiped is overtly presented as the cartesian product of the horizontal rectangle and a vertical height. No wonder it is interpreted this way by the subjects.

14.6–7. *Polar Coordinatisation*

14.6. Piaget's idea of mathematics is one picture of mathematics, unambiguously structured and moreover such that its structure is an image of cognitive development. Mathematicians are less dogmatic. They know about many a way to build mathematics and they choose according to needs. Piaget's starting point is the straight line as a model of the linear order; plane and space must be constituted as products of two or three lines, even if these products are in no way prestructured. As he noticed this cannot succeed but by prestructuring.

Piaget has intensively studied the linear but never — if I am not mistaken —

the cyclic order. There is no doubt that cyclic orders are early mental objects and arranging cyclically is an early mental activity,

> sitting around a table,
> standing or dancing in a circle,
> walking in a circle
> counting out,
> walking around a block,

and many objects suggests cyclic order,

> rims of dishes and cups,
> dials.

I have dealt with cyclic order earlier*, also in its relation to linear order, at such a length that didactically—phenomenologically I have little if anything to add. I summarise the main ideas:

> recognising,
> causing,
> inverting,
> analysing

the

> discrete cyclic order,
> continuous cyclic order;

further,

> embedding and mapping cyclic orders,

> relating cyclic and linear orders by counting on and rolling off respectively.

14.7. The motive for this digression is the cyclic order displayed by the system of directions at a point in the plane. This system, determined by the horizontal movements of the head, the horizontal turns of the body, and the possibilities of stepping forwards, must be counted among our first mental objects. It is structured by the front—behind and the right—left in a way that of course changes with the position of the body. The primordial and natural coordinatisation is the polar rather than the cartesian one, that is, as explained in Section 14.1, with the standpoint as the pole, and as coordinates the distance and direction, mathematised by the angle with a fixed direction. As we noticed, Piaget's experiments rather offer evidence in favour of the primordiality of this kind of localisation. Historically, too, polar coordinate systems precede cartesian ones.

---

* *Mathematics as an Educational Task*, pp. 465—494.

Mathematically the polar coordinate system means a structuring of the plane, after deleting one point (the pole) as the

product of the ray of the (positive) distances
by the cyclic system of angles (a circle).

Mathematically there is no reason why this structuring of the plane should be inferior to that by means of the product of two lines, which is the cartesian co-ordinatisation. It is true that if there is no need to do so, nobody will prefer the polar coordinatisation; the need, however, is often felt enough to be accounted for.

Earlier the coordinatisation of the spherical surface was mentioned, in particular that of the terrestrial surface. This coordinatisation can be considered as a restriction of the polar coordinatisation of space. After a point (the pole) has been deleted, space is structured as the

product of the ray of (positive) distances
by the system of directions,

which in turn can be structured, for instance, by meridians and parallel circles.

Another coordinatisation of space is by cylinder coordinates, that is, structuring space as

product of a straight line
by a plane with a polar structure.

14.8. *Algebraisation*

The global localisation by means of coordinatisation leads to the algebraisation of geometry. Whereas the polar coordinate systems used to describe the celestial vault and terrestrial surface primordially served to systematise the localisation, the cartesian coordinate systems became particularly effective for describing geometric figures and mechanical movements, and later on, mappings in general. A figure is thus algebraically translated into a relation between coordinates, a movement into a function of time, a geometrical mapping into a system of functions of a certain number of variables. In this course — historically viewed — the theory of conics is algebraised and at the same time extended, Newton's geometrical mechanics transformed into the more efficient analytical mechanics, and the theory of mappings is designed *a priori* algebraically—analytically. This will later be exposed more broadly.

14.9–23. *Orientation*

14.9. In the topographical context Section 10.3 dealt with "polarities". There are more polarities than the ones mentioned there. Quantified ones like gain—loss, assets—debts, heavy—light, hot—cold and non-quantified ones like good—bad, beautiful—ugly. Since plus and minus signs, "positive" and "negative" entered mathematics, we have got accustomed to mathematise polarities by plus—minus, positive—negative, to distinguish opposite forces by signs, as we call positive and negative the two "kinds" of static electricity, human attitudes and values, and in general use this "model" of polarity — quantified and non-quantified.

This indication of oppositions — in which ever sense — by "opposite" signs is indeed an important piece of model forming. We will study it more intensely when we deal with negative numbers and directed magnitudes. At the present opportunity of mathematising polarities from the topographic context, we restrict ourselves to such polarities as need no further quantification: front—behind, left—right, below—above, left turn—right turn.

They are, as I explained at the former opportunity, closely connected to the human body but have been detached from it already in the topographic context. As far as mathematising means detaching from reality, the bonds with the human body and topography must be untied, though in a way that it does not obstruct restoring the ties.

Between the polarities just mentioned, there are relations; we need not repeat circumstantially what has been said in Section 10.3 about them. These mutual relations will show the way to, and the need for, mathematising.

The first three polarities we recalled are linear, while the fourth may be called cyclic or circular. By this I mean that their obvious standard models are the straight line, and the circle, respectively, both of them travelled in a certain sense, and on the other hand that these polarities can be used to establish on straight lines and circles respectively, a sense of travelling. Of course, in this context the straight line may be replaced with any simple curve, open at both ends, the circle by any simple closed curve — ultimately also by a screw.

14.10. Let us first consider linear polarities. I *step upon a line* in order to look along it and to mentally travel on it from behind to the front. Or I *look upon a line in front of me* in order to mentally travel on it from the left to the right, or from below to above.

Behind—in front of, left—right, below—above are linear polarities, on straight lines which are situated in space or with respect to my body in a way that allow me to indicate a mental travel upon them. By change of perspective I can establish on *each* straight line in space a behind—in front of, left—right, below—above, just as I like it or need it, in order to indicate a mental passing over the line. In this process of mentalisation the *spatial* behind—in front of, left—right, below—above is replaced with the *temporal* past—future.

It is profitable to choose a neutral term to indicate this polarity, and as such "positive—negative" presents itself. The linear polarity can then be expressed as follows:

Globally: The straight line possesses two opposite senses of travelling, which can arbitrarily and according to the needs be distinguished as positive and negative. The line provided with this sense, is called oriented.

Locally (p): A point p on the straight line has two sides — it divides the line into two rays —, which can arbitrarily and according to needs be distinguished as positive and negative.

Connection between "Globally" and "Locally (p)": Running from p positively on the line, one gets the positive side of p.

Connection between "Locally (p)" and "Locally (q)": If q is on the positive side of p, then the positive side of q is *a fortiori* on the positive side of p.

The ultimate mathematisation of these phenomena is the

Linear order: A set S with a binary relation $<$ such that for all a, $b \in S$ one of the choices

$$a < b, \quad a = b, \quad b < a$$

is realised, and transitivity, that is

$$a < b \wedge b < c \to a < c$$

for all a, b, c, holds, is called ordered. The opposite order is the relation $<'$ with

$$a <' b \leftrightarrow b < a.$$

The straight line is by preference ordered by two mutually opposite order relations such that for the oriented line it holds that:

$$a < b \leftrightarrow b \text{ lies at the positive side of } a.$$

14.11. Though the polarities in plane and space bear a circular or screw-like character, when passing from the line to plane and space we are confronted with the straight lines and their orientations as our primary objects of attention, thanks to polarities, specified by the new situation, of behind—in front of, left—right, below—above. These polarities can be

topographically influenced

as far as the below—above is tied to the "upright", they can mentally get detached from topography by a change of perspective that nominates an arbitrary plane to become the

mental standing plane with its own below—above.

By change of perspective a

mental behind—in front of,
mental left—right

can replace the polarities determined by the actual countenance of the body.
The polarities can be related to an actual or mental plane

on which one is standing,
in front of which one is standing
in which one is standing (lying).

A change of perspective we have got so acquainted with in early childhood that we deal with it unconsciously as an obvious one, is that between

> the vertical plane from which, and
> the horizontal plane on which one copies,

where

> left—right is unchanged

and

> below—above is replaced by behind—in front of,

such that finally

> behind—in front of is even indicated by below—above.

If one renounces combining two polarities in a planar structure, space shows

> three polarities at a time,

which in a first step of mathematisation are described by three oriented (mutually orthogonal) lines through one point. Going on one can have an

> arbitrary triple of lines

play this part, where at most the arrangement of the axes (x-axis, y-axis, z-axis) and a sketchy figure reminds one of the original role distribution. Such a system of axes is used to coordinatise space. In physical applications the relatedness of the system of axes to the human body is felt. Physicists have the habit of distinguishing *lefthand* and *righthand* coordinate systems. It is not farfetched to suppose at the background of this terminology an impossibility — in principle or in practice — to detach the space used in physics, entirely from the topographical context influenced by the human body.

14.12. In order to deal with this problem let us return to two dimensions. Linear polarities separated from each other do not suffice to grasp the situation.

On the plane where I am standing, I can make a right or a left turn. At the plane which I look at, I can make a turn with, or counter to, the clockhands. In the plane I am leaning against I can have my arms swing from above to in front to below, to behind, and again to above, or the other way round. I can impose a turning sense to each plane whatever it may be. I need not restrict myself to horizontal and vertical planes to impose turning senses on them. Each simple closed curve — Jordan curve — in such a plane with a sense of circulating around it contaminates with its sense of circulation as it were all others. If in a plane two Jordan curves are given, one of which with a circulation sense, I can immediately at a glance fix what is the "same sense" on the other one. How is this decided?

How can I see it — this was my original question but possibly it is more feeling than seeing, a kinaesthetic rather than optical experience. Something is turning if I perceive a Jordan curve with an arrow; mentally I am running over the curve and while I do so, something is turning simultaneously, the surroundings of the curve, the whole plane. Anyway, this holds for Jordan curves that are circles or like circles. How does it happen?

I am fixing a point o, in the interior of the curve — if it is a circle, by preference the centre. The ray from o to a point p on the curve moves with p — a turning movement that involves the whole plane. To be sure with more complicated Jordan curves (Figure 195) it does not work that smoothly, the ray can swing back and forth but finally a whole turn, in the one sense or the other, will be accomplished.

Fig. 195.

Turns of the plane can take place around different centres. How do I know whether they are the same sense or the opposite one? This, too, seems primordially to be decided kinaesthetically. One carries with oneself a feeling for the turning sense, and so it is carried from one place to another. Mathematically this can be established more precisely by shifting the rotation that determines the turning sense by means of a translation. One can also look far away where it hardly matters any more which point is the centre of the turn.

This transport of the turning sense of the plane conversely takes with itself the sense of running around a Jordan curve. But the relation between the senses of two Jordan curves can be made in still another fashion: under a continuous transformation in which the two Jordan curves appear as the first and last of a continuous array of Jordan curves, the sense is preserved. The question as to whether in such a process a Jordan curve cannot be transformed in the opposite one, will not be asked in this context let alone be answered by a formal proof.

Let us summarise the phenomena discussed here and their mutual relations:

$Ru(J)$:   the two (left and right) senses of travelling a Jordan curve in the plane,

$Tu(o)$:   the two turning senses of the plane around o.

For two Jordan curves J, J' and two points o, o':

$Ru(J)$ and $Ru(J')$ are related by continuous deformation transfer,

$Tu(o)$ and $Tu(o')$ are related by parallel transfer, view from far away, or continuous transfer,

Ru(J) and *Tu(o)* are related by assuming *o* in the interior of *J* and having the ray from *o* to a variable point of *J* and the whole plane participate in the same turn.

There are two — mutually opposite — senses at issue. The decision about which one is called the left or the right, is related to the human body or somehow motivated by physical phenomena.

Just as the behind—in front of, the left—right, and the below—above on a line are determined by choosing an at least mental position on or with respect to the line, so is the left or right turn in the plane determined by an at least mental position on or with respect to the plane.

When standing on the plane, a right turn means that the vision line turns towards the right hand, that is, the direction that means front turns to the right — perhaps in order to return after four quarters of a turn to its original situation.

When standing in front of the plane (for instance a clock dial) a right turn means that the above turns to the right (the clockhand from 12 towards 3) — again a quarter of a turn which when repeated four times restores the original situation.

(When standing within the plane, the arm swinging from above to the front, or the body falling on the face, determines something that can be called a left or a right turn, depending on the side from which it is viewed. In other words: sitting on a bicycle I cannot tell whether the wheels turn left or right.)

Just as on the line, there is a need in the plane for complete mathematisation in the sense of detachment of the human body or the topographical reality of which right and left turn are derivatives, and just as on the line, there is a need in the plane for more neutral terms like

positive—negative

for the two mutually opposite senses. By assigning the predicate "positive" to one of the senses the plane is transformed into an

oriented plane.

The phenomena and relations expressed by means of the abbreviations

Ru(J) and *Tu(o)*

are then to be read by replacing

left—right

with

positive—negative.

14.13. By the preceding the explanation of what is a left or a right turn is reduced to the cooperation of two polarities — in the terminology of standing upon the plane:

at a left (right) turn the "front" swings to the left (right).

We could also say to the left (right) side. Indeed a straight line in the plane has two sides (half planes). If the line in the plane I am standing upon (in front of myself) is oriented, I can distinguish the sides as left and right; if the orientation of the line is reversed, left and right are interchanged.

This concerns the plane I am standing upon (and with the usual change of perspective, the plane in front of me). If travelling a Jordan curve in the plane or the turning sense of the plane is at stake, we can free ourselves from the bond with the human body or with topography by assigning arbitrarily to one of the senses the predicate "positive" and to the opposite one the predicate "negative". We can do likewise with our reformulation of what is a left and a right turn.

A point determines two sides on a line; one of which can arbitrarily be called positive and the other negative. Similarly I can arbitrarily determine a

positive turning sense

in the plane by assigning

to an oriented line l

in the plane

its positive side:

under this positive sense, some p situated on l at the positive side of o turns to the positive side of l (Figure 196).

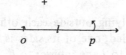

Fig. 196.

It is intuitively obvious that this definition does not depend on the choice of o and p on l (provided p is lying on the positive side of o).

By moving the oriented line l I can carry with it its orientation as well as the definition of what is its positive and negative side. It is likewise intuitively obvious that this does not influence the definition of the positive turning sense.

Summarising these phenomena and relations, I can say:

$S(l)$: the two (positive and negative) sides of an oriented line l in the plane,

$S(l)$ and $S(l')$ related by continuous transfer,

$S(l)$ related with $Tu(o)$: the positive turning sense turns the positive part of the oriented line l through o to the positive side of l.

Here we understand $Tu(o)$ in the way that left and right are replaced with positive and negative. It is, indeed, usual in mathematics to identify in drawings the positive turn with the left turn. This is a consequence of the fact that one draws the first axis "horizontally" and the second "vertically"; one is as it were standing mentally on the horizontal axis, while viewing to the positive (right) side; then the left side of this oriented line is lying "above"; the positive turn happens from horizontally left towards vertically above, which is against the clock and all that is called right turn in everyday life.

14.14. Like a straight line any curve extending from infinity to infinity has two sides, which if the curve is oriented can be distinguished as left and right. Or conversely: one assigns to one side the predicate positive and to the other the predicate negative and by this way defines the positive turning sense of the plane.

I will take this variant into account by allowing l in $S(l)$ to be a curve extending from infinity to infinity.

14.15. An even more striking case is the simple closed curve – the Jordan curve – which moreover has an interior and an exterior. Just as the positive (left) sense of J was fixed, so the interior is lying *left* of J. It is then quite natural to assume that this interior of J is its positive side.

Summarising:

$Int(J)$:   a Jordan curve determines in the plane an interior and an exterior; if J is being travelled in the positive sense, the interior of J is left of J.

If two Jordan curves J, J', lying outside each other, meet along an arc, then while this arc is being travelled, one of its sides is interior to J and exterior to J' and the other side exterior to J and interior to J'. Thus

$Int(J)$ and $Int(J')$ are under these circumstances related in such a way that if both J and J' are being travelled in the positive sense the common arc is being travelled in opposite directions.

This leads to

$Int(J)$ and $Int(J')$ related by combinatoric transfer:

Between J and J' one does not interpolate a continuous array for continuous transfer but a chain of Jordan curves such that each has just an arc in common with the next, and so transfers the sense from each to the next – for instance, between two oriented triangular circumferences a chain of triangles (Figure 197). In Section 13.27 this has been anticipated.

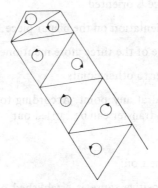

Fig. 197.

14.16. Whereas on the line the

global idea of orientation

is easily given a precise meaning, in the plane it starts so vaguely that it requires a profound analysis. We got acquainted with three ways to fix a positive orientation of the plane,

by defining a positive turning sense,
by defining a positive sense on a Jordan curve,
by defining the positive side of an oriented line.

This can be extended to polyhedra and curved surfaces, though it requires some caution. Let us explain this with a spherical surface.

A rotation around o is at the same time one around its antipodic point o', but for somebody standing on the surface at o' the right turn at o is a left turn at o'. With a Jordan curve on the sphere there is no distinction between interior and exterior, whereas in $Ru(J)$ the point o must taken in the interior in order to create a link with $Tu(o)$, if an orientation is to be imposed on the surface. There is no obstacle if one chooses the third way: a Jordan curve gifted with a positive sense, the positive side of which determines an orientation of the spherical surface. But if one passes to other kinds of surfaces, one has to be even more cautious: on the torus there are Jordan curves that do not divide it and consequently have one side only. As soon as one restricts oneself to "small" Jordan curves, one can distinguish two sides, and can stick to the method used on the sphere. Such difficulties arise with surfaces that *globally* differ from the plane — closed surfaces and surfaces on which Jordan's theorem does not hold. This is a reason for starting the business of orienting on such surfaces locally:

A surface is considered

as flat in the neighborhood of a point o —

polyhedra are so in a "general" point and curved surfaces can be approximated by flat pieces. The surface is oriented

by imposing an orientation on the planar piece,

which can be done in one of the three afore mentioned manners. By

continuous transfer to other points

the surface gets oriented at any point. According to Section 14.15 rather than in a continuous way this transfer can be carried out

combinatorically.

For instance, for a polyhedron:

for each face a positive sense is established on its border in a way that common segments of neighboring faces are travelled in opposite directions.

A curved surface can be dealt with similarly by division into small pieces.
There are, however, also

non-orientable surfaces,

as is the Möbius strip (Figure 198), a strip with the opposite short sides stuck together after inversion. Here the above procedure fails: after a trip around the strip, divided into triangles, one returns with a triangle bordered in the opposite sense.

Fig. 198.

14.17. Up to now I have neglected a phenomenologically important source of orientation: the reflections (planar ones if the plane is concerned). As a matter of fact the need for orientation as a mental object is primordially — or at least to a high degree — due to the existence of reflections. The 'd' and 'b' are the same and yet not the same — how can they be distinguished? The plate of Figure 199a does not fit into the hole of Figure 199b — why not? Curls and spirals are of two kinds — what is the difference? The S is easily confused with its mirror image. Space is even more intricate: the left shoe does not fit the right foot; in twin houses the stairs turn differently.

(a)   (b)
Fig. 199.

A reflection in one dimension, on the line, interchanges the two senses. A reflection in the plane one is standing on or in front of you interchanges right turn and left turn, thus both of its orientations however the plane is situated. The planar congruences are of two kinds: motions that preserve orientation, and slide reflections inverting it. Two figures that arise from each other by side reflections do not differ from each other in their internal properties: it is perhaps for this reason that the optical sense needs the support of the kinaesthetic one to constitute the difference between left and right and between left and right turn as such.

How this comes about, how and when things are made conscious that occurred in our phenomenological analysis — I cannot tell. I know that some children have difficulties with left—right, and with left—right-turn, but I did not observe learning processes related to it, and I do not know about people who did. Neither geometrical propedeutics nor advanced geometrical education — traditional or modern — pay attention to this complex of phenomena and relations, and so it is not to be wondered at that it has eluded the attention of psychologists and educationalists. This is just as true with respect to orientation of space, which will be our next concern.

14.18. I am starting with two striking phenomena. The first is what I would like to call the

стamp imprint effect,

or the

shop window inside—outside effect —

a right turn on a stamp is printed as a left turn, a right turn drawn inside looks like a left turn from outside.

The second phenomenon is exemplified by the existence of

screws, cork-screws, winding stairs

of two possible kinds — right and left ones.

Both phenomena are closely connected. Let us first consider the second, which is the most solidly materialised. All cork-screws and most screws show the same "turn"; it is quite natural and anyway usual to call it the right one. A striking geometrical feature: if the screw is standing in front of you the screw-thread runs from left below to right above, even though you view the screw from the backside or upside down.

About winding stairs — at least in old castles — there is a tale that most of them are left winding, for a good reason, as is claimed: so the defender of the castle who is facing downwards has more elbowroom for his right arm — the sword arm — than the assailant facing upwards. The minority of right winding stairs — it is claimed — were built by left-handed castle lords.

Other examples of right "screws": the hair crown of the human skull, the hop stalk, most of the snail-shells. Left screws: the bean stalks.

The ordinary — so-called clover leaf — knots can be distinguished according to right and left ones (Figure 200).

left knot   right knot

Fig. 200.

Running along the knot one has to consider how a new piece of the string turns around one that has been passed before. Righthanded persons have the habit of tying right knots.

Kinaesthetically the (right) cork-screw is characterised by a

    synthesis of right turn and inwards,

which is also experienced with the screw-driver.

One can turn a right screw into a thin plate — representing a plane — from one side or the other; this causes within the plane opposite turns depending on the side from which the screw enters, although both turns are viewed as right ones at the side where one is operating — this then is the shop-window effect — or, closely connected — the stamp effect.

The act of turning in or screwing in provides the turn or screw axis with a sense, which quite naturally may be defined as the positive one. The plane orthogonal on this axis has two sides — half spaces; operating the instrument I am looking from the negative to the positive side.

We notice the coupling between three phenomena:

    a screw sense in space,
    a turning sense in a plane,
    a distinction between the positive and negative side of the plane.

Each pair of them determines the third:

    given a screw (right or left) and an oriented plane if the screw is to be drilled into the plane in the positive sense, it is determined from which side it shall be done,

    given a screw (right or left) and the side from which it is entered, the turning sense in the plane is determined,

    given a turning sense in the plane and the side of the plane from which the screw is driven, it is determined whether it is a right or a left screw.

14.18. About a straight line or a plane in space I cannot tell any more what is left and right, left turn and right turn as soon as they are detached from their topographical context. I can arbitrarily call one orientation positive and the other negative, with no regard to polarities depending on the human body and topography. In order to state what is a right or a left screw, I can dispense with the topographical context, but I must know what is my right and left hand. If this knowledge is eliminated as not belonging to mathematics, then one is still left with the existence of two kinds of screws and too screw senses though the question which is right and which is left, becomes meaningless.

Among the methods of putting an orientation upon a straight line or a plane, there was one, proposing:

a line is oriented by assigning to a point on it a positive side,
a plane is oriented by assigning to an oriented line in it a positive side.

This can be continued as follows:

the space is oriented by assigning to an oriented plane in it a positive side.

It should be noticed that:

if the positive side of one oriented plane is fixed, it is so by continuous transfer of any oriented plane, in particular, if an oriented plane is transferred into its opposite, its positive sides are interchanged.

If the preceding definition is to be fed back to screws in real space, I have to settle how the plane is to be oriented in reality and which is its positive side. Then the orientation defined by the right screws is the negative one, just as for the positive orientation I have to use the left screws.

Summarising I state the following phenomena and relations

$Sc(P)$:   the space possesses two orientations (screw senses), which can arbitrarily and according to needs be distinguished as positive and negative. Fixing the positive orientation is equivalent to fixing the positive orientation and the positive side of a plane P in space.

$Sc(P)$ and $Sc(P')$ can be related by continuous transfer of the plane, its orientation, and its positive side.

14.19. In order to impose an orientation on space, we do not need full planes, rather we can be satisfied with little pieces. As a consequence we can replace the plane in the preceding by any curved surface if it extends to infinity, as does the plane. We can, however, do it also with an oriented closed surface and make good use of the fact that it possesses an interior and an exterior. As a convention we can consider its interior as the positive side, which then fixes an orientation of space by means of an oriented closed surface:

$C(S)$:   an oriented closed surface S determines an orientation of space with its interior as its positive side.

$C(S)$ and $C(S')$ can be related by continuous transfer.

If two such surfaces S and S' are touching each other from outside along something looking like a circular disc, one of its sides belongs to the interior of S and the exterior of S', and the other to the interior of S' and the exterior of S. Thus

> $C(S)$ and $C(S')$ are under these circumstances related by the opposite orientation borne by the common piece in S and S'.

Moreover

> $C(S)$ and $C(S')$ can be related by combinatoric transfer:

Rather than by a continuous array and continuous transfer, one joins S and S' by a chain of closed oriented surfaces where each touches the next along something like a circular disc, for instance, between two oriented tetrahedra, a chain of tetrahedra (Figure 201) each of which touches the next from outside along a whole face.

Fig. 201.

14.20. An oriented tetrahedron lends an orientation to each tetrahedron, by continuous or combinatoric transfer. Before introducing the usual symbolism for oriented tetrahedra, I shall return for a moment to lower dimensions.

The orientation of a line can be given by choosing two points a, b; the positive sense of the line (and the line segment) is indicated by

   $a b$.

The orientation of a Jordan curve can be given by choosing three points a, b, c, such that the curve is travelled from a via b to c; the orientation of the curve can be indicated by

   $a b c,$ $b c a,$ $c a b$

as you want it.

In particular the orientation of the border of a triangle can be indicated by its vertices, placed according to the orientation

   $a b c,$ $b c a,$ $c a b.$

The orientation of a tetrahedron with vertices a, b, c, d is determined by that of a face, say that with vertices b, c, d. Let its orientation be

   $b c d$ (or equivalently $c d b, d b c$).

Then by the rule of opposite senses of the common edge the orientation of the other faces is

a d c (or equivalently *d c a, c a d*),
a b d (or equivalently *b d a, d a b*),
a c b (or equivalently *c b a, b a c*).

By means of the oriented tetrahedron, space is oriented: the interior of the tetrahedron is its positive side. Thus: of each face the opposite vertex is on its positive side.

If one prefers a notation for the orientation of the tetrahedron by means of all of its four vertices, I might convene that the fourth vertex is added to the left of the triple. Then the oriented tetrahedron is indicated in twelve ways:

abcd, acdb, adcb,
badc, bdca, bcad,
cabd, cbda, cdab,
dacb, dcba, dbac.

The 24 permutations of *abcd* fall into two classes, the even and the odd ones, according to whether they are obtained by an even or an odd number of pairwise exchanges. The above lists just the even permutations of *abcd*.

Thus, the orientation of a tetrahedron can be indicated by its vertices in order, where

an even permutation describes the same orientation,

under the orientation *abcd* the faces are oriented as *bcd, adc, abd, acb* and each vertex is lying on the positive side of its opposite face,

in other words,

a on the positive side of *bcd*,
b on the negative side of *acd*,
c on the positive side of *abd*,
d on the negative side of *abc*.

Feeding back to the space of our kinesthetic experience we can distinguish a right and a left screw sense, respectively suggested by the left and the right drawing of Figure 202.

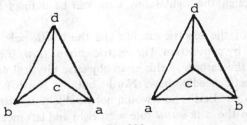

Fig. 202.

14.21. Starting from the orientation *abcd* of the tetrahedron (and thus of space), one can arrive at another description of spatial orientation, that is by means of an ordered triple of oriented lines

$$da, \quad db, \quad dc,$$

which can be identified with the triple of axes of a coordinate system: the third axis shows to the positive side of the plane oriented by the first and second axis. Fed back to the space tied to our body it is identifying the first axis with the direction left—right, the second with back—front, the third with below—above, as is often suggested by drawings: the turn from the first to the second axis combined with the progression along the third gives the left screw sense.

Another method to orient space is by means of a (non-ordered) pair of oriented skew lines: assuming on each two points in the order of orientation

$$ab \quad \text{and} \quad cd,$$

and orienting the space by the oriented tetrahedron *abcd*. (The order of the pair does not matter since *cdab* represents the same orientation.) One can connect with it the kinaesthetic experience of screwing *ab* into *cd*.

Feeding back into physical space, it becomes meaningful to talk about two oriented straight lines in terms of whether they cross each other left or right. This applies also to clover leaf knots: two pieces of the knot are considered as skew oriented lines; according to how they cross each other it is a left or a right knot.

14.22. The existence of two orientations or screw senses of space is a mathematical fact; it depends on knowledge about our body which is called left and which right — we said. Does it really depend? The screw sense is applied in physics: If electricity flows through a conductor above (or below) a magnetic needle, the needle turns as though it tries to stand orthogonal to the current. It is settled by the cork-screw rule in which sense it turns: One imagines the (right) cork-screw with its axis in the conductor drilled into the direction of the current; then the North pole of the magnetic needle moves as though it is hit by the handle of the cork-screw.

It looks as though the right screw sense can be defined by pure physics. Is this true?

The direction of the electric current and the North pole of the magnetic needle are defined by convention. The electric current flows if it is a dry battery from the zinc to the carbon — this is an objective physical datum. The north pole of the magnetic needle shows North, but the North is a topographical datum, fettered to our Earth. So I would not be able to inform extra-terrestrial beings by means of the cork-screw rule what right and left means among people on Earth.

However, there are other natural laws, related to the so-called weak force, which do allow us to characterise right and left screws as such.*

14.23. What was *didactical* phenomenology in the sections on orientation? Perhaps a few remarks at the end of Section 14.17, which bear witness to my ignorance on genetic and learning processes with respect to orientation, left—right, left turn—right turn; I can now add the screw senses to this list. I guess that these objects are mentally constituted with kinaesthetic support. This would be an interesting subject for research. Of none of the relations between back—front, left—right, above—below, left turn—right turn, left side—right side, interior—exterior, left screw—right screw, which I analysed phenomenologically, would I be able to tell, when, how spontaneously, how directed, how explicitly they can, may, must be learned, and how far the road from the mental object to the concept should be travelled. Though important enough, I do not consider it at present as a learning matter. I dealt with it circumstantially to stimulate research in this field.

---

* See H. Freudenthal, *Mathematics Observed*, World University Library, 1967, Chapter VII.

CHAPTER 15

NEGATIVE NUMBERS AND DIRECTED MAGNITUDES *

15.1. *An Apology*

I have hesitated for a long time about which order I should deal with the subjects indicated in the title. The chapters of this phenomenology are not logically ordered; even *a posteriori* I do not see any possibility of introducing such an order. I cannot do without anticipating and resuming. I do not write a mathematical treatise with a deductive structure that helps one to avoid logical circles. In general the mathematical ideas under consideration are supposed to be known. In the phenomenological approach I may sometimes suppose that the reader is familiar with mathematical subject matter which phenomenologically is delayed and dealt with later on.

15.2–3. *History*

15.2. Historically negative numbers are much earlier than directed magnitudes. If precursors, as in Hindu mathematics, are disregarded, negative numbers arose about 1500, though three centuries passed before they were wholeheartedly accepted; the directed magnitudes are an invention of the 19th century. This, however, does not say anything about their mutual relation.

The origin of negative numbers is of course the algebra of equations, such as they are symbolically written. Methods of solving such equations were developed and gradually automatised. As the automatisation progressed, one wanted to extend their domain of validity. One tried to remove obstacles; solving equations such as it had been automatised should go on under all circumstances. The equation

$$ax + b = c$$

should possess a solution x under all circumstances (of course provided $a \neq 0$) – we would say at present, but in history it was the quadratic equations (and equations reducible to quadratic ones) that created this need, say, something like

$$x^2 - x - 2 = 0$$

or

$$x - 1 = \frac{2}{x}.$$

* Directed magnitudes are dealt with in this chapter only in connection with negative numbers, thus with the stress on "directed" rather than on "magnitude".

The algorithmic procedure led to two solutions, one of which was "impossible" but would nevertheless be admitted, at first hesitatingly but then with an ever increasing degree of conviction. It is the same idea that led to imaginary numbers, in particular when solving the cubic equation: pressing the solving formula at any price. So it is not to be wondered at that the imaginary numbers did not enter that much later than the negative ones. The resistance against both of them lasted about equally long.

The first extension of the natural number concept, towards the fractions, had been much less problematic. From the first mathematical documents onwards we meet with fractions. Heart-searching in this domain was of a much later date, in Greek mathematics, where for philosophical reasons it was forbidden to break the unit. The Greek mathematicians replaced fractions with ratio while the calculators in commerce as well as in science continued to do it with fractions.

One can readily understand why the step from natural numbers to fractions was unproblematic. Fractions have to do with and are required for magnitudes, that is, wherever people are measuring and where continuous quantities are divided, and from the first calculating activities onwards people had to perform these activities.

Mathematics in the Greek sense is about numbers, and as far geometry is concerned, about magnitudes — a view that even mathematicians in more recent times tried to share, at least in theory. The negative numbers had originated from the formal algebraic need for the general validity of solving formulae, but not until the algebraisation of geometry (the so-called analytic geometry of former times) did they become effective — I mean content effective.

The idea of algebraically describing geometric figures and solving geometric problems is older than Descartes. We owe to Descartes the use of one coordinate system (to express it in modern terms), independent of the figure and the problem. Algebraisation and coordinate systems were dealt with in Chapter 14, where we tacitly presupposed negative numbers. Descartes had some trouble with them; numbers were introduced as magnitudes; letters indicated magnitudes, thus positive numbers. But those who applied Descartes' method, could no longer avoid having letters also mean negative numbers. If straight lines are to be described algebraically in their totality, if curves are to be described algebraically in any situation, one cannot but admit negative values for the variables. The need for

general validity of algebraic solution methods

to which the negative numbers owed their existence, is reinforced from the 17th century onward by the need for

general validity of descriptions of geometric relations.

The second need, more content directed than the formal algebraic one, is the most natural and compelling. It is properly responsible for the success story of negative (and also of complex) numbers.

If the negative numbers are introduced, it does not suffice to claim their existence — this is often didactically overlooked, as it happens also with the rational numbers. The negative numbers become operational by their use in calculations, which obey certain laws. As I have explained elsewhere* at great length, these calculation laws are uniquely determined as extensions of certain laws governing the positive numbers. I gave the idea of extending while preserving certain laws the name of

the algebraic principle.

It includes what I just called the

general validity of solution methods,

and virtually it is the same idea, albeit formulated in a broader view.

I recall a few examples of the algebraic principle.

$$(-3) + (-4) = -(3+4)$$

is proved by starting with the definition equations for $-a$,

$$(-3) + 3 = 0, \qquad (-4) + 4 = 0,$$

adding them formally, while using commutativity and associativity, in order to arrive at the definition equation

$$((-3) + (-4)) + (3+4) = 0$$

for $-(3+4)$.

Or: Starting with the same definition equations, one proves

$$(-3) \cdot (-4) = 4 \cdot 3$$

by multiplying distributively the first with 4 and the second with -3,

$$4 \cdot (-3) + 4 \cdot (3) = 0, \qquad (-3) \cdot (-4) + (-3) \cdot 4 = 0$$

and subtracting them from each other.

Or: With $\sqrt{a}$ defined as the x making $x^2 = a$, one gets

$$\sqrt{a}\,\sqrt{b} = \sqrt{ab}$$

by multiplying the definition equations

$$x^2 = a, \qquad y^2 = b$$

to get

$$(xy)^2 = x^2 y^2 = ab.$$

Similarly if operations are to be extended,

$$a^{\frac{1}{n}} = \sqrt[n]{a}$$

because both terms have the same nth power.

* *Mathematics as an Educational Task*, pp. 224, 275 sq.

With hindsight the term "algebraic principle" looks to me too colourless. In the sequel it will be the

algebraic permanence* principle,

if I mean the – in general unique – extension of algebraic structures with preservation of certain fundamental or fundamental looking or as desirable considered properties. As far as negative numbers are concerned, the problem is to introduce new objects beyond the positive numbers with operations such that the impossibility of

$$a + x = b \quad \text{with } b < a$$

is abrogated, while other laws, which can be specified, are preserved and such that these acquirements of preservation imply the uniqueness of extension.

In history the negative numbers (and later on the complex ones) have been invented according to the algebraic permanence principle. Initially people worked with negative numbers in a naive formal way. Later on arguments of content character were contrived to be adduced, which still subsist in newer arithmetic and algebra textbooks, although some of them are not quite convincing (positive-negative as capital–debt, gain–loss, and so on).

15.3. I have discussed earlier the algebraic permanence principle and its didactical implications at such a length that there is little to be added. I only stress once more its didactical soundness and the didactical weakness of the traditional objects against it.

I have propagated still another didactical means to introduce the negative numbers, the

induction extrapolatory method,**

using tables like

$$
\begin{array}{llll}
3 + 2 = 5 & 3 - 2 = 1 & 3 \cdot 2 = 6 & (-3) \cdot 2 = -6 \\
3 + 1 = 4 & 3 - 1 = 2 & 3 \cdot 1 = 3 & (-3) \cdot 1 = -3 \\
3 + 0 = 3 & 3 - 0 = 3 & 3 \cdot 0 = 0 & (-3) \cdot 0 = 0 \\
3 + (-1) = \ldots & 3 - (-1) = \ldots & 3 \cdot (-1) = \ldots & (-3) \cdot (-1) = \ldots \\
3 + (-2) = \ldots & 3 - (-2) = \ldots & 3 \cdot (-2) = \ldots & (-3) \cdot (-2) = \ldots
\end{array}
$$

If I do not deal with these methods now again, I do not mean to renounce them. I would rather view them more closely connected than I formerly did to the geometrical means of introducing negative number. Didactically I now see the algebraic permanence principle rather as a

geometric–algebraical permanence principle.

---

* Recalling Hankel's permanence of the calculation laws.
** *Mathematics as an Educational Task*, p. 281.

This fits better the historical course; I already mentioned that the negative numbers did not become really important until they appeared to be indispensable for the permanence of expressions, equations, formulae in the "analytic geometry". What I called the inductive extrapolatory method also fits into the geometric—algebraical context (in particular of mappings), in which justice will be done by the analysis I intend to carry out.

15.4. *Old Models*

Long before analytic geometry, numbers were used in geometry. Lengths of line segments were measured by numbers of a unit. This then was a way to introduce numbers and operations on numbers in a geometrically-based building of algebra. Numbers as coordinates were used in astronomy — with characteristic non-numerical extra data such as East and West, South and North.

Coordinatising a road by mile stones has already been exercised in antiquity. Each path or line can be measured in this way: measuring distances from a fixed point — the origin — and noting them down. Roads start somewhere (in antiquity of course in Rome, whereas in later times all of them led to Rome); even after arbitrary extensions they remain half lines. If the origin (say Rome) is passed, the half line can virtually become a whole one. Two points on such a track with the same distance from the origin can be distinguished as back and front, and in order to measure distances on such a path one has to distinguish pairs of points at the same side of the origin from those on different sides.

To make it more precise let us view a "horizontally" drawn line. The points can be determined by their distance, measured in a certain unit, from the "origin", with the extra datum "left" or "right". If the mutual distance between two points is to be expressed by means of such data, three cases are to be distinguished: both points left, both points right, one left and one right.

Once negative numbers are admitted, it is not a farfetched idea to count further on the number line beyond 0 with negative numbers, stepwise subtracting a unit, or straightaway placing $-a$ at the mirror image of a with respect to the origin. With capital and debts the term is black and red, reminding one of the old table computers and the print-outs of more modern ones. Or the centigrade thermometer they used to call it heat and cold and indicate it red and blue, but in meteorological surveys it is now plus and minus.

Lodging the negative numbers on the number line is of course not the thing that matters. Negative numbers are not only introduced; they are to be operated on, added, subtracted, and later on, also multiplied and divided, to be put into exponents and in all kind of functions. Let us first view adding and subtracting.

Models in the older literature to introduce negative numbers and their operations, are

> the game of gain and loss, also with debts,
> the stairs up and down, also into the cellar,
> the thermometer up and down, also below the freezing point.

The second and third are similar to the number line, and the first can also be illustrated on the number line. The arithmetical operations are in all cases related to events, the result of a game, a movement on the stairs, a change of temperature. The same holds with the model that is now preferred,

the number line, and jumping on it.

The negative numbers are reached as soon as

one loses more than one possesses,
one descends more steps than one was above street level,
the temperature drops more than "it was".

Or with the number line as path: if

one jumps back more than one had advanced.

These models make plausible

adding a positive number a to any given one,
subtracting a positive number a from a given one.

With the given point as a variable x, it means concretising the functions

$$x \rightarrow x + a$$
$$x \rightarrow x - a$$

with a *positive a*.

That is all these models can grant. In the older textbook literature more was tried, but the best they could do, was to decree honestly but with a brazen face:

adding (subtracting) a negative number is the same as
subtracting (adding) its opposite,

thus

$$x + (-a) = x - a$$
$$x - (-a) = x + a$$

for positive a.

This decree can hardly be justified by the models used. The most satisfactory is perhaps

negative gain is positive loss,
negative loss is positive gain,

but negatively the stairs up and down and negative temperature gains and losses are unworthy of belief.

Of course by formal algebra it can be justified. From

$$(-a) + a = 0$$

it follows, if commutativity is granted,

$$a + (-a) = 0,$$

and from this by associativity

$$x + (-a) = ((x - a) + a) + (-a) = (x - a) + (a + (-a)) = x - a,$$

and even more involved,

$$x - (-a) - ((x + a) - a) - (-a) = ((x + a) + (-a)) - (-a) =$$
$$(x + a) + (-a) - (-a) = x + a.$$

But this is a bit too much.

The fault of the models, dealt with so far, is the didactical asymmetry between positive and negative numbers. The positive numbers are more concrete in the sense of greater originality; so one can operate with them; the negative numbers are secondary, introduced as results of operations, which formerly were impossible, fit to be operated on if need be, but unfit to have operations performed with them. In other words, the positive numbers are active, the negative numbers only passive.

If rather than asking for a model, one is satisfied with the formalism of what I called the algebraic permanence principle, this difference is non-existent; as soon as one has decided about extending, the negative numbers have the same legal status as the positive ones; operating with negative numbers is formally justified and in no way distinguished from that with positive ones.

15.5—8. *New Models*

15.5. If rather than being satisfied with the algebraic permanence principle, one looks for more satisfying models than those dealt with so far, it is now clear that positive and negative numbers shall be given the same opportunity. The former models admitted a symmetry between

adding and subtracting

as inverses of each other —

the one undoes the other.

What is asked for is an equal status for

positive and negative numbers,

such that the

operations can be performed by means of this equal status.

A — non-geometric — example are

positive and negative counters,

a method proposed, if I am not mistaken, by Gattegno; a geometric model is the

reinterpretation of points on the number line as arrows.

15.6. One works with, say black and red, counters such that

a black and red one can annihilate each other,

or conversely that

a black—red pair can come into being from nothing.

It does not matter which colour is identified with positive and which with negative; only where it is needed for the notation we will identify black with positive. The annihilation rule allows us to identify with each other for instance

7 black and 3 red ones,
6 black and 2 red ones,
5 black and 1 red one,
4 black and 0 red ones.

Of course also

3 black and 7 red ones,
2 black and 6 red ones,
1 black and 5 red ones,
0 black and 4 red ones.

As a matter of fact it means considering an integer as

an ordered pair of natural numbers

with the equivalence relation

$$\ulcorner a, b \urcorner \sim \ulcorner c, d \urcorner \leftrightarrow a + d = b + c,$$

thus as an equivalence class, or at least operating with them in a way that mathematically conceptualised is known as equivalence class formation. The additions of type

$$7 + 3 \qquad\qquad (-7) + (-3)$$

can be performed with the black and red counters directly by

taking together.

$$7 + (-3) \qquad (-7) + 3$$

require

annihilation

of three black and red ones. The subtractions

$$7 - 3 \qquad\qquad -7 - (-3)$$

can be performed by

taking away,

but

$$7 - (-3) \qquad (-7) - 3$$

would require taking away three red (black) ones from a pile that contains black (red) ones only. This is the first opportunity to apply

creating from nothing:

in order 3 red (black) ones to be taken away which are non-existent, three red—black pairs are added to the present stock:

(10 black and 3 red ones) minus 3 red ones
        (10 red and 3 black ones) minus 3 black ones,

thus

$$7 - (-3) = 10 \quad (-7) - 3 = -10.$$

Finally one has the types

$$3 - 7 \qquad\qquad (-3) - (-7),$$

when 7 black (red) are to be subtracted from too small a quantity. So one completes the insufficient quantity by pair forming such that subtracting becomes feasible

(7 black and 4 red ones) minus 7 black ones
        (7 red and 4 black ones) minus 7 red ones,

thus

$$3 - 7 = -4 \qquad (-3) - (-7) = 4.$$

These are various cases — the enumeration is complete — but all of them are solved according to the same principle:

simplifying by annihilation,
pair forming if "it can't be done".

It recalls artifices in column arithmetic,

> replacing 10 units by a ten,
> dissolving a ten into 10 units,

and this analogy with

> pair annihilation,
> pair forming

is not at all superficial.

Multiplying integers cannot be dealt with in this model unless it is decreed by brute force that

> black times black = red times red = black
> black times red = red.

15.7. The geometric model I announced in Section 15.5 is that of arrows. A point a on the number line is at the same time interpreted as an arrow from 0 to a, that is, as an arbitrarily movable arrow, a vector. The objects are, mathematically understood,

> equivalence classes of arrows

with equality of length and direction as equivalence relation. Or if the arrow is hollowed out to become the pair "tail–head", an

> equivalence of pairs of points

with equality of distance and order as the equivalence relation.

A warning: Arrows — most often curved ones — are also used with the model of the number line such as mentioned in 15.2 to indicate operations; then an arrow of length a means a jump of width a along the arrow (to the left or right according to the arrow direction), that is, the operation

$$x \rightarrow x + a$$

or

$$x \rightarrow x - a.$$

As explained in Section 15.2 this leads no further than adding and subtracting positive numbers. It is true there are textbook authors who tacitly switch from that interpretation to this new view on arrow classes as numbers and by this way suggest more than the first interpretation can yield — a not unusual kind of intimidation.

In order to avoid this misunderstanding it is worth recommending that one puts the name of the arrow nor above but as its head (Figure 203). Moreover adding arrows must be defined explicitly:

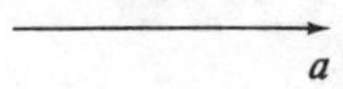

Fig. 203.

Adding arrows means putting them tail upon head, while using freely the movability of arrows (Figure 204).

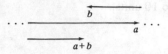

Fig. 204.

How to define subtraction? As the converse of adding? It is possible, but it easily leads into the temptation we warned against of interpreting the arrow as an operation rather than a number. It is better, and even easier, to define $a - b$ by

$$x = a - b \leftrightarrow a = b + x.$$

So one gets

$a - b$ as the arrow from the head of b to that of a

provided the arrows a, b are joined at their tails (Figure 205).

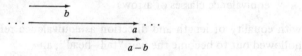

Fig. 205.

It is perhaps surprising that in this model the subtraction is better visualised than the addition. It is visually also obvious that $0 - a$ is the vector opposite to a. If the notation

$0 - a$ is to be replaced by $-a$,

this should be done *explicitly*. The definitions imply that

$$0 - 3 = -3 \quad \text{and} \quad 0 - (-3) = 3$$

but as such

$-(-3)$ is meaningless

unless it is defined, that is by

$$0 - (-3).$$

As soon as this definition has been accepted

$$a \rightarrow -a$$

is a meaningful notation for the mapping that carries each arrow in its opposite, and one gets

$$-(-a) = a.$$

Again it is clear that

adding and subtracting are opposites of each other.

The associativity of addition is conceptually obvious in this model, which cannot be asserted as confidently of the commutativity, but also the visualisation of commutativity is poor as shown by Figures 206 and 207.

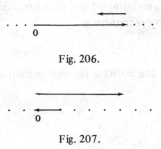

Fig. 206.

Fig. 207.

Not only of commutativity. The authors of various textbooks are wrestling, each individually and anew without benefitting from the lessons taught by former failures, with the visualisation problems for this model of operations on integers. Properly the arrows should be placed on the number line, but then they fall together and cannot be distinguished. A way out is to draw them at a little distance parallel to each other or to the number line, another one is to replace them with arcs from tail to head – it is not elegant but at least practical. It might work as long as one restricts oneself to the operations of adding and subtracting positive, in particular natural, numbers, and perhaps even in the general case. But a convincing visualisation may hardly be expected of this method.

After adding and subtracting, the next issues are multiplying and dividing. Let us restrict ourselves to multiplying.

$$b \to 2 \cdot b$$

is meaningful and visually recognisable as a

dilatation

(and similarly for other factors). This mapping can be continued on the negative side as a dilatation, which yields for instance

$$2 \cdot (-3) = -6.$$

Or otherwise: it is visually obvious that a dilatation behaves symmetrically on both sides of 0, thus

$$2 \cdot (-b) = -2 \cdot b.$$

Or still differently:

$$2 \cdot b = b + b$$
$$2 \cdot (-b) = -b + (-b) = -2 \cdot b.$$

In general

$$a \cdot (-b) = -a \cdot b,$$

which can be considered as meaningful and true for positive a.

Again this is extended by recognising

$$a \rightarrow a \cdot (-b)$$

as a reversing dilatation of the positive ray and as such continued on the negative side. Thus in general,

$$a \cdot (-b) = -a \cdot b$$

and similarly

$$(-a) \cdot b = -a \cdot b$$
$$(-a) \cdot (-b) = a \cdot b.$$

Of course there are fewer words and formulae needed to visualise this. This leads to a decent motivation of

plus times plus = minus times minus = plus
plus times minus = minus.

Commutativity and associativity of multiplication are almost obvious. Distributivity, however, requires either distinguishing a number of cases or a visualisation that is not very convincing though.

15.8. Translating both models of Sections 15.4–5, counters and arrows, with each other, one can parallelise

red counters	arrow tails
black counters	arrow heads
black-red pair	opposite arrows
annihilation and creation	displacing of arrows
equivalence classes of pairs of counters	equivalence classes of arrows.

The two models differ in that

counters represent integers	arrows represent arbitrary numbers
a counter is either red or black	each point can be both head and tail;

multiplication is

    an artificial   a natural

operation.

15.9–12. *Directed Magnitudes**

15.9. As P. M. van Hiele has rightly remarked — in order to draw the didactical consequences — the didactical malfunction of the arrow model is caused by its one-dimensional character. For a didactics of directed magnitudes — as arrows are — the dimension 1 is the most inappropriate, because in one dimension these objects and their operations lack badly needed relief.

Directed magnitudes (such as speed, force, acceleration) occurred earlier in mechanics than they did in geometry. The mechanical principle by which these magnitudes become operational is the so-called

    parallelogram of speeds, forces, accelerations,

as an expression of the way to

    compose and decompose

such magnitudes.

Adding vectors in geometry starts with another pattern: by bringing one vector with its tail to the head of the other, one indeed sweeps out a parallelogram, which moreover demonstrates the commutativity of the addition (Figure 208). As a matter of fact, such a figure can degenerate, if the two vectors fall into one line and the parallelogram collapses — a degeneration, which should not be neglected didactically though there is little need to exaggerate its importance.

Fig. 208.

The most comprehensive concept to fit these phenomena is *vector space* or linear space (as called in linear analysis) over a field, say that of real numbers. A brief definition: "V is a vector space" means:

V is a commutative group with respect to an addition '+', thus

$$x + y = y + x$$
$$(x + y) + z = x + (y + z)$$
$$x + 0 = x$$
$$x + (-x) = 0$$

---

* See the note at the title of the chapter.

for x, y, $z \in V$; and V allows for multiplication $\xi \cdot$ with real numbers ξ, thus for x, $y \in V$,

$$\alpha \cdot (x + y) = \alpha \cdot x + \alpha \cdot y$$

such that

$$(\alpha + \beta) \cdot x = \alpha \cdot x + \beta \cdot x$$
$$(\alpha\beta) \cdot x = \alpha \cdot (\beta \cdot x)$$
$$1 \cdot x = x$$

These axioms can be fulfilled by a visual geometrical model with

as elements of V the arrows from a point o,
as addition the tail by head composition after parallel displacement,
as α-multiplication the geometric multiplication from o.

A mathematical model fulfilling the axioms is *n-dimensional number space* R^n: the set of ordered n-tuples of real numbers with the operations $+$ and $\alpha \cdot$ such that

$$\ulcorner \xi_1, \ldots, \xi_n \urcorner + \ulcorner \eta_1, \ldots, \eta_n \urcorner = \ulcorner \xi_1 + \eta_1, \ldots, \xi_n + \eta_n \urcorner$$
$$\alpha \cdot \ulcorner \xi_1, \ldots, \xi_n \urcorner = \ulcorner \alpha\xi_1, \ldots, \alpha\xi_n \urcorner.$$

This model is in a sense exemplary. If the vector space R has a basis

$$e_1, \ldots, e_n,$$

each vector $x \in R$ can unambiguously be represented by

$$x = \xi_1 e_1 + \ldots + \xi_n e_n.$$

Assigning to each x the system of its coordinates with respect to the basis

$$x \to \ulcorner \xi_1, \ldots, \xi_n \urcorner$$

effects an isomorphic mapping of R onto R^n.

15.10. The axiomatic version of vector space is the final result of a historical process. In a mathematical *system* it is — rightly — a starting point. The n-dimensional number space R^n is a richer structure than the n-dimensional vector space R, which in order to yield R^n must be enriched by a chosen basis. Progressing from poorer to richer structures is a sound principle of mathematical systematism. It depends on circumstances whether it is didactically as sound a principle.

If one asks for didactical starting points, there are a few possibilities.

1. Arrows in plane or space with a point o as their tail, and as addition the tail by head composition after displacement of one of the summands along the other.

2. Equivalence classes of equal and equally directed arrows in plane or space with as addition the tail by head composition of appropriate class representatives.

3. The translations of plane or space with as addition the successive performance.

For 1 it is required that plane or space are first being enriched by an arbitrary point o which finally does not matter much; this can be dispensed with by 2 and 3.

Connections between

1 and 2: Each equivalence class according to 2 is unambiguously represented by an arrow starting at o;

2 and 3: An equivalence class of arrows represents a translation tail → head.

1 and 3: The translations according to 3 have the form $x \to x + a$, to be understood according to 1.

No doubt 3 is the most elegant and at the same time the least didactical approach. A grave objection against 3 is that the α-tuple of a translation does not possess a natural interpretation until a high level is reached.

An objection against 1 is the geometrically unmotivated choice of the point o as tail of all vectors. This objection, however, can be disregarded if the intention prevails that plane or space will anyway be pictured with a coordinate system of axes intersecting in o.

To 2 the *general* objection applies which holds against the explicit use of such logical complexities as equivalence classes. If such a terminology is avoided — as was done here in Section 15.7 — this objection may be dropped. The best choice seems a compromise between 1 and 2:

> One considers arrows in plane or space, agrees on considering equal and equally directed arrows as "the same" and givens and draws them by preference with the tail at o.

This is what I called * the intentional — versus the extensional — concept formation.

15.11. One can go a long way in mathematics with vector spaces before introducing a basis and coordinates; in the theory, in particular if it is functional analysis, it is often unsuitable or even obnoxious to pass to special coordinate systems. In geometrical vector spaces, however, the coordinate free approach can degenerate into a hobby — with respect to content as well as to didactics. Certain figures ask, in order to be represented algebraically, for special coordinate systems — in this way Apollonius already used coordinates. In order to understand linear equations and solving them in the context of vector spaces, I am compelled to enrich the space with a basis. Determinants as expressions for volumes do not arise unless I choose a basis in order to substitute the coordinates with respect to the basis in the expression for the determinant. To be sure, this use of bases and coordinate systems can be wrapped into an unimpeachable garment that is adapted to "the system": a coordinatisation of an n-dimensional

---

* *Mathematics as an Educational Task*, pp. 30–32.

vector space R is nothing but a linear mapping of R on the n-dimensional number space R^n, indeed. But unless we are obsessed by formalising, we may happily forget about it in the present context.

We were induced into the foregoing phenomenological analysis of vector spaces by certain didactics of negative numbers. In Section 15.9 we took a standpoint where negative numbers were already a didactical acquisition; whereas in Sections 15.2–8 we were concerned about acquiring negative numbers, with the final conclusion that directed magnitudes was a good approach, were it not that if restricted to one dimension they create more problems than they are called on to solve. The idea of passing to the vector plane was primarily suggested here by a need to algebraise not the plane, but the line in order to yield the number line. Primordially the plane will be an auxiliary means; if it can be more, it is a bonus; even if it should dominate, it could be accepted since sometime the plane will come about anyway. But if it will, can, must come up here intentionally, it has to do so as an algebraic, coordinatised plane. Not as an axiomatic plane that is afterwards coordinatised and exemplified by an algebraic model, but as a geometric object, as the drawing plane that will be as naively provided with an origin, axes, and coordinates, as happened in history. The historically more recent turn towards the axiomatic vector space, the inversion from the numerical to the axiomatic vector space should also didactically be of a later date — after the mastery of negative numbers and algebra rather than as a certainly didactically insufficient precondition for it.

15.12. In P. M. van Hiele's newest approach negative numbers arise in a two-dimensional frame. A number pair

$\ulcorner 3,4 \urcorner$	means 3 steps to the right,	4 steps upwards
$\ulcorner -3,4 \urcorner$	means 3 steps to the left,	4 steps upwards
$\ulcorner 3,-4 \urcorner$	means 3 steps to the right,	4 steps downwards
$\ulcorner -3,-4 \urcorner$	means 3 steps to the left,	4 steps downwards.

The left–right, up–down are those of the drawing plane, with horizontal and vertical axes on which the numbers of steps can be read in units.

Performing such operations in succession one describes or prescribes rectilinearly constructed drawings in the plane. Adding these vectors is nothing but performing these operations in succession.

$$\ulcorner 3,-4 \urcorner + \ulcorner -5,2 \urcorner$$

arises in a natural way and defines as naturally what shall be

$$3 + -5 \quad \text{and} \quad -4 + 2.$$

Let us consider a translation from his textbook*:

In Figure 209 a cat's head has been partially drawn. You should finish it.

* P. M. van Hiele et al., *Van A tot Z*, 1a, Purmerend, 1976, pp. 74–75.

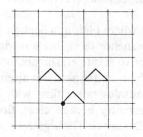

Fig. 209.

a   Draw Figure 209 on squared paper.
Go from the fat point one step to the left and one upwards.
This is indicated by ⌐– 1, 1⌐.

b   Draw then a line segment going 0 steps to the left and 2 up. We write it ⌐0, 2⌐.

c   Finish the picture by drawing one after the other the line segments ⌐1, – 1⌐, ⌐1, 0⌐, ⌐1, 1⌐, ⌐0, – 2⌐, ⌐– 1, – 1⌐, ⌐– 1, 0⌐.

d   Take a piece of squared paper, 9 steps wide and 6 steps high. Start at the point that is 4 steps from the left and 1 from below. Make a maze by drawing from this point as a start one after the other the line segments ⌐0, 3⌐, ⌐1, 0⌐, ⌐0, – 1⌐, ⌐– 2, 0⌐, ⌐0, 2⌐, ⌐3, 0⌐, ⌐0, – 3⌐, ⌐– 4, 0⌐, ⌐0, 4⌐, ⌐5, 0⌐, ⌐0, – 5⌐, ⌐– 3, 0⌐.

e   Take a piece of squared paper, 8 steps wide and 8 steps high. Start at the point exactly in the centre. Make a mill by drawing from this point one after the other the line segments ⌐1, 0⌐, ⌐2, – 1⌐, ⌐– 3, 0⌐, ⌐– 2, 1⌐, ⌐2, 0⌐, ⌐1, 2⌐, ⌐0, – 3⌐, ⌐– 1, – 2⌐, ⌐0, 3⌐.

In the two-dimensional model both the associative and the commutative law are visually obvious; the same is true of subtraction as adding the opposite. Likewise the multiplication is well visualised by a geometrical multiplication as is the distributive law. Gradually, in order to attain the algorithmic automatisation, the bonds with the visual model are loosened. Meanwhile the visually acquired insight into calculating in the extended number domain is again visually used to extend functions formerly introduced by tables or else imperfectly such as

$$x \to x - 2,$$
$$x \to 2 - x,$$
$$x \to 2x,$$
$$x \to \tfrac{1}{2}x.$$

It is an intentionally many-sided approach to calculations in the extended number system.

15.13–15. *The Geometrical-Algebraical Permanence*

15.13. I take up the thread where I left it in Section 15.2. The negative numbers were invented to lend permanence to algebraic solving methods, and the operations with them were defined to grant permanence to their properties. Afterwards they entered geometry, which as far as it was algebraic, had from olden times been concerned with magnitudes rather than with directed magnitudes. Soon their indispensability in geometry became clear, that is, in an algebraised geometry that developed after Descartes. The negative numbers would have remained a nice plaything, and the operations, motivated by algebraic permanence, rules of a game, which could have been fixed in another way, were it not that geometry had seized upon them. The negative numbers are needed to describe the whole plane by coordinates and planar figures in their whole extension by equations. The simplest figures in the plane, lines, are then translated by the simplest equations, those of the first degree, called linear because of their relation to straight lines; circles and other conics are fitted by second degree equations. I think that both in phenomenological analysis and didactics too little emphasis is laid onto this fact:

> the justification of the numerical operations and their laws by the simplicity of the algebraic description of geometrical figures and relations.

Briefly said: algebra is valid because it functions in geometry.

It is strange that so far this insight has not or not strongly enough been pronounced. In my view it is one of the objectives of algebra instruction to convince the learner of the validity of the operations and their property so forcefully that he cannot but accept them. The most convincing argument is to show him the operationality of algebra in geometry. This, I believe, should be our policy with teaching negative numbers.

Here "geometry" does not mean an axiomatic structure but what is visually obvious or conceptually follows from what is visually obvious – a visuality that neither requires involved explanations in the vernacular nor sophisticated replenishments. The one-dimensional medium, the straight line, has not enough visual structure, two dimensions is the minimum that is required, and with a view to the graphic possibilities the most appropriate medium.

15.14. I am going to elucidate the preceding claims, and I will do so by the way of a didactical sketch.

Functions like

$$x \rightarrow x - 3$$
$$x \rightarrow x + 3$$
$$x \rightarrow 3 - x$$
$$x \rightarrow 2x$$

restricted to the positive realm, show when put into a graph, the image of a *part* of a straight line. We wish their extension to be prescribed not by algebraic but by geometric algebraic permanence. Let us start with

$$x \to x - 3$$

which before the introduction of negative numbers is meaningful only for $x \geq 3$, where it is represented rectilinearly (Figure 210). The rule of extension is obvious: rectilinearity. This means geometrically and in a table (recall the inductive extrapolation)

$$x \quad = 4, 3, \quad 2, \quad 1, \ldots$$
$$x - 3 = 1, 0, -1, -2, \ldots$$

Fig. 210.

where the negative numbers are being put on the axes, albeit as abbreviations of $0 - 1$, $0 - 2$, ... Of course there is no need restricting oneself to integers. On the contrary, the replenishment with

$$x \quad = 3\tfrac{1}{3}, \ 2\tfrac{1}{3}, \quad 1\tfrac{1}{3}, \ldots$$
$$x - 3 = \tfrac{1}{3}, -\tfrac{2}{3}, -1\tfrac{2}{3}, \ldots$$

can add much to understanding.

So far negative numbers occurred as results only. If

$$x \to x + 3$$

is considered (Figure 211), geometric algebraic permanence requires negative x to be admitted to fit the algebraic expression to the geometric image

$$x \quad = 1, 0, -1, -2, \ldots$$
$$x + 3 = 4, 3, \quad 2, \quad 1, \ldots$$

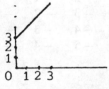

Fig. 211.

In fact the same step is required with $x \to x - 3$ as soon as the graph is to be extended left of the "vertical axis" (Figure 212). In

$$x \to 3 - x$$

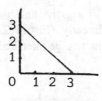

Fig. 212.

there is a need for extension on both sides since the graphical image is only a line segment unless negative numbers are being introduced. Maybe by this very fact this function if especially appropriate to be used to start with:

$$x \quad = 2, 3, \quad 4, \ldots$$
$$3 - x = 1, 0, -1, \ldots$$

on the one side and on the other

$$x \quad = 1, 0, -1, \ldots$$
$$3 - x = 2, 3, \quad 4, \ldots$$

By the preceding exposition addition and subtraction in the extended realm are completely defined, at least if the commutativity of addition is assumed. The introduction looks somewhat disintegrated but the operations can be integrated in tables:

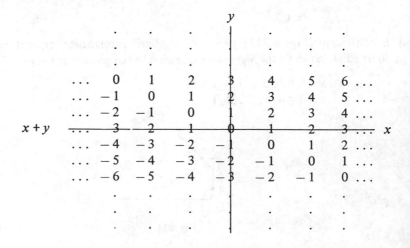

					y					
	...	0	1	2	3	4	5	6	...	
	...	−1	0	1	2	3	4	5	...	
	...	−2	−1	0	1	2	3	4	...	
$x+y$	...	−3	−2	−1	0	1	2	3	...	*x*
	...	−4	−3	−2	−1	0	1	2	...	
	...	−5	−4	−3	−2	−1	0	1	...	
	...	−6	−5	−4	−3	−2	−1	0	...	

			y			
·	·	·		·	·	·
·	·	·		·	·	·
... −6	−5	−4		−2	−1	0 ...
... −5	−4	−3		−1	0	1 ...
... −4	−3	−2		0	1	2 ...

$x - y$

... −2	−1	0		2	3	4 ...
... −1	0	1		3	4	5 ...
... 0	1	2		4	5	6 ...
·	·	·		·	·	·
·	·	·		·	·	·
·	·	·		·	·	·

x

(In the case of $x - y$ it is recommended to place no numbers upon the axes.)

From this representation it is a small step to arrive at nomograms for $x + y$ and $x - y$ (Figure 213) which with necessary interpolations define addition and subtraction completely.

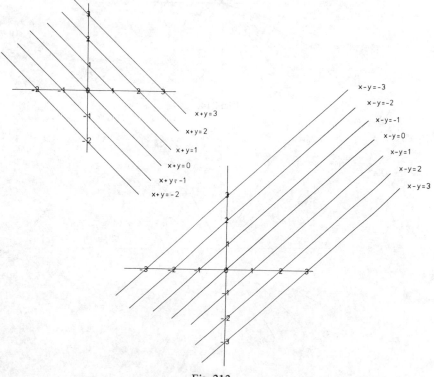

Fig. 213.

The relation

$$x - y = x + (-y)$$

is illustrated by the opposite slope of the nomograms for $x + y$ and $x - y$.

The multiplication can be dealt with in a similar way.

$$x \rightarrow 2x \quad \text{and} \quad x \rightarrow \frac{1}{2}x$$

initially are defined only for positive x and thus graphically represented by half lines. Geometric algebraic permanence requires that the representations is extended (Figures 214–215) such that

$$2 \cdot -1 = -2, \qquad 2 \cdot -2 = -4, \quad \ldots$$

$$\frac{1}{2} \cdot -1 = -\frac{1}{2}, \qquad \frac{1}{2} \cdot -2 = -1, \quad \ldots$$

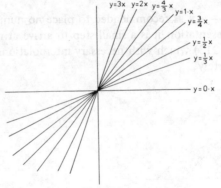

Fig. 214.

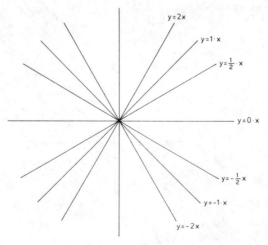

Fig. 215.

or in general, for positive a and b,

$$a \cdot (-b) = -a \cdot b.$$

This leads to a nomogram of the multiplication by positive numbers, which again according to geometric algebraic permanence can be filled up with lines in the other quadrants.

This then yields the general validity of

$$a \cdot (-b) = -a \cdot b.$$

What about the insight into the arithmetical laws in this approach? The commutativity of the addition is visually obvious, the associativity appears from the nomogram of $x + y$:

substituting y by $y + z$ means
shifting the lines over z,

of course specified by numerical values of z.

Associativity and commutativity of multiplication, once supposed for positive factors, cause no problem. The commutativity is illustrated anew by the symmetry of the nomogram for $x \cdot y$ (Figure 216).

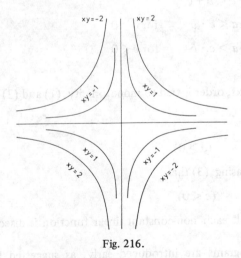

Fig. 216.

15.15. Rather than leaving it here, I would immediately put the rectilinearity of the graphs of

$$x \to ax + b$$

(for paradigmatic numerical values of a and b) into this context and deal with the continued harmony between geometry and algebra as a teaching matter as well as a means of motivation. I would do so each time in the sequel if a simple figure is graphically represented by a simple equation.

15.16. *Order*

So far in this chapter I have neglected the order relation. Without a model, say

$$-3 < 5,$$

can reasonably be argued,

$$-5 < -3$$

is more difficult unless it is based on

$$-3 - (-5) = 2 > 0,$$

that is requiring in general

$$a < b \leftrightarrow b - a > 0.$$

In each geometric model the algebraic order is satisfactorily motivated by the geometric order of below–above (on the thermometer or in the staircase) or left–right (on the number line). Likewise the behaviour of the order relation under adding and multiplying

(1) $a < b \rightarrow a + c < b + c$

(2) $a < b \rightarrow c \cdot a < c \cdot b$ for $c > 0$

(3) $c \cdot a > c \cdot b$ for $c < 0$

is visually obvious.

In the graphic context, order is tied to monotonicity, (1) and (2) express that the functions

$$x \rightarrow x + c$$
$$x \rightarrow c - x \qquad (c > 0)$$

are monotonically increasing, (3) that

$$x \rightarrow c \cdot x \qquad (c < 0)$$

is decreasing. In general, each non-constant linear function is discovered to be monotonic.

If graphs and nomograms are introduced early, as suggested by Sections 15.13–15, there is another opportunity to demonstrate the algebraic-geometric harmony, that is defining sets by inequalities: besides the line represented by the equation

$$2x - 3y = 5,$$

the half planes represented by the inequalities

$$2x - 3y > 5,$$
$$2x - 3y < 5.$$

15.17. *Reasoning Back from the Goal*

After the preceding analysis we should keep in mind that the instruction on negative numbers, if successful, should lead to automatisms. How does a person who has become familiar with negative numbers, manipulate them, how does he solve problems? I think the answer is so obvious and so unambiguous that I may draw general conclusions from my own procedures to cover those of other people and even of computers.

Let us first classify the possible additions and subtractions paradigmatically.

$$7 + 3 \qquad 3 + 7 \qquad 7 - 3 \qquad 3 - 7$$
$$7 + (-3) \qquad 3 + (-7) \qquad 7 - (-3)$$
$$-7 + 3 \qquad -3 + 7 \qquad -7 - 3$$
$$-7 + (-3) \qquad -3 + (-7) \qquad -7 - (-3)$$

The first line consists of problems which, except for the last, are already meaningful before the extension of the arithmetical domain; the next lines contain systematic variations, which become meaningful only after the extension with negative numbers. The transformation rules I apply here, are

$$\ldots + (-a) \to \ldots - a \qquad \text{for } a > 0$$
$$\ldots - (-a) \to \ldots + a \qquad \text{for } a > 0$$
$$a - b \to -(b - a) \qquad \text{for } 0 < a < b$$
$$-a + b \to -(a - b) \qquad \text{for } 0 < b < a$$
$$-a + b \to b - a \qquad \text{for } 0 < a < b$$
$$-a - b \to -(a + b) \qquad \text{for } a > 0, b > 0$$

with the common pattern

$$a - b = -(b - a).$$

And of course for the multiplication

$$(-a) \cdot b \to -a \cdot b \qquad \text{for } a, b > 0$$
$$a \cdot (-b) \to -a \cdot b \qquad \text{for } a, b > 0$$
$$(-a) \cdot (-b) \to a \cdot b \qquad \text{for } a, b > 0.$$

These are not excessively many, nor complicated, transformation rules, particularly if seen in the common pattern of the quartet of addition and subtraction. It is almost nothing compared with the rules a child must learn in order to master column arithmetic. What does this mean for the didactics of this field?

That after all it would be the most simple and efficient way to programme the learner with these six rules? This is all he needs, or isn't it? It is a question I could have asked at many opportunities before and could ask as often in the sequel. Though, there are reasons why it is asked right now. Elementary arithmetic is blessed with a cornucopia of such transformation rules, from arithmetic below 10 with all its connections between additions, subtractions, and equations via column arithmetic to the various operations with fractions.

Some of them are of the kind that every child — not only the bright learner — can discover them if he is allowed to do so. Other rules are adapted to more or less steered reinvention. But among them there are also a few that depend to such a degree on appropriate and appropriately motivated definitions that they cannot be understood unless their definitory character is made explicit, with or without motivation. This holds in particular for multiplying and dividing fractions.

It goes without saying that we strive for automatisation where it is required. The problem of whether automatisms are better pursued by goal directed programming or as a result of learning by insight, is not at all original. There are plenty of answers to this question, by way of theoretical expositions, of textbooks, of teacher guides to textbooks, and by the ways individual teachers practice teaching. Theoretical expositions and teacher guides no doubt emphasise learning by insight even if — as it happens sometimes — they are belied by the underlying textbooks. The actual policies in the classroom are certainly covering a broad scale, from downright programming to learning by insight to attain automatisation. "First doing then understanding" is the catchword of one of our textbooks for secondary mathematics. To put it the other way round may be a honest intention, or a cheap philosophy, or a swindle. "First doing then understanding" can mean that some never achieve more than — imperfect — doing, while understanding is an objective reserved for the privileged ones. "First understanding then doing" can amount to verbalistic understanding that does not lead and cannot lead to doing.

It is a good thing to fix the attention on three points while discussing the learning of automatisms:

first, are the intended automatisms
    worth being learned,

second, once learned,
    how well do they function,

third,
    how much value is the *process* of learning them?

These three questions are closely connected with each other. Clearly it is no use learning automatisms

    that are seldom applied and soon forgotten,
    that are too complicated to be safely manipulated,
    that can easily be confused with each other

unless the way of the check by insight of those automatisms remains open. On the other hand too much insight during the learning process can block the automatisation.

Is it in the era of the handheld calculators still meaningful to learn column arithmetic? Well, if at the critical moment the calculator fails, what is one to do? Then automatisms learned by insight might help better than programmed ones.

Tables of multiplications are most often learned by insight: the learners are taught to build them by successive additions. After memorising, however, the insight that multiplying comes down to successively adding, fades away. Why? Because multiplication as repeated addition is only used to build up the tables, in order to be repressed afterwards if column multiplications are learned and exercised.

The consequences of column multiplications learned by being programmed – rather than by insight – are particularly striking with, say 100, as first or second factor. But the same features can be observed if multiplying was learned by insight and has been repressed in the process of automatising.

Learning by insight, if restricted to an *introduction* by insight, does not deserve this name; the didactic of teaching automatisms should rather be turned to not repressing. This requires a more broadly flowing process rather than the narrow one of learning algorithms. In the process that finally leads to acquiring algorithms, more and other things should be learned than only that which can aid the acquisition of automatisms.

Once more, why are these questions asked here? According to the tradition negative numbers are not part of the teaching of arithmetic. Arithmetic is usually taught and learned in narrow goal directed learning processes, though sometimes a bit broadened by – again too narrow – applications. A pupil who has gone successfully through these learning processes, may not, in general, be supposed to have experienced them consciously enough to be able to reconstruct them. This kind of learning arithmetic has created a mentality where being programmed dominates insight. It takes enormous efforts to break open the automatism to have it understood by insight. Some textbooks at the secondary level try it; at teacher training institutions one exerts oneself to have the future primary school teacher experience consciously the processes that can promote learning by insight.

After this long introduction, what can be said about the pros and cons of learning arithmetic of negative numbers by being programmed or by insight? The answer depends on the place negative numbers occupy in the total learning process. If, after years of instruction in arithmetic negative numbers is the first opportunity to learn meaningful, not yet algorithmised mathematics – which is the normal situation – the answer must be different from what it might be if the learner has already acquired a mathematical attitude.

The operations on natural numbers are usually learned by insight though it is a fact that often because of the lack of relations with reality the original insight is suffocated by automatisms. Fractions is an opportunity to justify the operations and the rules governing them, which, however, is most often disregarded. Mathematics – at least algebra – is too often presented as a system of working according to prefabricated and then imposed rules. There is everything to be said in favour of preventing this impression as far as the learner is concerned.

Certainly one is allowed to interpret mathematics *also* in this way, as a formal system, as a game played according to certain rules, or even as a language constructed by arbitrary syntactic rules. This is a view of mathematics that under certain circumstances — foundations of mathematics — may be useful. It is a view worth being noted by somebody who masters mathematics at a high level. It is a *view* of mathematics, yet it is no *mathematics*, and certainly it is no mathematics to start with as a learner.

However one proceeds in extending the number concept, it is a necessity that the fact and the mental process of extending are made conscious, that it is made conscious why one has the extension take place in this, and in no other way, and that as a background of the compulsory, the arbitrary in the definitions becomes clear. To get this done — I think — negative numbers offers a good opportunity, at least if others have been missed. Indeed, the rules of operating on negative numbers are on the one hand relatively simple, wheras the constraint to work with negative numbers just this way, is remarkably strong, provided this rather narrow learning process is embedded in a broader current, like the one I have sketched before.

With these remarks I conclude this chapter. Again I apologise that I dealt with a question of mathematical principle not in a general context but at the opportunity of some special subject.*

---

* Compare Section 6.19.

CHAPTER 16

THE ALGEBRAIC LANGUAGE

16.1–16.4. LANGUAGE IN THE BROAD SENSE

16.1. *Language, Languages, Translation*

The singular "language" is to obscure the fact that there are no two people speaking the same language. Even one individual can use various languages – not only if he masters foreign tongues, but also depending on with whom and how he communicates, orally or in writing, talking, calling, crying, whispering, singing. Of course that individual will also listen to various languages. Learning one's mother language is communicating with one's environment in two widely differing languages, with adaptations which in the course of the time increase at the side of the learner and decrease at that of the people he communicates with.

Learning to read means initially becoming acquainted with a language cut into letters, syllables, sentences, which compared with the spoken and heard language is extremely poor with regard to form and content, until the printed text starts carrying on words and constructions unknown and unheard in the spoken language: as far as the level is concerned, the printed language overtakes the vocal one.

When I put it – a bit provocatively – that no two persons speak the same language, I used the word "language" in an improper sense, that is, not as I do if I oppose, say, English and French. Between these two meanings of "language" there is a scale of gradations depending on the means of expression – spoken and written language – or on the environment – dialect and educated language, childrens' language, boys' language, girls' language, bookish language, teachers' language, church language, thieves' language, secret languages – and on the subject matter – lawyers' language, chemistry language, mathematics language.

Rendering the same content in another language is called translating, though in the relation between spoken and written language it is called writing and reading, in that between a language and a crypto language it is called coding and decoding, and in the relation between people speaking "the same language" it is "transforming". Transforming is partly a lexicographic concern – a word or combination of words replaced with a synonymous one. Partly it is structural and this is where so-called transformation grammar comes in.

16.2. *Transformation Rules*

In the course of individual development the first linguistic transformations are determined by what I have called change of perspective. The example

461

to the question "where do you live?" a child (2; 8) answers "I lives [sic] there" (showing with her finger)

demonstrates a transformation of person description: you → I. These transformations form a whole bunch: I, me, my → he, him, his are examples, which must be replenished with proper names, other pronouns, personal descriptions by means of appellatives, or of a relative structure like

> the thief,
> the man who has stolen the bike.

The first transformations by change of perspective are as such formal, the more advanced ones show at least some formal structure.

By change of perspective

> here and there,   inside and outside,   above and below

change their meaning according to who pronounces them or is considered to pronounce them, and correspondingly they are to be interchanged in communicating.

Of course these are not the only transformation phenomena: parts of clauses, replaced with others, singulars with plurals and conversely, present tense with past tense, active with passive, creating comparatives and superlatives, transforming verbs into nouns, and so on — these are just a few examples, which could be multiplied *ad lib*. In particular the transformation into the interrogative is quite productive at a certain age: it can be a pleasant game to have every sentence of the partner followed by one that starts with "why", "how", "what", "when".

Our languages are so irregularly constructed that formal transformation rules cannot be purely formal — "no rule without an exception" holds in particular for this kind of rule. It is a mystery as to how the learner finds his way in this labyrinth, how he learns to speak well and judge a language — at least if one does not estimate the influence of formal rules on learning to speak (and to act) at its true rate, if one does not consider the rules within their proper frame. Languages are being learned within a *factual* context, in close connection to the facts of the human, objective, and literary environment. This is a rich context, and it is particularly so in the course of learning one's mother language; when learning "new languages", in whatever of the earlier senses, the learner can rely, by translation or transformation, on previously learned languages. The wealth of context can extend from the human and objective environment to the literary one. A more complex language — more complex with regard to vocabulary or syntax or transformation rules — is required to be learned in a richer context than a simple language. The more formal a language is the poorer the context can be, were it not that a large number of transformation rules demands, to be handled, a reorientation towards the factual context.

16.3. *Formal Languages*

According to the wealth and kind of structure there is a broad scale of languages. The language of pictograms such as used by, say, the Netherlands Railways on the stations does not exhibit any structure; the language of traffic signs contains a few structural elements, for instance the speed signs, by combining a general pattern with a particular number; flections, conjunctions, sentence structure are structural devices of what is commonly called language, with moreover the punctuations in the written, and the intonations and pauses in the spoken language. At the other end of the scale are the totally formalised languages, which will be dealt with later on.

Pictograms are symbols which by their exterior betray what they mean (or at least by the designer are judged to do so). In order to grasp the meaning of words and sentences one needs linguistic experience; structural elements in the language make it possible to understand linguistic utterances, words, clauses, sentences, never met with before. On the other hand useful reproduction, production and creation of linguistic utterances is only possible if one understands what the meaningful linguistic elements mean and masters the functioning of the structuring elements. It is true, however, that while observing the structural requirements one can produce linguistic utterances that do not mean anything nor are intended to do so, and this is the easier the more structure the language possesses.

A language is purely formal if its utterances can be handled, imitated and tested to see whether they are correct (that is, exhibit the required regularity), without paying attention to their meaning, which is perhaps even absurd. Under this label of formal language much can be comprised that in the usual terminology does not deserve the predicate of language: for instance the game of chess with the possible chess positions as linguistic utterances and the rules of the games as transformation rules, which in each thinkable position define what is a new admissible position obtainable from the given one by a move. The rules do not tell what is a *good* move in a given position as little as the rules of grammar give you any information about the value of a linguistic utterance.

There is a lot of theory in formal languages. Rules describe how words and sentences are built from elementary signs, how from a sentence accepted as true (or otherwise accepted) new true (or accepted) sentences are derived. No concepts need be associated with the sentences; the mere form of signs, words, sentences determines what can and may be done with them — a work that can be performed by a computer.

16.4. *Arithmetical Language*

This reminds one of the arithmetical work done by handheld calculators or, if you prefer it, by well-trained human calculators. Arithmetic is to a certain extent, which depends on the calculator, such a formal language. Problems have to obey certain rules.

$$7 + 5 =$$

is admissible while

$$7 + = 5$$

is not. The counting sequence is a formal system — as it were a story where each "word" produces the next by necessity. There are formal rules to tell you how to translate a vernacular numeral into an algorithmic one, an "Arabic" into a "Roman" one, and conversely — at least for not too large numbers.

But there is more to it. Beyond the fact that arithmetical problems are formal data the form of which can be tested, solving such a problem is a formal business, that is, one that goes on according to certain rules. Of course not from the start onwards.

$$7 + 5 = 12$$

is acquired by insight and finally engraved on the memory to such a degree that it becomes an automatic linguistic utterance such as continuing "I pledge a . . ." with "llegiance". And so it happens with the tables. To be sure

$$37 + 25 =$$

is another case. The units and tens are added according to rules acquired by insight or decree until these rules too are engraved on the memory. Solving has then become a purely formal process. This extends similarly to column arithmetic. In order to calculate safely and quickly one is advised not to attach any meaning to symbols and operations. According to fixed rules well-shaped utterances of arithmetical language are transformed into others. A lot of training is required to attain this goal. The rules are not that simple — in particular for long division and fractions — the insight can be of great help to reconstruct, distinguish and correctly apply them. The computer can do it without this kind of insight. Because of their small number its switches function better than the associations in the human brain; moreover unlike the human brain it is intentionally designed to perform certain programs.

The human calculator, however, is prepared for tasks that are not the electronic one's virtue, that is, use opportunities to calculate handily, and though computers can be programmed to perform the same kind of tricks, one will in general refrain from it, because it is unnecessary or unnecessarily expensive. According to their size and means of being programmed, computers have a restricted capacity and however this is increased, one can always create situations where man uses better his likewise restricted capacity. A special case is word problems; even if a whole dictionary is stored into the memory of the computer, one can dress the simplest arithmetic problems in a fashion that the computer cannot handle them. As soon as applications are at issue, arithmetic starts ceasing to be a formal language.

Many one will not agree if I interpret figuring out such problems as transforming the given expression into its result. An oldfashioned view — many one will say. Of course I know — and I stressed it before — that the equality sign is understood to mean an identity: at its left and right are names of the same thing. It is just the same whether I say

$$7 + 5 = 12 \quad \text{or} \quad 12 = 7 + 5.$$

It is still the same if I say

$$7 + 5 = 5 + 7 \quad \text{or} \quad 12 = 12, \quad \text{and so on.}$$

Well, it is an objective fact that $7 + 5$ and 12 are the same thing, as for Edgar Allen Poe it was an objective fact that the killer of the rue de la Morgue was an ape. The only thing that mattered was to discover this objective fact. Someone who does not yet know that $7 + 5$ and 12 are the same thing, understands the equality sign in another way, to wit, as a command to write behind

$$7 + 5 =$$

something that, for more or less intelligible reasons, will be approved. $5 + 7$ will certainly not, though it is correct, and $6 + 6$ will not either. This then — being approved — is the meaning of the equality sign in this kind of arithmetical language. On this level arithmetic is not yet purely formal. Formally one could write after the equality sign $5 + 7$; it yields a correct equation. In the arithmetical language the equality sign invites a transformation procedure, and one has to know which one. Right of the equality sign is the place of the "result". But what does it mean: result? In

$$7 + \cdot = 12$$

the "result" is 5, whereas the figure behind the equality sign is 12. The fact that a few pupils are able to grasp this, may not obscure the even more important fact that the majority is not, at least at the age where it is being taught.

For many pupils the result of a division is the remainder. As late as the seventh grade I came across pupils that answered $8 \div 4$ by 0. No doubt this is a domain where the traditional arithmetical language shows serious defects.

I will try to sharply outline the problematic of arithmetical language.

16.5–16.8. LANGUAGE AS ACTION

16.5. *Performing Tasks — Answering Questions*

The learner performs tasks he is given by others or by himself. I order him to count, and the oral or written number sequence means performing the task, the activity of counting is documented by its linguistic result.

I let him count *something*. The task can be formulated either "count the marbles", or "how many marbles is this?" In the second case the task is implicit, at least if the child knows that in order to answer the question, he must count. Anyway the task is performed by counting. The linguistic document can show variants: the produced counting sequence or the number of marbles counted.

Operations is a similar case. A *task*

add 3 to 4,   take away 4 from 7,

or the question

how much is 4 + 3,   how much is 7 − 4?,

most often abridged to

4 + 3 =,   7 − 4 =

with, if it is asked orally, in the "equals" the intonation of interrogating. One should expect

after the task: the performance,
after the question: the answer.

In the first case the command "add" or "take away" should indeed stimulate the activities of adding and subtracting. But the performance of the task can also be documented by the result of the problem, and in general the one who sets the task will be satisfied unless there are special reasons why he would ask the pupil to perform the task explicitly, for instance if the alleged result is wrong or if he would like to check whether the pupil has not guessed it, or if he is for theoretical reasons interested in the pupil's procedures.

It depends on the pupil's level what follows on the question '4 + 3', '7 − 4'. If he has not yet memorised the problems, he is obliged to instruct himself to add 3 to 4, to take away 4 from 7 − with concrete material, or mentally.

A six-year old who had trouble with arithmetic, communicated at a certain moment to her father the trick she had found out to do 4 + 3 − counting further from 4: 5, 6, 7. She considered it as a ruse, since she pretended she knew it whereas in fact she figured it out.

All that we demonstrated with counting, counting something, and simple arithmetic, repeats itself again and again at every higher level and with more complicated material. The task that is given or that is performed to answer a question can be composed of more or less explicit partial tasks; which in turn can be subdivided − a long division with a divisor of more than one digit is an example of an extremely involved task, with branchings depending on the circumstances. At present such tasks are schematised by flow diagrams, for instance in order to program people or computers.

But let us return to our starting point: the double dichotomy

task   − performance
question − answer.

Task → performance is the most primitive relation. The task will mostly be given by a linguistic utterance though it may also be a gesture, whereas the performance need not contain a linguistic element other than the comprehension of the task. If, however, this linguistic element is present, it may consist of the documentation of the performance by means of the answer to a question substituted for the task.

Question → answer can often be a direct reaction, for instance the answer to the question "How old are you?", "How much is 10 × 10?", "How many days does the week have?". The questioner may mean a task — "May I get 10 postcards?" — or he is sure that the question illicits a task — for instance in the arithmetic question "How old were you, when the war began?" or "How much is 37 × 37?". The question can be understood as a *question*, requiring an answer obtained via a task and its performance; it can also happen that the question is interpreted as a *task* to be performed while the result of the performance is not translated into an answer — a well-known phenomenon, if after the question "How many marbles?" the child has counted without formulating an answer.

16.6. *Knowledge of Facts and of Procedures*

No cognitive teaching area is less concerned with the problem of what is called ready knowledge in instruction than is mathematics. Questions about a phone number, about the way to go to some place, about the spelling of a word, about a year, about the name of a plant, about the meaning or translation of a text, can more often be answered by looking up, consulting sources of information than is the case with questions of mathematical origin or in which mathematics is involved. Well, even in such cases it may be an art to know where to look up, which sources of information to consult, how to use the means of information, what to do with the information retrieved and how to know whether you can trust it. On the other hand mathematics too cannot dispense with ready knowledge or knowledge from sources of information — the elementary additions, the tables of multiplication, formulae, tables of all kind of functions, and computer solutions. But as far as this is ready knowledge, it is easy to be tested, and as far as sources of information are concerned, reliability and standardisation are so strong that consulting plays a minor part in the process of problem solving.

As a matter of fact, if in mathematics ready knowledge is called upon, it is the solution procedures that really matter. Though such things are difficult to estimate, I would guess that outside mathematics and its applications ready knowledge of facts has by far the upper hand of knowledge of procedures, while in mathematics the procedures look relatively simple, though one is easily inclined to underestimate their length and depth.

16.7. *Procedures as Linguistic Transformations*

In Section 16.3 we already discussed a peculiarity of mathematics as compared to any other subject. Mathematics can be more adequately formulated even in linguistic respect than any other knowledge. Of no physical or mental objects can such a structurally simple and precise description be given, which moreover disregards the diversity of languages, as can be done for the mathematical objects; nowhere can operations on objects as adequately rendered linguistically as those on numbers. Even the mental objects of geometry can, though structurally not as simply, be described with all desired precision by the conventional terms used to indicate them in the various languages. This is one of the reasons why mathematics is very often identified with its linguistic expression. Both understanding and misunderstanding about what is mathematics, feelings of power as well as of helplessness can be the consequence.

If a child has understood elementary arithmetic, it enjoys mastering it; the ease of answering questions invites the desire to have them posed. He soon forgets how much trouble the acquisition of this ability has cost. Meanwhile the subject matter is extended: ever more experience is gathered about transformation rules that turn queries into tasks, by which tasks are split into partial tasks, to be performed, in order finally to answer the questions. The data are of linguistic character and gradually the whole solving procedure gets the character of a linguistic transformation or a sequence of linguistic transformations, which are mostly simpler and anyway more formal and regular than the transformation rules of the vernacular.

16.8. *Formalising as Means and Aim*

The ease of formalising is a striking, though at the same time misleading feature of mathematics, a virtue that, as with other virtues, can turn into a vice. The ease of formalising gives us a feeling of power that can hardly be overestimated and certainly not be disregarded with impunity, though it is one that can turn against mathematics. For the majority who have got into contact with mathematics, it is mastering (or in fact not being able to master) formal rules. What to do about it? Desisting from teaching mastery? This would be a preposterous solution.

The rule of "multiplying by 100 by appending two zeroes" (which has to be modified if decimals are concerned), the rule of so many places after the decimal point if decimal fractions are multiplied, which competes with that for adding decimal fractions, the mechanism of long division with all its ins and outs — shouldn't one be glad if this functions well? Whoever was able to learn and apply this correctly, is certainly able to find out where these rules come from and how they can be justified. Let us be cautious: maybe he would be able to. Yet even the faintest idea that such a question admits of, let alone, asks for a rational answer, is absent in the mind of the great majority of those who master these formalisms. Even student teachers who are expected in the future to teach such formalisms often do not grasp the aim of racking one's brain about formalisms one masters perfectly (or thinks so). After a long division, performed in all detail, the question what is a division, is shelved as irrelevant. The usual explanation is that they have learned it as a mere trick and that is all they know about it. Is it really true that all who react this way have learned the formalisms as mere tricks?

This is a rhetorical question. I have observed, not only with other people but also with myself — I mentioned it in Section 15.17 — that sources of insight can be clogged by automatisms. One finally masters an activity so perfectly that the question of how and why is not asked any more, cannot be asked any more, and is not even understood any more as a meaningful and relevant question.

This looks a natural course of affairs, and nevertheless one would like somehow to change it. It is a problem that does not concern only arithmetic. Somebody once said: "When calculating starts, thinking finishes." No, it need not be true, unless one blocks the way back to insight.

The didactical mistake resides in the principle of once learning by insight, and then irrevocably going forward to the automatisms.* A variant is: back to the start by insight as soon as something goes wrong. This variant is better, though still unsatisfactory. Even if the formalism functions reasonably, the teacher or the one who defines the instruction should avail himself of each opportunity to return to the source of insight. Such opportunities exist on each level, for instance. As regards column arithmetic, if procedures of measuring are discussed or (in algebra) if powers are dealt with, or if the technique of brackets is taught, or if arithmetical laws are formulated: write numbers and operations in the form

$$(a_m 10^m + \ldots + a_0) + (b_n 10^n + \ldots + b_0),$$
$$(a_m 10^m + \ldots + a_0) \cdot (b_n 10^n + \ldots + b_0).$$

There is no need to pass to other number systems in order to restore preformalist insight. Other number systems are *ad hoc* creations, which can be viewed by children as irrelevant games or dead ends rather than as an intention to lead them back to the sources of insight.

The didactical necessity, which I stressed, is a consequence of the high degree of formalising of the language of arithmetic. It is as strongly felt in the didactical phenomenology of other formalised languages. In the normal use of less formalised languages, such as mother language and foreign languages, the form is so strongly attached to the content that it can permanently be checked by means of the content; in general it would even be impossible to handle the linguistic form without content support.

Here I disregard cases of autonomy of the linguistic form, such as cultivated in poetry and philosophy as phenomena of a use of language which does not aim at, or is not appropriate for, unambiguous communication. I also exclude as an aim cases of formalising, where language itself becomes a matter of study. I restrict myself intentionally to didactical cases where learning formalising

* See also Sections 6.19 and 15.17.

and formalisms is a didactic necessity, though a necessity among, and for the benefit of, others.

A characteristic feature of mathematics is a line of progressive formalisation. Transformation rules which have, or have not, been acquired by insight, are generalised in order to solve problems with a greater efficiency while using the acquired apparatus, while by this activity new more complex transformation rules are formed and again generalised, which continues the same way; progressive formalisation leading to ever and ever more radical shortcuts.

16.9–16.15. CHARACTERISTICS OF THE LANGUAGE OF MATHEMATICS

16.9. *Algorithmic Construction of Proper Names*

The most important source of progressive formalisation is the algorithmic construction of vocabulary, the first and foremost example of which is the construction of proper names of natural numbers.

Algorithmic features are not unusual in the vocabulary and the syntax of whatever language, though they are incidental and unsystematic: plurals, past tenses, comparatives, word compositions, sentence patterns to be filled out. None of them approaches remotely the systematic structure of the *numerals*. Certain historical remnants of irregularity — mostly with respect to small numbers — have wholly been eliminated in the *digital* language of the decimal system. Already the pre-positional systems knew a decent regularity; in a positional system one can, starting with a small stock of digital symbols represent all natural numbers according to strict algorithmic rules. This trend continues if the number concept is extended; synonymy of fractions is tied to rigid transformation rules.

Likewise for the names of tasks or statements such as "add three to four" (in the form $4 + 3$) or "four is greater than three" ($4 > 3$) the algorithmic construction dominates, though there may be a variety of names for the same task or statement.

According to how complex objects, tasks, statements are, the names assigned to them need a certain structure, which will now be discussed.

16.10. *Punctuations*

The natural languages have developed a large number of structuring devices such as prepositions, conjunctions, affixes, suffixes, subordination of clauses, and so on. Moreover *spoken* language uses for structuring such devices as pauses and intonation, whereas in the written version structure is more or less adequately indicated by punctuations. The most explicit structuring element in the language of mathematics are brackets of various kinds. Moreover there is a lot of implicit structure: in performing a task or reading a statement some operations take precedence over others — multiplication over addition.

Explicit punctuations can usually be dispensed with in the vernacular. Often criteria of content rather than formal ones decide the structure. In the sentence

There were aged ladies and children in the bus

the "aged" refers to the ladies only, whereas in the formally almost equivalent sentence

There were aged ladies and gentlemen in the bus

the "aged" is very likely to include the gentlemen. (Spoken language is more sophisticated than the written one; the two sentences are probably pronounced with a different melody.) In

We visited Dutch towns and villages

and

We got lessons in Dutch history and mathematics,

the "Dutch" has different domains. A well-known example is

pretty little girls schools

which according to the places of the − lacking − brackets can have 17 different meanings.

In mathematics more care is practised.

5 times . . . 3 plus 7

must be distinguished from

5 times 3 . . . plus 7

and this distinction is formalised by putting in the first case the 3 plus 7 between brackets. I should do the same in the second case with the 5 times 3 were it not that a convention says that multiplying takes precedence over adding and subtracting. I already pointed out that spoken language can be more sophisticated than written one. In the last two examples I tried to render pauses and intonation by dots; in the written language no similar devices are available. We must rely on understanding the content in order to grasp the syntactical structure. Rather than from the structure of the linguistic utterance, I grasp what "aged" and "Dutch" include, from the content that has to be meaningful.

In the language of mathematics meaning is not a reliable criterion. In fact,

5 times . . . 3 plus 7

is as meaningful as

5 times 3 . . . plus 7;

it must be perfectly clear which is meant if the expression is to be used. This is attained by strict punctuation rules, strict indications as to how an expression should be bracketed and read. If the mathematical rules were to be adopted, one would write

> (aged ladies) and children,
> aged (ladies and gentlemen).

This, however, does not happen, which is one of the big differences between the vernacular and formalised language.

I pointed out that the language of mathematics possesses more devices of structuring than bundling brackets, in particular the precedence of some operations over others if a task is to be performed or a statement to be read. These too are formal devices of structuring, in contradistinction to those of content used in the vernacular. I do not explain details, which I actually have done earlier.* What matters here is to map out the differences between vernacular and the language of mathematics.

I must, however, stress that the system of structural rules in the language of mathematics is not as simple as one would believe at first sight. Troubles that learners experience with the language of mathematics, can be explained at least partly by the lack of insight of textbook authors and teachers into the complexity of the matter. As those who are responsible for the instruction of this matter are not conscious enough of these details, they lack the insight into the possible sources of mistakes.

It is a remarkable fact that children do learn their mother language from people who have never thought about the structure of this language; no theoretical analysis of the mother language is required in the intercourse with children who are to learn their mother language; it is even thinkable that an adult does not become conscious about such features as, say, weak and strong verbs until he hears children commit mistakes; that he is not puzzled by polarities such as "yesterday—tomorrow" until he observes the uncertainty of children about this pair.

The case of the language of mathematics looks a bit more like that of learning foreign languages: in order to guide the learner, the teacher should master the matter more consciously than the parents do with the vernacular that they are to deliver — this will be clear without specific examples.

This, however, does not mean that the highly involved system of mathematical-linguistic structure rules should be made conscious to the learner. In remedial teaching this can become a necessity, in particular if strongly ingrained habits have to be driven out. Primordially the rules for applying structuring devices (as well as transformation rules) should be learned by *using* them — consciousness is required for fighting *misuses*.

Though I do not repeat my former phenomenological analysis of structuring

* *Mathematics as an Educational Task*, pp. 304—311.

devices in mathematical language, I do not wish to have it disregarded. On the contrary, I would urgently recommend textbook authors to pay attention to it and to realise and show that learning this linguistic element requires a more conscious didactics.

16.11. *Variables in the Vernacular*

The long step from arithmetic towards algebra (from primary to secondary instruction in mathematics) is calculating with letters rather than numbers; "letters reckoning" used to be a familiar term. It looks a superficial phenomenology to put it this way, but in phenomenology profundity starts at the surface.

How letters entered mathematics, how halfconscious of what they did, people started calculating with letters, how the increasing consciousness influenced didactics of mathematics in the present century, how these precious attainments have been smashed to smithereens by the set theory rage of New Math and what is required to clear away the rubbish — I have told this in the past* though not as a connected story, but dispersed over various places according to which detail just mattered. So I have to start anew.

In the last few sections I used the term "name" in a somewhat broad sense: I attributed names not only to objects but also to acts, tasks, statements, even though such names could have the linguistic form of sentences. In traditional linguistics one knows the term "noun" and nouns are distinguished as proper names and generic names (appellatives). It is difficult to draw a sharp borderline between both of them. "John" and "Dad" can be proper names in a family, but it is easy to think of situations where they are used as appellatives. "My Peugeot" and "the lady across the street" can be proper names, which, however, show a more involved structure than simple nouns. " $2 \times 2 = 4$ " is the name of a certain statement, but in the context "it is as sure as two times two is four" it looks rather as the "Tom" in "Tom, Dick and Harry".

Day by day we have to communicate about individual objects — physical and mental ones, about processes, acts, desires, which are objectivated, and in order to do so we need names — names which in fact, though they do not look this way, are proper names. How can we invent ever and ever new proper names for this incredible variety of objects? The answer is: by manipulating skilfully with generic names, by tying generic names in changing situations to changing individuals. For such names one knows in mathematics the term "variable" and for fixing a variable in a given situation the term "binding".

According to an old story Adam's first task in the paradise was to name the creatures. He himself had got the name "Adam", that is "man", and after him his offspring were called Adam, that is man. He gave names to all animals that passed in review, and that one he had called lion transmitted this name to his

---

* *Mathematics as an Educational Task.*

offspring. So "lion" became a name with which you could name each particular
lion; in order to distinguish lions from each other, one could speak of this lion
and that lion, of the old and the young lion, the lion in the Amsterdam Zoo
and the lion in the London Zoo. It would not be feasible to invent brandnew
proper names for each mouse, chair, bicycle; we are familiar with the use of
ambiguous, or rather polyvalent, names — one name for many objects. When my
daughter was at the age when children play the game of "what does this mean?"
and I asked her what is "thing" she answered: Thing is if you mean something
and you do not know what is its name. "Thing" is a name that fits an incredible
variety of objects, chairs, bicycles, trees, and so on. Words as "here" and "now"
are also polyvalent names: "here" as name of the place where one says "here",
"now" for the moment when one says "now". In

> the cricket is an insect

"cricket" is a generic name, the name of a species, but with

> each cricket is an insect

one can maintain that "cricket" is a polyvalent proper name that fits each
particular cricket. In La Fontaine's 'La cigale et la fourmi' "cricket" is the
unambiguous name of a fable animal. In

> the mouse is a rodent

"mouse" is the name of a species, in

> the mouse is in the trap,

pronounced in a quite special situation, "mouse" will mean a particular mouse
that by these special circumstances is supposed to be well-known and well-
defined. In

> we have again a mouse

or

> we have mice,

the name "mouse" is given to one individual or to a few about which not much
more is known than the existence.

   In order to function as proper names, variables must be bound. Variables
can be bound independently of any context, by linguistic logic devices, or in
dependence of a context. Logical means are

> the universal quantifier
> > a mouse is a rodent — for every x, if x is a mouse it is a rodent,

> the existential quantifier
> > we have got a mouse — there is an x such that x is a mouse and we
> > have got x,

the article
> our mouse; the mouse we have — the x such that x is a mouse and we have got x,

the set former
> our mice — the set of x such that x is a mouse and we have got x,

the function or species former
> the species mouse — the property of being a mouse

the interrogative
> which mouse? — which x such that x is a mouse?

Context depending devices of binding are

the demonstratives
> this mouse, that mouse, the mouse in the trap, the mouse I hear rustling.

Many variables assume another form according to the way of binding, for instance the variable of place

universal	:	everywhere,
existential	:	somewhere,
article	:	the place where,
set forming	:	the places where,
function forming	:	the place of,
interrogative	:	where,
demonstrative	:	here, there.

Or the variable of time

universal	:	always,
existential	:	some time,
article	:	the moment when,
set forming	:	the time in which,
function forming	:	the time of,
interrogative	:	when,
demonstrative	:	now, today, yesterday, tomorrow.

Similarly the variable indicating persons

universal	:	everybody,
existential	:	somebody,
article	:	the one who,
set forming	:	those who,
function forming	:	being a person,
interrogative	:	who,
demonstrative	:	I, you, this one, that one.

16.12. *Variables in the Language of Mathematics*

It is worthwhile keeping in mind this variegated use of variables in the vernacular in order to confront it with the more regimented and at the same time emaciated use of variables in mathematics.

The variable "mouse" can be used for mice only, a variable like "here" only for places, a variable like "I" only for persons — in a certain measure they are already bound by content. In mathematics symbols for variables are usually letters, sometimes combined with each other or with numbers $(A_1, A_2, \ldots, A_n)$. Such symbols are unburdened and the variables denoted by them are unrestricted with respect to their domain or only *ad hoc* restricted.

The use of letters for variables stems from — Greek — geometry. In a period of exclusively oral communication of knowledge one was very likely to reason about a figure while speaking of this point and that point, thus demonstratively. For written communication a more practical way of description was required: in a figure one numbered the points under consideration with the letters of the alphabet, which in numerical texts were indeed used as numerals. The original terminology "the point at A" — that is beside A — was later shortened to "the point A". In a similar way lines were named, triangles, quadrilaterals, by letters or by combinations of letters, which in turn indicated points.

There we are on the threshold from the demonstrative description to the use of genuine variables. In fact the variables are bound demonstratively by the figure to spots on the drawing material. On the other hand the points and figures are arbitrary paradigms. All points are the same, statements on a triangle mean all triangles, however drawn or indicated. By this fact the variables look as though they are bound by logical rather than demonstrative means. Indeed the figure may even be forgotten.

> If ABC is a triangle the orthogonal bisectors of AB, BC, CA pass through one point

is such a statement where A, B, C may be considered as point variables bound by the universal quantifier.

In geometry the use of letters for variables did not lead to the birth of an algorithmic language; except for the indication of line-segments by pairs of end points, there is hardly any algorithmic vocabulary nor is there any formalisation of operations, which at most takes place in the vernacular serving as meta-language.

In algebra letter variables are even of a later date. Linear and quadratic equations and general solving methods are as old as the oldest cuneiform texts, but always in a numerically paradigmatic setting. In the oral transfer of knowledge the unknowns could have been named in an informal way — thingummy — and this actually happens in the later development. In Greek the letters also meant numbers, which fact may have prevented Greek mathematicians from using them for unknowns. There was, however, one more impediment to transfer the geometrical use of letters for variables directly to algebra: whereas all points are "the same", numbers have a well-distinguished individuality. Finally, in the Hellenistic period, in Diophant's work, there is at least a symbol for the unknown, an abbreviation of arithmos (number). In the middle-ages cosa (Ital. thing) becomes the name of the unknown — the "cossists" developed a whole symbolism for powers of the *unknown*.

The decisive step toward a more useful algebraic notation was taken by Vieta (about 1600), who indicated also the *indeterminate* magnitudes, the variables in algebraic expressions, by letters. This notation is the proper start of the development of an algebraic language, which gets more and more detached from the vernacular. Letters are first used to indicate arbitrary numbers, but soon as well for arbitrary functions. At present we use letters for all kind of mathematical objects — sets, relations, propositions, spaces, metrics and all kind of structures — and if the need is felt, we take them from all kind of alphabets.

This should be kept in mind: letters in mathematics mean something: as symbols they represent something. At other opportunities* I have signalled and analysed the misapprehensions demonstrated by New Math at this point: letters that mean nothing or themselves, and mathematics viewed as a meaningless game with symbols. It is true that this latter aspect can be consequentially elaborated, with the aim to do foundations of mathematics, rather than mathematics itself. Then meaningful mathematics is taken as a subject to put its meaning as it were between parentheses and pay attention to its form only — the formalism. This, however, happens with and for the benefit of a meaningful mathematics. As early as the language of arithmetic I have discussed both the usefulness and the didactical danger of this procedure if a mathematical attitude is to be developed. No doubt in the case of the language of algebra usefulness and danger of rigid formalisation are as great or even greater. Miscomprehension of what is mathematics is often generated by blind algebraic calculations with letters. New Math has contributed to spread these misconceptions even among teachers.

16.13. *The Equality Sign*

If in Section 16.4 I stressed the transformation character of the language of arithmetic, I feel now obliged to attenuate it:

$$4 + 3 =$$

is primordially read as a *task* or a *question*, I said. The equality sign looks as if it were asymmetric. One side is given and the other is to be filled out. The problem

$$4 + \cdot = 7$$

---

* *Mathematics as an Educational Task*, Chapter XV.

fits less into this picture — for learners a reason to get confused. If numbers are replaced with letters, the confusion is even bigger. Putting an equality sign in

$$a + b = ,$$

what should it mean? Well, it can be meaningful in an exercise like

how much is $a + b$ if $a = 4$ and $b = 3$? ,

thus if numbers are to be substituted for a and b. In contradistinction to

$$4 + 3$$

one cannot consider

$$a + b$$

as something to figured out. It is rather the algebraic expression of a number the value of which depends on those of a and b. The expression $a + b$ does not suggest a task or a question, but it names a number depending on a and b. The notation of fractions was the point where this turn took place in history. With

$$\frac{3}{4}$$

there is nothing to be figured out. It is as such a new number represented in a conventional way by means of the previously known number symbols 3 and 4.

Of course the expression $a + b$ can occur with an equality sign behind it, such as in

$$a + b = c,$$
$$a + b = b + a,$$

in the first case either to introduce a new symbol for $a + b$ or to explain c by means of a and b or to require a, b, c to assume only values that lead to an equality, and in the second case to state something that is true for all a and b.

By this turn the character of the equality sign of arithmetical instruction has changed; it is "symmetrised". It is the intention that left and right of the equality sign the same thing occurs, and this intention can be actualised as a fact or as a demand.

The symmetric "is" plays hardly any part in the vernacular, even not in

Amsterdam is the capital of the Netherlands,

which betrays another intention than

The capital of the Netherlands is Amsterdam,

answers as it were to different questions, not unlike those asked by the pair

$$4 + 3 = \cdot$$
$$7 = 4 + \cdot$$

Even less is this equality related to the "is" in

Socrates is a man,

or with its plural in

Apes are mammals —

in the first case it rather reminds one of the symbol $\in$, in the second case of $\subset$. Again another thing is the "is" in

The apple is ripe,

namely part of the predicate being ripe, and so is the "is" in

There is time left

as part of the logical quantifier there being.

Traditional arithmetic knows equality signs in divisions with remainder like

$$16 \div 3 = 5 \, r \, 1$$

If this were the symmetric equality sign, it would imply

$$16 \div 3 = 21 \div 4$$

German didacticians of arithmetic have paid much attention to the problematic around this notation.* Is it worth the trouble? Yes and no. It is worth being seriously discussed but is not worth turning arithmetical instruction upside down.

I confessed earlier that mathematicians have convened to use the equality sign in the way that things left and right of it are the same, albeit indicated by different names. $4 + 3 = 7$ is allowed because with more or less ease one can verify that $4 + 3$ and 7 are indeed the same things. $\frac{4}{6} = \frac{2}{3}$ is allowed because by definition the two fractions mean the same rational number. Actually equality in mathematics is often a matter of definition. At a certain moment it can happen that a set of things so far viewed as different might better be considered as the same, and in order to do this with a good mathematical conscience, one introduces a new thing, the class of the things to be considered as equal and avails oneself of the names of these individual things, like $\frac{2}{3}$, $\frac{4}{6}$, . . . as various names for the new thing, the class — a convenient and therefore allowed linguistic misuse.

But let us forget for a moment about this sophisticated mathematics. In fact even mathematicians play it fast and loose with their rules about the equality signs, $\pi = 3{,}14159 \ldots$ — what does the righhand term mean? And now $\pi = 3{,}1415926 \ldots$ Does this mean that $3{,}1415926 \ldots = 3{,}14159 \ldots$? Or $g = 9{,}81 \pm 0{,}003 \ \mathrm{msec}^{-2}$. How about this? Well, one can object that this is applied mathe-

---

* See the excellent paper by H. Winter, 'Zur Division mit Rest' in: *Der Mathematikunterricht* 4/78, pp. 38–65.

matics. But number theory is pure mathematics, where nevertheless one writes in cold blood $16 \equiv 1 \mod 3$. (If number theory had been invented later, one would write $16 \sim 1 \mod 3$.) I agree this is not an equality sign, it is kind of equivalence symbol. But which equivalence? This is told only afterwards (as is done with the dots behind 3,14159 and the $\pm$ 0,003 at 9,81, which tell afterwards what kind of equality sign was meant.)

If one decides to place an equality sign between two expressions, one must first know what kind of things one has in mind to name by them. What kind of thing is $16 \div 3$ in the division with remainder? In higher mathematics such a colon is unusual. ($16 : 3$ can mean a ratio, but that is not what is meant here.) What kind of thing is $16 \div 3$? Of course no natural number. It is an ordered pair, consisting of the "quotient" 5 and the "remainder" 1. Or if you like more mathematical sophistication:

$$a \div b = \left\ulcorner \left[\frac{a}{b}\right], \ a - b\left[\frac{a}{b}\right] \mod b \right\urcorner,$$

a pair consisting of the "whole" $\left[\frac{a}{b}\right]$ and a residue class mod b, indicated in its "reduced" form.

This is a highbrow interpretation. As a matter of fact in arithmetical instruction we do not meet $16 \div 3$ but

$$16 \div 3 =$$

a task such as

$$4 + 3 = .$$

As early as elementary arithmetic there is a need besides the meaning of

$$4 + 3$$

as a task to pave the way for the meaning of

$$4 + 3 = 7$$

in the sense of $4 + 3$ being another expression for 7. In fact problems like

$$4 + \cdot = 7$$

is the first opportunity. Would it not be reasonable to adhere to this interpretation from the start onwards? Traditional arithmetic possesses symbolic devices to formulate tasks by means of dots. This use could systematically be extended even to the type

$$4 + 3 = .$$

Modern textbooks prefer squares,

$$4 + 3 = \square$$
$$4 + \square = 7.$$

There are more symbolic devices to formulate tasks, for instance in the arrow language

$$4 \overset{+3}{\frown} \cdot \,,$$

$$\cdot \overset{+3}{\frown} 7,$$

and so on. Here the equality sign is explicitly renounced and reserved for a more symmetrical use.

Let us reconsider for a moment the division with remainder.

$$16 \div 3 = 5 \, r \, 1$$

requires a lot of reinterpretation to be considered as containing a mathematically meaningful equality sign, but in fact '16 ÷ 3' is a task. If one sets out to switch early from the equality sign as a signal for a task to its static meaning, one can choose another pattern for the division with remainder, for instance a form to be filled out like

... divided by	... goes	... times with remainder	...
16	3	5	1
37	5	7	2
.	.	.	.
.	.	.	.
.	.	.	.

where in plain language the same is told as we did before in obscure mathematical language.

If one thinks that by the intervention of algebra the equality has been abolished as a signal to perform a certain task, one is mistaken. In the traditional school algebra it returns in full bloom in problems like

$$(a + b)(a - b) =$$

or

$$a^2 - b^2 = ,$$

where the pupil is expected to fill out behind the equality sign something depending on a general assignment that is or is not made explicit. Algebra as a system of linguistic transformation rules leads automatically to an asymmetric interpretation of the equality sign as unilaterally directed towards a "reduction".

"Reducing" is indeed a term characterising school algebra and, more general, automatised mathematics. According to certain rules expressions are "reduced" in one sense or the other, that is transformed into others, where the intention that meanwhile they remain the same, may readily be forgotten. This pattern of behaviour is reinforced by the application of "reductions" not only on algebraic expressions but also on equations, in order to solve them. The sequence of steps

$$x^2 - 3x + 2 = 0,$$
$$x^2 - 2 \cdot \frac{3}{2}x + \frac{9}{4} + 2 - \frac{9}{4} = 0,$$
$$(x - \frac{3}{2})^2 - \frac{1}{4} = 0,$$
$$(x - \frac{3}{2})^2 = \frac{1}{4},$$
$$x - \frac{3}{2} = \frac{1}{2} \text{ or } -\frac{1}{2},$$
$$x = \frac{4}{2} = 2 \text{ or } 1,$$

is assigned the same character as that of the problem of dissolving

$$a^2 - 3ab + 2b^2,$$

where then the various steps are justified by equality signs:

$$a^2 - 3ab + 2b^2 =$$
$$= a^2 - 2 \cdot \frac{3}{2}ab + (\frac{3}{2}b)^2 + 2b^2 - (\frac{3}{2}b)^2 =$$
$$= (a - \frac{3}{2}b)^2 - (\frac{1}{2}b)^2 =$$
$$= (a - \frac{3}{2}b + \frac{1}{2}b)(a - \frac{3}{2}b - \frac{1}{2}b) =$$
$$= (a - b)(a - 2b).$$

One is tempted in the first example to make the connection between the subsequent lines explicit by a mental, vocal or written equality sign. There, however, the thing that remains equal at the various steps is not a numerical but a truth value, which fact is registered by a $\iff$ sign, or in the asymmetric case, that is, if the truth value increases by $\Rightarrow$.

16.14. *Formal Substitution*

Replacing in

> Socrates is a man

"Socrates" by "I" is not allowed; one had simultaneously to change "is" into "am". Replacing in

> It snows

the hidden time variable "now" with "yesterday" or "tomorrow" requires a change of the verb tense into the past or the future. The substitution processes that lead from

$$3 + 4 = 4 + 3$$

to

$$a + b = b + a$$

or conversely, are more formal. By formally substituting 3 for x one can verify whether

$$x^2 - 3x + 2 = 0$$

is satisfied by $x = 3$.

Formal substitution, however, extends further. From

$$(a + b)(a - b) = a^2 - b^2$$

one obtains

$$(a + c + b + d)(a + c - b - d) = (a + c)^2 - (b + d)^2$$

by replacing

a with $a + c$, b with $b + d$,

that is, variables in an expression are replaced with more complex expressions, which can in turn involve variables.

A powerful device — this formal substitution. It is a pity that it is not as formal as one is inclined to believe, and this is one of the difficulties, perhaps the main difficulty, in learning the language of algebra. On the one hand the learner is made to believe that algebraic transformations take place purely formally, on the other hand if he has to perform them, he is expected to understand their meaning. If in

$$\ldots - b$$

I have to replace

b with $b + d$,

it becomes

not $- b + d$ but $-(b + d)$.

The minus sign of b is expected to extend its activity to the whole of $b + d$.

This then is the inevitable consequence of the interpretation of

$$b + d$$

as

sum of b and d

rather than the task

add d to b,

which holds whether the b and d are genuine variables or already specified constants. Thus

the difference of ... and b

becomes after substituting $b + d$ for b

> the difference of . . . and the sum of b and d.

The learner is expected to read formulae with understanding. He is allowed to pronounce

$$a + b, \quad a - b, \quad ab, \quad a^2$$

as

> a plus b, a minus b, a times b, a square.

Yet he has to understand it as

> sum of a and b,   difference of a and b,   product of a and b, square of a.

The action suggested by the plus, minus, times, square and the linear reading order must be disregarded. The algebraic expressions are to be interpreted statically if the formal substitution is to function formally indeed.

The formalism of algebra could have been designed more rigidly by requiring that in substitutions the substitute is only accepted within brackets; in the case of substituting $a + c$ for a and $b + d$ for b in

$$(a + b)(a - b) = a^2 - b^2$$

this produces

$$((a + c) + (b + d))((a + c) - (b + d)) = (a + c)^2 - (b + d)^2,$$

where parentheses that we do not like can be eliminated afterwards, and in this way we are likely to instruct a computer, indeed. But since in this way algebra would become a wearisome business, we appeal to understanding, even if we teach algebra. It is a fact of didactic experience that this appeal falls on deaf ears — at the same time asking for formalist acting and content directed under-standing is too much. What to do about it didactically?

After the preceding phenomenological analysis the advice is easy. In order to teach the language of algebra, in particular the formal substitution, we have to make an appeal to intelligent reading, and this appeal should be well-aimed and if necessary made explicit. It would be unnecessarily complicated to read

$$(a + b)(a - b)$$

again and again as

> product of the sum of a and b and the difference of a and b,

but in order to act as though this is said, it can didactically be necessary to recall it to consciousness. An intermediate stage, kind of flow diagram, would be:

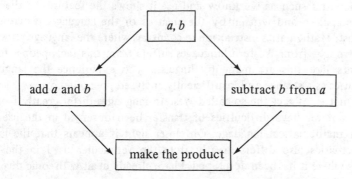

How and when this should be done, depends on the total organisation of formalising in the didactical sequence, in particular on the way how, and the degree to which, thinking in functions is going to be formalised.

If the teacher can afford to be patient, the formal substitution may arise selfreliantly. A personal experience with a 14-year old underachiever:

I allowed her to solve equations such as

$$\frac{3}{4}x = 7$$

as she liked, in two steps ("times 4, divided by 3) until she had found out how to do it in one step. I had her calculating products like

$$a^4 \cdot a^3$$

by writing each power as a product until she became tired. When with types like

$$(a + b)^2$$

my efforts to have her combing both mixed factors, fell flat, I switched to

$$(a + b)(a - b)$$

where I succeeded, and then back to $(a + b)^2$, which became a success too. This was the start of a slow progress on the road towards the formal substitution.

Patience is expensive; the teacher must know whether the price is worth paying.

The formal substitution is so important because its validity extends much further than substituting into given expressions. For solving certain problems the algebraist has rigidly formalised solving procedures at his disposal. Solving linear and quadratic equations and systems of equations follows patterns which may be numerically introduced, then generalised by formal substitution and again applied in the generalised form by means of formal substitution. Numerous algebraic strategies function thanks to the possibility of formal substitution in strategic patterns.

16.15. *Idiom*

In the last ten or twenty years it has become a fashion to assert that the language of mathematics such as we know and use it shows the features of the natural languages spoken and written by the creators of the language of mathematics in the past. Mathematics instruction in countries where the language structurally differs strongly from Western languages suffers from this fact, people tell you. Statements like this are made by linguists who experience the language of mathematics as a strange and insufficiently analysed phenomenon.

As a matter of fact the so-called Western languages differ greatly from each other, but never have difficulties of transfer been observed in the use of the common mathematical language. On closer sight it appears that the language of mathematics also differs greatly from the vernacular used in the various countries where it has been developed. I have already dealt with some divergences of principle and I will add a few more.

Divergences between a natural language and the language of mathematics can indeed cause learning difficulties. The most striking and most serious is probably the deviation from the rules of order according to the positional system in the construction of numerals — in many languages up to 20, and in a few even up to 100. We have got used to most of the other deviations to such a degree that we are not anymore aware of it.

$$4 + 3$$

is, one closer inspection, a strange manner to write down the

> task "add three to four"

and an almost absurd one for

> the sum of four and three

as it should properly be read. Nevertheless everybody reads it cooly as

> four plus three

which is not English — as little as the French, German, Italian . . . "translations" of it are French, German, Italian.

Subtractions were up to the beginning of this century in German and certainly in many other languages formulated according to the pattern

> vier von sieben   (four from seven)

for $7 - 4$. When pedagogues got fed up with it, they made it

> sieben minus vier

for the schools of the elite and

> sieben weniger vier   (seven less four)

for the common people, both of which are not normal German — a fact noticed today by nobody. The Dutch terminology for $12 \div 3$ was and often still is

> dividing three on twelve

which is normal Dutch, though now in favour of the international mathematical language and without tears replaced with

> twelve divided by three.

Never in former times has the equality sign been read as "equals".

$$4 + 3 = 7$$

was read as

> quatre et trois font sept,

or similarly with *facit, macht* and so on in other languages.

"Three times" can degenerate into "one times" and "nought times" both of which are all but English. One reads — in all languages similarly

> a^2 as a square,

instead of

> square of a

as required by decent linguistic habits. In expressions is

> log a

the order of the natural language is at least preserved in the mathematical language, but nobody complains about the lack of something like the word "of" between the two components as far as it is required and possible in the surrounding language.

Let me restrict myself to these examples, which have no other pretension than to make clear how strongly the language of mathematics differs from the languages in which it has come into being, and that on points which as a consequence of habituation we do not notice any more.

16.16–25. *Algebraic Strategies and Tactics*

16.16. A more profound study than I have been able to undertake so far, should be required to enumerate if not exhaustively then at least representatively, and at the same time systematically, algebraic strategies and tactics. Let me produce that which just crossed my mind and repeat the warning that it is neither complete nor well-organised.

16.17. *The algebraic principle* – I used the term earlier,* but in Section 15.3 I changed it into the more colourful

algebraic permanence principle,

with the shade (Sections 15.13–15) of the

geometric–algebraic permanence principle.

It means the idea of extending operations and relations, wherever needed, in a way that

certain appreciated properties are preserved.

As a creative principle as well in its didactic function it has been discussed so often that no further explanations are necessary. I will only add some more colour to the originally neglected geometric shade by repeating expressively:

the justification of the numerical operations and their laws by the simplicity of the algebraic description of geometric figures and connections.

16.18. *The (formal) substitution* – discussed as early as Section 16.14 has numerous tactical aspects. It can yield

specialisation – if variables are numerically fixed;
   (example: verifying whether $\Sigma_{i=1}^{n} (2i - 1) = n^2$ is correct);
generalisation – if numerical terms are replaced with variables;
specialisation – if a solving schematism is *applied*;
generalisation – is a solving schematism is *extended*;
structural simplification – if in a given expression partial expressions are replaced with variables
   (example: replacing $x^4 + 2x^2 - 3$ the x^2 with y);
structural complication – if in a given expression variables are replaced with expressions
   (example: when solving $x^3 + px + q = 0$ replacing x with $u + v$ in order, by having $3uv + p$ vanishing, to later simplify the left side);
elimination – of a variable involved in a substitution;
restructuration – in order to recognise patterns that admit of applying certain algebraic laws
   (example: replacing in $x^2 + 2x + 3$ the x by $y - 1$);

* *Mathematics as an Educational Task*, pp. 228–241.

On a higher level substitution is involved

> if operations and relations are replaced with their defining terms;
>> (examples: multiplication as repeated addition,
>> raising to a power as repeated multiplication,
>> subtraction as inverted addition, and so on,
>> and the converses)

16.19. *The algebraic translation* of properties, connections, and problems — in the most original example, characterised by the advice "call the unknown x". In order to describe a problem in its context algebraically,

> knowns, unknowns, parameters and relations between them are tracked down and translated by means of variables, equations and inequalitites.

In order to describe properties (sets) algebraically,

> variables are introduced
>> (example: "odd number" algebraically described by "$2n - 1$ with integral n").

It is surprising that even pupils in the higher grades of the highschool solve problems like

> A fraction with value $\frac{2}{5}$ where numerator and denominator differ by 45.
> Mary is six years older than John, and eight years ago she was thrice John's age,
> The length of a rectangle is $4\frac{1}{2}$ times its width and its surface is 288 cm^2

by trying; they do not hit on the idea "to do it with x".

16.20. *Solving equations and inequalities*

> by systematic trials,
> by transferring all the terms to one side,
> by distributing the terms appropriately on both sides,
> by applying the same operations to both sides,
> by complication to prepare a simplification;

in systems with several unknowns

> by substitutional elimination of unknowns, if needed, prepared by other transformations.

16.21. *Considering an expression as a function,*

> a function as composite of functions,
> a function as the inverse of a function.

16.22. *Change of perspective*,

> by considering a datum as an unknown,
> by considering an unknown as a datum,
> by considering data as solutions of a system of conditions,
> by replacing inequalities with equalities,
> by replacing equalities with inequalities.

16.23. *Exploiting symmetries*

> in expressions, equations, inequalities, functions.

16.24. *Stating the positivity of some expression*

> by trying to regard it as a sum of squares.

16.25. *Exploiting analogies*

> in an informal way;
> in functional connections,

such as the analogy between addition and multiplication informally or by means of exponential and logarithmic functions.

CHAPTER 17

FUNCTIONS

17.1. *Variables*

In Sections 16.11–12 we subjected the variable as a linguistic phenomenon in the vernacular and in mathematic language to a comparative analysis. Indeed what we discussed there, was language, the need for names, proper names as well as

polyvalent names,

which according to certain procedures may supply proper names. This need was felt early on in mathematics: to distinguish this point and that point as A and B, the three vertices of a drawn triangle as A, B, C, which at the same time can name the vertices of any triangle and by this very fact occur as polyvalent names for points, and combined, for triangles.

Polyvalent names are a means to formulate general statements, that is statements that hold for all objects they name:

the mouse is a rodent

applies to every mouse,

$$\overline{AB} + \overline{BC} \geqq \overline{AC}$$

to every triple of points in the plane,

$$a + b = b + a$$

to every pair of numbers. Polyvalent names can, as we pointed out, be fixed in various ways, "bound" in order to be used in statements on objects.

The mathematical habit of calling polyvalent names variables is of a rather recent date. Originally "variable" meant something that really varies, something in the

physical, social, mental

but as well in the

mathematical

world, that is

perceived, imagined, supposed

as varying, that is in addition to

> the time that passes,
> the path that is covered,
> the aim that changes,
> the moon that is waxing,
> the temperature that oscillates,
> the wind that is changing,
> the days that lengthen,
> the mortality rate that decreases,
> the progressive rate of income tax,

also the

> variable mathematical objects

by which these phenomena are described. From the

> variable physical, social, mental

phenomena one is led to

> variable numbers, magnitudes, points, sets,

in general

> variable mathematical objects.

Locutions like

> the number ϵ approaches (converges to) 0,
> the point P runs on the surface S,
> the element x runs through the set S,
> the number e is approached by the sequence $\left(1 + \dfrac{1}{n}\right)^n$ if n goes to infinity,

witness this kinematic aspect of the "variable". It is true that in the course of, say, the past half century such locutions have been outlawed by purists. Indeed one can dispense with them,

> x_n converges to 0

can be written

> $\lim_n x_n = 0$

and be defined, with no kinematics involved, by

> for every $\epsilon > 0$ there is an n_0 such that $|x_n| < \epsilon$ for $n \geqq n_0$;

> x runs through the set S

can very simply be written as

$$x \in S.$$

Well, one can dispense with that kind of kinematics provided one has once been in its possession, learned to use and then to eliminate it — this didactical feature will be dealt with later on.

The effort to suppress the kinematics of the variable goes hand in hand with the annexation of the term "variable", stripped of its kinematic undertone, for the benefit of what we just came across as "polyvalent names". In our former examples

the mouse is a rodent,

$$\overline{AB} + \overline{BC} \geqq \overline{AC},$$
$$a + b = b + a$$

"the mouse", and the various letters are primordially polyvalent names by means of which statements can be made, which, if not restricted, hold for all bearers of these names. Variable objects is not the original idea here as it is in the examples of

the time that passes,
the number that approaches 0,
the point running on the surface.

Lumping concepts of various origin together, using one name for things which stripped of mere frill boil down to the same, this is one of the characteristics of mathematical activity. We have met here again such an example:

polyvalent name

and

variable object

are being related with each other in the term

variable,

but they pay for it with the loss of important phenomenological properties, which mathematically look like mere frill, though phenomenologically they cannot be dispensed with. What happens if they are renounced can be described linguistically and ontologically:

Linguistically: the variable even ceases to be a name, it becomes a place-holder for names of a certain kind of object, which are to be put into the place — that is, in a certain context for the same variable at each occurrence the same name, while one context may require different kinds of place holders.

Ontologically: The variable does not indicate an object but rather the opportunity to evoke a certain kind of objects — a virtual rather than an actual variability.

Expellas naturam furca, tamen usque recurrit — nature though driven out
with the fork, nevertheless returns. The mathematical purism — of high value
within mathematics — is a forced and less sufficient language as soon as one steps
out of mathematics. The abundance of variable objects of the half-way mathematised vernacular can be eliminated by linguistic sophistication but by this
linguistic measure they are not disposed of. And — even more important — in
order to eliminate them by linguistic tricks, one must once have experienced
them. Indeed this is the only way to guarantee that one is able to restore them
as soon as one falls back on the vernacular in order to do something with mathematics in the real world. The world is a realm of change, describing the world
is describing change, and to do this one creates variable objects — physical,
social, mental, and finally mathematical ones. There exist many languages of
description, or rather many levels of describing. At a high level of formalisation
the variable mathematical objects can be forgotten, but at less formalised levels
they are a genetically and didactically indispensable link with the physical,
social, and mental variables, which for their part are indispensable.

17.2. *Dependence (or Connection)*

In the phenomenological approach the "variable" is more than an instrument of
formalised mathematical language and even more than something one uses by
way of speaking. This should be completely plain before I start dealing with
functions. Indeed, the very origin of the function is

> stating,
> postulating,
> producing,
> reproducing

dependence (or connection) between variables occurring in the

> physical, social, mental world,

that is, in and between these worlds. Particularly important are

> mathematical variables
> > mutually related or related with the others.

The dependences themselves can in turn be objectivated, that is given the status
of mental objects. On the way towards this objectivation such a dependence can
be

> mentally experienced,
> used,
> provoked,
> made conscious,
> experienced as an object,
> named as an object,
> placed in larger contexts of dependences.

The precision with which such a dependence is objectivated can vary from

> classifying according to sorts of dependences
>
>> to relating orders with each other (the more this the more that)
>> to relating in a more or less precise, possibly numerical way with each other,

and depending on this precision the dependence can be given

> a generic name,
> a proper name
>> of *ad hoc* character
>> of algorithmic character.

Examples:

A body falls. There is a dependence between witnessed time and place of the body. The dependence is more or less consciously experienced, as free fall (of the body), which is the sort in which this dependence is placed. The longer the body falls the faster — an order relation, it falls according to a formula, the law of free fall, by which the dependence is placed within the larger context of uniformly accelerated motions.

Two elastic bodies collide. In the game of billiards the dependence between the pairs of velocity vectors before and after the collision is experienced, used, provoked; it is consciously encountered in a larger complex of experiences, and in its sort described by the term "collision". The dependence is more precisely described by a formula and a whole theory, which also accounts for the spin of the billiard balls.

For dependences in which time plays a part, one knows a number of generic names, such as

> movement, growth, decay, process, course, trend,

some of which are also used metaphorically when time is replaced with

> change of standpoint, direction of vision, and so on.

For numerous physical dependences a generic name is

> causal connection,

made more precise by such terms as

> by attraction, friction, heat transfer, oxydation, collision, refraction, optical representation, and so on.

From this one should distinguish the

> automatic (or programmed) connection,

for instance

> between touching a key on the piano or typewriter and the production
> of a sound or a typographical sign,
>
> between turning a switch and certain mechanical or electrical phenomena,
> between aiming and hitting.

A three-year old draws with chalk on a blackboard. "Doll growing bigger, chalk growing smaller".

The dependence can be described by the generic name "chalk consumption by drawing" and made more precise by a numerical connection between the size of the drawing and the piece of chalk.

"When the days begin to lengthen, the cold begins to strengthen" – a dependence between day length and cold (which holds only for a restricted period, and even so in a stochastic way). Quality pays, the more they have the more they want, the sooner the better, the longer it lasts the worse it gets – many sayings and locutions aim at dependences that are somehow described by the relation between orders "the . . . the . . .".

17.3–4. *Function – Terminology*

17.3. The function is a special kind of dependence, that is, between variables which are distinguished as

> dependent and independent

– an old-fashioned looking terminology, which, however, stresses the phenomenologically important element:

> the directedness from something that varies freely to something that
> varies under constraint.

Mathematised: the function from A to B as

> act* that assigns to each element of A an element of B.

Functions are all around in mathematics and its applications, albeit labelled in various ways:

> mapping, transformation, permutation, operation, process, functional,
> operator, sequence, morphism, functor, automaton, machine,

which are used according to needs and opportunities:

> *Function* is preferred if the set of values is numerical,
> *mappings* and *transformations* come from geometry but serve as well, with
> certain attributes added in algebraic structures such as
> *morphisms*, prefixed with certain prepositions or adjectives,

---

* I now prefer "act" to the more statical "law", which I used formerly.

functors, acting on morphisms;

permutation is the term for a one-to-one mapping on itself, in particular, if studied in a group theory context;

operation or *process* is the term used with certain simple standard functions (addition, root extraction).

If *A* itself is a set (or space) of functions, then in order to avoid repetitions or ambiguities,* the terms

functional and *operator* are used for functions from *A* to *B* — the first if *B* is a set of numbers and the second if it is also a set of functions.

The term
sequence is usual, if not obligatory, for mappings from **N**.

Automaton or *machine* are didactical equivalents of function.

Only to mention it:

argument as a term for the independent variable in a function, which is

falling in disuse.

17.4. Functions can be considered as special relations. A relation from *A* to *B* is any subset of the cartesian product $\ulcorner A, B \urcorner$. Such a relation *f* is called a function from *A* to *B* if

for every $a \in A$ there is exactly** one $b \in B$ such that $a, b \in f$.

This definition is *logically* equivalent with the former; phenomenologically it is not, and didactically phenomenologically not at all. It obscures the essential action of assigning directed from *A* to *B*, which — in order to repeat the terms at the start of Section 17.1 — is of a

stating, postulating, producing, reproducing

character. One can oppose these two definitions to each other as

dynamic versus static

if one does not care to use outworn terms.

From the fact that all mathematics can be reduced to set theory, one may not conclude that it should be done and even less that it is a didactic necessity and possibility. Switching

from $\ulcorner a, b \urcorner \in f$ towards $fa = b$

* In a similar fashion class, family, system are welcome synonyms for "set".
** According to some authors: at most.

is mental gymnastics for mathematically trained brains, which can be too much for less advanced people.

I have discussed this question more profoundly in the past.* At that opportunity I have also refuted the fairy tale as though subset of $\ulcorner A, B \urcorner$ is to be preferred above "law of assignment",** because of its greater solidity. Didacticians around 1960, who experienced set theory as a new revelation, believed they had to impose this revelation to mathematical instruction, a tyranny from which instruction has not yet wholly recovered.

17.5–9. *Functions – Terminology*

17.5. A *direct* pattern to describe the connection between independent and dependent variable is the

function table

$x = 1$	2	3	...
$y = 2$	4	6	...

with a progression of assignment suggested by the dots. A similar device is the

function graph

which suggests a connection

$$x = \ldots 2 \ldots$$
$$y = \ldots 3 \ldots$$

visually (Figure 217).

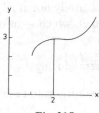

Fig. 217.

The function table can be used for a complete definition in the case of small finite sets, for instance

$x = 1$	2	3	4	5
$y = 2$	1	4	3	5

---

* *Mathematics as an Educational Task*, pp. 387–390.
** Now I would even say: act of assignment.

or

$$\begin{pmatrix} 1 & 2 & 3 & 4 & 5 \\ 2 & 1 & 4 & 3 & 5 \end{pmatrix}$$

for the permutation that interchanges 1 with 2, 3 with 4, and leaves 5 invariant.

A more general and more generally applicable device to indicate the connections between independent and dependent variable is the pattern

... as a function of ...

Here the function can describe a *factual* connection such as in the given graph

the ordinate as a function of the abscissa,
the slope of the tangent as a function of the abscissa,
the area extending from a certain abscissa onwards;

or at a given curve

the curvature as a function of the arc length from a given point onwards,
the torsion as a function of the arc length from a given point onwards;

or on a given surface

the normal vector as a function of the point,
the total curvature as a function of the point;

or with spheres

the area as a function of the radius,
the volume as a function of the radius;

or in the movement of a mass point

the path as a function of the time,
the velocity as a function of the time;

or at a certain place

the temperature as a function of the time,
the atmospheric pressure as a function of the time;

or at a certain moment

the temperature as a function of the place,
the atmospheric pressure as a function of the place;

or combined

the pair ⌜temperature, pressure⌝ as a function of the pair ⌜place, time⌝;

or

the state of a system as a function of the time,
the expansion of the mercury as a function of the temperature,
the attractive force as a function of the distance,
the period of oscillation as a function of the length of the pendulum,
the intensity of some radiation as a function of wave-length and temperature;

or in a certain national economy

the price index as a function of the time,
the income tax as a function of the income;

or in a certain society

the death rate as a function of the age,
the average income as a function of education,
the life expectance as a function of sex, social class, and so on,
the fertility as a function of the social class.

In all such cases there might be more causes of variability while one of them is stressed. "Sleeping variables" which at some time can be "wakened up" are termed

parameters

— their significance will be dealt with later on.

17.7. The pattern ". . . as a function of . . ." also serves to describe more formal connections, like

with long objects
        the length measure as a function of the measuring unit,
in the act of measuring
        the precision as a function of the measuring procedure,
in the act of calculating
        the precision as a function of the calculating procedure,

and purely formal connections, like

$$x^2 - 3x + a \qquad\qquad \text{as a function of } x,$$

$$\frac{1}{\sqrt{2\pi}\sigma} \cdot e^{-\frac{1}{2}x^2/\sigma^2} \qquad\qquad \text{as a function of } x,$$

$$x^2 + y^2 \qquad\qquad \text{as a function of } \ulcorner x, y \urcorner.$$

Shorter ways of notations are: the function f defined by

$$f(x) \quad = x^2 - 3x + a,$$

$$f(x) \quad = \frac{1}{\sqrt{2\pi}\sigma} e^{-\frac{1}{2}x^2/\sigma^2},$$

$$f(x, y) = x^2 + y^2$$

respectively, or

$$x \qquad \rightarrow x^2 - 3x + a,$$

$$x \qquad \rightarrow \frac{1}{\sqrt{2\pi}\sigma} e^{-\frac{1}{2}x^2/\sigma^2},$$

$$\ulcorner x, y \urcorner \rightarrow x^2 + y^2$$

respectively, or

$$Y_x \, (x^2 - 3x + a),$$

$$Y_x \, \frac{1}{\sqrt{2\pi}\sigma} e^{-\frac{1}{2}x^2/\sigma^2},$$

$$Y_{\ulcorner x, y \urcorner} \, (x^2 + y^2).$$

In two of these cases the function moreover "depends on a parameter a or σ"; it is a function that is itself a variable. If the dependence of this function on the parameter is to be made explicit, then in the first notation the parameter must be made explicit in naming the function

$$f_a \text{ defined by} \quad f_a(x) = x^2 - 3x + a,$$

$$f_\sigma \text{ defined by} \quad f_\sigma(x) = \frac{1}{\sqrt{2\pi}\sigma} e^{-\frac{1}{2}x^2/\sigma^2}$$

in order afterwards to define a function f by

$$f(a) = f_a,$$
$$f(\sigma) = f_\sigma,$$

respectively. According to the second kind of notation it would be

$$a \rightarrow (x \rightarrow x^2 - 3x + a),$$

$$\sigma \rightarrow \left(x \rightarrow \frac{1}{\sqrt{2\pi}\sigma} e^{-\frac{1}{2}x^2/\sigma^2} \right),$$

though I never came across this notation. According to the third kind of notation it is

$$\underset{a}{Y}\underset{x}{Y}\,(x^2 - 3x + a),$$

$$\underset{\sigma}{Y}\underset{x}{Y}\,\frac{1}{\sqrt{2\pi\sigma}}\,e^{-\frac{1}{2}x^2/\sigma^2}.$$

In general, an expression

> $T(\ldots x \ldots)$ as a function of x can be briefly written as the function f for which $f(x) = T(\ldots x \ldots)$

or

$$x \rightarrow T(\ldots x \ldots)$$

or

$$\underset{x}{Y}T(\ldots x \ldots).$$

If this expression involves parameters

$$T(\ldots x \ldots a \ldots)$$

then it becomes

> the function f_a for which $f_a(x) = T(\ldots a \ldots x \ldots)$, and the function f for which $f(a) = f_a$,
>
> $a \rightarrow (x \rightarrow T(\ldots x \ldots a \ldots))$,
>
> $\underset{a}{Y}\underset{x}{Y}T(\ldots x \ldots a \ldots)$,

respectively.

In all these cases the variability can be restricted by means of subscripts

$$x \in \ldots, \qquad a \in \ldots.$$

17.8. Another pattern comes from indicating functions by means of

> the ... of

and suchlike schemes — in linguistic terms, the genitive or genitive-like constructions.

Examples to illustrate this:

the mother of ...	(a person),
the birthday of ...	(a person),
the address of ...	(a person),
the value of ...	(a coin),
the rise of ...	(a celestial body),
the start of ...	(a process),
the target of ...	(aiming),

the moment of ...   (a happening),
the place of ...   (an object),
the tip of ...   (an iceberg);

or from mathematics

the midpoint of ...   (a line-segment),
the median point of ...   (a triangle),
the interior of ...   (a sphere),
the area of ...   (a plane figure);

the half of ...   (a magnitude),
the triple of ...   (a magnitude, quantity),
the square of ...   (a number),
the square root of ...   (a positive number);

the sine of ...   (an angle, arc),
the maximum of ...   (a function),
the limit of ...   (a sequence),
the greatest common divisor of ...   (a set of natural numbers),

the union of ...   (a set of sets),
the complement of ...   (a set),
the number of elements of ...   (a set),
the set of subsets of ...   (a set);

the sum of ...   (two numbers),
the set of divisors of ...   (a natural number),
the number of divisors of ...   (a natural number);

the degree of ...   (a polynomial);

the operation of multiplication of ...   (a group),
the factor group of ... by ... ;

the solution set of ...   (a condition),
the domain of ...   (a function),
the derivative of ...   (a function);

the closure of ...   (a set in a certain topology),
the topology of ...   (a linear space);

the function belonging to ...   (a graph),
the graph of ...   (a function);

the equation of ...   (a conic),
the curve belonging to ...   (an equation);

the truth value of ...   (a proposition),
the disjunction of ...   (a system of propositions),
the system of well-formed formulae of ...   (a formal language),
the system of true statements of ...   (a formal system).

17.9. Some functions have fixed notational names in the mathematical language. Usually it is a function symbol with on its right side — between parentheses or not — the thing on which the function is acting. Examples:

$$\sin, \cos, \tan, \log, \sqrt{\ }, \max, \sup, \lim, \gcd.$$

Less usual

$$\#: \quad \text{number of,}$$
$$\mathscr{P}: \quad \text{set of subsets of}$$

If a finite or infinite sequence is interpreted as a function acting on

$$\{1, \ldots, n\} \quad \text{or} \quad \mathbf{N},$$

then in

$$a_i \quad (i = 1, \ldots, n \text{ or } i \in \mathbf{N}, \text{ respectively})$$

a can be considered as a function symbol with the thing on which it acts put lower down rather then level. As regards the exponential function

$$e^{\cdots}$$

one could maintain that the variable has climbed, but the exponential function as such is better indicated by

$$\exp.$$

In the sum as a function of two summands the function symbol appears between the variables,

$$a + b \quad \text{instead of} \quad +\ulcorner a, b\urcorner.$$

A similar notation is that of

$$\text{factor group of} \ldots \text{by} \ldots \quad : \quad \ldots / \ldots$$

In the notation of

$$\text{square of} \ldots \quad : \quad \ldots^2$$

it is the function symbol that has climed (to the right), as well as in

$$\text{derivative of} \ldots \quad : \quad \ldots'.$$

In

$$\text{closure of} \ldots, \quad \text{conjugate complex of} \ldots \quad : \quad \overline{\ldots}$$

the function sign is floating above that on which it acts.

In

> absolute value of . . . : |...|

the function sign split into two parts surrounds the variable acted on, as it is the case in

> whole part of . . . : [...].

These, however, are mere interpretations which can hardly be maintained. If there is a need for symbols of such functions, one is better served by

$$\ulcorner a, b \urcorner \to a + b \quad \text{or} \quad \mathsf{Y}_{a,\,b}\, a + b,$$
$$\ulcorner G, H \urcorner \to G/H \quad \text{or} \quad \mathsf{Y}_{G,\,H}\, G/H,$$
$$x \to x^2 \quad \text{or} \quad \mathsf{Y}_x\, x^2$$
$$f \to f' \quad \text{or} \quad \mathsf{Y}_f\, f'$$
$$A \to \bar{A} \quad \text{or} \quad \mathsf{Y}_A\, \bar{A}$$
$$x \to |x| \quad \text{or} \quad \mathsf{Y}_x\, |x|$$
$$x \to [x] \quad \text{or} \quad \mathsf{Y}_x\, [x]$$

The differential operator is more favourably placed; in

$$\frac{\mathrm{d}}{\mathrm{d}x}\,(xe^{-\frac{1}{2}x^2})$$

the $\frac{\mathrm{d}}{\mathrm{d}x}$ can be regarded as a function (operator) symbol that describes a transformation of a function of x into a function of x.

The integral operator is more troublesome; in order to consider

$$\int_a^x f(t)\, \mathrm{d}t$$

as a function of f, and certainly with

$$\int_a^b K(s,\, t) f(t)\, \mathrm{d}t$$

as a function of f, descriptions like

> g defined by $g(x) = \int_a^x f(t)\, \mathrm{d}t$ as a function of f

or

$$\mathsf{Y}_f \mathsf{Y}_x \int_a^x f(t)\, \mathrm{d}t$$

and

> g defined by $g(s) = \int_a^b K(s,\, t) f(t)\, \mathrm{d}t$ as a function of f

or

$$\mathsf{Y}_f \mathsf{Y}_s \int_a^b K(s,\, t) f(t)\, \mathrm{d}t,$$

respectively, are to be preferred.

17.10–14. *Sorts of Functions – Terminology*

17.10. Beside explicitly given functions and functions indicated by algorithmic or proper names, there is a need for ambiguous names for functions in order to deal with functions affected with, or subjected to, variability. Such devices have already repeatedly been used in the preceding sections. In most cases the function indicated by the letter f was specified, but in Section 17.9 letters such as f were also used to indicate rather arbitrary functions. The need for ambiguous names for functions is felt as soon as statements (definitions included) are to be made about a whole class of functions, the need for variable functions as soon as acts are performed on, or yield, functions – the function as independent or dependent variable.

In Section 17.3 we enumerated synonyms for the term "function"; according to the kind of functions under consideration there is some preference for a special synonym. The following expositions about terminology with regard to sorts of functions somehow overlap that of Section 17.3.

17.11. One interprets

functions of, say, m variables

as simple functions by nominating

an ordered m-tuple of variables

as independent variable, thus

$$f(x_1, \ldots, x_m) \equiv f(\ulcorner x_1, \ldots, x_m \urcorner).$$

An

ordered system of n functions

with the same independent variable can be considered as a simple function by nominating

an ordered n-tuple of variables

as dependent variable, thus

$$y_i = f_i(x) \, (i = 1, \ldots, n) \equiv \ulcorner y_1, \ldots, y_n \urcorner = f(x).$$

In this way we get for instance

mappings out of m-space into n-space,

$$\ulcorner y_1, \ldots, y_n \urcorner = f \ulcorner x_1, \ldots, x_n \urcorner.$$

Such mappings are also used to describe

curves, surfaces, and so on
paths, deformations,
movements.

Used in this way, the independent variables are also called parameters — a point to which we will pay attention later on.

17.12. Other aspects according to which functions (mappings) from A to B are distinguished are

> peculiarities of A,
>
> peculiarities of B,
>
> peculiarities of original sets of elements of B, for instance a one-to-one (injective) mapping,
>
> peculiarities of the factual image, for instance a mapping on (surjective),
>
> smoothness of the mapping, for instance continuity, differentiability, and so on.

Some school textbooks distinguish between functions and mappings in the way that functions from A to B need not be defined in all A, wheras mappings from A to B are defined in all A. This diverging school text terminology is not to be recommended. If one wishes to indicate functions (mappings) to B without specifying the part of A in which they are defined, one may prefer the terminology

> function (mapping) out of A into B.

I used it before (Section 17.11).

17.13. Sorts of functions can be defined implicitly by imposing conditions, for instance

> operations on or in certain sets,

required to fulfill certain conditions, which lead to a certain structure,

> group, ring, field, linear space, and so on.

Mappings of certain structures can be subjected to requirements of preservation of structure, the

> morphisms

(with various kinds of prefixes).

17.14. If a function acting on $\mathbf{N}$ is considered a sequence, one gets in accordance with the variability of the dependent variable

> number, point, function, and so on, sequences.

A function acting on $\mathbf{N}$ must not necessarily be considered as a sequence. One speaks of the *sequence* of natural numbers: $\curlyvee_n n$

of even numbers   : $\curlyvee_n 2n$

of squares   : $\curlyvee_n n^2$

of prime numbers : $\curlyvee_n$ nth prime number

yet of the

function φ of Euler

(that is the number $\varphi(n)$ of numbers $< n$, relative prime to n), and the

function μ of Moebius

($\mu(n) = 0$ if n is a square, otherwise $+ 1$ or $- 1$ according to whether the number of prime divisors is even or odd), though in the context of the function of s

$$\Sigma \mu(n)n^{-s}$$

one would be inclined to speak of the sequence of coefficients — this expression is as it were structured as the inner product of two sequences, that of the $\mu(n)$ and that of the n^{-s}.

Similarly in

$$\Sigma a_n x^n$$

one speaks of the sequence of coefficients and the sequence of powers of x.

17.15–16. *Parameters*

17.15. The term parameter as synonym for variable is used with three meanings.

First, a secondary — as it were sleeping — independent variable, which, if need be, can be accounted for — as it were wakened up — for instance in order to get a system

$$f_t$$

of functions by the variability of the "parameter" t.

Second, a variable that according to its origin is dependent but according to its appearence is independent, and which serves to distinguish figures, structures, and so on, from each other within their sort, for instance

the radius of the circle as a function of the circle, by means of which circles are distinguished according to size,

the parameter p of the parabola (equation $y^2 = 2px$) by which non-congruent parabolae are distinguished — this is the very origin of the term "parameter",

the halving time as a parameter of a decay process.

17.16. As a third occurrence one knows the so-called

parameter representation

of curves, surfaces, and so on.

Examples:

The unit circle in the $\ulcorner x_1, x_2 \urcorner$ plane (equation $x_1^2 + x_2^2 = 1$) is parametrised by means of the arc length s from a fixed point,

$$x_1 = \cos s,$$
$$x_2 = \sin s.$$

The unit sphere in the $\ulcorner x_1, x_2, x_3 \urcorner$ space (equation $x_1^2 + x_2^2 + x_3^2 = 1$) is parametrised by means of two parameters s and t (corresponding to geographical longitude and latitude)

$$x_1 = \cos s \cdot \cos t,$$
$$x_2 = \sin s \cdot \cos t,$$
$$x_3 = \sin t.$$

In general, what is phenomenologically experienced as curve or surface, is parametrised by one parameter, or two, respectively. These parameters arise as dependent variables (arc length, longitude and latitude), dependent on the point of the curve and surface. *A posteriori* they are used as independent variables in order to

represent the curve and surface parametrically

(1) $x = f(s),$

(2) $x = f(s, t).$

Then the curve (surface) appears as the image of a number (pairs of numbers) set.

An important aspect is the liberty to change parameters,

the parameter transformation.

With the new parameters the curve (surface) is described by (Figure 218)

$$x = g(u),$$
$$x = g(u, v).$$

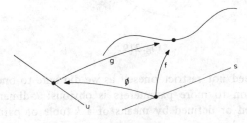

Fig. 218.

Old and new parameters are coupled to each other by the points of the curve (surface) they describe — a one-to-one mapping, respectively:

$$\varphi: s \to u$$
$$\ulcorner \varphi, \psi \urcorner: \ulcorner s, t \urcorner \to \ulcorner u, v \urcorner,$$

where, respectively:

$$g(\varphi(s)) = f(s),$$
$$g(\varphi(s, t), \psi(s, t)) = f(s, t).$$

Afterwards by a change of perspective the mental object curve (surface) is mathematised:

A curve (surface) in R is a sufficiently smooth mapping out of the one-dimensional (two-dimensional) parameter space into R,

where mappings that arise from each other by parameter transformations are considered as the same.

More precisely:

A curve (surface) in R is an equivalence class of (sufficiently smooth) mappings out of one-dimensional (two-dimensional) space (the parameter space) into R with the equivalence relation: $f \sim g$ if and only if there is a one-to-one mapping φ (out of the parameter space into itself) such that $f = g \circ \varphi$.

By this change of perspective it is made possible to consider curves (surfaces) with multiple points (lines) really as curves (surfaces) and to count multiple points as simple or multiple according to circumstances: parameters on the curve (surface) are introduced which take different values in the multiple points according to the direction one comes from, and by means of these parameters the curve (surface) can be described *as a map or a mapping* — in the map the multiple point is counted simply, in the mapping multiply (Figure 219).

Fig. 219.

Of course one need not restrict oneself as we did here to one or two parameters. The extension to more parameters is obvious: k-dimensional hypersurfaces are described or defined by means of a k-tuple of parameters — this indication may suffice here.

17.17–21. *Function – What is their Meaning and Use?*

17.17. An overwhelming phenomenological variety of what is called function! But is it not the same with all mathematical concepts? Mathematical activity often aspires at unification, with the consequence of a vision directed from the sublime endpoint of the activity to the cloudy sources – a vision that reveals itself mathematically and mathematical-didactically.

So far, however, we have stuck to an enumeration of the phenomenological appearances of the function. What do you do with functions? – this question has hardly been tackled, except perhaps in the last section where the phenomenon "parameter" required some illustration. But such an enumeration does not mean much as long as the function does not become "functional". If this point is arrived at, it may appear that what is unified under the term of function is not that much a union.

"Set" too is ascribed a unifying task, structuring mathematics itself. The universe of sets with its relations of intersection, union, complement, cartesian product, subset, and so on, is a rich structure yet because of lack of interpretation poor mathematics. The true mathematical wealth is created by the perspective of function. We have displayed a rich variety of objects which are understood as functions. But it is one thing to bestow on an object the title of function and another to do something with it, to work with functions in a way that is characteristic of functions. Is function a name I can attach to all that fulfills certain requirements or rather a signal how to act in certain contexts? Does one call a thing a function in order to do something with it, and if so, what? It will appear that "function", which by its pure definition looks a unifying concept, is much too variegated as regards its operational virtue to be profoundly unifying. Moreover it will appear that the need for objectivation – that is promotion to an object – of the function is quite often no more than a need for unification. The intellectual pleasure of unifying is certainly not a bad thing. There are enough examples to prove that this pleasure can create precious mathematical products. But beyond this pleasure to structure the phenomenon that is called mathematics by unifying means, one would expect other incentives. Anyhow if they exist, phenomenology is obliged to account for them and to analyse them.

17.18. The question "is this really worth being called a function?" will less often be asked of the examples 17.6–7 which illustrate the pattern

   . . . as a function of . . .

and more often of quite a few of the examples 17.8, which fill out the function forming pattern

   the . . . of . . .

What kind of context should justify the part played by functions like

> the mother of . . .   (a person),
> the address of . . .   (a person),
> the birthday of . . .   (a person),
> the place of . . .   (an object),

and what part do they actually play? It is perhaps possible to contrive such context but as long as this has not been done, I prefer to affirm that they have been invented only in order to give a few more examples of what can be called a function.

The more mathematical examples joining that list are a bit different. In contradistinction to the preceding list it may happen that in many, if not in all, of these cases one meets or introduces some symbol for the function under consideration, whereas function symbols for "mother of . . ." look farfetched. Some usual function symbols, belonging to this list, have been mentioned in Section 17.9. The list is not large; one can perhaps add some symbols but even then it would not become imposing. Even algorithmic name giving is unusual for many among these functions. The midpoint of a line-segment is readily called M, and if it really matters one can add in parentheses the line-segment that is actually meant, but does this promote $M(. . .)$ to the rank of a function? And the same holds for "median point of . . .", "interior of . . ." and many more.

Giving the topology of a space R the name $T(R)$ may be a pleasure that is restricted to such contexts as "by $T(R)$ I mean a system of subsets of R with the following properties . . ." and to the trick of comparing two different topologies T_1 and T_2 of the same R by making use of a possible inclusion relation between $T_1(R)$ and $T_2(R)$. Yet does this lend T the status of a function?

For the closure of a set S there is even a fixed notation $\overline{S}$, but who then imagines $\overline{S}$ as a function of S (and of the topology that defines the closure)? Though

$$\ulcorner x, y \urcorner \to x + y$$

or

$$\curlyvee_{\ulcorner x, y \urcorner} x + y$$

is clearly recognisable as a function, nobody is aware of the fact that this function has an established name,

> sum of . . . and . . .

For the solving set of the condition $F(x, y, z)$ one has the notation

$$\{ \ulcorner x, y, z \urcorner \mid F(x, y, z) \}$$

but it requires quite special circumstances to consider this set as a function of the condition $F(x, y, z)$, for instance that of a differential equation where one is interested to know how the solution depends on the initial conditions.

The widely diverging degree of operationality of the concept of function will be illustrated with three paradigms.

17.19. A quite usual notation is

$$C(M, r)$$

for

>   the circle with centre M and radius r.

Here C seems to be a function that produces from a pair ⌐point, number⌐ a circle. But is it really a function? I do not think so, it is just a notation. Of course, one knows that a circle in the plane is determined by its centre and radius, but one's attention is not attracted by this dependence as a mental object. It is just a notation such as PQ for the line determined by the points P and Q. Nobody asks more profound questions. The C of circle is not properly a function symbol. The notation is just a formal imitation of the genitive construction

>   the sine of . . .
>   the logarithm of . . . ,

an abbreviation by analogy of

>   the circle with centre M and radius r.

"C" awakes the impression to be a function symbol, but it is not operational as such.

But let us now restrict the liberty of point M to a curve K and r to a fixed value. Immediately the function character of C becomes patent. M on K determines a circle $C(M, r)$ and if M runs along the curve K a nice system of circles with an envelope — by pure set theory* determined as the set

$$G(r) = \{P \mid \text{distance } (P, K) = r\}.$$

If moreover r is made variable and considered as the time t, then one can interpret

> $C(M, r)$ as a wave front

of a wave motion and the envelope of the circles $C(M, t)$ $(M \in K)$ as

>   wave front G emitted by K.

"G" is now a true function symbol, that is for

>   the wave motion G,

which at the moment t produces

>   the wave front $G(t)$.

---

* A few conditions must be fulfilled if indeed the envelope has to coincide with this set.

What does this example mean? That which originally was introduced as a pure notation, bears in itself the potency of a function, though — at least in the present case — a potency that could be actualised without speaking of functions. Indeed, I could have told the same story with the same or perhaps even more convincing power without using the word "function" even once.

17.20. For the symmetric group of the n permutands $1, 2, \ldots, n$ one knows the notation $\Sigma(n)$. Is this more than a notation or does it really mean Σ (symmetric group) as a function of the *number* of permutands? The answer is of course: it is not.

Yet as soon as one considers the symmetric group $\Sigma(A)$ of the (finite) set A (the group of all permutations of A, that is of all one-to-one mappings of A on itself), one heads for a more essential use of the function notation. For instance: for the aim to prove that a one-to-one mapping of A in B induces an isomorphism of $\Sigma(A)$ in $\Sigma(B)$.

I took care to introduce $\Sigma(n)$ as the symmetric group of $\{1, \ldots, n\}$. The usual terminology is

the symmetric group of n permutands,

as though the nature of the permutands did not matter. Indeed, often it does not. Yet sometimes it does. For instance if I set out to state that two full permutation groups with the same number of permutands are isomorphic.

If $\Sigma(n)$ is interpreted this way, that is, as

the symmetric group of n permutands,

then Σ as a function symbol does not assign one group to n but a whole class of groups that are isomorphic with each other in a well-determined way. Then

G is a symmetric group of n permutands

should rather be written as

$G \in \Sigma(n)$.

17.21. Let us dive more profoundly into this example. Let $\Sigma(n)$ again be defined as was done in the beginning of the preceding section. As far as function symbolism is concerned, the notation $\Sigma(n)$ does not count much if compared with all that is required and advantageous for understanding permutations as mappings. Well, I can dispense with function symbols as long as I deal with a single permutation, say for $n = 5$ the permutation that replaces

$$1 \quad 2 \quad 3 \quad 4 \quad 5$$

with

$$3 \quad 2 \quad 1 \quad 5 \quad 4,$$

respectively. On this level I can even perform given permutations in succession and invert them and even arrive at some understanding of, say, Σ (5). Of course this is quite a job: considering a permutation of objects as an object, beside, and cooperating with, other objects. Just because this is a hard thing, one feels the need for a notation of functions and operations with functions. This then leads us into the alley, or rather on the highway of formalising.

Let A and B be two equipotent sets. Their Σ (A) and Σ (B) then are isomorphic. More precisely, each one-to-one mapping from A on B produces an isomorphism from Σ (A) on Σ (B). Or, more generally:

A one-to-one mapping φ from A in B induces an isomorphism of Σ (A) in Σ (B),

to wit according to the pattern

the permutation f of A determines a permutation g of B: for each $\varphi(a)$ $(a \in A)$ we put by definition

$$g(\varphi(a)) = \varphi(f(a))$$

and for each $b \in B \backslash \varphi A$:

$$g(b) = b.$$

In order to express how this g depends on f we need some function symbolism, say,

$$g = \varphi(f),$$

and only if this has been settled one can go further, for instance to state that

$$\varphi(f_1) \varphi(f_2) = \varphi(f_1 f_2)$$

— no profound fact, but anyway a fact that in order to be enunciated requires the function as a mental object with the full function notation.

17.22—36. *The Historical View*

17.22. From enumerating what is called function and how this is done, we were led to the question "why?", from stating facts to looking for causes — causes that, as it happens in mathematics and mathematical terminology, are necessities.

The rich variety we have displayed, emerged within a wholly mathematised context. Phenomenological analysis should start earlier. With the question whether this "earlier" implies an earlier place in this chapter, as well as with the total arrangement of the subject matter, I have unceasingly struggled — the present chapter has been written and rewritten five or six times.

The phenomenological analysis should start earlier, but this "earlier" must be exposed later. "Earlier" can also be understood historically, and in such an investigation, history can be a guide-line that is not to be despised, albeit one with twists that need not be followed though in the past some people believed that the individual repeats the history of mankind — in fact education and instruction take care that this neither does nor need happen.

17.23. How much can the present investigation profit from history depends on the profundity allowed the historical look. One can simply state that the word "function" emerges in the correspondence of Leibniz and the Bernoullis and that not until at least Euler does the use of this word start resembling what we are accustomed to. We can also state that Euler and d'Alembert were the first to use indeterminate function *symbols*, and then by preference the familiar f and φ. We can signal concept and term "mapping" (Latin: representatio) in Euler's paper (1777) where he studies mappings of the sphere on the plane in a cartographical context. Such mappings are of course analytically described by systems of functions of systems of variables ("parameters"), yet this perspective is changed in the beginning of the 19th century when one starts considering — more or less explicitly — such systems of functions as mappings. Mappings, if not produced analytically, are of geometrical origin. They become an important device in set theory and an object of investigation in topology. Meanwhile function theory — real and complex — has grown enormously. Not until the 20th century do these two streams — function and mapping — merge in one bed.

17.24. In many respects this historical account is too simplistic. Two streams are not enough, there are more of them which can historically and must phenomenologically be distinguished. Moreover they can be traced back to more ancient sources. The history of a mathematical concept starts long before it is given a name.

Tables often represent functions and about such tables we know from Babylonian mathematics onwards; it is as the background of tables that the function reveals itself most patently in astronomy — tables well-known from Babylonian and Greek astronomy, mathematically constructed from empirical data. The Babylonians — at least since the 5th century B.C. — proceeded in this work more phenomenologically than the Greeks. They observed periodicities in the *positions* of celestial bodies and interpolated their *course* in between linearly, or rather piecewise linearly, taking into account the observed periodical variations of velocity. The functions by which they attempted to describe the celestial motions and by means of which they calculated their tables, are consequently superpositions of periodic zig-zag functions (Figure 220). The Greek

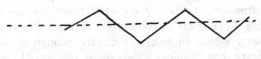

Fig. 220.

proceeded less phenomenologically. They used kinematic models: centric and eccentric circular movements and epicycles, that is, circular movements dragged on by others (Figure 221). Mathematically this means that they described the celestial movements as superpositions of trigonometric functions — a model that maintained itself up to Kepler. Then circles were superseded by ellipses — the sun in a focus and in equal times equal surfaces swept out by the vector from sun to planet.

Fig. 221.

Interpreting celestial motions in functions — this is indeed a historically natural invitation to mentally constitute functions and continuity. One observes the rotation of the celestial globe and with it the motions of sun, moon, and stars as function of the experienced time, making this subjective time more precise by means of the more objective looking, suggested by the kinematics of the celestial globe. One sees the monthly backward course of the moon outlined according to longitude and latitude between the fixed stars and learns to express it numerically by means of the "objective time", the kinematics of the celestial globe — to be sure only pieces of the course are actually observed and the remainder is smoothly filled out. Less easily is the yearly backward course of the sun along the stars observed — directly by the setting and rising of stars of the zodiac in the morning and evening dusk, or indirectly by the full moon as the opponent of the sun. One states "inequalities" in these motions and attempts to describe them by superpositions of piecewise linear or trigonometric functions and, as to the latter — to explain them by mechanical models. The forward and backward movements of the planets are as many invitations to constitute and analyse functions — continuous functions of time. There are more of them: the shadow directions and lengths as functions of the time — time objectivated on the sundial — the meridian altitude of the sun as function of the geographical latitude and noon as function of the geographical longitude.

17.25. So far we have functions of time — linear zig-zag or trigonometric functions. But Greek astronomy knew trigonometric functions also under another aspect. Place finding by numerical coordinates in the celestial globe requires spherical trigonometry which in fact — strangely enough as we might think — has preceded plane trigonometry. We cannot tell exactly when this had happened, but in Ptolemaeus' astronomical work, the "Great Syntax" we find it in the state where it remained for better than a millenium. Spherical triangles have sides and angles, as the plane triangles do, but their sides are circular arcs and hence also measured by angles. Between sides and angles of a spherical triangle there exist relations resembling those of plane trigonometry, for instance the sine theorem of spherical trigonometry

$$\sin a : \sin b : \sin c = \sin \alpha : \sin \beta : \sin \gamma.$$

Greek astronomers and mathematicians knew what we call trigonometric functions, though in another setting. In Ptolemaeus' work we find *chord tables*, the subtended chord as function of the angle in a circle, the radius of which is divided in 60 and the circumference in 360 parts. The connection between sines and chords is clear (Figure 222)

$$\text{chord } \alpha = 2 \sin \tfrac{1}{2}\alpha$$

Fig. 222.

The Hindu word for chord, phonetically adapted by Arabs and in their language interpreted as bosom, led to the Latin translation "sinus" (the Latin equivalent of bosom).

17.26. Even older than the sine is the tangent, which can be traced back to Babylonian mathematics. As a measure of inclination and as the shadow length belonging to a given length or angle of incidence, the tangent is quite natural. Trigonometric functions, in particular the tangent, are used in Archimedes' Circle Measurement, the calculation of π.

17.27. I started this historical exposition with trigonometric rather than with the more elementary and natural functions like linear, quadratic, cubic ones, which in fact can also be traced back to Babylonian mathematics. The Greek term for linear dependence means proportionality: two variable magnitudes, most often of geometrical origin, behave according to a fixed ratio. The areas of circles, the volumes of spheres are in the ratios of the squares and cubes, respectively, of the radius — a terminology, which maintains itself up to Kepler, and even longer: according to his third law the squares of the times of revolution of the planets are to each other as are the cubes of the long axes of their orbits.

The Greek had even names for inhomogeneous linear functions: a magnitude if compared with another is "a given magnitude bigger (smaller) than in proportion"[*] — because of the lack of appropriate symbols a cumbersome verbalisation of the relation between x and y:

$$y = ax \pm b.$$

Even more involved is the terminology to describe quadratic functions. It is a geometric terminology which serves to formulate, classify quadratic equations — known before to the Babylonians — the so-called *application* of areas: applying a given area F as a rectangle to a line-segment a (Figure 223),

plainly, that is

$$ax = F,$$

such that a square falls short

$$(a - x)x = F,$$

such that a square exceeds

$$(a + x)x = F.$$

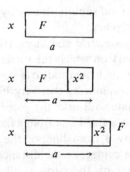

Fig. 223.

These three cases are distinguished with the Greek words for agreement, falling short, and excess as

parabolic, elliptic, hyperbolic

application. This, then, is via Apollonius the origin of our terms for the conics: if F is given as a square with side y, then the above equations become

[*] Euclid's Data.

$$ax = y^2,$$
$$(a-x)x = y^2,$$
$$(a+x)x = y^2,$$

respectively, which are indeed the equations of parabola, ellipse, hyperbola.

In this terminology a quadratic function in x is described as an area applied to a given line-segment (the independent variable) with deficiency or excess. This is the way Archimedes expresses himself* when he considers what we would call the graph of a quadratic function in order to perform what we would call an integration.

17.28. The Greeks were familiar with describing curves and surfaces by means of equations; in the case of the conics the equations even preceded the curves. The equations, however, were not written in an algebraic but in a geometric language: addition and subtraction are operations on line-segments, areas, and other geometric magnitudes; the product is obtained as a rectangle of line-segments or avoided by positing proportions between geometric magnitudes. The Greek mathematicians had sound reasons to

express algebraic relations geometrically,

which I will not explain here. The change of perspective to

express geometric relations algebraically

was performed by Descartes (and simultaneously by Fermat); it became possible by Vieta's creation of letter algebra.

Indeed, it is first of all geometric and kinematic-geometric magnitudes and relations (such as the cycloid) on which the predecessors and inventors of the analysis test the strength of their method. If a point is running along a curve, one perceives a great many magnitudes that vary in dependence of each other: not only abscissa and ordinate, but also the piece cut out by the tangent or normal in a variable point on a fixed line from a fixed point, or the piece of the tangent or normal cut off by a fixed line, or the radius of curvature, or the coordinates of the curvature centre, or the distance from a fixed point, or the arc length from a fixed point on the curve — all of them "functions" in the original Leibniz—Bernoulli terminology. Magnitudes vary in dependence on each other — the picture present in Newton's mind is that all of them together depend on something like time — and these dependences can, but need not, be expressed by formulae.

If u and v are such magnitudes, then thanks to the connection that — ceteris paribus — exists between them, an "infinitesimal" growth of u by du corresponds to an "infinitesimal" growth of v by dv while neither of them is indicated as the dependent and independent variable. The quotient du/dv yields the growth

* See for instance: E. J. Dijksterhuis, *Archimedes*, 1956, p. 124.

ratio — in Newton's notation the ratio of "fluxions" $\dot{u}/\dot{v}$. If the connection between u and v is specified, du/dv can be more closely determined, for instance if

$$u^2 + v^2 = \text{const.},$$

then

$$udu + vdv = 0,$$

thus

$$du/dv = -v/u.$$

I have explained elsewhere* that this view has maintained itself in physics, that is, in those parts of physics that have sprung from classical mechanics. If a mass point moves, its place, velocity, acceleration are functions of time, the velocity is the derivative of place by time, the acceleration the derivative of the velocity,

$$dx = vdt,$$
$$dv = adt$$

but there is no need to make the functions explicit, as we have learned to do with function symbols,

$$x = f(t),$$

in order to express a special dependence between time and place, which implies that of velocity and acceleration,

$$v = f'(t),$$
$$a = f''(t).$$

Function symbols are of a later date than functions in the Leibniz–Bernoulli sense, but to date the confusion between symbols for magnitudes and for functions has persisted — I discussed it in the earlier mentioned place.

In the original approach there is nothing like an independent and dependent variable. There is dependence of variables on each other, which can be turned into linear dependence between differentials. According to the needs some of the variables may be stamped as independent and the others as dependent.

The need can be prompted by necessity, and this happens as soon as one passes to differentials of higher order. From the second order onwards one knows the asymmetry between numerator and denominator in such expressions as

$$d^2u/dv^2.$$

The inventors of differential calculus had a hard time before they understood that second differentials are not meaningful unless certain first differentials are "kept constant", namely those of the variable declared as independent.

* *Mathematics as an Educational Task*, pp. 553–559.

I apologise for this side-leap, which may be less intelligible to outsiders. I had somehow to explain why the easy looking game with *mutual* dependence did not satisfy in the long run; at certain moments one must decide which are dependent and which are independent variables if successful analysis is at stake. It is strange that people who believe they can replace calculus by "non-standard analysis" refuse to learn this from history — a point that should be discussed at greater length.

17.29. The original view that each variable (or each system of mutually independent variables) can be chosen as the independent one (ones) had momentous consequences for the development of Calculus. One could work with du/dv without specifying or at least symbolising the actual dependence and in the same breath one could assert that

$$dv/du = 1/(du/dv).$$

One could switch to and fro between the available variables, for instance simplify integrals

$$\int y \, dx$$

by passing from the independent variable x to a more appropriate t with

$$dx = \frac{dx}{dt} \, dt.$$

(In fact such an integral transformation was the great "aha-experience" in Leibniz' discovery story.)

Knowing how x and y and how y and z depend on each other one could state by elimination the kind of dependence of x and z, and conversely by inserting between x and z an intermediary y one could break the mutual dependence of x and z into one between x and y and one between y and z, moreover with regard to the differentials one could establish the useful fact that

$$dz/dx = (dz/dy)(dy/dx).$$

17.30. This is an appropriate place to uncover the roots of the explosive growth of the analysis and in particular of the tremendous success that was meted out to Leibniz' notation — no sooner than in the course of this phenomenological contemplation have the ideas that I am going to explain become clear to me in all their pregnancy.

In the function notation the transition from one variable t that is given as independent to another u that is chosen as independent means a substitution

$$t = \varphi(u).$$

Deriving from the dependence between x and y and that between y and z the one between x and z means in the function notation composing the function f and g that describe the dependences

$$y = f(x), \qquad z = g(y)$$

to a function h

$$z = h(x)$$
$$= g(f(x)).$$

In the dependence between x and y passing from x to y as the independent variable, means inverting the function f that maps x on y.

Up to the rise of the new analysis, algebra was dominated by the typically algebraic operations of adding, subtracting, multiplying, dividing, raising to power and extracting roots. All these operations can also be performed with functions. But functions open a new perspective:

> substituting into each other and inverting.

This created on the one hand a never before known wealth of new objects — functions as wild as one wants to contrive; on the other hand the opportunity to break up and invert functions in order to handle them more easily, and for all these operations simple rules to put differentials and integrals in relation to each other.

In Sections 17.17–21 we asked the critical question "functions — what is their meaning and use?" Partly this is answered in the last paragraph, albeit in a historical perspective: the strength of the function concept is rooted in the new operations — composing and inverting functions — which create new possibilities.

17.31. Let us return to Section 17.28 where we interrupted the course of the exposition: the viewpoint where the dependence between variables was not yet explicitly indicated but implicitly determined by the geometric, kinematic, physical relation at the background. Sooner or later this relation itself is made explicit; the dependence is described by an equation, say,

$$f(x, y, z) = 0.$$

Considering f anew as a function, of three variables, is the next step, which by itself creates the opportunity to consider arrays of dependences

$$f(x, y, z) = \text{const.},$$

according to the choice of the constant.

The dependence between variables, say,

$$x_1, \ldots, x_n$$

can be given by a k-tuple of equations

$$f_1(x_1, \ldots, x_n) = 0, \ldots, f_k(x_1, \ldots, x_n) = 0,$$

and here too one gets the opportunity to consider $f_1, \ldots, f_k$ as functions and

$$f_1(x_1, \ldots, x_n) = y_1, \ldots, f_k(x_1, \ldots, x_n) = y_k$$

as a description of a system of dependences, according to the choice of the $y_1, \ldots, y_n$. On the other hand this becomes the description of a mapping out of $\ulcorner x_1, \ldots, x_n \urcorner$-space into $\ulcorner y_1, \ldots, y_k \urcorner$-space; conversely such a system of equations can be interpreted as a mapping.

17.32. This does not interfere with the need, if an equation

$$f(x, y, z) = 0$$

or a system of equations

$$f_1(x_1, \ldots, x_n) = 0, \ldots, f_k(x_1, \ldots, x_n) = 0$$

is given, not to be satisfied with the mutual dependence but to consider

one variable (a few variables) dependent from the others,

which means

solving with respect to one variable (a few variables)

the equation (the system of equations).

Equations most often have more than one solution. If it is the equation

$$x^2 = 9$$

one shall make explicit both of them, by $x = \pm 3$, or $x_1 = 3, x_2 = -3$, or $x = 3$ or -3. In the case

$$x^2 = 2,$$

and certainly in the case

$$x^2 = a$$

one will be inclined to be satisfied with the notation

$$x = \sqrt{2},$$
$$x = \sqrt{a}$$

to indicate both solutions — root symbols with an ambiguous meaning.

With this kind of multivalued symbol one has indeed worked and struggled during centuries in algebra. In fact one can work with them and even quite smoothly as long as one takes care and sees to it that a multivalued symbol (even as a variable) keeps the same meaning in the same context. It is true this may be not that easy. For instance, if one uses the formula

$$\sqrt{a}\sqrt{b} = \sqrt{ab}$$

in order to replace $\sqrt{a}\sqrt{b}$ with $\sqrt{ab}$, one may not forget to check whether $\sqrt{ab}$ did not occur earlier in the same context, and then perhaps in another meaning, and this indeed may require too high an intensity of watchfulness.

The knot has finally been cut, though afterwards it has lasted almost a century until this gordian solution was generally accepted:

the — by definition — univalent function.

But blood is thicker than water: in order to globally "save" complex analytical function such as the nth root or the logarithm, which cannot be dispensed with, one invented *ad hoc* domains where they could behave univalently — the Riemann surfaces — but this is a question I had better skip.

17.33. Originally functions are given — implicitly or explicitly — by analytic expressions. Yet as soon as he had to solve the differential equation of the vibrating string, d'Alembert came across "arbitrary" functions, the amplitude of the string at a certain moment could "arbitrarily" be prescribed. Solutions of partial differential equations in general depend on arbitrary functions. As a matter of fact this phenomenon led to the function notation, the ambiguous names f and φ for a function. During the 18th century one has continuously but unsuccessfully struggled with the arbitrary functions. Even more confusion was created by the phenomenon that the differential equation of the vibrating string could be solved by means of arbitrary functions as well as by what we now call Fourier series, which look analytically nice but can represent strange functions, even with discontinuities.

This knot was cut by Dirichlet, when in 1829 he studied Fourier series: the function, with no regard to its generation, which to each point (of an interval) assigns an arbitrary number. This was so revolutionary an idea that still a century later older mathematicians, if they meant an arbitrary function, for safety reasons added "in the sense of Dirichlet".

This is the essential feature of the function concept as it has evolved in history

univalent   and   arbitrary.

The arbitrariness is no longer limited — one may imagine a function equal to 1 for each rational number and equal to 0 for each irrational number.

Well, if it is needed, one will restrict the arbitrariness. But even continuous functions can be arbitrary. I could not find out who was the first to interpret the daily course of temperature at a certain spot as a function (of time). Which was earlier, the graph of an empirical happening or the locution that this represents a function of time?

17.34. The adjective "arbitrary" on "function" aimed in this historical exposition at the kind of dependence that the function constitutes between the variables, not at the variables themselves. The variables were numbers or otherwise continuous magnitudes — real and complex ones, possibly bundled in n-tuples, in order to describe points in geometry or systems in mechanics. If at present we speak of a function from A to B we admit in principle arbitrary sets A and B. The road to this goal was paved with the one-to-one mappings of finite sets in themselves or others, which were known under names like permutations, combinations, substitutions. The finite sets concerned were primordially — with Lagrange, Ruffini, Abel, Galois — the roots of algebraic equations, and from the start composing permutations was an essential element in these investigations. Long before group theory, groups of permutations (performed in succession) were implicitly involved in these investigations. Composing and inverting is even here the most relevant aspect of what appears as a function to our view.

17.35. From the self-mappings of finite sets the road leads to mappings, in particular one-to-one mappings, of arbitrary sets, the instrument by which Cantor compares infinite sets. And now all the fences are down. Functions, under any name one can imagine, are called up to restructure old domains. An instructive example is calculus of variations. From the dawn of analysis one knew maximum and minimum problems for functions to be solved by calculus: maximising or minimising a function depending on certain variables. Beside this there were the variational problems: the curve of the fastest descent (between two points in a vertical plane though not on a vertical line), the revolution body with the least resistance, the shortest line between two points on a surface are a few examples. Here the variables are not numbers or points, but curves, surfaces, functions, and that what should be maximised or minimised by an appropriate choice of the variable is a number depending on the variable curve, surface, function — a functional, as such a function, of a higher order as it were, will be called later on.

And so it goes on: variables which are in fact functions, or operations on functionals, operators, and so on do the job of functions. And of course variables which in fact are operators, acted on by functions called by various names.

17.36. Geometrical mappings, whether algebraically represented or not, were the principal motor of geometry as it developed in the 19th century, and again composing and inverting such mappings was the structuring feature, which finally resulted in the explicit investigation of the implicitly much older geometrical groups. In topology, mappings — continuous and topological ones — become both the instrument and object of investigation. Even if they were not called functions the preferred letter f bears witness to the function at the background of the mapping. The unification is announced before it is pronounced.

17.37. The present author has lived through and undergone a part of this development, but it should require cumbersome heart searchings to find out the dates when for the first time he sensed, accepted, applied the various mathematical objects as functions. It would be worthwhile identifying their first occurrence and their sporadic or regular or definitive use in the literature, though of course literature is a belated witness of what occurred in minds.

Sequences of numbers, points, functions as functions from **N** to a set of numbers, points, functions — when did it come up, when did it become conscious to myself, when did I make it conscious to others? Coordinates as functions of points (and bases in vector spaces), the i-th element of an ordered n-tuple as a function of i, a predicate as a function from individuals to propositions, a binary operation in G as a function from $\ulcorner G, G \urcorner$ to G, a function of n variables as a function of an n-tuple of variables, the limit of a convergent sequence as a function of convergent sequences — I could repeat all the above questions with respect to these subjects and I would as little be able to answer them.

The closure $\bar S$ of a set S as a function of S — perhaps I saw it always this way, perhaps it first crossed my mind when I saw Kuratowski's axiomatics of topological space, which postulates for each set S an $\bar S$ such that

$$(1) \quad S \subset \bar S \qquad (2) \quad \overline{S \cup T} = \bar S \cup \bar T \qquad (3) \quad \bar O = O \qquad (4) \quad \bar{\bar S} \subset \bar S$$

because not until (4) where the iteration of the closure emerges, does closure become operational as a function.

Similarly "mother of ...", "father of ..." become operational only in the context of genealogy where functions of this kind are iterated and composed to "grandmother of ...", "grandfather of ..." and with eliminations to "brother of ...", "sister of ...", "uncle of ...", and so on.

17.37a. *Lessons of History*

There were functions prior to the use of this name, and there are and will be functions before they are named as such and before you have a name for them.

Systems have been developed to name functions algorithmically.

"Variables" sprung from two sources: variable object (physical, social, mental, mathematical) and ambiguous name. Both sources, though covered by the modern view of "variable as placeholder" are nevertheless still active.

Functions — explicitly — entered as relations between variable magnitudes, the variability of which was infinitesimally compared.

The liberty to change variables, from dependent to independent, and between independent ones, led to a new kind of operation with functions: composing (substituting) and inverting (accompanied by eliminating).

This new operational wealth has caused the success of the function.

The necessity of distinguishing independent from dependent variables led to stressing functions rather than relations.

Notwithstanding what is suggested by algebraic and analytic expressions, the development tended towards the univalent function.

From the description of visual data by functions a change of perspective led to visualisation of functions by graphs.

Though somewhat younger than the algebraically and analytically expressed function, the arbitrary function arises as early as variational calculus and solving partial differential equations.

This "arbitrary" can mean the character of the functional relation as well as that of the variables: numbers, number tuples, points, curves, functions, permutands, elements of arbitrary sets.

The functions of analysis, the geometric transformations, the permutations of finite sets and the mappings of arbitrary ones flow together, in order to generate the general function concept.

This concept is used to comprise a great variety of things: algebraic operations, functionals, operators, even sequences, coordinates, logical predicates.

It can happen that what looks like a functional symbol is a mere notation, which, however, can become operational as a functional symbol.

The operationality of what is called function expresses itself often in the activities of composing and inverting.

17.38–51. *The Genetic View*

17.38. *Variability as a Source of the Function*

The function enters history explicitly rather late and then by means of new and advanced mathematics — Calculus — but with a view to its much earlier virtual presence, one may expect similar discrepancies in the individual developmental rise of the function.

In mathematical instruction functions have moved downwards from Calculus, via graphs and supported by equations to the primary school, even to its lower grades, where they are concretised by imaginary machines and expressed and symbolised in table and arrow language. The fact that this is feasible provides a certain susceptibility for functions at a young age. What does this susceptibility involve?

At a certain age children are susceptible to learning reading and writing; the question of at what age they might be able to invent an alphabet is as meaningful as the similar question we can ask about the wheel. We cannot check it because it is imposed by the social environment, as is the case with language. Would they be able to invent number selfreliantly? The numerals are imposed on them by the environment and even if they refuse to accept them (see Bastiaan 4; 3*) and pursue an autonomous development, words such as "that many" and "how many", blown in from the environment, are as many milestones along the road to number, halfway or even nearer the goal. The early ability to operate with magnitudes — at least with lengths — has to wait for many years to be made explicit, if at all, in arithmetical instruction — an irresponsible lag.

What about the function? For its constitution, we noticed a few components are required:

> variables,
> mutual dependence,
> independent and dependent variable,

---

* *Weeding and Sowing*, p. 281.

where the variability may be

  physical, social, mental

in order to be mathematised, or

  mathematical,

and the same holds for the dependence, which can be

  mentally experienced,
  used,
  provoked,
  made conscious,
  experienced as an object,
  named as an object,
  placed in larger contexts of dependences,

in particular in that of

  composing and inverting.

Humans – even animals – are from the start of their development confronted with variabilities and dependences: the mother reacts on crying, with the hand in front one perceives a thing, while moving one pushes something away, one collides with something, the place is changed, turning a switch causes light or dark – qualitative dependences, which are later made more precise by order relations: the harder one cries, the stronger the sound, the harder one tries, the higher one jumps, the more blocks the higher the tower and the louder the noise if it is knocked down.

17.39. *Qualitative Variability*

The simplest case of mappings of sets is – trivialities excluded – that where both sets have two elements, which for our convenience we will name 0 and 1.

Let A, B, C be three sets like this and let the functions from A to B be indicated by f_i, those from B to C by g_i, and those from A to C by h_i such that

x	$f_1(x)$	$f_2(x)$	$f_3(x)$	$f_4(x)$
0	0	0	1	1
1	0	1	0	1

and similarly for g_i $(i = 1, \ldots, 4)$ and h_i $(i = 1, \ldots, 4)$. Then there is a composition table

∘	g_1	g_2	g_3	g_4
f_1	h_1	h_1	h_1	h_1
f_2	h_1	h_2	h_3	h_4
f_3	h_4	h_3	h_2	h_1
f_4	h_4	h_4	h_4	h_4

or if A, B, C are identified with each other

∘	f_1	f_2	f_3	f_4
f_1	f_1	f_1	f_1	f_1
f_2	f_1	f_2	f_3	f_4
f_3	f_2	f_1	f_4	f_3
f_4	f_4	f_4	f_4	f_4

Physical models of this are systems with two states and appropriate couplings to realise the functions.

Example: A: Two pushbuttons at a sun curtain with the signs ↓ and ↑; B: curtain down and up; f_1: press ↓ ; f_2: do nothing; f_3: press ↓ or ↑ according to what causes changes; f_4: press ↑.

Example: A, B are two lamps with states "on" and "off", f_i a certain circuit.

The examples are artificial, it is not easy, indeed, to integrate all four functions in one system. The following pattern looks more natural:

Suppose with respect to some action (for instance stimulating somebody)

A: performing or not (0 or 1)

and with respect to the effect

B: taking place or not (0 or 1).

Then the meaning of the functions f_i is as follows:

f_1:   not being able to cause,
f_2:   being able to cause by doing,
f_3:   being able to cause by not doing,
f_4:   not being able to prevent.

This scheme reflects certainly early experiences, but is it somehow experienced as a system of functions? Asking this question means answering it in the negative. Or it requires much empirical experience and a high level of experiencing functions in order to place this in the context of functions. The right place for schemes like the preceding one is the end rather than the beginning of our knowledge about functions.

I chose this pattern of functions as being the simplest possible. Yet logical simplicity does not mean developmental primordiality.

17.40. *Richer Structure*

Let us proceed to broader connections, structured in a richer way. As an example we choose: fitting a set of objects into more or less predetermined holes; which in a highly structured and completely determined way is realised by a jigsaw puzzle. Children educated in our environment are quite early (from about four years onwards) able to fit flat figures into characteristic holes and to complete simple jigsaw puzzles; moreover after the first success they repeat it with the same puzzle without any hesitation. The ever greater complication is conquered by ever more exercise. What exercise means for acquiring this ability, is shown when children who have grown up in greatly differing environments are confronted with such tasks; they lag far behind and slowly, if at all, catch up.

The jigsaw puzzle pieces form a set with – depending on colour and shape characteristics – a more or less clear connection structure; arranging the pieces is indeed a mapping that assigns to each piece its correct place. The criterion of the correct assignment is fitting; if it is done according to a model, it is identification with the model.

The question is again: is this action experienced as a function? I would say: what looks like an action, is in fact enforced; there is not much to be experienced like the act of assignment that characterises the function.

But there is no reason to restrict ourselves to the rigid determinism of the jigsaw puzzle. A less rigid situation is that of the circle, triangle, square shaped holes and flat objects fitting them, not necessarily according to a one-to-one assignment, but in a more general relation. The factual placing then produces a one-to-one connection, albeit between *part* of the pieces and a *part* of the holes – a function arising within the relation of fitting. Is this experienced as a function? It depends on the total context: are two functions of this kind compared with each other, are they related with each other by permutations (thus by composing with other functions), is there perhaps even a view on the totality of these functions? One can imagine natural contexts to realise it and didactical ones to promote it.

17.41. *Structured by Counting Sequence*

The same can be asserted about functions arising from counting a set of objects, or counting a required number of objects out of a given set of objects. The case looks like the preceding one, the freedom of choice may even be greater: the allocation is replaced with the counting process, the holes with the numbers. The act of assigning is entangled with that of structuring (for the benefit of counting) the set of objects to be counted or counted out, which in this way is functionally adapted to the counting sequence. Is this counting or counting out experienced as a function? Counting and counting out are so much product oriented, that is towards determining the number or producing a set of a given number of objects, that the act is hardly experienced as such, unless this natural and didactical context is stressed to make it conscious.

To a higher degree the function can be experienced in the equipartition process (dealing cards, sharing sweets, and so on) which produces functions that need not be any more one-to-one as in the preceding examples. But here too much depends on the context as to whether the functional experience is made possible or not.

17.42. *Function and Cardinal Number*

At an earlier opportunity* I have reported learning experiences with the "drawer principle" and have asked the question which mental conditions must be fulfilled in order that it works. It was the well-known question as to whether there exist two (not bald) people on the world with the same number of hairs on their heads, or rather a didactical sequence of this kind of question. The solution can be mathematised by means of the function that assigns to each person the number of his hairs; since the number of hairs on whichever head is certainly smaller than that of people, this function cannot possibly be one-to-one. I linked the successes of ten-year olds and the failures of eight-year olds to the ability or non-ability to constitute certain sets and mappings. As soon as there is more structure to profit from, this can be easier. On the same occasion** I reported certain questions to test the mental possession of the cardinal number: A vase, or its picture, decorated with a girdle of alternating suns and moons (only partly visible) and the question "which is more, suns or moons?". Or a more sophisticated one about a wall-paper pattern, or a long open chain of beads, alternating red and blue — or its picture — with the complication that the ends can be equally and unequally coloured. Or the question of whether each layer of a —drawn — wall contains the same number of bricks, or how a game should be played so that each gets the same number of turns. Children in the first grade can not only answer such questions by "yes" and "no" but also motivate the answer meaningfully, that is, by arguments that reveal their experiences with the relevant one-to-one mappings — mappings that are inherent to the given structure but not forced upon the user as happens with the jigsaw puzzle and which differ from those occurring in counting and counting out by the fact that they can be made conscious by the query "why?".

* *Weeding and Sowing*, pp. 210–214.
** P. 199.

17.43. *Linguistic Structure as a Source of the Function*

The possibility of experiencing functions cannot but be determined to a high degree by the given structure of the underlying sets and of the functional relation — a structure that may be geometric or at least visual. Even linguistic structure can be influential. The functional connection between units and tens by means of "... ty" is at least by some children experienced before they know written numbers and after they become acquainted with written numbers before they can identify this function with the function "ten times . . .". For that matter the language creates numerous — not always regular enough — opportunities like this one: suffixes for comparatives and superlatives, tenses, plurals, action and actor nouns, and so on — none of them as convincing as the one we related to "10 times . . .", "100 times . . .", and so on, though experienced earlier. As early or even earlier can geometric or geometrically structured connections be experienced as functions — I will deal with them at a later stage, in order to tap meanwhile a particularly rich source of experiencing functions.

17.44. *Geometric Mappings as a Source of the Function*

There can be little doubt that geometrical mapping if picturally structured, in particular faithful copies, but also mappings in scale, are experienced as functions — in Section 14.5 I cogently refuted Piaget's postulate that this would require the cartesian structuring of the plane. I have extensively dealt with the mental constitution of all kind of geometric mappings in Section 12.6. All kinds of — that is, not only pointwise and geometrically heavily structured ones but also objectwise ones — as in the compository reproduction (Section 8.9) — which are geometrically less structured. It is an advanced stage, though with geometric mappings more easily accessible than with others, that such mappings are not only passively and actively experienced but also objectivated and named. A rare case:

In his holiday resort, Bastiaan (7; 3) has made, along a piece of a brooklet, a model of the North Sea coast of The Netherlands, Germany and Danmark, with islands and so on, as far as known to him. He calls it a miniature [North Sea] — a word he probably knows from golf. His grandmother takes a picture of his work. He says: "This becomes a double miniature".

The mapping here is so strongly felt as a mental object that it expresses itself in the composition of mappings, and the verbalisation of the activity of composing.

17.45. *"The . . . the . . ." as a Source of the Function*

In Section 17.38 and even earlier I placed the pattern "the . . . the . . ." into the context of the function. Rightly, I think, since this relation in a way expresses a property of functions, called

   monotonicity:

If A, B are supposed to be ordered sets, the function f from A to B is called monotonic increasing if

$$a < b \to f(a) < f(b)$$

and monotonic decreasing if

$$a < b \to f(a) > f(b).$$

(A weaker form of monotonicity requires only $f(a) \leqq f(b)$ and $f(a) \geqq f(b)$, respectively, but this kind of sophistication does not matter now.)

Monotonicity is certainly experienced at an early age, but is this then a monotonicity that presents itself as property of functions? One would be inclined to say if monotonicity is a property — exclusively — of functions then empirical and mental experience and use of monotonicity presupposes that of the function. This, however, would be too rash a conclusion. In our mathematical formulation of monotonicity A and B were ordered sets. One may doubt whether in the monotonicity experienced in the pattern "the . . . the . . .", the underlying domains (the sets A and B) are really experienced as ordered sets or even as sets.

In Section 3.12a I have discussed order structures and explained what is required for the constitution of order in a set as a mental object. Comparatives like those occurring in the pattern "the . . . the . . ." can be restricted to local comparing; they need not indicate the mental presence of a set, let alone an ordered set or even something like a scale. "Doll growing bigger, chalk growing smaller" quoted in Section 17.2, is a surprisingly early utterance, which seems to reveal something about the constitution of variables in ordered sets and of a monotonic dependence. *Seems* to reveal — in fact it can be interpreted merely locally, that is, as the gradual transfer of chalk on the drawing board.

Moreover one has to admit that with comparatives regarding spatial (or perhaps temporal) dimensions in the pattern "the . . . the . . ." we are left in a rather concrete sphere. Comparatives of intensities like

> heavier,* hotter, sweeter, dearer, nicer

bear witness even less to the mental existence of sets or, let alone to the ordered sets of the qualities they are supposed to order.

I would require for "the . . . the . . ." to bear witness to an experienced function that the order structures in both "the"-domains are more than only locally experienced and that the same should hold for the relation connecting them.

Order and monotonicity are weak structures. In stronger structures the function could possibly be more strongly experienced. This strengthening can take place in quantities and magnitudes. Then the pattern "the . . . the . . ." can acquire greater precision. This may happen in different ways:

---

* In the development "heavy" is initially indeed intensive, as at a very early stage are even "long", "far", and so on.

additive — subtractive
multiplicative — inverse.

The additive–subtractive variant could in a merely mathematical context easily be neglected, though in the development it might be the original one. The multiplicative kind gets its first expression in ratios and proportionality, not only of magnitudes but also of their powers, positive and negative ones.

We pursue the exposition in both directions.

17.46. *"The . . . the . . ." Additive–Subtractive*

One of the earliest functional specifications of the pattern "the . . . the . . .", experienced by children, is the functional dependence

between ages in the course of time,

an additive dependence, which lends itself even for

functional composition and inversion.

x years older

has as its inverse

x years younger.

"A is x years older than B" and "B is y years older than C"

are composed to get

A is $x + y$ years older than C.

Because of the different birthdays it is not a clean function, though

age is a (discontinuous) function of time,

which, however, is experienced as such not until a more advanced stage, as will be the connection with the earlier mentioned functions.

Beside this additive specification there are — perhaps not that early in the development — subtractive ones:

On the way from A to B,
In the course of time from A to B
the farther from A the nearer to B

made precise by the knowledge about spatial and temporal distances and their additivity,

distance from B = distance AB — distance from A,

an equality which can more or less be experienced precisely. There are a lot of
realisations of the subtractive "the ... the ...":

> the string through a slit in the door,
>> the more inside the less outside,
>
> pouring from one vessel in the other,
>> the more in the one the less in the other,
>
> on the seesaw,
>> the higher the one the lower the other,
>
> playing with marbles,
>> the more the one wins, the more the other loses.

Additive and subtractive specification of the pattern "the ... the ..." can
precede — at least as far as magnitudes are concerned — the additive and subtractive arithmetic. It is a didactical problem as to whether the experience of these
functions helps, or is killed by, arithmetical instruction — something to be
recalled.

17.47. *"The ... the ..." in Proportion*

Even less does the multiplicative specification of "the ... the ..." require the
support of arithmetical multiplication. "Multiplicative" is even not a good
term for what I actually mean. It is "multiplicative" if viewed from an already
constituted linear function; on the road towards this constitution the correct
term is "ratio" or "proportion", which shows up in the title of the present
section.

I have dealt with ratio and proportion extensively in Chapter VI. This very
fact indicates that I reserve a particular place for ratio, not only in the development but also in didactics — contrary to modern trends.

It is a commonplace that ratio and proportion have to do with the linear
function. In antiquity (and not long ago even at school) much was described
in the language of ratio, which is now done with linear functions — if need
be with certain positive or negative powers attached. Nevertheless I have claimed
a special place for ratios and proportions as opposed to fractions and linear
functions, a kind of phenomenology, which must have descriptive consequences
in developmental and normative ones in didactical respect.

In the development, ratio comes prior to the linear function, perhaps even
to function at all, and this origin should leave didactical traces. I mention
"relatively larger, smaller, sweeter, more expensive, and so on" (Section 6.8),
which precede ratio as a mental object and only by mathematisation on a high
level can be fitted to the pattern of the linear function. I also mention the exposition and composition states (Section 6.4), which are proportionally compared
with each other with no intervention of linear functions. If the linear function

is experienced as a connection between variable objects — as is natural — the appeal to linear functions for norming and renorming (Sections 6.9–11) requires a mental switch to *measuring numbers* instead of objects, which is only possible in an advanced stage. The visualisations (Section 6.13) of position and composition states differ widely from the usual visualisations of the linear function. Criteria for linearity (Section 6.16) will more often be derived from corresponding ones for preservation of ratio than conversely.

Closer connections between proportionality and linearity are found in the structure states (Section 6.5), particularly in those of geometrical origin. If internal length ratios in a piece of reality are compared with those in their images, or internal length ratios in two images with each other, the dominatingly experienced function is the more global geometric mapping; there is little need for a separate experience of the length increase or reduction as a function, unless it is instigated or — in an advanced stage — it is used for explicit *calculations* with scales. On geographical maps the centimeter on the paper *is* the kilometer or so in reality with no intervention of the function that transforms centimeters into kilometers. Experiencing such a function could even cause damage to the operationality of the correspondence of distances. More strongly than the action of comparing does that of constructive increasing or reducing work in favour of experiencing functions.

17.48. *"The . . . the . . .", Proportionality and Function*

In Section 17.44 we pointed out the conditions which the structure involved by a pattern "the . . . the . . ." should fulfill in order to witness the experience of the function. Even then this experience might remain qualitative as long as the pattern "the . . . the . . ." is not sufficiently specified. Such specification can tend towards the additive–subtractive or the multiplicative-inverse. Mutual dependence of ages in the course of time is expressed by an additive constant — a function that becomes familiar to children at an early age. As stressed before, this does not at all mean that the multiplicative interpretation of the pattern "the . . . the . . ." must wait for the acquisition of the arithmetic multiplication or that it is necessarily supported by it. The natural development seems rather to take place along proportionality understood as equality or at least qualitative agreement of internal ratio. I recall the example of Bastiaan (see Section 6.7) who reproduced the ratio altitude of two kinds of clouds by distances of his hands from the ground — the strong need for ratio models leads to a first step towards the linear function. Recognising and reproducing topographic data on scale, comparing well or less well arranged mixtures (red and blue beads in a chain or some other pattern), distribution problems according to given or wanted ratios require no multiplicative techniques, though at a later stage they can be formulated and solved multiplicatively, by means of linear functions. Here too the specification of the pattern "the . . . the . . ." by means of the linear function is preceded by that of proportionality.

17.49. *Aiming and Hitting as Source of the Function*

The link between aiming and hitting is in the experience of the function perhaps the earliest. Monotonicity plays an important part in this, somewhat refined by more or less rough quantification. Hitting too high elicits aiming lower, too far to the right elicits a correction towards the left, and conversely, and this can be modulated by adverbs like much, a bit, a little bit, a bit more, a bit less. Marksmanship in general is conditioned by a wonderful feedback system in the brain but this does not exclude opportunities to consciously steer the aim variable and observe the hit variable. As a matter of fact aiming and hitting is not restricted to a spatial context. Times, colours, sound pitches and volumes can be an aim, which is to be hit.

Even if there is no quantification, successful hitting supposes as a precondition: knowledge about the kind of monotonicity in the relation between aiming and hitting. By this one gets enabled to steer the right way. The frequency of qualitative errors in this field — even among adults — is rather high, even if there are no misleading cues as with the behaviour before the mirror, looking through an astronomic telescope, backward driving with a trailer. If feedback is lacking or weak the reaction often goes the wrong way. A well-known example: if a lever is to be balanced, young children react by charging the heavy side even more, in particular if the extra charge does not change the — unstable — state of the lever. If older ones perform better it is not a symptom of progressive development as some psychologists claim but empirical experience. All the same serious psychological research into the reasons for the wrong way reaction would be useful.

Repeatedly* I have recommended the aim-hit-relation as a paradigm for continuity: little causes yield little consequences. Continuity is not being discussed now; we are far away of any problematic regarding continuity. What counts here, is the dependence itself between aiming and hitting, which is used "to score a bull's eye".

Aiming and hitting cuts in two ways as a source of experiencing the function:

> globally and locally.

Globally means — if space is concerned — the mapping of the

> cone of aim directions on the target,

thus a mapping in two dimensions, which can serve as a useful paradigm for mappings. The local way is operationalised in

> scoring a bull's eye,

* For instance 'Les niveaux de l'apprentissage des concepts de limite et de continuité.' Atti del Convegno Internazionale sul tema Storia, Pedagogia e Filosofia della Scienza; Pisa, Bologna, Roma 7–12 Oct. 1971. Accad. dei Lincei 1973, Quaderno 184.

that is choosing the aim variable appropriately to get a wanted value for the hit variable. Here this activity has been placed in the context of "the . . . the . . .". Indeed what happens if one repeatedly aims at a goal is answering deviations in a perceived sense by corrections in the opposite one.

Both global and local function experience are integrated in aiming at and hitting of a moving target. Here we come again across a phenomenon we have met before without paying explicitly attention to it: compensation.

17.50. *Compensation in the Pattern "the . . . the . . ." as a Source of the Function*

Most often one independent variable is concerned if a dependent variable is to be controlled. Control of a dependent variable may mean keeping it constant: hitting a fixed target, balancing a lever or some other contraption, and so on. The variable z I want to stabilise depends on the variable x I can perceive and the variable y I can directly control, and the dependences are monotonic — at least in the local domain I am concerned with. The direction of monotonicity must be known in order control to be feasible and even somehow specified quantitatively in order the influence to be more effective.

In order to stabilise z I have to compensate for the perceived or anticipated influence of x on z by a change of y. This activity can be

preprogrammed or exercised

to such a high degree that the compensation is

practically perfect

or perfection can be more or less

approximated

by

aiming and hitting.

Yet the compensation can also

theoretically be steered

by experiencing certain lawlike connections such as

compensating for the length in the width in order to keep the area of a rectangle constant,

compensation of lever variables to keep equilibrium.

In all cases the influence of x on y that stabilises the variable z can be experienced as a function — a function where the composition via z of

$$x \to z, \quad z \to y$$

may or may not be an explicit component of the experience.

17.50a. *Inter- and Extrapolation as Sources of the Function*

> Meteorological data read thrice a day,
> the precipitation of the last 24 hours registered at 8 AM,
> the morning and evening temperature of a patient,
> the number of traffic deaths each month in The Netherlands,
> the gold treasure of the National Bank at the end of each month,
> the cost of living at the first of each month,
> the electricity consumption of the Netherlands in a calendar year,
> the (estimated) world population in subsequent years,
> grades on successive reports of a pupil,
> the number of storks breeding in The Netherlands each summer —

these are primarily discrete data, often presented by line graphs, and nominated to be functions of time. They are rightly so, if intermediate values can be ascribed to them though they have neither been measured nor estimated as such; hypothetically some of these functions can be extended beyond the limits of measuring and estimating, to the past or future. Between the measuring points it is done by linear or somehow smooth interpolation; outside the measuring interval by continuing an apparent or patent trend.

In some of these cases intermediate states are out of the question. The births, traffic deaths, the electricity consumption have already been accumulated monthly; the grades of a pupil express his performance over a period; the storks leave in the winter. This does not prevent one from constructing functions.

The weight of a more or less coarse matter as a function of the quantity, its price as a function of the weight are primarily functions of a discrete variable, continuously interpolated. Sometimes the meshes of the independent variable are indeed so narrow that interpolation is natural and admissible, even if the matter consists of indivisible units. This continuisation of the discrete extends even to mathematics, for instance to the number of primes below n, called $\pi(n)$, if it is estimated by $n/\log n$.

17.51. *Piaget*

The function has also been one of Piaget's themes. The work where this is exposed* is utterly disappointing. A quarter of the booklet, by Piaget & Grize, is a kind of epistemology: generating functions by what is alternately called "structuring functions", "constitutive functions", "coordinators", with — in particular by Grize — the display of mathematical terminology that is out of order, and formulae that look cabalistical rather than mathematical. If rather than five times one would try and read this part fifty times, one would perhaps

---

* J. Piaget, J.-B. Grize, Alina Szeminska, and Vinh Bang, *Epistémologie et psychologie de la fonction* (Etudes d'épistémologie génétique XXIII), Paris, 1968. Translation: *Epistemology and Psychology of Functions*, Dordrecht, 1977.

find out what the authors intended to convey with this theory, but it is a comfort to notice that the collaborators, responsible for the experimental part (three quarters of the book) did not understand it either, or at least did not show that they cared to. The terminology of the epistemological part is only used, but hardly elucidated, in Chapter 3 and then in research where the relation with functions is farfetched if existent at all.

The relation with functions is more often lacking. In some cases a phenomenological appearance of the function is belaboured in a manner that everything that has to do with functions is extracted: the subject is asked to perform artificial, aimless and useless, operations, the sediment as it were of what was left after the extraction. (For instance Chapter 1 on permutations.)

Another objection can fairly be illustrated by the experiments of Chapter 2. They are what in Section 17.40 we called the jigsaw puzzle type, albeit with a certain flexibility. What is tested, however, is whether children can manage the puzzle rather than whether the activity of the more or less enforced assignment is experienced as a function. The same objection can be raised more or less as criticism of the other test series. For instance, in Chapter 5 the — subtractive — connection between the two pieces in which a string is divided by a roll over which it is conducted; in Chapter 6 the connection between the two parts in which a quantity is divided; in Chapter 7 the connection between the levels of the same quantity of liquid in glasses of different shape; in Chapter 8 the two sides of a rectangle with a given circumference; in Chapters 9–12 with proportions. It is always the same: about something that is viewed by the experimenter as a function, asking the subject to predict for a given value of one variable the corresponding one of the other. What is not done or not in a sufficiently structured way is: finding out whether the connection is experienced by the subject as a function. Of course, somebody who experiences the connection as a function, has a better chance to score high on such tests. But conversely, a wrong answer can be caused by a function experience, which happens to be wrong. Intentional experiments could show which is the one and which the other. Unfortunately the experiments are often artificially complicated to such a degree that the experimenters themselves were unable to describe them decently. Protocols of interviews are often lacking or restricted to examples, which do not allow one to check how the experimenters had instructed the subjects who were expected to perform the complicated tasks.

The investigation on proportions is also too little tuned on functions; the tasks are needlessly complex.

The lack of phenomenology there is not compensated by what the authors call epistemology.

17.52–57. *The Didactics of the Function Experience*

17.52. *Experiencing*

In this phenomenology the present chapter is the first where I used the term "experiencing". I should say I am not too happy with this term. In German I would say *erleben*, which is just what I mean. In English I tried sometimes "lived through" or "felt" instead of "experienced", but this does not work too well except in rare cases. In French I would choose *éprouvé* though the undertone of "suffered" can be misleading. The advantage in German is that there are two words *erleben* und *erfahren*, which in English are covered by one. What is the difference? A shade? Or isn't it more? Translating is a good opportunity to analyse what one knows by intuition — no, by use. German *Aha-Erlebnis* is translated by aha-experience, there is no *Aha-Erfahrung; erleben* is more emotional than *erfahren*. Older people are wiser by *Erfahrung* and sadder by *Erlebnis* — is it this way? I think I had better stay close to mathematics. Time, space, infinity if *erlebt* can be empty, if *erfahren* they are likely to be full. It is easier to tell people about *erfahren* than *erleben; erfahren* is easier to be verbalised than *erleben*.

I would like to understand experiencing in this less pregnant way though I do not know how to do it. One remembers that from the start onwards I pushed concept formation far away. Constitution of mental objects was the first that matters. Small cardinals are early constituted as mental objects, yet one number is not the other (while any point can be any other); the constitution of big natural numbers happens within that of *the* natural number, which moveover can be differentiated according to different aspects. The point, the straight line, the circle, the sphere, however, are mutually constituted in all their generality without any previous constitution of specialisations (except that in the case of the straight line horizontal and vertical ones can initially play a special part). In the case of magnitudes, real numbers, ratio, proportion I did not feel the need for an explicitly indicated period of experiencing preceding that of constituting the mental object, though it would not have been farfetched to do so.

The mental constitution of the *function*, in an advanced stage, can hardly be distinguished from concept formation. I stressed earlier that so much and so many things are comprised by the function concept that all this hardly corresponds to one mental object. Moreover beside mental objects of vastly diverging origin and character, they are linguistic terminological structures, lumped together conceptually — by means of an explicit definition. Investigations up to the highest grades of secondary education* have pointed out that this conceptual definition does not work, that is, that in instructional practice it has neither replaced nor overarched the various mental objects.

Various aspects of the functions can grow to become mental objects, though in an early stage and not without directed didactical support. As a mental object the function is more complex than number, than geometrical objects, even than ratio, though one aspect may be more easily accessible than the other.

---

* See, e.g., R. Barra and J. J. Pensec, 'Un défaut de notre enseignement à propos de fonctions', *Bulletin APMEP* 57 (1978), pp. 357–362.

Therefore a broader margin should be left for experiencing, allied to using and provoking, of functions, before they are made conscious, accepted as an object and named. This of course does not exclude didactical attempts at level raising.

This leads us to the problem of how to justify having and letting functions be experienced. There is not the slightest need for functions in order to base the cardinal number on them, or to introduce little finite groups. One could, however, suppose and even maintain that the function, made explicit or not, can fulfill an important strategic task. Repeatedly I pointed out that mathematicians avail themselves of bundles of strategies, which for a long time could have been operational before somebody hit on the idea to condense them in a comprehensive theory. Even at the highest levels of mathematical education it can be didactically recommendable in this matter to pursue somehow the historical course, that is, maintaining and teaching as incidental strategies things that will later be overarched by a theory.

17.53. *Geometrical Mappings*

According to the title this could be a repetition of Chapter 12. But there geometrical mappings ranged in the context of geometry whereas here they are meant in that of functions. Nevertheless much could be repeated. I will try to avoid this.

It is in the visual realisation that the function is the easiest accessible. How strongly can it be experienced in this approach? Reproduction of a piece of (genuine or imagined) visual reality is often compository in character: objects from the piece of reality are reproduced by partial pictures, united in a composition to one picture of that reality, and then according to structure rules that are arbitrary or difficult to find out — a structural relation, which can perhaps be interpreted but hardly experienced as a function.

Perspective reproduction on paper of a three-dimensional piece of reality is, at least in its initial stage, as we pointed out earlier, rather a kind of composition; perspectivity as mapping by means of vision lines is a rather advanced stage.

If we speak of geometrical mappings, we mean a pointwise rather than an objectwise assignment, which can most strongly be experienced in the relation between an objective situation and a picture or model of it or between two pictures or models of the same objective situation. This then is — at least in principle — the pairwise mapping which can be

> given in order to be stated, or
> asked to be made or completed.

A special case is

> given or asked for symmetries

in a picture or model.

It is not required that such a mapping be a similarity; it can be a subject of search or discussion to find out the rules governing the mutual ratios (if there are any). What matters is

experiencing the act of assigning as a pointwise one

and moreover

experiencing the composition and inversion of mappings,

as soon as more pictures of the same objective situation are at stake. In a natural way this happens if

the same drawing is copied from the blackboard by a number of pupils,

or if

pupils are given copies of the same demonstration picture.

The composition of the mappings

copy of pupil $A \to$ copy of pupil B
copy of pupil $B \to$ copy of pupil C

and their inversions can here be experienced. In the case of similarities

scale factors and their behaviour under composition and inversion

can be stated or observed. But even with mappings of other sorts

characteristics and their behaviour under composition and inversion

are worth being taken into account.

I considered here pictures and models of an objective situation because they lend themselves excellently to suggesting pointwise assignment. Geometrical figures as objects have been dealt with in Chapter 13 with this intention. The assignment there is in fact restricted to a finite number of points (for instance the vertices of a tetrahedron) in order to be extended according to geometric rules — a procedure at a higher level than is viewed here. Even if the geometric figure is strongly prestructured, for instance by means of a lattice or like the number line, the experience of the pointwise assignment is more demanding than in the case of naturally structured pictures and models of an objective situation. In fact it requires, as pointed out above, a knowledge of geometric relations to perform exact geometric constructions or a kind of aim—hit procedure by which supposed assignments are systematically corrected.

17.54. *Classifying by Parametrisation*

Classifying is a popular subject in developmental psychology. Two trends of research can be distinguished: research whether subjects can perform a certain classification or a certain kind of classifications — for instance according to categories like living and non-living, solid—liquid—gaseous, size, weight, age — or investigating whether subjects master classifying as a formal activity, that is, using formal criteria which can be imposed on classifications. Both kind of abilities get sometimes entangled with each other in the course of such experiments, with the consequence that one does not know for sure what is actually being tested.

Functions have to do with classifications because they can be used in classifying. A function from A to B classifies in A by partitioning A into classes of originals of the various elements of B, where moreover a possible structure in B can be transferred to the system of classes. For instance the function that assigns to everybody one's age, divides the population into age classes, which can be ordered according to the ages they represent.

Classifying as understood and investigated by psychologists is a more primitive and rough conception than that which is performed as such in mathematics. This is particularly true with respect to the operations acting on and between classes and the connection with functions (as far as ever considered by psychologists). As a matter of fact we will also view the connection between classifications and functions in a special direction. Rather than in *a posteriori* classification by means of an *a priori* given function we are interested in the construction of a function for the benefit of, and as a tool for, classification.

Classifying can be done by *comprehension*, that is, by arbitrarily combining into classes or by *discrimination*, that is by distinguishing characteristics that are interesting in themselves. The second manner is by far the most important.

A characteristic of classification, as intended, can be *colour*. The function generating this classification is "colour of", that is, a function that to everything assigns its colour. In a didactical context, aiming at experiencing functions, it is a wrong example. As far as the colour can or must serve as a distinguishing characteristic, one can be satisfied with this use rather than dragging in a function, which is a needless duplication of the distinguishing characteristic.

The title of the present section sounds a bit strange, perhaps by its brevity, but even with more words I could not have made the intention clearer. So I will start the exposition with an example, which is rich enough to be paradigmatic, and brief enough to be clear.

I am considering a rhombus, realised by the combination of four equally long rods which are hinging in joints (Figure 224). The rhombus can take different shapes: a square, more or less stretched or pressed, up to coincidence of the

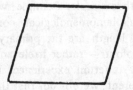

Fig. 224.

sides. It is a whole system of geometric figures — though all of them rhombuses — produced by this instrument. The term "classification" in the title aims at the need for mutual distinction of all these rhombuses. Distinguishing can be done by characteristics such as mentioned before: square, more or less stretched. Yet "more or less" can be made more precise. The term "parametrisation" in the title means the choice of a parameter that can be used to distinguish various rhombuses. If such a parameter is found, then the

> shape of the rhombus as a function of the parameter

is a natural experience of a function.

In fact such a parameter is itself a function, to wit of the rhombic shape it describes. The choice is among a number of possible parameters,

> the angle of two rods,
> the "height" of the rhombus,
> the "width" of the rhombus,
> the distance between opposite vertices (diagonal),
> the area,

a system of variables that can also be viewed in their mutual dependence

> qualitatively or more or less specified.

Not only

> shape as function of a parameter

but also

> parameters as functions of other parameters

can be experienced, as well as the mutual relatedness by means of

> composition and inversion.

As a matter of fact the change of perspective

> from shape to a required parameter

towards

> shape determined by a parameter value

is worth being experienced.

I could have started with a simpler example, which, however, would have lacked the rich didactical-phenomenological context. For instance, with Cuisinaire rods colour and length are the primary criteria of classification. I already pointed out that colour — rather irrelevant in the present case — can hardly be an incentive for function experience. Length, measured with the shortest unit, could be different, were it not that the system of rods is too soon identified with the parameter: the rod of length 3 *is* 3. This quasi identity is an impediment in the way of experiencing a non-trivial function. The number line, right or curved, is not much different: it is just the intention that the points are identified with the numbers at the points, which trivialises the assignment required for the function. Number lines on different scales is another case, which, however, belongs rather to the geometrical mappings dealt with in Section 17.53.

Squares, cubes, circles, spheres classified up to congruence is a better example, though less rich than the rhombuses we dealt with formerly. Parameters, which introduce themselves, are diameter, side or diagonal, and perhaps area and volume. They are appropriate as examples for the function experience, but to begin with, a richer example such as that proposed, looks more appropriate.

A rich example is also that of rectangles classified up to similarity. Parameters that present themselves are

> the angle of the diagonals,
> the angle between diagonal and side,
> the ratio of sides,
> the ratio of side and diagonal,

where the occurrence of ratio in the last example requires broader previous knowledge than was supposed in the former examples.

A similar, perhaps even richer, example is triangles classified up to congruence or similarity.

A more elementary case is rectangles with a given circumference, classified up to congruence, with classifying parameters

> length of a side,
> length of a diagonal,
> angle of side and diagonal,
> angle of the diagonals,
> area.

A sophistication we did not pay attention to so far: are the considered rectangles supposed to possess well-distinguished (numbered) vertices, or are the "lying" and "standing" rectangle, if congruent, to be consiedered as equivalent? According to this decision certain parameters shall be more precisely specified; certain function can become two-to-one rather than one-to-one.

The preceding examples do not exhaust the possibilities. Rather than with respect to congruence or similarity one can classify up to translations and, for instance, choose a

> system of circles with the centre on a fixed line and the radius constant
> or depending on the centre,

or the

> system of equilateral triangles inscribed to a fixed triangle

or the

> system of squares inscribed to a fixed square,

systems which can nicely be realised by a kinematic contraption (Figure 225):
A closed string as the periphery of the fixed figure with three or four nooses,
respectively, attached at points that divide the circumference of the string in
equal parts, and through the nooses an elastic string.

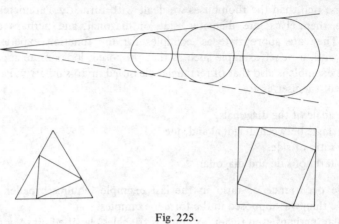

Fig. 225.

There are a great many such mechanisms. To terminate this section with an
example of a string construction we mention that of the ellipse: the point of
the ellipse as a function of one of the radii vectores or of one of the related
angles or all of them as functions of each other; and the shape of the ellipse
as a function of focal distance and long axis.

17.55. *Time as the Independent Variable*

In Sections 17.53—54 geometry was dominating. There was no need for visualising since the data and their mutual relation themselves were visual. An effective
visualisation of the function experience is function graphs. Of course a number
of functions we came across in Section 17.54 can also graphically be visualised
and for a better understanding this can be useful in some cases. Various approaches towards graphical visualisation of functions will be reviewed later on.
One of them, and perhaps the didactically most effective, is the one where the
"horizontal" axis is interpreted as a time axis (or time log).

The development of time experience is perhaps even more mysterious then
that of space experience. Though little children have trouble with the distinction
between "yesterday" and "tomorrow", one may ask whether this fact really
proves a disability in distinguishing past and future, or is it rather a linguistic
disability comparable to the confusion of colour *names* while colours as such
and the difference of colours is mastered at a quite early age. Whatever the

reason may be, in the development of time experience, periodicity of the happenings in time must play an important role, the change of light and dark, waking and sleeping, school times and Sundays — celebrated in the church or not —, vacations, birthdays, Santa Claus. By the events in time, and certainly by the periodic events, time is structured. The days of the week, the hours of the day determine an order structure, the "unwinding" of a primordially cyclic structure, and even more: a mental — rough — measurement of time.

Length and time measurement differ greatly. Lengths can be compared by direct transfer, with an increasing precision. In order to compare and measure time intervals as reliably one needs more or less complicated instruments. Or aren't we in this respect spoiled by the possession of such instruments?

It seems that in his experiments on the inclined plane Galileo measured time intervals, by singing a melody. So it is perhaps not that difficult to keep a fixed rhythm to use it for time measurement — as a matter of fact the duration of a *pater noster* or *ave Maria* was from olden times a not unusual time measure.

At a kind of bowling (in rails) where the number of tumbling skittles depends on the initial ball velocity only, I succeeded after a few observations in predicting exactly the result — which shows that velocities can be observed with a reasonable precision.

Time is a magnitude as is length. But, other than length, time is not primordially a magnitude. Lengths can be compared and displaced within the plane and space. Time is first of all experienced, that is, the thing experienced is the distinction between moments and the order structure. Duration is secondary.

Time as a linearly ordered system precedes time measurement and time interval. With length it is just the other way round. Straight lines can be structured or unstructured. One line is not the other, but to acquire the mental object of straight line, identifications are required. Time as such is unique and structured by living in it. As a matter of fact the microstructure of time is more easily accessible than its macrostructure. Globally time is structured by history. Elswhere* I pointed out that for 8—9-year olds the past may still be unstructured though accessible to structuring.

Structuring of time — the order structure as well as the metric one — can intentionally be supported by visualising: the spatial translation of time into the time axis. It is meaningful to use this "model" of time from the beginning onwards on *various scales* with the structural units of

century, year, month, week, day, hour, minute, second,

to accentuate the structuring by

history, life history, season's events, school events, day division, clock division,

* *Weeding and Sowing*, p. 290. *Zentralblatt Didaktik d.Mathematik* 10 (1978), No. 2, pp. 76—77.

and illustrate the last example by

> unwinding the dial upon the line,

a geometrically active illustration.

Not surprisingly does the introduction of the time axis historically coincide with that of function graphs (Oresmus in the 14th century), that is, graphs of simple functions. Even now we feel obliged to introduce time as a function variable by certain contemplations on time as a magnitude and on the time axis.

Didactically it can be useful to have dependence on time experienced as a function, and the most natural objects for this are

> growth processes.

First of all a somewhat improper but nevertheless effective example, the growth of age:

> age of a person as a function of calendar time

or

> age of one person as a function of that of another,

in both cases graphically represented (if scales are equal) by

> a line, inclined 45°,

provided age is interpreted as a continuous variable. If age is rounded in the usual way, that is downwards, then the

> age of a person as a function of calendar time

is a step function.

Age graphs are a means to answer questions like

> when was A double the age of B?
> when will A be double the age of B?

at a glance.

The condition "A double the age of B" can also be visualised by a function graph; problems that involve this condition can be formulated as a pair of equations in two variables and solved by intersecting graphs. The function "A double the age of B", however, is of an entirely other character than that which connects the ages of A and B. The first is as it were of logical, the second of physical origin. I realised this important fact when I noticed with higher grade students the confusion caused by the difficulty of grasping and appreciating this difference: they tended to reject the function "A double the age of B" because it is not preserved in the course of time. It looks a strange attitude which, however, would be worth investigating more closely.

Age as a function of time is an improper growth process. Growth processes rather regard such variables as

length, weight, product, number, energy consumption

of

persons, animals, nations, population, industry

as a function of time. I used the word "growth" which, though not in the vernacular but by convention, also means

loss as negative growth,

a new example of the habit, since the adoptation of "plus" and "minus", to mathematise polarities by positive—negative (see Section 14.9). A certain object becomes in time more or less according to some numerical property because it gains or looses something.

Another kind is

air temperature, air pressure, prices

as a function of time.

Again of another character, though no growth processes:

path, velocity, acceleration

as functions of time.

The algorithmic operational force of most of these function examples, as understood in Section 17.37, is not great. Most of them derive their practical and didactical value from the visualisation by graphs, which demonstrate not only

growth and loss,

but also

the strength of growth and loss.

Experiencing growth functions certainly includes the recognition of such features. It is quite another thing to experience functions and classes of functions by

laws of growth.

Certainly experiencing and even verbalising of a

linear law of growth

is possible early. An

exponential law of growth

— however strongly experienced by trained persons — requires much more. One would guess that handheld calculators should make an early, that is prealgorithmic, experience of the exponential law of growth possible, but I do not know of any experiments or observations that would confirm this hypothesis.

Special didactic attention what regards experiencing functions, should be paid to

 time—distance graphs.

Such a graph, which indeed describes a motion, easily conveys the misleading suggestion of picturing the path covered by the moving object.

Pupils who — in grades 5 to 6 — are for the first time taught graphs and well the time—distance type, for instance graphs of the course of trains along a certain traject, or of a skating competition, are inclined to interpret intersections of this kind of graphs as intersections of paths.

It seems didactically commendable not to introduce time—distance graphs, and for this reason also distance as a function of time, unless the pupil has beoome familiar with graphs of functions in other contexts.

On the other hand richer opportunities are offered by time—distance graphs and functions than by any other kind, thanks to the special meaning of

 intersections of such graphs in one picture

as well as to the particular character of related measures of growth, such as

 velocity and acceleration,

witnessed by proper names, which can provoke the experiencing of

 new functions and graphs.

It is one of the approaches to

 derivatives

which may be possible and useful at an early stage.

Closely connected to the time—distance graphs are what I would call

 time—path graphs.

A closed circuit is as it were cut open and mapped on the "vertical" axis, which causes unavoidable discontinuities deriving not from the function but from the cut. An example: the graph of the motion of a clock-hand on the dial (Figure 226) or the (apparent) motion of the Sun or the Moon around the Earth, or one of many laps of the cycling track. This kind of graph becomes operational in problems of overtaking and meeting such as:

 When do hours and minutes hands cover each other?

 Why does an eastbound or westbound globe traveller win or lose a day, respectively (that is, sees one sunrise less or more)?

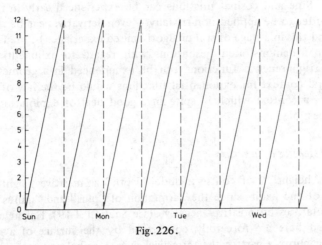

Fig. 226.

Distance is the most striking and first experienced motion parameter, but it is not the only one. We already mentioned velocity. As regards closed circuits it is elucidating to link the image of an irregular circuit and the velocity graph, where minima and maxima are determined by bends (Figure 227).*

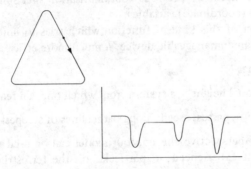

Fig. 227.

Another important motion parameter as a function of time is

projection on an axis.

The uniform motion of a point on a circular orbit, projected on the "vertical" axis (Figure 228) is a harmonic vibration, or in other words, a trigonometric

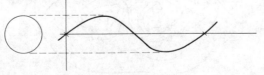

Fig. 228.

* See p. 572 footnote (Janvier).

function. Sine and cosine functions can be experienced early in this graphical way in order to be applied non-trivially — even derivatives can be read from the figure, and this in a stage of still unalgorithmised experience of the function.

Earlier I demonstrated hesitations with regard to exponential functions. Whereas trigonometric functions can be experienced in a geometric context, the natural context for exponential functions would be the law of exponential growth — a context which in spite of a good deal of exploration is still not easily accessible.

17.56. *Height as a Function*

If I say "height" I of course include "depth" as negative height — again an example of the habit, since the adoptation of "plus" and "minus", to mathematise polarities by positive—negative (see Section 4.19). Height and depth as understood here are forcefully concretised by the surface of a mountain or a sea-floor above a part of the terrestrial surface or below the sea level, respectively. Originally I chose "functions of two variables" as the title of this section. It would have been a wrong choice because a pair of variables need not be the means to experience height as a function of the underlying relevant variable. The ground plane needs no previous coordinatisation that splits the one (point) variable into two (coordinate) variables.

A visualisation of this kind of function which tries to imitate graphs, is the concrete model, an unmanageable device. A much more efficient method is

     contour lines

that is, lines of equal height — a system from which one can read

     peaks, valleys, passes, ascent along paths, lines of steepest descent.

By change of perspective the contour model can be used to have functions other than height experienced, in particular on the terrestrial surface (Figure 229),

     temperature, air pressure, density of population

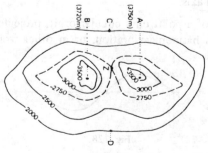

Fig. 229.

as functions of the place, with the corresponding

isolines,

though the terrestrial surface as function domain can also be replaced with a plane as such,

unstructured or structured (by a pole or as product of two lines).

17.57. *Elementary Operations as Functions*

By elementary operations I mean here adding, subtracting, multiplying, and dividing. They are binary operations, thus involving three variables: an "improper one" that is fixed in order to characterise the operator, while the two others are considered as independent and dependent, thus the kind of functions which are algebraically denoted by

$$x \rightarrow x + 3$$
$$x \rightarrow x - 3$$
$$x \rightarrow 3 - x$$
$$x \rightarrow 3 \cdot x$$
$$x \rightarrow 3/x,$$

and perhaps more complicated ones, composed from these.

There are many contexts in which to introduce

function tables

of these functions. First of all I mention a not so trivial one under its IOWO didactics term,

the double decker,

that is: a fixed number of persons is divided over the upper and lower deck which yields the function table

above	0	1	2	. . .
below	40	39	38	. . .

of the function $x \rightarrow 40 - x$. A great many variables as convincing as this can be thought of:

pages of a book: read and still left,
distribution of a fixed number of marbles between two persons,
distance covered and still left,

and so on.

There are many more opportunities to draw up function tables — I will leave it at that one example, which otherwise would perhaps escape the attention.

A modern linguistic device, which is intended to stress the functional character of elementary operations is

$$\ldots \overset{+3}{\frown} \ldots, \quad \ldots \overset{-3}{\frown} \ldots, \quad \ldots \overset{\times 3}{\frown} \ldots, \quad \ldots \overset{\div 3}{\frown} \ldots.$$

They can also be used to formulate the traditional "dot problems" more forcefully: in

$$\ldots \overset{\cdots}{\frown} \ldots$$

two values are being given and the third is asked. On the other hand the use of the arrow creates an opportunity to relieve the equality sign from improper use and to stress its static character.

A more concrete, that is, geometric feature is added to the arrows language by placing

arrows on the number line.

More involved operation patterns can be built with arrows (Figure 230).

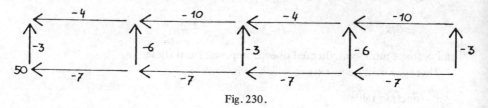

Fig. 230.

A recent — though rather verbal — concretisation of the function character of operations is

machines

with an input and output and labels characterising them, which moreover can be composed in various ways such as (Figure 231)

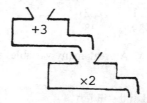

Fig. 231.

or even more involved, for instance with several inputs and outputs. This linguistic device can also replace the traditional "dot problems", if again two values are being given and the third is asked. A further schematisation leads to

flow diagrams.

The didactician will not stick to one device. He will rather have the same function experienced

> in a story, as a table, in arrow language, on the number line, as machine, as flow diagram

where stress will be laid on the

> isomorphism of the various function experiences.

17.58–63. *The Algebraisation of the Function*

17.58. *Approaches to Algebra*

Algebraised functions were already discussed in Chapters 15 and 16 (in particular Sections 15.9–12, 16.21). In the past it was a popular controversy which approach to algebra is to be preferred

> via *identities*, that is, where letters indicate *indeterminates*

or

> via *equations*, that is, where letters indicate *unknowns*.

At present one would add a third,

> via *functions*, that is, where letters indicate *variables*.

The didactic goal in the latter case is: letters as variables, and algebraic expressions — in the broadest sense — as functions of one variable or several variables, corresponding to the letters. It is quite natural to start a didactical sequence with the expedients of function experience I exposed in Section 17.57, thus with a

> translation of these expedients of function experience into algebra.

17.59. *From the Elementary Operations to Functions*

I will illustrate this with a few examples:

(a) Think of a number, add 5, multiply by 2, subtract 5. What is the result? . . . Then you had . . . in mind.

(b) Think of a number, multiply by 2, add 10, divide by 2. What is the result? . . . Then you had . . . in mind.

(c) Think of a number, add 5, square it, subtract 20. What is the result? . . . Then you had . . . in mind.

(d) Think of a number, multiply by 2, add 10, divide by 2, subtract the number you first thought of. The result is 5, isn't it?

(e) Think of a number, add 5, square it, subtract 25, divide by the number you first thought of. What is the result? . . . Then you had . . . in mind.

The "riddles" (a), (b), (c) are of the type that directly springs from the use of "machines". In (d) and (e) the number in mind is fed once more into the process. This means that in (a), (b), (c) the original number can be retrieved by inverting the process as a whole:

(a): $\ldots\overset{+5}{\frown}\ldots\overset{\times2}{\frown}\ldots\overset{-5}{\frown}$ yields $\ldots\overset{+5}{\frown}\ldots\overset{\div2}{\frown}\ldots\overset{-5}{\frown}\ldots$

(b): $\ldots\overset{\times2}{\frown}\ldots\overset{+10}{\frown}\ldots\overset{\div2}{\frown}$ yields $\ldots\overset{\times2}{\frown}\ldots\overset{-10}{\frown}\ldots\overset{\div2}{\frown}\ldots$

(c): $\ldots\overset{+5}{\frown}\ldots\overset{sq}{\frown}\ldots\overset{-20}{\frown}$ yields $\ldots\overset{+20}{\frown}\ldots\overset{\sqrt{}}{\frown}\ldots\overset{-5}{\frown}\ldots$

In (d) and (e) this does not work. Retrieving the original number seems to lead into a vicious circle, such as a millenia old prototype of this kind of queries:

A stone weighs one pound more than half its weight, what is its weight?

It seems that in order to calculate the weight of the stone one must first know half its weight, which would be true, indeed, in

A stone weighs the double of half its weight. What is its weight?

What is the difference between these problems? More profound insight leads to an answer.

Solving by starting at the end, as with type (a), (b), (c) is to my view the indispensable predecessor for learning to prove algebraic identities and solving equations: the new aspect of the circular looking problems becomes thus clearer and the search for new methods gets more strongly motivated.

The search can first mean trying:

(d): $1\overset{\times2}{\frown}2\overset{+10}{\frown}12\overset{\div2}{\frown}6\overset{-1}{\frown}5$

      $2\overset{\times2}{\frown}4\overset{+10}{\frown}14\overset{\div2}{\frown}7\overset{-2}{\frown}5$

      $3\overset{\times2}{\frown}6\overset{+10}{\frown}16\overset{\div2}{\frown}8\overset{-3}{\frown}5$

      .

      .

      .

(e): $1\overset{+5}{\frown}6\overset{\times6}{\frown}36\overset{-25}{\frown}11\overset{\div1}{\frown}11$ solved by $11\overset{-2\times5}{\frown}1$

      $2\overset{+5}{\frown}7\overset{\times7}{\frown}49\overset{-25}{\frown}24\overset{\div2}{\frown}12$ solved by $12\overset{-2\times5}{\frown}2$

      $3\overset{+5}{\frown}8\overset{\times8}{\frown}64\overset{-25}{\frown}39\overset{\div3}{\frown}13$ solved by $13\overset{-2\times5}{\frown}3$

      .

      .

      .

The next step would be introducing a variable at the place of the number in mind:

(a): $x \overset{+5}{\curvearrowright} x+5 \overset{\times 2}{\curvearrowright} 2(x+5) \overset{-5}{\curvearrowright} 2(x+5)-5$

(b): $x \overset{\times 2}{\curvearrowright} 2x \overset{+10}{\curvearrowright} 2x+10 \overset{\div 2}{\curvearrowright} (2x+10) \div 2$

(c): $\underline{x} \overset{+5}{\curvearrowright} x+5 \overset{\times(x+5)}{\curvearrowright} (x+5)^2 \overset{-20}{\curvearrowright} (x+5)^2 - 20$

(d): $x \overset{\times 2}{\curvearrowright} 2x \overset{+10}{\curvearrowright} 2x+10 \overset{\div 2}{\curvearrowright} (2x+10) \div 2 \overset{-x}{\curvearrowright} (2x+10) \div 2 - x$

(e): $x \overset{+5}{\curvearrowright} x+5 \overset{\times(x+5)}{\curvearrowright} (x+5)^2 \overset{-25}{\curvearrowright} (x+5)^2 - 25 \overset{\div x}{\curvearrowright} ((x+5)^2) - 25) \div x$

Simplifications underway would lead to a final

(a): $2x+5$

(b): $x+5$

(c): x^2+2x+5

(d): 5

(e): $x+2$

which in the cases (a), (b), (c), (e) would be compared with the "result" in order to figure out x; in the difficult case (c) the way back would be shown by the way from the number in mind to the "result".

17.60–62. *The Algebraisation of the Inversion and Composition of Functions*

17.60. The sequence of algebraisation of the function has been started with the function such as experienced. The algebraisation of the function, however, includes as well

the algebraisation of inverting and composing

of functions, which in the didactical sequence should finally get entirely algorithmised. This is a well-known thorny aspect of the algebraisation of the function. It is quite natural to require that the algebraisation of inverting and composing continues the

experience of inverting and composing

of function, in machine or arrow language, such as sketched in Section 17.57 and elaborated in Section 17.59.

The "final" notation* of the kind of functions we came across is

$$x \to x+5, \quad x \to 2x, \quad x \to 2x+5, \quad \ldots$$

* The reader will notice that in the sequel I will not mention the $\curlyvee$-notation I proposed to supersede the – rather deficient – arrow notation.

or

$$f: x \to x + 5, \qquad g: x \to 2x, \qquad h: x \to 2x + 5.$$

The difference, compared with the former arrow language, is not only that the arrows have been straightened out and the specification of the operation above the arrows has been dropped. In the definitive notation the original experience of inversion and composition is lost. The gulf between these two notations is broad and not easily bridged, and certainly not within the frame of the usual didactics today, which after the rise of New Math has got addicted to the one symbol x for the variable (or x and y if there are two of them).

Even well-trained pupils are exposed to confusion by the multiple appearance of the letter x if functions like the above $f, g, h, \ldots$ are to be inverted and composed. This is a serious handicap to the algebraisation and finally algorithmisation of these activities.

In the notation

$$x \overset{+5}{\curvearrowright} x + 5 \qquad\qquad x \overset{\times 2}{\curvearrowright} 2x$$

one can clearly recognise

$$x \overset{-5}{\curvearrowright} x - 5 \qquad\qquad x \overset{\div 2}{\curvearrowright} \tfrac{1}{2}x$$

as respective inverses. Moreover composition of operations in this or that order is clear in

$$x \overset{+5}{\curvearrowright} x + 5 \overset{\times 2}{\curvearrowright} 2x + 10 \qquad \text{or} \qquad x \overset{\times 2}{\curvearrowright} 2x \overset{+5}{\curvearrowright} 2x + 5.$$

The need for brackets might be experienced in this context and their use might be exercised. Beware of the didactical problems that arise for inverting and composing functions f, g, h, given in their "definitive" notations!

17.61. To face these difficulties one has to break the monopoly of the variable x. The transition from

$$\ldots \overset{+5}{\curvearrowright} \ldots \qquad , \qquad \ldots \overset{\times 2}{\curvearrowright} \ldots$$

to

$$x \overset{+5}{\curvearrowright} x + 5 \qquad , \qquad x \overset{\times 2}{\curvearrowright} 2x,$$

should be accompanied by that to

$$y \overset{+5}{\curvearrowright} y + 5 \qquad , \qquad y \overset{\times 2}{\curvearrowright} 2y,$$

$$a \overset{+5}{\curvearrowright} a + 5 \qquad , \qquad a \overset{\times 2}{\curvearrowright} 2a,$$

$$A \overset{+5}{\curvearrowright} A + 5 \qquad , \qquad A \overset{\times 2}{\curvearrowright} 2A,$$

$$* \overset{+5}{\curvearrowright} * + 5 \qquad , \qquad * \overset{\times 2}{\curvearrowright} 2*$$

and so on and to

$$x \overset{+5}{\frown} y = x + 5 \quad , \qquad x \overset{\times 2}{\frown} y = 2x,$$

and so on, and it should be understood that all of these are notations for the the same function. Not until this has been achieved, can the definitive notation be tried and then not only in the form

$$x \rightarrow x + 5, \qquad\qquad x \rightarrow 2x, \qquad\qquad x \rightarrow 2x + 5,$$

but as well in the form

$$y \rightarrow y + 5, \qquad\qquad y \rightarrow 2y, \qquad\qquad y \rightarrow 2y + 5,$$
$$a \rightarrow a + 5, \qquad\qquad a \rightarrow 2a, \qquad\qquad a \rightarrow 2a + 5.$$

and so on, and in the form

$$x \rightarrow y = x + 5, \qquad\qquad x \rightarrow y = 2x, \qquad\qquad x \rightarrow y = 2x + 5,$$
$$y \rightarrow z = y + 5, \qquad\qquad y \rightarrow z = 2y, \qquad\qquad y \rightarrow z = 2y + 5,$$

and so on. Not until this has been attained, can one count on algebraic inversion and composition of functions in a way which in the long run would be accessible for algorithmisation. In order to compose

$$x \rightarrow x + 5, \qquad\qquad x \rightarrow 2x$$

in this order, one makes implicitly present variables explicit by putting

$$x \rightarrow y = x + 5, \qquad y \rightarrow z = 2y,$$
$$x \rightarrow z = 2y = 2(x + 5) = 2x + 10,$$

and similarly in order to invert

$$x \rightarrow 2x + 5$$

one writes

$$x \rightarrow y = 2x + 5$$

for the same function, in order to solve

$$y = 2x + 5$$

for x,

$$x = \frac{1}{2}(y - 5),$$

and to get the function

$$y \rightarrow x = \frac{1}{2}(y - 5),$$

which finally can also be written

$$x \rightarrow \frac{1}{2}(x - 5).$$

I think it is the way adult mathematicians do it. It is unbelievable that by restricting the independent variable to the one symbol x, one creates didactical problems for the learner which an adult mathematician would never risk incurring.

17.62. Earlier we pointed out that — in particular in applications from physics — certain letters are stuck to certain magnitudes. This should not conflict with the need for free exchange between letters that indicate variables. This need is felt not only in

　　　composing and inverting

functions but also in

　　　interpreting functions as inverted or composed ones,

whether such functions are

　　　specified or general.

I mean expressions like

　　　$\log \sin 2x$

which must be interpreted as functions, in order to be tabulated or graphically represented, differentiated or integrated. It is the classical procedure to split it into

$$x \to y = 2x,$$
$$y \to z = \sin y,$$
$$z \to u = \log x.$$

It requires routine, for instance in differentiations, to perform this splitting purely mentally, that is, without introducing new variables. Devices which are familiar in Calculus — the introduction of new variables — should be not only allowed but even promoted at lower levels.

17.63. *Once More: the Function Experience*

Composing functions is in a primitive way experienced by coupling "machines" in a row. A more sophisticated form is the arrow language. It is algebraised in the formal substitution. But even at the level of algebraisation, there are aspects of the functions which remain a matter of experiencing.

I again start with two examples:

Given the graph of $x \to \sin x$, what does that of $x \to \sin (x + \frac{1}{3} \pi)$ look like, what that of $x \to \sin 2x$?

It is of course the intention to have the second and third function to be interpreted as composed from the sine function and $x \to x + \frac{1}{3} \pi$, and $x \to 2x$, respectively.

More generally: Given the graph of f (Figure 232), what does that of fg look like, where g is a nice function, linear or at least monotonic.

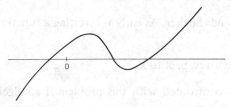

Fig. 232.

What happens with the graph of f if the x in $f(x)$ is replaced with $x + a$ or with ax? If f has a maximum or a minimum or takes the value c at x_0, where do $f(x + a)$ and $f(ax)$ do the same?

Answer: at $x_0 - a$, and x_0/a, respectively.

In composing f with $x \to x + a$ or $x \to x/a$

the graph of f is shifted horizontally by $- a$,
              shrunk horizontally by the factor a,

respectively.

More general: if g is monotonic, the graph of fg arises from that of f by subjecting the horizontal axis to the transformation g^{-1}.

A wave movement in time to the right with the profile f is described by

$$f \to f \circ (x \to x - t)$$

a uniform swelling by

$$f \to f \circ (x \to x/t),$$

in other words,

$$f(x - t) \qquad \text{as a function of time } t$$

represents a wave progressing in the positive sense,

$$f(tx) \qquad \text{as a function of time } t$$

a decompression.

To this kind of experiencing of the composition of functions more didactical attention should be paid.

17.64–65. *Composing Implicitly Given Functions*

17.64. After this side-leap let us return to the theme of algebraisation of the function. So far our view was directed to

explicitly given functions,

that is, algebraic expressions (in the broadest sense) with a variable,

> interpreted as a function of that variable.

The algebraisation extends farther. As early as inverting a function f one requires, at least in principle,

> solving $y = f(x)$ with respect to x.

More generally one is confronted with this problem if an algebraic expression involving more variables is given,

> one of which is considered (implicitly) as a function of the other,

for instance in

$$x^2 + y^2 = 1$$

the y as a function of x. This leads to the interpretation problems we sketched in the historical context. In general

$$f(x, y) = 0$$

yields a relation between x and y: the solution set

$$\{ \ulcorner x, y \urcorner \mid f(x, y) = 0 \},$$

which can possibly be split into function graphs, in the case

$$\{ \ulcorner x, y \urcorner \mid x^2 + y^2 = 1 \},$$

those of

$$x \to \sqrt{1 - x^2}, \qquad x \to -\sqrt{1 - x^2}.$$

If univalency of functions is to be maintained, the implicit definition

$$f(x, y) = 0$$

is to be refined by domain restricting conditions

$$x \in A, \qquad y \in B$$

in the case of

$$x^2 + y^2 = 1 \qquad \text{for instance by} \qquad y \geqq 0.$$

Then the solution set appears to be split into two implicitly defined function graphs, to wit

$$\begin{aligned} x^2 + y^2 = 1, & \qquad y \geqq 0, \\ x^2 + y^2 = 1, & \qquad y \leqq 0. \end{aligned}$$

On the other hand it might be possible to describe the solution set of $f(x, y) = 0$ by

parametrisation,

in the case

$$x^2 + y^2 = 1$$

by

$$t \to \ulcorner x, y \urcorner = \ulcorner \cos t, \sin t \urcorner,$$

from which one can return by

elimination of t

to the previous, parameter free, equation. In other words the connections between x and y described by

$$t \to \ulcorner x, y \urcorner = \ulcorner \cos t, \sin t \urcorner$$

and by

$$x^2 + y^2 = 1$$

are equivalent.
    The

elimination

is the pattern that also serves in composing implicitly given functions. Let by

$$f(x, y) = 0, \qquad g(y, z) = 0$$

functions

$$x \to y, \qquad y \to z$$

by given. The composed function

$$x \to z$$

can be obtained by eliminating y from

$$f(x, y) = 0, \qquad g(y, z) = 0,$$

which may be a purely formal action as far as algebraic functions are concerned. As an example take the elimination of y from

$$x^2 + y^2 = 1, \qquad y^3 + z^3 = 1$$

by means of

$$y^2 = 1 - x^2, \qquad y^3 = 1 - z^3$$
$$(1 - x^2)^3 = (1 - z^3)^2.$$

This formal acting springs from experiencing the implicitly given functions: from the equations

$$f(x, y) = 0, \qquad g(y, z) = 0$$

an equation, free from y,

$$h(x, z) = 0$$

is derived, which "thus" determines the connection between x and z. In fact, one has to prove the equivalence of both of them:

as counterpart of the elimination of y from

$$f(x, y) = 0, \qquad g(y, z) = 0 \quad \text{to } h(x, z) = 0,$$

the parametrisation by means of y

of $h(x, y) = 0$ to $f(x, y) = 0, \qquad g(y, z) = 0.$

In general this will require the choice of appropriate domain restricting conditions

$$x \in A, \qquad y \in B, \qquad z \in C.$$

In the previous example, for instance

$$-1 \leq x \leq 1, \qquad 0 \leq y \leq 1, \qquad 0 \leq z \leq 1.$$

As a matter of fact, elimination is reduced to substitution if an explicitly given function

$$x \to y = f(x)$$

and an implicit one

$$y \to z \quad \text{with} \quad g(y, z) = 0$$

are composed, that is, to

$$x \to z \quad \text{with} \quad g(f(x), z) = 0.$$

17.65. What is the meaning of this exposition in a didactical context? My answer is: to give a fresh chance to the natural operations with implicitly given functions in opposition to the dogmatism of formalist hobbyists.

Implicitly given functions are at present didactically banished. Things like

$$x \to y \quad \text{with} \quad f(x, y) = 0$$

and

$$y \to z \quad \text{with} \quad f(y, z) = 0$$

are nowadays rejected as function definitions. It *must* be

$$x \to y = \varphi(x),$$
$$y \to z = \psi(y),$$

which then may be composed to get

$$x \to z = \psi(\varphi(x));$$

and this is imposed and required, contrary to one's better knowledge, that is, even if explicit expressions for φ and ψ are not available while elimination of y readily yields the wanted connection between x and z. Users of mathematics, in particular physicists who do not understand these formalist hobbyism, rightly stick to their own methods that bear witness to insight into what variables and functions mean, compared with a formalism detached from its context. I considered it useful to stress here the content aspect of working with variables and functions.

17.66—68. *The Algebraisation of the Pattern "the . . . the . . ."*

As we know, the pattern "the . . . the . . ." is a vigorous source of the function. It is mathematised in various ways — additively, subtractively, by ratio, or more general, by monotonicity — monotonic increasing or decreasing.

Additive and subtractive "the . . . the . . ." are algebraised, starting from the function table, the geometric picture or the graph by

$$x \to a + x \quad \text{and} \quad x \to a - x,$$

or implicitly written

$$x \to y \quad \text{with} \quad y - x = a, \quad \text{and} \quad y + x = a, \quad \text{respectively.}$$

The paradigm of the ratio "the . . . the . . ." is the uniform motion or growth:

in equal times equal distances,
in equal times the same increase.

Or, more general, with two variables,

the growth of y is independent of the initial state determined by that of x.

In a — high level — formula:

$$y - y_0 = \varphi(x - x_0) \quad \text{for all } x_0, x.$$

This supposition, implies that

to the double, threefold, and so on, increase of x corresponds the double, threefold, and so on, increase of y.

(See Figure 233.) In formulae, with

$$x_1 = x_0 + h, \qquad x_2 = x_0 + 2h, \ldots,$$
$$\varphi(h) = \varphi(x_1 - x_0) = \varphi(x_2 - x_1) = \ldots$$
$$y_1 = y_0 + \varphi(h), \qquad y_2 = y_1 + \varphi(h) = y_0 + 2\varphi(h), \ldots.$$

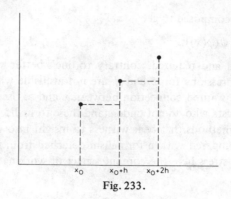

Fig. 233.

Thus, the growth function φ satisfies

$$\varphi(2h) = 2\varphi(h), \qquad \varphi(3h) = 3\varphi(h), \ldots .$$

Likewise

> to the half, the third, and so on, of the increase of x corresponds the half, the third, and so on, of the increase of y.

(See Figure 234.) In formulae

$$\varphi(h) = \varphi(2 \cdot \frac{1}{2} h) = 2\varphi(\frac{1}{2} h),$$

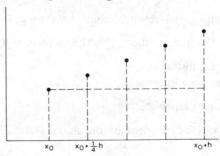

Fig. 234.

thus

$$\varphi(\frac{1}{2} h) = \frac{1}{2} \varphi(h),$$

and similarly

$$\varphi(\frac{1}{3} h) = \frac{1}{3} \varphi(h) \quad \text{and so on.}$$

If

> to the threefold increase of x corresponds the threefold increase of y

is combined with

> to half the increase of x corresponds half the increase of y

one gets

to the $\frac{3}{2}$-fold increase of x corresponds the $\frac{3}{2}$-fold increase of y,

and in general for any fraction $\frac{m}{n}$

to the $\frac{m}{n}$-fold increase of x corresponds the $\frac{m}{n}$-fold increase of y.

In a formula

$$\varphi(rh) = r\varphi(h) \qquad \text{for (positive) rational } r.$$

This leads to an explicit form for φ as soon as the increase of x, say for $h = 1$, is supposed to be known (Figure 235 and 236): for $\varphi(1) = c$

$$\varphi(r) = cr,$$

which yields φ for all rational values of r.

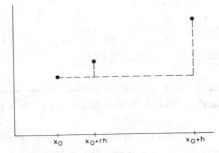

Fig. 235.

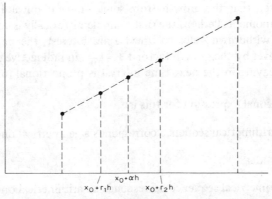

Fig. 236.

The pattern "the ... the ..." requires monotonicity of the growth function φ. Each real number α can arbitrarily closely be confined between two rationals

$$r_1 \leqq \alpha \leqq r_2.$$

Since φ is monotonic increasing or decreasing,

$$\varphi(r_1) \leqq \varphi(\alpha) \leqq \varphi(r_2),$$
$$cr_1 \leqq \varphi(\alpha) \leqq cr_2,$$

respectively

$$\varphi(r_1) \geqq \varphi(\alpha) \geqq \varphi(r_2),$$
$$cr_1 \geqq \varphi(\alpha) \geqq cr_2$$

from which by having r_1, r_2 approach α (and each other) one gets

$$\varphi(\alpha) = c\alpha.$$

Uniform growth means that the increase of y is a fixed multiple of that of x. The fixed factor means, depending on the circumstances, a velocity of motion, of growth, and so on. Uniform growth is algebraically described by a linear function

$$y = a + cx,$$

called linear because its graph is a straight line.

17.67. Uniformity means that growth is viewed as an additive phenomenon: in equal times the same is added, which implies a growth proportional to the time elapsed.

Another view of growth is introduced by the concept of growth factor; of the two values of a variable at two moments the *quotient* is taken to tell us how much the variable has increased in between. It is a most natural idea as soon as it is posited that the growth from some moment onwards depends on the state at that moment. Perhaps the first example historically is compound interest: interest not withdrawn helps to breed again interest, the part of the additive $p\%$ is taken over by the growth factor $1 + \frac{p}{100}$. In radioactive decay the number of atoms decaying in the next time interval is proportional to the quantity that is left.

The traditional expression for this is:

> to an arithmetical sequence corresponds a geometrical one.

The converse idea

> to a geometrical sequence corresponds an arithmetical one

once led to the invention of logarithmic tables as a means to replace time-wasting multiplications by time-saving additions. The slide-rule is an offshoot of the same idea: thanks to the logarithmic scale multiplications can be performed by the additive operation of adding line-segments.

To the arithmetical sequence, say,

$$a, a + d, a + 2d, \ldots$$

corresponds a geometrical one, say,

$$c, cq, cq^2, \ldots$$

thus the assignment, for $n \in \mathbf{N}$ is

$$a + dn \to cq^n.$$

However, the growth, observed discretely, takes place continuously — even compound interest could be bred continuously. Though built stepwise the table of logarithms pretends to continuity.

Corresponding to the arithmetical sequence there is a geometric one. In arithmetic sequences one can interpolate, halving the step d, or dividing in ten parts, or even smaller ones, say, N parts, where N may be as large as you wish. The corresponding geometric sequence is well-determined. But how?

One needs a multiplicative counterpart of halving, dividing into ten parts, into N parts. Euclid's terminology was continual proportion: switching, say $N - 1$ terms between two magnitude values A and B:

$$A : C_1 = C_1 : C_2 = C_2 : C_3 = \ldots = C_{N-1} : B.$$

Additively dividing into N parts is the inverse of multiplying by N. Multiplicatively dividing into N parts is the inverse of raising to the Nth power. $9 = 3 \cdot 3$ is a directly visible multiplicative halving of 9 and similarly $1024 = 2^{10}$ a multiplicative dividing into 10 parts of 1024.

In this course root-extraction offers itself as the inverse of power-raising:

$$\text{asked for the } r \text{ such that } r^N = q,$$

the Nth root of q, denoted by

$$r = q^{1/N}$$

because

$$r^N = (q^{1/N})^N = q^{N/N} = q$$

extends a law for whole exponents, as does the definition

$$q^{M/N} = q^{(1/N)M}.$$

This then is the way to interpolate the assignment

$$a + nd \to cq^n$$

of a geometrical to an arithmetical sequence to

$$a + rd \to cq^r.$$

for rational r and — repeating the monotonicity arguments of Section 17.62
— to

$$a + dx \rightarrow cq^x$$

for real x — perhaps even negative ones; or somewhat simpler:

$$x \rightarrow CQ^x$$

with C and Q, which can easily be calculated. This then is an exponential function, describing what is nowadays called exponential growth (or decay).

17.68. Questions as to whether

power raising is invertible,
$q \rightarrow q^r$ is defined for rational r,
$x \rightarrow CQ^x$ is meaningful,

are not put here in their didactic context, since I did so earlier* quite circumstantially. One aspect, which was then still of minor importance, should be added: the use of handheld calculators in what I would call

imparting meaning to existence queries by operationalisation:

the

power, root, exponential, logarithmic functions

being effectuated by computing them.

Properly I should have amplified my arguments at many points by an appeal to handheld calculators. However, I could not collect sufficient experience with calculators — either by working didactically with them or by consulting literature. If I ever catch up with this lag, didactical phenomenology of the handheld calculator ought to be a separate chapter.

17.69–76. *Graphs*

17.69. Graphs have been mentioned several times in the present chapter (for instance Section 17.50a, 17.55, 17.63); I even promised to deal with graphs more systematically. Without entering into details, I would like to draw the attention to research by C. Janvier,** which unites phenomenological and experimental aspects.

I restrict myself here to line graphs while disregarding histograms and similar

---

* *Mathematics as an Educational Task*, Chapter XIV.
** C. Janvier, 'The interpretation of complex cartesian graphs representing situations — studies and teaching experiments'. Thesis, Nottingham, 1978 — Accompanying brochure including appendices. Shell Center for Mathematical Education and Université de Québec à Montréal.

means of visualisation, which have been discussed earlier (e.g., Section 16.13). With regard to the functions I stick to numerical ones of one variable (mappings out of **R** into **R**). Taking these restrictions into account I state a phenomenological reciprocity between functions and graphs:

functions, visualised by graphs,
graphs interpreted as functions.

17.70. Janvier put the graphs into a larger frame of devices to describe connections between variables. The media of description are put into a double entry table with the means of translation from one medium to the other in the compartments — even those in the diagonal can be meaningfully filled out.

to from	verbal	table	graph	formula
verbal				
table				
graph				
formula				

17.71. I recall from the beginning of the present chapter the phenomenology of variables from the

physical, social, mental, mathematical

world with related dependences which as functions can be

stated, postulated, produced, reproduced.

This can indeed happen in a verbal description. Even in the world of mathematics this is not usual ("the chord as a function of the angle", "the area (of a circle) as a function of the radius"). The temperature (at a certain place) as a function of the time can have been put into a table where it can be read, or a thermograph can have recorded it in the form of a graph.

A formula pretends to have more precision than a table or a graph. The table explains the values of the dependent variable only on a discrete set of the independent variable, though this can be done with any precision that is relevant to the problem; moreover the mesh of this discrete set can be chosen as is required by relevance. The precision of graphs, however, is limited by the unavoidable thickness of lines. As regards precision no graph of the logarithmic function can compete with the simplest logarithmic table, but in many phenomena described by functions one can be satisfied with the precision of the graph.

The often decisive feature by which the graph can compete with table and fomula is its visualising power. The computer, which produces long lists of numbers, is assisted by a plotter, which lends perspicuity to this chaotic material. On the other hand we will come across the graph as a device to visualise functions that have only been subjected to qualitative requirements.

17.71. Data for a graph construction can be:

      (a)   an explicit or implicit equation that expresses a dependent by an independent variable,

      (b)   empirical numerical data,

      (c)   general instructions on the behaviour of the function.

Examples

      (a) $x^2 + y^2 = 1$, $y \geq 0$

      (b)   the temperatures at de Bilt during 1978.

      (c)   the function is monotonic increasing, decreasing, convex, possesses a certain number of maxima with minima in between.

Explanation of (c):

Graphs of this latter, less determined type are used in econometry. One speaks about demand and supply as functions of the price of a utility. These, however, are — useful — fictions. What happens in the economy depends on too many and too closely entangled variables to be accessible to a successful isolation of dependences between a pair of variables; fixing a moment the other variables is a — sometimes useful — illusion. Demand and supply functions are quite simplistic models. Nevertheless there is something like a trend: according to the increase of the price of a utility, the demand may decrease and the supply increase — the degree to which demand and supply react on the change of the price, called *elasticity* by the economist, can be visualised by the slope of a graph; for extremely low or extremely high prices the curve can behave deviantly — so cheap that no consumer cares, so expensive that nobody dares to produce. The behaviour in general or in special cases can be sketched by a graph. Such a picture can for instance show the equilibrium of demand and supply at the intersection of the demand and the supply graph (Figure 237).

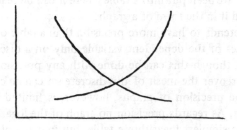

Fig. 237.

Another example (by Janvier, see Section 17.55) is the speed of a race-car (Figure 238) as a function of the path — it is clear that the function must have a steep minimum and two less steep minima with horizontal pieces in between — only the distances between the minima can quantitatively be more precisely specified (Figure 239).

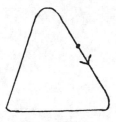

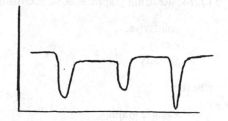

Fig. 238.   Fig. 239.

17.72. In order to put data of the type a and b (see Section 17.71) into a graph, pointwise plotting is the appropriate method. The pointwise given graph is completed by interpolation to yield a linear zigzag function if the "in between" is meaningless or irrelevant, or a so-called smooth curve. If the character of the function that is to be constructed, is established (or accepted, or postulated) *a priori* up to a number of parameters, then a corresponding number of graph dots suffices to determine the graph — if it is a linear function, two points are needed, which are joined by means of a ruler; if it is a polynomial of degree n a $(n + 1)$-tuple suffices. But apart from such theoretical condition, there is a certain feeling for what is a smooth curve, which in the practice of the conscientious draughtsman is backed up by the use of models with a multifariously varying curvature. It is impossible to say, and it does not seem to have been investigated, what is the origin of this feeling for adapting a curve to graphical data. The need and feelings for smoothness of a graph constructed from empirical data can lead to rejecting misfits. If it may be assumed that the empirical data are affected by observational errors, a cloud of measuring points is replaced with a smooth line either guided by subjective feelings or according to an objective method of adaptation to a functional type depending on a number of parameters.

17.73. Plotting a graph is preceded by the choice of a scale or scales for the variables. The graph must fit the dimensions of the paper — if it is the graph of a function given by a formula, at least so far that certain characteristics can be visualised or a proposed problem becomes graphically solvable (for instance intersections of graphs become accessible). The scales of independent and dependent variables can differ from each other. For instance this divergence of scales is badly needed for drawing the frequency curve of the normal distribution if its characteristics are clearly to be displayed; an exaggeration of the ordinate in the ratio 10 : 1 is not unusual. In a graph of the vibrating string this exaggeration might even be 100 : 1.

Beside linear scales one knows logarithmic and double-logarithmic ones; exponential growth processes can slip away if plotted on a linear scale. By a logarithmic scale for the independent variable exponential growth processes are apparently featured by linear ones in order to be visualised and compared.

17.74. Reading graphs is done according to the needs

> pointwise,
> locally,
> globally,

whether

> a single graph

is concerned or several graphs are

> compared separately or in the same picture.

"Pointwise" means that for a given value of the independent variable that of the dependent one is looked up or conversely for a given value of the dependent variable those of the independent one. This is usually done by scanning the graph

> vertically — horizontally,
> horizontally — vertically,

while interpolating, if need be, on the axes.

Locally — this means comparative investigations of the graph behaviour near a point:

> positive, negative,
>
> increasing, decreasing, constant, minimal, maximal,
>
> steeply or less steeply increasing or decreasing, up to a discontinuous jump or fall,
>
> rectilinear or bent in the convex or concave sense,
>
> more or less bent in one sense or the other,
>
> sharply or less sharply bent at a minimum or maximum,
>
> changing from one type of bending to the other (turning point).

This qualitative behaviour can quantitatively be specified by considering

> a measure of increase or decrease,
> the height of minimum or maximum,
> a measure of curvature (in general or at an extreme value),

in particular by plotting graphically

the derivative, or second derivative

of the function — a didactically useful activity, which in Calculus instruction is not paid the attention it deserves.

Globally — this can mean

searching for
and comparing

local characteristics in order to find

intervals of positivity, negativity, confinement between certain bounds,

intervals of increase, decrease, constancy,

minima, maxima,

points of steeper, less steep, steepest, least steep behaviour,

jumps,

intervals with certain curvature,

turning points,

points of sharper, less sharp, sharpest, least sharp bending.

Reading graphs globally beyond the search for the comparison of global characteristics takes place if one states

monotonicity,

oscillating behaviour,

asymptotic behaviour of various kind,

periodicity,

damped, excited, slightly or more seriously disturbed periodicity,

linearity, quadratic or otherwise algebraic character,

exponential character,

sinusoidal character.

Reading globally a function given in a certain restricted interval can lead to

extrapolation based on certain characteristics mentioned earlier.

If the graphs of two or more functions are confronted with each other, it is in particular

function values,
slope measures,
places of minima and so on

that are compared with each other. If families of functions depending on parameters are discussed

>the diverging asymptotic behaviour

for different parameter values can play a part.

17.75. Properties and peculiarities such as those summarised here, can more
easily be recognised in the graphical picture than in formulae, and this is one
of the reasons why graphs are so important.

Even the restriction to one variable is not as absolute as it seems. Graphs
are pictures of function-relations between an independent and a dependent
numerical variable. Functions of more variables can be mastered by declaring
some of the variables as parameters, and plotting functions corresponding
to various parameter values besides or below each other as if they were moving
pictures. Another graphical means to picture functions of more, in particular
two, variables, has been shown in Section 17.56: level lines with the function
value as level.

LIST OF THE AUTHOR'S PUBLICATIONS
ON MATHEMATICS EDUCATION

1. 'De algebraische en de analytische visie op het getalbegrip in de elementaire wiskunde', *Euclides* **24** (1948), 106–121.
2. 'Kan het wiskundeonderwijs tot de opvoeding van het denkvermogen bijdragen. Discussie tussen T. Ehrenfest-Afanassjewa en H. Freudenthal' (Publicatie Wiskunde Werkgroep van de V.W.O.), Purmerend 1951.
2a. 'Erziehung des Denkvermögens' (Diskussionsbeitrag), *Archimedes* Heft **6** (1954), 87–89. (This is a translated extract from 2.)
3. 'De begrippen axioma en axiomatiek in de Wis- en Natuurkunde', *Simon Stevin* **39** (1955), 156–175.
3a. 'Axiom und Axiomatik', *Mathem. Phys. Semesterberichte* **5** (1956), 4–19.
4. 'Initiation into Geometry', *The Mathematics Student* **24** (1956), 83–97.
5. 'Relations entre l'enseignement secondaire et l'enseignement universitaire en Hollande', *Enseignement mathématique* **2** (1956), 238–249.
6. 'De Leraarsopleiding', *Vernieuwing* **133** (1956), 173–180.
7. 'Traditie en Opvoeding', *Rekenschap* **4** (1957), 95–103.
8. 'Report on Methods of Initiation into Geometry', ed. H. Freudenthal, (Publ. Nederl. Onderwijscommissie voor Wiskunde), Groningen, 1958.
9. 'Einige Züge aus der Entwicklung des mathematischen Formalismus, I, *Nieuw Archief v. Wiskunde* **3** (1959), 1–19.
10. 'Report on a Comparative Study of Methods of Initiation into Geometry', *Euclides* **34** (1959), 289–306.
10a. 'A Comparative Study of Methods of Initiation into Geometry', *Enseignement mathématique* **2**, 5 (1959), 119–139.
11. 'Logica als Methode en als Onderwerp', *Euclides* **35** (1960), 241–255.
11a. 'Logik als Gegenstand und als Methode', *Der Mathematikunterricht* **13** (1967), 7–22.
12. 'Trends in Modern Mathematics', *ICSU Review* **4** (1962), 54–61.
12a. 'Tendenzen in der modernen Mathematik', *Der math. und naturw. Unterricht* **16** (1963), 301–306.
13. 'Report on the Relations between Arithmetic and Algebra', ed. H. Freudenthal (Publ. Nederl. Onderwijscommissie voor Wiskunde), Groningen, 1962.
14. 'Enseignement des mathématiques modernes ou enseignement moderne des mathématiques?' *Enseignement Mathématique* **2** (1963), 28–44.
15. 'Was ist Axiomatik, und welchen Bildungswert kann sie haben?' *Der Mathematikunterricht* **9**, 4 (1963), 5–19.

16. 'The Role of Geometrical Intuition in Modern Mathematics', *ICSU Review* **6** (1964), 206–209.

16a. 'Die Geometrie in der modernen Mathematik', *Physikalische Blätter* **20** (1964), 352–356.

17. 'Bemerkungen zur axiomatischen Methode im Unterricht', *Der Mathematikunterricht* **12**, 3 (1966), 61–65.

18. 'Functies en functie-notaties', *Euclides* **41** (1966), 299–304.

19. 'Why to Teach Mathematics so as to Be Useful?' *Educational Studies in Mathematics* **1** (1968), 3–8.

20. Paneldiscussion, *Educational Studies in Mathematics* **1** (1968), 61–93.

21. 'L'intégration après coup ou à la source', *Educational Studies in Mathematics* **1** (1968–1969), 327–337.

22. 'The Concept of Integration at the Varna Congress', *Educational Studies in Mathematics* **1** (1968–1969), 338–339.

23. 'Braces and Venn Diagrams', *Educational Studies in Mathematics* **1** (1968–1969), 408–414.

24. 'Further Training of Mathematics Teachers in the Netherlands', *Educational Studies in Mathematics* **1** (1968–1969), 484–492.

25. 'A Teachers Course Colloquium on Sets and Logic', *Educational Studies in Mathematics* **2** (1969–1970), 32–58.

26. 'ICMI Report on Mathematical Contests in Secondary Education (Olympiads)', ed. H. Freudenthal, *Educational Studies in Mathematics* **2** (1969–1970), 80–114.

27. 'Allocution au Premier Congrès International de l'Enseignement Mathématique, Lyon 24–31 août 1969', *Educational Studies in Mathematics* **2** (1969–1970), 135–138.

28. 'Les tendances nouvelles de l'enseignement mathématique', *Revue de l'enseignement supérieur* **46–47** (1969), 23–29.

29. 'Verzamelingen in het onderwijs', *Euclides* **45** (1970), 321–326.

30. 'The Aims of Teaching Probability, in L. Råde (ed.), *The Teaching of Probability & Statistics*, Almqvist & Wiksell, Stockholm, 1970, pp. 151–167.

31. 'Introduction', *New Trends in Mathematics Teaching*, Vol. II, Unesco, 1970.

32. 'Un cours de géométrie', *New Trends in Mathematics Teaching*, Vol. II, Unesco, 1970, pp. 309–314.

33. 'Le langage mathématique. Premier Sém. Intern. E. Galion, Royaumont 13–20 août 1970', OCDL, Paris, 1971.

34. 'Geometry between the Devil and the Deep Sea', *Educational Studies in Mathematics* **3** (1971), 413–435.

35. 'Kanttekeningen bij de nomenclatuur', *Euclides* **47** (1971), 138–140.

36. 'Nog eens nomenclatuur', *Euclides* **47** (1972), 181–192.

37. 'Strategie der Unterrichtserneuerung in der Mathematik', *Beiträge z. Mathematikunterricht*, Schroedel, 1972, 41–45.

38. 'The Empirical Law of Large Numbers, or the Stability of Frequencies', *Educational Studies in Mathematics* **4** (1972), 484–490.

39. 'What Groups Mean in Mathematics and What They Should Mean in Mathematical Education', in *Developments in Mathematical Education, Proceedings of the Second International Congress on Mathematical Education* 1973, pp. 101–114.

40. *Mathematics as an Educational Task*, Reidel, Dordrecht, 1973.

40a. *Mathematik als pädagogische Aufgabe*, Band 1, 2, Klett, Stuttgart, 1973.

41. 'Mathematik in der Grundschule', *Didaktik der Mathematik* **1** (1973), 2–11.

42. 'Nomenclatuur en geen einde', *Euclides* **49** (1973), 53–58.

43. 'Les niveaux de l'apprentissage des concepts de limite et de continuité', Accademia Nazionale dei Lincei, 1973, Quaderno N. 184, 109–115.

44. 'De Middenschool', *Rekenschap* **20** (1973), 157–165.

45. 'Waarschijnlijkheid en Statistiek op school', *Euclides* **49** (1974), 245–246.

46. 'Die Stufen im Lernprozess und die heterogene Lerngruppe im Hinblick auf die Middenschool', *Neue Sammlung* **14** (1974), 161–172.

47. 'The Crux of Course Design in Probability', *Educational Studies in Mathematics* **5** (1974), 261–277.

48. 'Mammoetonderwijsonderzoek wekt wantrouwen', University Newspaper "U", State University of Utrecht, June 1974.

49. 'Mathematische Erziehung oder Mathematik im Dienste der Erziehung', Address 21 June 1974, University Week, Innsbruck.

50. 'Kennst Du Deinen Vater?' *Der Mathematikunterricht* **5** (1974), 7–18.

51. 'Lernzielfindung im Mathematikunterricht', *Zeitschrift f. Pädagogik* **20** (1974), 719–738; *Der Mathematikunterricht* **23** (1977), 26–45.

52. 'Sinn und Bedeutung der Didaktik der Mathematik', *Zentralblatt für Didaktik der Mathematik* **74**, 3 (1974), 122–124.

53. 'Soviet Research on Teaching Algebra at the Lower Grades of the Elementary School, *Educational Studies in Mathematics* **5** (1974), 391–412.

54. 'Ein internationaal vergelijkend onderzoek over wiskundige studieprestaties', *Pedagogische Studiën* **52** (1975), 43–55.

55. 'Wat is meetkunde?' *Euclides* **50** (1974–1975), 151–160.

56. 'Een internationaal vergelijkend onderzoek over tekstbegrip van scholieren', *Levende Talen*, deel 311 (1975), 117–130.

57. 'Der Wahrscheinlichkeitsbegriff als angewandte Mathematik', *Les applications nouvelles des mathématiques et l'enseignement secondaire*, C.I.E.M. Conference, Echternach, June 1973 (1975), 15–27.

58. 'Wandelingen met Bastiaan', *Pedomorfose* **25** (1975), 51–64.

59. 'Compte rendu du débat du samedi 12 avril 1975 entre Mme Krygowska et M. Freudenthal', *Chantiers de péd. math.*, June 1975, Issue 33 (Bulletin bimestriel de la Régionale Parisienne), 12–27.

60. 'Pupils' Achievements Internationally Compared – the I.E.A.' *Educational Studies in Mathematics* **6** (1975), 127–186.

60a. 'Schülerleistungen im internationalen Vergleich', *Zeitschrift für Pädagogie* **21** (1975), 889–910. (This is a translated extract from 60.)

61. 'Leerlingenprestaties in de natuurwetenschappen internationaal vergeleken', *Faraday* **45** (1975), 58–63.

62. 'Des problèmes didactiques liés au langage', pp. 1–3; 'L'origine de la topologie moderne d'après des papiers inédits de L.E.J. Brouwer', pp. 9–16. Lectures delivered at the University, Paris VII, in April 1975 (offset). (With Krygowska).

63. 'Variabelen (opmerkingen bij het stuk van T. S. de Groot', *Euclides* **51**, 154–155), *Euclides* **51** (1975–1976), 349–350.

64. 'Bastiaan's Lab', *Pedomorfose* **30** (1976), 35–54.

65. 'De wereld van de toetsen', *Rekenschap* **23** (1976), 60–72.

66. 'De C.M.L.-Wiskunde', interview, *Euclides* **52** (1976–1977), 100–107.

67. 'Valsheid in geschrift of in gecijfer?' *Rekenschap* **23** (1976), 141–143.

68. 'Studieprestaties – Hoe worden ze door school en leerkracht beinvloed? Enkele kritische kanttekeningen n.a.v. het Colemanreport', *Pedagogische Studiën* **53** (1976), 465–468.

69. 'Rejoinder', *Educational Studies in Mathematics* **7** (1976), 529–533.

70. 'Wiskunde-Onderwijs anno 2000. Afscheidsrede IOWO 14 Augustus 1976', *Euclides* **52** (1976–1977), 290–295.

71. 'Annotaties bij annotaties, Vragen bij vragen', *Onderwijs in Natuurwetenschap* **2** (1977), 21–22.

72. 'Creativity', *Educational Studies in Mathematics* **8** (1977) 1.

73. 'Bastiaan's Experiment on Archimedes' Principle', *Educational Studies in Mathematics* **8** (1977), 3–16. (This is a translated extract from 64.)

74. 'Fragmente', *Die Mathematikunterricht* **23** (1977), 5–25.

75. 'Didaktische Phänomenologie, Länge', *Der Mathematikunterricht* **23** (1977), 46–73.

76. (Review, with J. van Bruggen) 'The Psychology of Mathematical Abilities in Schoolchildren by V. A. Krutetskĭ', *Proceedings of the National Academy of Education* **4** (1977), 235–277.

77. (Review, with J. van Bruggen) 'Soviet Studies in the Psychology of Learning and Teaching Mathematics (6 volumes), ed. by J. Kilpatrick and I. Wirszup', *Proceedings of the National Academy of Education* **4** (1977), 201–234.

78. *Mathematik als pädagogische Aufgabe* I, Klett, Stuttgart, 2. Aufl. 1977. (Cf. 40a).

79. 'Die Crux im Lehrgangentwurf zur Wahrscheinlichkeitsrechnung', *Didaktik der Mathematik*, 436–459 (ed. Steiner) Wiss. Buchgeselschaft Darmstadt, 1977. (Translated from 47.)

80. 'Brokjes Semantiek', *Pedagogische Studiën* **54** (1977), 461–464.

81. 'Cognitieve ontwikkeling – kinderen geobserveerd', *Prov. Utrechts Genootschap, Jaarverslag* **1977** (1978), 8–18.

82. 'Teacher Training – An Experimental Philosophy', *Educational Studies in Mathematics* **8** (1977), 369–376.

83. 'La sémantique du terme Modèle', *La Sémantique dans les Sciences, Colloque de l'Académie Intern. de Philos. des Sciences 1974, Archives de l'Institut Internat. des Sci. Théor.* **21** (1978).

84. 'Bastiaan meet zijn wereld', *Pedomorfose* **37** (1978), 62–68.

85. 'Address to the First Conference of I.G.P.M.E.', 29 Aug. 1977, *Educational Studies in Mathematics* **9** (1978), 1–5.

86. 'Modern wiskunde-onderwijs? Goed wiskunde-onderwijs', *Intermediaire 28 April 1978.*

87. *Weeding and Sowing – A Preface to a Science of Mathematics Education*, Reidel, Dordrecht, 1978.

88. 'Change in Mathematics Education since the late 1950's – Ideas and Realisation, an ICMI report', *Educational Studies in Mathematics* **9** (1978), 75–78.

89. 'Soll der Mathematiklehrer etwas von der Geschichte der Mathematik wissen?' *Zentralblatt Didaktik Mathematik* (1978), 75–78.

90. 'Change in Mathematics Education since the late 1950's – The Netherlands', *Educational Studies in Mathematics* **9** (1978), 261–270.

91. *Vorrede zu einer Wissenschaft vom Mathematikunterricht* (Original of 87), Oldenbourg, München, 1978.

92. 'Nacherfindung unter Führung', *Kritische Stichwörter*, ed. D. Volk, Fink Verlag, 1979, pp. 185–194.

93. 'Rings and String', *Educational Studies in Mathematics* **10** (1979), 67–70.

94. 'Lessen van Sovjetrekenkunde', *Pedag. Studiën* **56** (1979), 57–60.

95. 'Onderwijs voor de kleuterschool – cognitief, wiskunde', *De Wereld van het Jonge Kind* **1979**, 143–147, 168–172.

96. 'Introductory Talk, Congresso Internationale . . . 7–15 gennuio 1976', *Accad. N. Lincei* **326** (1979), 15–32.

97. 'Konstruieren, Reflektieren, Beweisen in phänomenologischer Sicht', *Schriftenreihe Didaktik der Mathematik*, Klagenfurt **2** (1979), 183–200.

98. 'How Does Reflective Thinking Develop?' *Proceedings Conference IGPME Warwick* **1979**.

99. 'Ways to Report on Empirical Research in Education', *Educational Studies in Mathematics* **10** (1979), 175–303.

100. *Mathematik als pädagogische Aufgabe* II, 2. Aufl., 1979.

101a. 'New Math or New Education?' *Prospects Unesco* **9** (1979).

101b. 'Mathematiques nouvelles ou éducation nouvelle?' *Perspectives Unesco* **9** (1979), 339–350.

101c. '? Matemáticas nuevas o nueva educación?' *Perspectivas Unesco* **9** (1979), 337–348.

102. 'De waarde van resumerende en tweede hands informatie', *Pedag. Studiën* **56** (1979), 323–326.

103. 'Un'experienza di insegnamento dello lege di Archimede', *Instituto della Enciclopedia Italiana* n. 1, 245–246. Roma (translation of 96; see 73).

104. *Weeding and Sowing*, 1980 (= 87).

584 LIST OF THE AUTHOR'S PUBLICATIONS

105. 'Invullen – Vervullen', *De Pedagogische Academie* **10** (1979), 197–200.
106. 'Invullen – Vervullen', *Euclides* **55** (1979/80), 61–65 (= 105 abridged).
107. 'Onderzoek in onderwijs', *Pedomorfose* **12** (1980), no. **47**, 57–66.
108. 'Wiskunde en onderwijskunde in de leraarsopleiding', *Resonans* **12** (1980), 20–21.
109. 'Lernprozesse Beobachten', *Neue Sammlung* **20** (1980), 328–340.
110. 'Examenpakket en wiskundige kiespijn', *Weekblad NGL* **13** (1980), 438–439.
111. 'Four Cube House', *For the Learning of Mathematics* **1** (1980), 12–13.
112. 'IOWO – Mathematik für alle und jedermann', *Neue Sammlung* **20** (1980), 633–654.
113. 'Kinder und Mathematik, Didaktik des Entdeckens und Nacherfindens', Wiskunde op de basisschool, *Grundschule* **13** (1981), 100–104.
114. 'Flächeninhalt in phänomenologischer Sicht', *Mathematiklehrer* 1981–1, 5–10.
115. 'Major Problems of Mathematical Education', *Educational Studies in Mathematics* **12** (1981), 133–150.
116. 'Verslaggeving over empirisch onderwijskundig onderzoek ten onzent', *Pedag. Studiën* **58** (1981), 141–143.
117. 'Hovedproblemen for matematikkundervisning', *Normat* **29** (1981), 49–66. (Transl. from 115.)
118. 'Should a Mathematics Teacher Know Something about the History of Mathematics?' *For the Learning of Mathematics* **2** (1981), 30–33. (Transl. from 89.)
119. 'Roltrappen', *W. Bartjens* **1** (1981/2), no. 1, 3–4.
120. 'Binnen is 't minder', *N. Wiskrant* **1** (1981), no. 2, 18–20.
121. 'Mathematik, die uns angeht, IOWO', *Mathematiklehrer* **2** (1981), 3–4, 44.
122. 'Wat is er met het aftrekken aan de hand?' *W. Bartjens* **1** (1981/2), no. 2, 4–5.
123. 'Fifty–fifty', *N. Wiskrant* **1** (1982), no. 3, 7–8.
124. 'Winkel messen und berechnen', *Mathematiklehrer* **1** (1982), 4–5.
125. 'Taalfetisjisme', *Euclides* **57** (1981/2), 297–291.
126. 'Mathematik – eine Geisteshaltung', *Grundschule* **14** (1982), 140–142.
127. 'Kinder und Mathematik', in *Grundschule* **14** (1982), Arbeitskreis Grundschule, Beiträge . . . 50 (= 113).
128. 'Ik was moeder en ik doe boodschappen', *W. Bartjens* **1** (1981/2), 131–133.
129. 'Differentialen – ja of neen en zo ja, hoe?' *N. Wiskrant* **1**, No. 4, 15–18; **2**, No. 1, 16–18 (1982).
130. 'Studentenhaver', *W. Bartjens* **2** (1982), 214–215.
131. *Mathematika kak pedagogičeskaja zadača*, Prosveščenie, 1982. (Translation of 78.)
132. = 127 = 113. Beltz Praxis 1982.

133. 'Fiabilité, validité et pertinence', *Educational Studies in Mathematics* **13** (1982), 395–408.

134. 'Een visie op onderwijskundig bezig zijn', *Utrechtse Pedagogische Verhandelingen* **5** (1982), No. 1, 3–11.

135. 'De kortste weg', *N. Wiskrant* **2**, No. 2 (1982), 19–21.

136. 'Inderdaad een oud probleem', *Euclides* **58** (1982/3), 65–68.

137. 'Die Entwicklung des mathematischen Denkens der 10–14 Jährigen', *Mathematiklehrer* **1982**, No. 3, 5–13.

138. 'Ganzeborden – andersom', *W. Bartjens* **2** (1982/3), No. 1, 18–19.

139. 'Kortste Wegen op een krom oppervlak', *N. Wiskrant* **2** (1983), No. 3, 9–11.

140. 'Wo fängt die Geometrie an?' *Mathematiklehrer* **1983**, 1, 2–4.

141. 'Pourquoi de la géométrie dans l'enseignement primaire?', *Colloque international sur l'enseignement de la géométrie*, Mons, 1982.

INDEX